Jesse David Wall
Elender Wall

INTRODUCTORY
PHYSICS

A Problem-Solving
Approach

Illustrated by Paul G. Hewitt

SECOND EDITION

ANALOG
PRESS

Introductory Physics: A Problem-Solving Approach

Copyright © 1997 by Analog Press

First edition:
Copyright © 1988 by Insight Press
Copyright © 1978 by D. C. Heath and Company

Library of Congress Catalog Card Number: 97-72862
ISBN 1-890493-04-X

Printing: 5   4   3

Text design: Kajun Graphics, San Francisco
Composition: Proctor-Willenbacher
Printer and Binder: McNaughton and Gunn
Publisher: Analog Press

Published simultaneously in Canada.

Printed in the United States of America.

Analog Press
225 Edna Street
San Francisco, CA 94112
415-239-2338

Contents

Chapter **11** **SOUND AND OTHER OSCILLATIONS**

Chapter **12** **ELECTRICITY**

Chapter **13** **ELECTROMAGNETIC INTERACTIONS**

Chapter **14** **MODERN PHYSICS**

Preface

General Objectives

INTRODUCTORY PHYSICS: A PROBLEM-SOLVING APPROACH is written for people who have never studied physics before. It is designed as a one semester college course to provide background and problem solving skills for those who would like to take additional courses in physics and related sciences.

The purpose of this book is to take the fear out of physics, to make it possible for people otherwise frightened of the sciences to succeed in subsequent classes. Physics has the reputation of being a difficult, almost impossible, course for many students. Otherwise intelligent people, who bring insight to discussions of art, literature and current affairs, stumble when they get to general college physics. Part of the reason is the wide diversity of their backgrounds. Some students taking general college physics have had a problem-solving physics course for two semesters in high school. Some have had advanced placement physics after that. Sadly, however, the majority of students entering college have had no previous experience in physics.

That is who this book is written for. We have found that people in college without a high school background in physics are on unequal footing with the students who have. A semester of *Introductory Physics* in college, however, can make the difference between success and failure for these students. For many people, changing their life's goals to go into science would be unthinkable without a way to do well in college physics.

Physics is a large body of knowledge, best mastered at several levels. Such ideas as force, acceleration, energy, momentum, entropy and voltage are sophisticated. Most people need to work with them, play with them, and think about them before they can claim a reasonable understanding of these concepts. This book provides a first exposure to these challenging ideas.

The approach needed by college students differs in several respects from that found most useful on the high-school level. The most obvious difference is a question of pace. The material must be highly condensed and carefully selected if it is to be covered in one semester. All of that good material found in the usual high-school physics book has been boiled down, then boiled down again, so that only the best and most essential remains.

Another difference is in reading level. For the college student, greater attention has been given to making the material forceful and interesting than to making it simple. Both the pace and reading level have therefore been adjusted to the needs of the college student.

To the Student

You already know almost all the physical principles dealt with in this book. You have learned them by just living in our modern society. If you didn't know them, you wouldn't be able to tell magic from ordinary reality. When you watch a magician, you know that he is tricking you in some way when he pretends to do things that violate physical principles. This text will convince you that you really do understand physical principles better than you thought you did.

What you may not have learned yet is the structure upon which to hang these principles. We will provide that structure. The general principles of the way things work have been distilled by physicists over the years into a surprisingly small number of statements. Once you see these distilled ideas and play with them for a little while, you will see that you knew them all along. A convenient way to express these ideas is in mathematical equations, and we do that in this book. You should always keep in mind, however, that these equations are shortcut notations for thoughts that were, in almost all cases, first developed by real experiments and observations before being expressed mathematically.

To give you a chance to play with these general ideas, we will provide you with some puzzles to think about. They are called problems. You will probably find that you actually enjoy working problems when you see how these puzzles yield to the rules of logic you have learned.

We assume that you are already reasonably competent in the rules of logic called algebra. If you are not comfortable with that level of math, you may wish to turn to the Algebra Review in Appendix I. By Chapter 3 you will also need to know how to manipulate a couple of simple ideas from trigonometry. The few things that you need to know are summarized in the Trigonometry Review in Appendix II. We will put off laying the foundation for the course for a couple of chapters in case you need to catch up. Optics is taken out of the usual order so that you have time to review.

Starting with Chapter 3, the course starts to build. The chapter on Newton's Laws of Motion, Chapter 5, requires an understanding of the previous two chapters on force and acceleration. Everything else depends on an understanding of Chapter 5. Again, there are a few mathematical skills you must acquire, so if you feel weak in mathematics, you had better start to work now.

In using this book, you will gain insight into the world around you and the kind of background you will need to enjoy taking additional courses in physics. You will find that you actually like physics if you take the time to master the skills presented in this book. Physics is fascinating. We hope that after you have taken this course, you will be able to see the ideas of physics in practice by observing the world in which you live. Your understanding of the way things work may forever deepen your appreciation of the world.

Problem Sets

Problems at the end of each chapter fall into four sections: Example Problems, Sample Solutions, Essential Problems and More Interesting Problems.

The first section, **Example Problems**, is a set of problems selected to illustrate each of the physical principles covered in the chapter. A substantial amount of thought has gone into selecting these problems. We believe that an understanding of these examples is the key to success in this course. You should attempt your own solution to the Example Problems before turning to the Sample Solutions.

If you were able to work the problems by yourself, you can compare your solution to the **Sample Solutions**. It is likely that there will be some differences between your solutions and the sample solutions since there are usually several ways to solve each physics problem. Reading through the sample solutions, therefore, gives you a chance to see the problems you just solved in perhaps a different light. You need not think of the sample solutions as better than your own, as long as you agree with the final results, but you should be able to see parallels between yours and ours.

If you were not able to work an example problem, you should make your best attempt and then look at the Sample Solution. Study it so that you will be able to work the problem by yourself. After you have worked as many examples as you can and have read the solution to the ones you couldn't do, go back and see if you can now work the ones you couldn't do on the first try. Many will have become easy, but the ones that are still hard deserve special note for future study.

An example that you cannot work all the way through without looking at the solution should be of the most interest to you. Read and understand our solution. If you read it and still don't understand, then that is a really good problem, one to ponder and discuss with your friends. That is one to ask about in class.

If you have given the examples and solutions proper study, you should be able to easily work the **Essential Problems**. They are different words and different situations, but they are worked exactly the same as the examples. Use them as a test. If you have a problem with the third Essential Problem, for example, you didn't understand the third example. They are sister problems. The answers to all the essential problems are given at the end of the chapter so that you can be certain that you understood them.

The last section is called **More Interesting Problems**. These are problems that are interesting either because they are a bit more difficult or because they illustrate still other applications of a physical principle. In either case, these problems are not worked exactly the same way as the examples. They allow you to try your hand at solving problems without being told exactly how. Often these problems are harder only because they don't call for all the steps you need to get the answer. You will have learned enough from the examples and essential problems to select the proper intermediate steps you will need.

To summarize: At the end of each chapter you will find a set of problems and one way to work them. Then you will find another set of problems worked exactly the same way. Following is another set of problems worked only generally the same way. Last, you are given all the answers to the Essential Problems and the answers to half the More Interesting Problems. (You may find some similarity between the problems for which you have answers and the problems for which you do not.)

The structure of these problem sets makes this book easy to use. The structure of these problem sets also makes this book easy to misuse. If you skip over the Example and Sample Solutions, you are misusing the book. You will spend an excessive amount of time doing your homework. Don't do it. In military tactics there is a maxim: Never leave a castle to you rear. If you skip over the examples, the ideas you leave behind will sneak out and catch you.

To the Instructor

Rush right along; don't get bogged down. The point of this book is to serve as a basic introduction while covering a broad range of topics. For those of us accustomed to teaching general college physics, some topics just seem to be such a nice fit with others that we are tempted to sneak our favorite stories, demos, and derivations into this course.

The result is a course at about the same depth as general college physics without calculus, but only covering mechanics. That may help your students in the first half of general college physics, but it puts them at a disadvantage in the second half of the course.

Several features of this book will allow you to streamline your course if you take advantage of them. The problems are the simplest types that still illustrate the concepts treated. The only picture-frame force problems used, for example, are those in which the upward angles are equal. Although students are presumed at this level to be able to handle two equations in two unknowns, they are not asked to do so. Topics that are not critical to the central development have been omitted even though they might be simple enough in themselves. As another example, the fact that the index of refraction of a medium equals c/v is so nice and neat that it was omitted from the present development only after considerable pain and suffering.

Some instructors may find it impossible to treat force equilibrium problems without also mentioning rotational equilibrium. If you find yourself incorporating additional topics that seem to naturally fit with those presented in the book, be careful to either note them as asides or be prepared to sacrifice other material. As it is, you will need to rush to cover the material presented, and you may not get to modern physics at that.

You will notice that Optics is taken out of its usual order and presented in Chapter 2. This is done in order to help the student who is weak in algebra and trigonometry, perhaps a third of your class. Rather than spend a week on a math review and lose the forward momentum of the class, a math review is built into optics. There is plenty of good physics to engage the math whiz while providing an opportunity for those weak in math or those taking trigonometry concurrently to brush up.

The foundations of mechanics, starting with Chapter 3, can thus be postponed until the third week of the semester. It could be postponed still another week by taking Fluids out of order, as was done in the first edition of this text. The disadvantage of that approach is that the treatment of units is messy without first establishing the relationship between mass and weight. The advantage is that it gives an extra week for those students taking trigonometry concurrently to learn the things they will need to know.

Insist that your students solve problems in a suitable format. Many of your students will follow the lead of the sample solutions and work the problems with algebra before inserting data. Others will be more comfortable plugging and then chugging. That bad habit can only be discouraged by a grading system that gives credit for an algebraic general solution whatever numbers and units are put in. The format of the problem set should help in this regard, since the similarity between Example and Essential problems, and odd and even numbered More Interesting Problems will encourage algebraic solutions. Students will soon learn that they can verify their algebraic solution to even numbered problems by changing the given data and checking the answers to the odd numbered problems. In so doing, they will be exposed to a fresh application of a physical principle.

The problems in this text are stated in concrete terms, each to serve as an application of physical principles to realistic situations. Well, maybe sometimes surrealistic situations. In this book, no nameless objects are acted on by mysterious forces. You may wish to add some generic objects of your own to the homework assignments if you favor putting them on exams.

Nameless objects are easier to deal with when writing problems, but there is a second level of skill needed to tie such abstractions to the concrete. We believe that students at the introductory level need to be grounded at the lowest cognitive level if their foundation is to be solid. If you choose to test them at a higher level, although it is an easier level for more formal thought, you may want to give them some practice at your style of problem.

Because the problems in the text are concrete applications, they tend to be long and wordy. If most of your students are going to take general college physics or other science courses, and thus have an opportunity to further develop their problem-solving skills at a later time, you may wish to limit the level of skill they need to be responsible for in this course. One way to limit the course is to use only homework problems for exams— word for word with only the numbers changed. This is a good way to take the fear out of physics. Your students know exactly what is expected of them. They need to learn how to work all the problems in the text, but that is all they need to know. They can get through their first course in physics on pure diligence. You might think that this procedure would ruin your grade distribution, but it turns out that diligence falls into a Gaussian distribution like everything else. If you grade exam papers for such desirable characteristics as algebraic general solutions, correct use of units, and, where appropriate, diagrams, you will find that students who do well on your exams are well prepared for general college physics.

You may find that the entry level physics course becomes a favorite course to teach. Since they want to do well in subsequent classes, the students at this level are easy to motivate and are more interested in mastering the subject matter than in the grade they get. This may be the only course where a student, after discussing with you their solution to a problem on an exam, will not be particularly interested in your improving the grade. "Oh, I don't care about that," they are likely to say, "I just wanted to know how to work the problem."

Acknowledgments

We would like to express special gratitude to Paul Hewitt for inspiring both of us. He showed the father how to teach and convinced the daughter that she was a physics major. His pioneering of the idea that physics should first be understood conceptually before it is analyzed mathematically has been the guiding principle to our style. He also did all of the drawings in this book, both first and second editions. His drawings best illustrate his teaching maxim— "Physics is Phun!"

Special thanks for the second edition go to our friend Sue Broadston at Cabrillo College. Her unflagging enthusiasm for this book and for the writing of a second edition helped to make it possible. She also contributed important ideas for changes to be made to this edition and corrections to the final text. Other people who helped with corrections were Mary Hannah Hawkins, a conscientious City College student, and Father Tom McShane of Creighton University who class tested the book. Friends without whom this second edition could not have been produced are Karen Tuttle, who spent months student testing the first draft; Robert Bodlak and David Lewak, who worked problems; Linnea Johnson, who consulted on layout and continuity; Meg Bernstein, who when she was not doing expert copy editing, was cooking us delicious dinners.

We would also like to thank Pat Koren and Rad Proctor, our book designer and typesetter at Kajun Graphics in San Francisco, for putting together a book that is visually appealing and accessible. Also of great help in the layout were Rose Egar and Doug Cowan. Sylvia Rorem did the indexing.

Several students deserve special thanks for inspiration, criticism and help. Mary Jew did a beautiful job in hand lettering the example solutions in the first edition, and her suggestions as to form and content still live in the second edition. Martha Stasinos, Mary Jo Callahan, Julie Racz, and Sharon Akutagawa helped substantially on the first edition and many of their comments and suggestions are reflected in the second edition.

Among the many of our friends and colleagues who have helped over the years, special thanks are due to Lewis Epstein, Mario Iona, Jackie Spears, Jim Court, as well as other friends in the Physics department at City College of San Francisco.

Ellen Wall, our wife and mother, respectively, and her sister, Joyce Cauthen, helped to get this book started many years ago. Ellen deserves even greater appreciation for encouraging a second edition. It would never have been possible without her.

About the Authors

Jesse David Wall, author, has been teaching physics at the City College of San Francisco for more than a quarter of a century. He wrote *Introductory Physics: a Problem Solving Approach* in 1977 because no text then existed for his favorite course—a one-semester preparatory course for general college physics. He helped develop this course because of the great diversity of academic backgrounds of City College physics students. The course proved itself again and again as an essential step for many students in order for them to succeed in general college physics. The course deserved a text book.

Elender Wall, editor and contributor to the second edition, is Jesse David's daughter. She grew up thinking that Physics was *Phun,* an impression that was confirmed when she took Paul Hewitt's course in Conceptual Physics at City College of San Francisco. After finishing a Bachelor's degree in Physics at San Francisco State University, she decided to revise her father's book. She had used the book when she was in college, and loved it, but felt it needed a general updating. She expanded the problem sets, revised text, and added new topics.

Paul Hewitt, illustrator, is a renowned physics teacher and author. An instructor for many years at the City College of San Francisco, he now shares his teaching schedule between San Francisco and the University of Hawaii, Hilo. He is the author of the college text *Conceptual Physics,* the leading physics textbook for non-majors for the last quarter century. He is also the author of the text *Conceptual Physics: A High School Program,* and is the co-author, with daughter Leslie Hewitt and nephew John Suchocki, of *Conceptual Physical Science,* all published by Addison-Wesley-Longman. Before he found his calling as the world's leading missionary of conceptual

physics, he worked for many years as a sign painter, where he honed his artistic skills.

The authors wish you, the student, an exciting introduction to problem-solving physics. We hope you have as much fun with physics as we do.

CHAPTER 1 *Measurement*

Why Measure?

SCIENTIFIC KNOWLEDGE IS TENTATIVE. Truth in matters of science is considered to be that which has been demonstrated, corroborated and verified. Since scientific truth must be able to stand up against future testing, there always remains the possibility that future refinements will disclose weaknesses or limitations in what we presently accept as truth. Indeed, progress in science depends upon that kind of testing. The tentative nature of scientific truth turns out to be its very strength.

Scientific knowledge is also simple in the sense that results can be expressed in numbers. Sometimes people fall in love with this simplicity and become possessed with a feeling that scientific truth is somehow superior to other forms of knowledge. A famous quote from Lord Kelvin (1824–1907) says,

> When you can measure what you are speaking about, and express it in numbers, you know something about it; but when you cannot measure it, when you cannot express it in numbers, your knowledge is of a meager and unsatisfactory kind: it may be the beginning of knowledge, but you have scarcely, in your thoughts, advanced to the stage of science.

Lord Kelvin was an intellectual giant of his time. He made substantial contributions to theoretical and experimental thermodynamics and electrical instrumentation. The Kelvin temperature scale, which we will use when we deal with the ideal gas law, is named to honor him. Although the scientific opinions of this nineteenth century scientist continue to be greatly

Figure 1-1
Lord Kelvin

respected, few would agree with his smugness in characterizing non-scientific knowledge as meager and unsatisfactory.

Most scientists today are humble enough to recognize that they have become clever mostly by learning to simplify reality. Most realize that knowledge in such fields as art, music, history, and politics is not so much meager and unsatisfactory as the fields themselves are complex. Even Lord Kelvin's own judgment on the meagerness of non-scientific knowledge is itself made of the very stuff he judges to be meager and unsatisfactory. The degree to which knowledge is meager cannot be measured and expressed in numbers because the nature of knowledge is too complex to allow such simple treatment.

Questions of superiority aside, there remains a clear difference between things that can be measured in terms of numbers and things that cannot. The peculiar strength of scientific knowledge lies in its ability to be verified by measurement. Truth in this realm can be tested. The numbers can be checked. Science evolved from magic, the ancestor of science, mainly by asserting this very premise.

Science also differs from magic in another respect. Magic makes people powerful, but the power is personal. The magician can do things that other people cannot. Science also makes people powerful, but the power is impersonal. Results depend more on *what* is done rather than *who* does it. Only a magician can do magic, but anyone can do science.

Because it is impersonal, science gains strength by being shared. The adage "publish or perish," advice to the young academic of modern times, tells how to become powerful and survive in modern academic pursuits such as science. This was unheard of by magicians of old. Secrets of magical crafts lost strength by being shared. With science, it is the other way around.

New discoveries in science are thrown into the arena of peer scrutiny and judgment. The rush to test new theories and verify new results is what entertains and amuses the scientific community. This shared activity also provides inspiration for still newer theories and discoveries.

Figure 1-2

Only a magician can do magic

Units of Measurement

Any joint human endeavor requires language for communication. In science, a natural part of language is mathematics. Mathematics is more, however, than numbers. Naked numbers command fascination by themselves, but for them to mean anything there has to be some agreement of what they stand for. It is not helpful to tell people that an airplane flies 100 high, or the desk is 5 long. Units of measurement are essential for expressing measurements in numbers.

Modern science requires great precision in defining units of measurement. The need for some precision is, however, as old as language itself. The oldest written records, coming to us from the edge between historic and prehistoric times, are tax records and calendars. We can read them only if we understand the ancient units.

Early length measurements, dating from prehistoric times, were naturally enough based on the human body. Inventing standards as they needed them, the builders of antiquity used their own fingers, thumbs, hands, and arms as measuring implements.

We see this still in the British Imperial units, the foot and inch. The foot, by its very name, is clearly a unit based on the human body. So is the inch. As does the ounce, the inch derives its name form the Latin *unus*, meaning one, or unity. In Old English it was spelled "*ynch*" and was based on the thickness of the thumb. The unique and most powerful finger on the hand, was the one to measure with in old England.

To measure something by the *ynch,* use the thumbs of both hands. If you hold one thumb in place while moving the other, you can get very repeatable results. As you place the thumb you are moving, you can roll it slightly to adjust its position until you can feel it just barely touch the stationary thumb. Always overlap your thumbs the same amount so that the bases of the thumb nails line up.

Physics being an experimental science, you might wish to pause in your reading to take the *ynch* measure of the height of the printed column on this page. Do not try to get the "right" answer. That will be different for people with different size thumbs. Try instead to get the **same answer** each time you do it. We will soon see how to convert your answer into standard units. Your personal *ynch* will be with you for a very long time.

It is a good thing for a measuring unit to be repeatable. With a little practice, the *ynch* measure will serve in that regard. Another good thing is for the measuring unit to find wide agreement by different people. There the *ynch* measure leaves something to be desired.

There exists in antiquity a measuring unit that shares with the *ynch* measure the virtues of being highly repeatable and of being based on a body part, but that had the further advantage of being calibrated to a widely agreed standard. It is even older. It is the Egyptian Royal Cubit.

When the Greeks built the Parthenon, the Egyptian Royal Cubit was already ancient. From the time of building the pyramids, about 2500 BC, through thousands of years of the developing Egyptian civilization, the Royal Cubit, equal by our measure to about 52.3 cm, remained constant to within 0.3 cm. Taken up by the Greeks and spread by the conquering armies of Alexander the Great, it became the most widely agreed upon standard in Western civilization. The Greek foot, defined as three-fifths of an Egyptian cubit, remained the standard in medieval England up to the twelfth century.

The Royal Cubit owed its stability to the fact that it had a built in calibration factor. It could be adjusted to fit different people of different size. Like the long cubit, which was a forearm plus a hand, it literally added something to the short cubit.

Figure 1-3

Measuring a page in "ynches"

Figure 1-4

Precision of measurement is needed to build a cathedral.

The short cubit, which was probably the Biblical cubit, was just the length of the forearm from the elbow to the tip of the fingers. Add the width of four fingers of the other hand to the short cubit and you have the long cubit. With the long cubit you could hold your place with your hand while you moved your arm. Just as with the *ynch* measure discussed above, using the sense of feel gave the long cubit remarkable repeatability.

The Royal Cubit added to the long cubit a standardization practice. Short people might add their thumb. Tall people would use fewer fingers. Since there are about 28 finger widths to a long cubit, this alone allows for a calibration to one part in 28. Different combinations of different fingers improves calibration still further.

The precision with which a measuring unit must be defined depends on what is being measured and for what purpose. Always the idea is that different people making the same measurements should get more or less the same result. If by "more or less" we mean that the agreement is good enough to build pyramids, then clever use of arms and fingers is good enough.

If by "more or less" we mean that the agreement is good enough to measure farm land, then clever use of arms and fingers is too good. Feet are good enough and are attached to a more convenient end of the body. If, on the other hand, we mean the agreement must be good enough to build cathedrals, then arms and fingers aren't good enough.

For the great cathedrals, far greater precision was needed. The rules and squares of the stone masons became important symbols of their craft. As the need for precision spread with the development of manufacturing, so did the tools, standards, symbols and practices of the building trade. England developed the system of British Imperial units, still in use in a few parts of the world, most notably the United States. The rest of the world has gone metric.

The Metric System

The modern measuring system, based on the *meter* as a unit of length, dates back to slightly before the French revolution. Standards of weight and measure had become a big problem by then. Manufacturing and trade had developed in France to the point that the petty cheating and misunderstandings that resulted from variations in local measuring customs had become intolerable. The yard measure for cloth, for example, had at least thirteen different definitions in one province of France.

Times were unsettled. France was at war and King Louis XVI was in trouble. Revolution was brewing. The Bastille had been stormed, but the king still sat upon the throne. The king was bankrupt financially, but whenever he mentioned the idea of raising taxes to pay for the war, people would grumble about the inconsistent weights and measures.

In an attempt to deal with the problem of weights and measures, Louis XVI started reforms suggested by his radical but nimble-footed

Bishop Talleyrand. The bishop wanted France to adopt a new unit of length that everyone on earth could agree upon. It was similar to a unit that had been first suggested a century earlier by the Abby Gabriel Morton, Vicar of Lyons, and was based on the size of the earth itself.

The good vicar's idea had been to make a yard measure based on the distance from the North Pole to the equator. It was known that this distance was about 10.9 million yards. Maybe a little more, maybe a little less depending on whose yard measure you used. By a little larger measure, the pole to equator distance would be exactly 10 million.

This appealed to both the bishop and his king for a number of reasons. Not only was it a nice even number, but it had the international appeal of being based on the whole world.

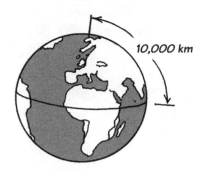

Figure 1-5

The original definition of the meter was that 10,000,000 m should be the distance from the earth's north pole to the equator.

Probably even more important, however, was the fact that this unit of measure was just a bit bigger than those in common use. Other unit changes had gotten smaller. To the people, this meant that they got less for their money. Smaller units therefore push economic inflation, and inflation has always been unpopular. The king was already unpopular enough. Louis XVI knew he had to go bigger so that people got more for their money.

The king had Bishop Talleyrand call a council of advisors including the leading scientists and mathematicians of the realm to approve the bishop's plan. They got into an argument over the size of the earth. To make everyone happy, the king commissioned a new survey of France so that the size of the earth could be determined with accuracy better than ever before.

This made everyone happy. The king's advisors particularly liked another part of the bishop's plan. The results of this new survey could be used to redefine units for mass. They took a cube that measured one meter on each side, filled it with water to get a new **tonn**, spelled with an extra "n" so that people would get the point that it was newer, bigger and better than the old ton.

Just about ten percent larger than the old ton, based on a cubic yard of dirt, the new tonn was in fact better. To improve the standard measure they could purify the water used to make the measurement. It is hard to agree on how to purify dirt.

They also came up with a name for a thousandth part of a cubic meter. Take a cube 10 cm on a side, and you have a liter, by the original definition. You can see that it takes a thousand liters to fill a cubic meter by lining up ten on a side in each of three dimensions.

The original gram was defined as the mass of water needed to fill a smaller cube, a cube one centimeter on a side, where a centimeter is one hundredth of a meter. Just as a liter is a thousandth part of a cubic meter, a cubic centimeter is a thousandth part of a liter.

All of these wonderful ideas for improving the king's measuring standards were worked out by his council of advisors before the survey team set out. Unfortunately, for the king, they were too late. Before the survey team got back to Paris, the good, honest, but ineffective King Louis XVI was sent by the mob to visit his less good, less honest ancestors. Fortunately, for the metric system, the revolutionaries thought this new metric system was just great and proceeded to take credit for the new standards.

Many people would like to see the metric system used universally, without exception. Others like the British Imperial measure with which they

HIGHLIGHT
The History of the Meter

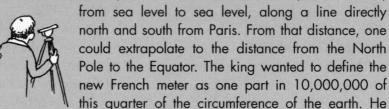

In 1792, King Louis XVI of France issued a Royal Warrant commanding provincial governments to give all assistance to a survey team consisting of J. B. J. Delambre and P. F. A. Mechain. The original idea was to refine and extend the measurements taken in an earlier survey of 1739. The objective of the survey was to measure the distance from Dunkirk, on the north coast of France, to Barcelona, from sea level to sea level, along a line directly north and south from Paris. From that distance, one could extrapolate to the distance from the North Pole to the Equator. The king wanted to define the new French meter as one part in 10,000,000 of this quarter of the circumference of the earth. He thus sent out his survey team on a job that was supposed to take a couple of years.

The royal survey team encountered substantial difficulty. Many of the landmarks had been lost or even moved, and the surveyors found they had to do almost everything from scratch. By far the greatest obstacle they had to overcome, however, was the Royal Warrant they carried. The king had become so unpopular that, the year after issuing this warrant, he lost his head. It is a great tribute to the survey team's commitment toward the ideal of establishing standards of measurement that they pushed forward, traveling across France during the Reign of Terror. What with having their equipment impounded and even having themselves jailed with monotonous regularity, the survey team took eight years to complete their measurements.

The effort cost Mechain his health, and he died a few years later. The science of measurement, too, has its martyrs and heroes.

The results of this survey were, however, truly revolutionary. Not only were the results of the new survey used to standardize length measurements, they were used to redefine units for mass and volume, as well. Based on the density of water, mass was defined in terms of water in such a way that everything in the new revolutionary units of measurement was to be governed by factors of ten.

So great was their enthusiasm for discarding the old that the revolutionaries even defined a unit of time based on factors of ten. Instead of dividing days into factors of two, twelve, and sixty, the new revolutionary time went by days, tenths of days, and thousandths of days. The revolutionaries were ready to throw out all of the clocks in France.

Clocks are expensive, however, and had become works of art. Moreover, time is not taxed in the same way as goods and lands. For a short time in France it was indeed a dangerous thing to measure by twelve and sixty. There were even clocks made with a public dial in metric time and a private dial on the bottom in common time. The urge toward metric time passed in less than a year, however, and the old clocks survived. In 1795, the idea of decimal clock reform was tabled by the revolutionary government. It remains tabled still.

Although the deciday and milliday never quite caught on, the new meter did. So did the rest of the metric system based upon the new meter. The length and mass standards established during the French Revolution became the official standards used all over the world with few exceptions.

Figure 1-6

During the French revolution decimal clocks were made dividing the day into ten parts. Sometimes a second face, hidden from public view, gave common time.

are more familiar and in which their machine tools and measuring instruments may be calibrated. The controversy has been with us for many generations and is not likely to go away soon. For example, the metric system was first officially adopted in the United States over a hundred years ago, in July of 1866. In the United States, however, the British Imperial System still seems to be hanging in there.

This controversy has little to do with physics, however, and much to do with the economics of a global economy. We could just as well use cubits as long as everyone agreed. We landed on the moon in feet and we put up a space telescope in meters. Except to the extent that they contributed to communication—or lack of it—on those missions, the system of units didn't matter.

That being said, nowadays scientists exclusively use the **International System** of units, or SI Units, which is a system based on the metric system. Metric units have a great advantage over the British Imperial system in that all units are related by some multiple of ten. This makes metric units useful for measuring things that are very small and very large, and reduces some conversion calculations to a matter of moving the decimal place.

Because metric units are used so universally in the scientific community, this text will use SI units almost exclusively. Our few references to British units will be to contribute to a conceptual understanding for the benefit of readers who have not yet become bilingual in this regard.

BASIC SI UNITS

A measure of:	Unit	Abbreviation
Length	meter	m
Time	second	s
Mass	kilogram	kg
Current	ampere	A
Temperature	kelvin	K
Luminous intensity	candela	cd

Table 1-1

The basic SI units

SI Units

The international system of units known as SI units, abbreviated from the French name "Le Systeme International d'Unites," is based on the metric system but is expanded to include all scientific properties. It is based on seven basic units, which can then be combined to measure other properties. The unit of length is the **meter**, the unit of mass is the **kilogram**, the unit of time is the **second**, the unit for electric current is the **ampere**, the unit for thermodynamic temperature is the **kelvin**, the amount of substance is the **mole**, and the luminous intensity is the **candela**. In this text, we will use all these units except the candela.

The meter of length was originally defined, as we just discussed, in terms of the distance from the North Pole to the equator. It is now defined in terms of the second of time and the speed of light in a vacuum. In one second of time, light in a vacuum will travel a distance of exactly 299,792,458 m, that now being the definition of a meter.

Before October 20, 1983, the definition of the meter was that the reddish-orange light from a krypton lamp has a wavelength of 1,650,763.73 m. With that definition, the best measurement of the speed of light was 299,792,458 m/s plus or minus a little bit. The Conférence Générale des Poids et Mesures rounded off the uncertainty and let the velocity of light be exactly 299,792,458 m/s.

The second of time was originally defined as a pulse beat, a natural measurement of time based on the rhythm of human body. By the time of the French Revolution, it was a certain fraction of the 24-hour clock as calibrated by the solar day. It is now defined in terms of an atomic clock, calibrated with microwaves tuned to an atomic resonance with cesium vapor.

The kilogram of mass was originally defined in terms of the gram, which was the mass of a cubic centimeter of water as discussed above. It was a good definition and still works for most purposes. A kilogram is a liter of pure water, and that's good enough until you start making measurements accurate to a few parts in a hundred-thousand. When you do that, the definition needs to be turned around. After that, a liter is a kilogram of pure water. Instead of mass being defined by volume, volume is defined by mass.

The change had to be made when people started measuring with so much precision that they could see a difference between samples of pure water prepared in exactly the same way. A natural sample of water contains a mix of oxygen and hydrogen isotopes. Water made with the heavy isotope of hydrogen is commonly called heavy water. Any natural sample of water contains some heavy water. Different samples will contain slightly different amounts of heavy water.

The difference is tiny—and we mean *really* tiny. The difference in mass of naturally occurring samples of pure water is about as much smaller than the mass of the samples as the mass of ink of a comma is smaller than the mass of the book you are reading.

For measurements more precise than that, you need to worry about the true definition of a kilogram. The problem with the true kilogram is that there is only one of them. It is a block of platinum and iridium alloy that is kept in a vault at the International Bureau of Weights and Measures at Sèvres, near Paris, France.

Other SI Units will be defined in the text as we use them.

A measure of:	Unit	Abbrievation	What it is:
Force	newton	N	$1\ N = 1\ kg \cdot m/s^2$
Work, energy	joule	J	$1\ J = 1\ N \cdot m$
Power	watt	W	$1\ W = 1\ J/s$
Frequency	herz	Hz	$1\ Hz = s^{-1} = \left(\dfrac{1\ cycle}{s}\right)$
Pressure	pascal	Pa	$1\ Pa = 1\ N/m^2$
(Electrical) Charge	coulomb	C	$1\ C = 1\ A \cdot s$
(Electrical) Potential	volt	V	$1\ V = 1\ J/C$
(Electrical) Resistance	ohm	Ω	$1\ \Omega = 1\ V/A$
(Electrical) Capacitance	farad	F	$1\ F = 1\ C/V$
Magnetic field	tesla	T	$1\ T = 1\ N/A \cdot m$

Table 1-2

Selected Combination SI Units

Metric Prefixes

One of the strongest features of the metric system is the use of a single set of prefixes to relate larger and smaller units of whatever it precedes. A microgram is a millionth of a gram, a micrometer is a millionth of a meter, and a microvolt is a millionth of a volt. A micro- stands for a millionth of whatever. The same prefix works for all. Since a kilogram stands for a thousand grams, it follows that a kilometer stands for a thousand meters. It also follows that a kilovolt stands for a thousand volts—no matter what a volt might be. We might say that these metric prefixes give us all a consistent method of multiplying both our knowledge and ignorance alike.

Of the nine metric prefixes most students should know, the largest is the **giga**-, standing for a thousand millions, or 10^9 times. It is abbreviated with an uppercase G. For example, a gigameter, or Gm for short, is a thousand million meters. The moon is about a third of a Gm from the earth.

The **mega**- is the next largest prefix in common use. It stands for a million, or 10^6 times, and is abbreviated with an uppercase M. A megameter, or Mm for short, is a million meters. The earth's radius is 6.37 Mm.

The **kilo**- is the next smaller by three orders of magnitude. It stands for a thousand, or 10^3 times, and is abbreviated with a lowercase k. A kilometer, or km for short, is a thousand meters. The span of San Francisco's Golden Gate Bridge is 1.28 km.

Most common metric prefixes differ from each other by three orders of magnitude. The **centi-** is the exception. Meaning a hundredth, or 10^{-2}, it is abbreviated with a lowercase c. A centimeter, or cm for short, is a hundredth of a meter. The British Imperial inch is exactly 2.54 cm as now defined in the United States.

Prefix	Abbreviation	Decimal Fraction	Power of ten
peta	p	10^{15}	1,000,000,000,000,000
tera	T	10^{12}	1,000,000,000,000
giga	G	10^{9}	1,000,000,000
mega	M	10^{6}	1,000,000
kilo	k	10^{3}	1,000
centi	c	10^{2}	100
deci	d	10^{-1}	0.1
milli	m	10^{-3}	0.001
micro	μ	10^{-6}	0.000001
nano	n	10^{-9}	0.000000001
pico	p	10^{-12}	0.000000000001
femto	f	10^{-15}	0.000000000000001

Table 1-3

Common metric prefixes

The **milli-** gets back into the three order-of-magnitude swing of things. It stands for a thousandth, or 10^{-3} times, and is abbreviated with a lowercase m. A millimeter, or mm for short, is a thousandth of a meter. The uppercase letters on this page, printed in ten point type, are 3.53 mm tall.

Notice that the abbreviations are case sensitive. The larger ones are abbreviated with uppercase letters, such as G and M, while the smaller ones are abbreviated with lowercase letters, such as c and m. There is a big difference between a Mm, short for megameter, and a mm, short for millimeter. A Mm is a billion times a mm.[1]

The **micro-** is the next smaller by three orders of magnitude. It stands for a millionth, or 10^{-6} times, and is abbreviated with the lowercase Greek letter mu, pronounced "mew" and written "μ." A micrometer, or μm for short, is a millionth of a meter. The wavelength of yellow light is 0.5 μm. In older literature, the micrometer is called a micron. It is at the limit of what can be seen with an optical microscope.

---

[1]In physics, we often make a distinction between upper- and lowercase. Be careful about this.

A **nano-** is still smaller by three orders of magnitude. It stands for a billionth, or 10^{-9} times, and is abbreviated with a lowercase n. A nanometer, or nm for short, is a billionth of a meter. The double helix DNA molecule of molecular biology is made of two strands that, when wound together, have a thickness of about 1.1 nm.

A **pico-** is still smaller by three orders of magnitude. It stands for a millionth of a millionth, or 10^{-12} times, and is abbreviated with a lowercase p. A picometer, or pm for short, is a millionth of a millionth of a meter. The covalent radius of an oxygen ion is 75 pm.

A **femto-** is the smallest of the common metric prefixes. It stands for a millionth of a billionth, or 10^{-15} times, and is abbreviated with a lowercase f. A femtometer, or fm for short, is a millionth of a billionth of a meter. The diameter of a carbon nucleus is, for example, 6.0 fm.

Of these prefixes, it is essential to learn right away four of them: M, k, m, μ

CHECK QUESTION

What is:

(a) a megavolt times a kilometer?

(b) a millisecond times a nanometer?

(c) 10^6 phone?

Answer: (a) a gigavolt-meter (b) a picometer-second (c) a megaphone[2]

Precision and Significant Figures

> *GIGO: /gi:'goh/ [acronym] 1. 'Garbage In, Garbage Out'—usually said in response to users who complain that a program didn't complain about faulty data. Also commonly used to describe failures in human decision making due to faulty, incomplete, or imprecise data. 2. 'Garbage In, Gospel Out': this more recent expansion is a sardonic comment on the tendency human beings have to put excessive trust in 'computerized' data.*
>
> *—FOUND ON INTERNET*

Through science, the rules that govern numbers can be brought into play in understanding and predicting results in the real world. That things of science are simple enough to be measured and expressed in numbers is the very fact that gives science its beauty and elegance. As we proceed into the study of physics, the most basic of sciences, we need to be wary of certain forms of smugness, however, that tend to creep into our thinking when we deal with simple things.

---

[2] Guilt for this pun lies with our friend Paul Tipler. *Physics for Scientists and Engineers*, Worth. 3rd Edition. Good book otherwise.

The rules of logic, which we know as mathematics, are so powerful that we sometimes tend to forget the limitations of the meaning of the numbers. It is one of the ways in which people sometimes come to believe that they know more than they really do. Pushing buttons on a little black box and then believing everything the little black box says, one can easily become over-confident in the truth of a result even when perfectly aware of the limitations of the data with which one starts.

Since there is a limit to the accuracy with which one can measure anything, scientists have a way of writing numbers so that they state quantities only to the number of numbers that they are sure about. This is called **scientific notation**.

Large and small numbers have an ambiguity to them. If something is 1,200 meters long, has it been measured extremely carefully to be exactly 1,200 meters, or has it been roughly measured to be 1,200 meters plus or minus 50 meters? It is hard to tell whether the zeros are there because we know their exact value, or whether they are there to hold the decimal place.

The method that scientists have found to solve this problem is to write the number in scientific notation. This is done by expressing the number as a single digit, followed by a decimal point and the proper number of figures, all multiplied by ten to a power. In this case, if we have just paced off a field and it is about 1,200 meters long, we would write this:

$$1,200 \text{ meters} = 1.2 \times 10^3 \text{ meters}$$

If, on the other hand, we had used advanced surveying techniques and measured the field to within half a meter, we would write:

$$1,200 \text{ meters} = 1.200 \times 10^3 \text{ meters}$$

There would be little doubt that the zeros after the decimal point were there for the sake of accuracy, since that is the only reason to write them.

The number of figures that it takes to write something in scientific notation is called the number of **significant figures** in a quantity. In the first example, the number 1.2×10^3 had two significant figures and the number 1.200×10^3 has four significant figures. More broadly, the number of significant figures in a number is the number of digits precisely known disregarding place-holding zeros. Thus, 166,288 has six significant figures, 0.00678 has three and 100,020 has five.

CHECK QUESTIONS

Write in scientific notation and give the number of significant figures:
(a) The Equatorial radius of the earth is about 6,371 kilometers.
(b) The radius of a droplet of fog is 0.0000006 meters.
   Answer: (a) 6.371×10^3 km, four; (b) 6×10^{-7} m (or 0.6 μm), one

A typical smugness that creeps into problem solving is the tendency to report data to more significant figures than the data will support. Say you hold a ruler up to one of your kitchen plates and find that it has a diameter of about 25.5 centimeters. You know that the circumference of a circle is given by the equation:

$$C = 2\pi R = \pi D$$
$$\text{Circumference} = 2\pi \cdot \text{Radius} = \pi \cdot \text{Diameter}$$

So you take out your trusty calculator, plug in your measurement for the diameter, multiply it by your calculator's value for pi and get:

$$C = \pi \cdot D = \pi \cdot 25.5\text{cm} = 80.110613\text{cm}$$

Figure 1-7

Measuring a kitchen plate

But wait! If you wrote down this answer, it would imply that you know the circumference of your plate to the closest micro meter! A micro meter is something like the size of a germ. You couldn't possibly know the measurement of the circumference to that degree of accuracy. The circumference of the plate shouldn't depend on how sterile the plate is!

Since you only measured the diameter of the plate to the nearest tenth of a centimeter, you can only know the circumference to the nearest tenth of a centimeter. Your answer, then, is:

$$\text{Circumference} = 80.1\text{cm}$$

The other numbers that showed up in your calculator after those digits are calculator garbage. They would be meaningful if we were dealing purely with numbers with no physical quantity attached. In the world of physics we always have to keep in mind that the numbers that we work with represent physical things or ideas, and every measurement has some sort of limitation.

There are some guidelines for dealing with how many figures to keep in a calculation. We will call them the **garbage rules**.

The first garbage rule is a rule of thumb. Generally speaking, you should keep only as many significant figures in your calculations as your least accurate data. If you know that a rectangular lawn is exactly 3.2500 meters wide and about 25 m long, you only know the area to two significant figures, or to within one meter of accuracy:

$$\text{Area} = \text{length} \times \text{width} = 3.25 \text{ m} \times 25 \text{ m} = 81.25 \text{ m}^2 \cong 81 \text{ m}^2$$

where the symbol $\cong$ means "is approximately equal to."

The second garbage rule is that half the time a little garbage is good. In the midst of calculations, you usually want to keep an extra significant figure to avoid introducing a round-off error to your calculations. Round-off errors occur when you do not take into account the next digit that you are discarding. For example, the number 5.59 is closer to the number 5.6 than it is to the number 5.5. Sometimes when working with several numbers they will multiply in such a way that the extra values become significant. So, in the course of calculations, keep an extra number in your data, and when you state the answer at the end, "round up" if the digit after the last significant figure is 5 or more.

Lest you think we are nit-picking here, remember that a new version of a common computer chip made front page news when it got a round-off error in one of its applications. Correcting the program error cost the manufacturer at least ten million dollars.

The most important thing when deciding how many significant figures represent real data is to keep in mind that you are dealing with physical quantities. You should try to have a conceptual idea of the accuracy of your numbers. When determining accuracy, it is helpful to understand the concepts of analog versus digital quantities and the idea of error.

CHECK QUESTIONS

Report to the proper number of significant figures:

(a) $(1.250 \, s \times 10^{-2} \, s) \cdot (2.0 \, s) =$

(b) $2200 \, m/s \cdot 3 \, m/s =$

   Answer: (a) $2.5 \times 10^{-2} \, s^3$; (b) $7 \times 10^3 \, (m/s)^2$

Theory of Error, Uncertainty

Physical quantities can either be **digital** or **analog**. Digital quantities can be expressed in exact numbers. Analog quantities are quantities that have a limited accuracy, and can often be reported with the amount of **error**, or uncertainty, to which they were measured.

Some data is digital. If a bunch of people were to go to the zoo and count the number of monkeys in a cage, everyone should get exactly the same number unless, of course, somebody made a mistake. In counting monkeys, there is plenty of room for mistakes, but there is no room for error. The nature of monkeys is that their number is digital.

Measurements tend to produce data that is analog. If you carefully measure the height of this page with a ruler, it is easy to measure it to the nearest millimeter, 25.3 cm. You can improve your measurement by estimating between millimeter marks to get 25.34 cm. Using ruler technology, that is as accurate as you can report the data. Since you estimated the last number, a more accurate way of reporting might be the distance plus or minus a couple of tenths of a millimeter: 25.34 ± 0.02 cm. The "± 0.02 cm" is an estimate of what we call the reported *error* in the measurement. Not *mistake,* mind you, but error.

Even with the most sophisticated measuring tools around, there is no way to measure the length of your textbook with absolutely no error in the measurement. On a microscopic level, the edges of the physics textbook are uneven. As you look closely at any solid object, it becomes less solid looking. On the atomic level there is actually a physical principle which puts a limit on accuracy of measurement; that is the Heisenberg Uncertainty Principle, which says that it is impossible to tell the position of an electron in an atom, not because we can not see it, but because it actually exists in more than one place in any given moment. We will learn about this in Chaper 14.

Some units are digital by definition. When we say that there are 60 seconds in a minute, 60 minutes in an hour, and 24 hours in a day, we do not mean these numbers to be approximate. We mean them to be exact. Calculate the number of seconds on a day, and you get:

$$\left(\frac{60 \text{ s}}{1 \text{ min}}\right)\left(\frac{60 \text{ min}}{1 \text{ hour}}\right)\left(\frac{24 \text{ hours}}{1 \text{ day}}\right) = 86,000 \text{ s/day}$$

If we didn't know any better, we could apply the garbage rules to say that the result is only good to one or two significant figures. Since we know these numbers are digital, however, we know that the number is exact. It is correct to any number of significant figures.

The Metric Earth

As an example of error propagation, let us return to considering the size of the earth. We recall that the meter was first defined by king Louis XVI as one ten millionth of the distance from the North Pole to the equator, assuming the earth is round. We can use the original definition to get the radius of the earth, and we will be exactly right.

Well—we will be exactly right to the degree that earth is exactly round. Everyone knows the earth is not really round. It has mountains, valleys and oceans, but the earth is pretty big and the irregularities are relatively small.

We might also worry about the survey. Sloppy work on part of the king's survey team would also keep us from being exactly right. We will see, however, the survey team did pretty good work.

Treating the distance from the North Pole to the equator as a quarter circle, where the full circle is the circumference of the earth, the circumference of the earth would be exactly 40,000 km if, as the king intended, the quarter circle were to be exactly 10,000,000 m. To find the radius of this circle, we can use the definition of π.

$$\pi = \frac{C}{D} = \frac{C}{2R}$$

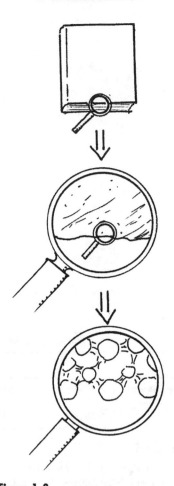

Figure 1-8

This is what the edge of a textbook looks like (a) normally, (b) microscopically and (c) atomically

Since π is defined as the circumference C of a circle divided by its diameter D, it is equal to the ratio of the circumference to twice the radius R. Sometimes people remember this definition as a formula

$$C = 2\pi R$$

Since the circumference is 4,000 km, we can calculate the radius of the earth to be

$$R = \frac{40,000 \text{ km}}{2\pi} = 6,366.19772 \text{ km}$$

In the perfect world of arithmetic, this is true. Your calculator cranks out these numbers with great authority. Bringing arithmetic to the real world, however, we know better than to trust all of those numbers. All of the numbers after the decimal point are fractions of a kilometer, and mountains are several kilometers high. To believe everything up to the last 2, we would need an earth that was round to the closest centimeter. Kick the dirt with your shoe, and you would mess up that number.

If we check our calculation with the real world, we can see how well the original definition holds up. The calculated value of 6,366 km puts the king's perfectly round earth only 5 km short of the real earth's average radius.

It might seem that 5 km is a big error, that being about the height of Mt. Blanc, the tallest mountain in France. On the other hand, we see that the equatorial radius is larger than the average radius by over 7 km. In all, the king's survey team came back with a value that is better than the earth is round. In fact, if you take a numerical average of the earth's equatorial and polar radii, you get a value that is only 1.5 km more than that of the king's survey.

The conclusion should be that the original definition of the meter holds to three significant figures, where the earth stops being round. In fact, it almost holds to four significant figures, where the real garbage begins.

	km	± km
Average	6371.315	0.437
Equatorial	6378.533	0.437
Polar	6356.916	0.437

Table 1-4

Radius of the Earth. (Source CRC Handbook, 53rd Edition. pg. F-150)

Limits of Perfection

If we want four significant figures of perfection, it is clear that we need a better earth. Our earth is too fat and not tall enough. You might wonder if the king's survey team knew about that.

As it turns out, they did.

Not only did they know about the bumps and dents in the earth, they knew that it should spin out and get wider at the equator. A perfectly spherical earth was something they knew to be only a first approximation to reality. But they decided to go with that approximation anyway.

As they measured the distance from Dunkirk, on the north coast of France, directly south to Barcelona, which is back at sea level in Spain, they kept track of their altitude. To get the accuracy they did, they needed to measure the circumference of an imaginary sphere down at sea level. They decided, however, not to account for the change in shape of that sphere due to the rotation of the earth. They decided to think in terms of a perfect sphere.

There is nothing wrong with that. Thinking in terms of a simplified model of reality is an essential skill in analytical thinking. It's something we have to do.

Nothing is perfect. Only our ideas of it are perfect, and they are imperfect to the extent that they don't match up with reality. It is important to know that. It is important to recognize those troublesome difficulties you set aside in your mind and be aware of how they can limit the perfection of your answer.

Even something like π, the familiar number that describes the ratio of the diameter to the circumference of a circle, cannot be perfect. The story of how we know the value of this number as accurately as we do could fill volumes. In fact, the value of the number itself could fill volumes. It has been calculated to millions of places and never found to start repeating itself.

By the time of the French revolution, the value of π had been calculated to fifteen places. They knew how to calculate it better than that, but they didn't have calculators. For reference, we will give the first 26 places.

$$\pi = 3.14159\ 26535\ 89793\ 23846\ 26433\ ...$$

Some people do take pleasure in knowing trivia. From time to time, you will encounter a person who has bothered to memorize π to this or greater accuracy. These numbers are not quite garbage, as they do represent some fact of mathematical significance. The interest in π to such accuracy does not, as we shall see, have anything to do with calculating the circumference of the earth.

For many purposes, the common approximation of twenty-two sevenths is good enough.

$$\pi \cong \frac{22}{7} = 3.1428\ 57143\ ...$$

We can see that the first three digits are correct and the next one is close. If the French team had used twenty-two sevenths, their result would have only been disturbed in the fourth significant figure.

$$R = \frac{40,000\ \text{km}}{2(22/7)} = 6,363.63636\ \text{km}$$

Instead of being 5 km short of the average radius of the earth, they would have been 8 km short. Again, the result is better than the earth is round.

There are, of course, better approximations to π. An easy one to remember is 355/113.

$$\pi \cong \frac{355}{113} = 3.14159\ 29203\ ...$$

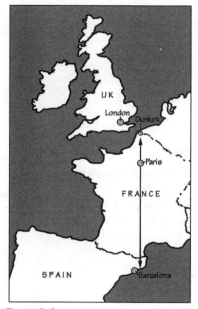

Figure 1-9

The original meter was based on a survey from Dunkirk to Barcelona through Paris.

This one is nice because of the doubling of 1, 3, and 5. It also gives π to almost eight significant figures.

To make full use of this approximation, however, we would need an earth that is round to at least seven figures of accuracy. Since the first four figures describe the radius of the earth in kilometers, the next three describe the radius of the earth in meters. We would need an earth that had no mountains more than about waist high.

Table 1-5

The radius of a mathematically perfect earth to the closest femptometer. Such an object could not have mountains as big as protons.

	km	m	mm	μm	nm	pm	fm
R=6371.	315	000	000	000	000	000
4 sig figs	7	10	13	16	19	21

To make full use of the ten significant figures of π found in many calculators, we would need an earth that is round to millimeters. Another three figures of accuracy, and we would need an earth round to the closest micrometer. That would be an earth that had no mountains more than waist high on a germ.

To justify the need for π to 19 figures of accuracy, our earth would need to be round to the closest picometer. Since atoms are on the order of hundreds of picometers in diameters, they would make mountains too high for such a perfect earth.

Finally, to justify using π with an accuracy of twenty-six significant figures for the purpose of calculating the circumference of an object the size of the earth, one would need to know the radius of that object to at least twenty-five significant figures. The object would need to be made out of some material much finer than nuclear particles.

Mental Models

> **Common sense is the collection of prejudices acquired by the age of eighteen.**
> —*ALBERT EINSTEIN*[3]

Figure 1-10

The act of thinking involves the making and using of mental models.

The idea of a perfectly spherical earth is an example of something we call a mental model. It is a geometrical model, so we can visualize it. Some of the mental models we will encounter are more algebraic and more difficult to visualize. This will be nothing new to you, however, because you are used to dealing with mental models much more sophisticated and abstract than any of the mathematical models we will deal with.

In fact, inductive and deductive logic amounts to little more than the creation and testing of mental models. We all need this kind of thinking just to get along in society. Each time you make a new friend, you form a

[3] *Scientific American* Feb 1976

mental model of that person. With that model you predict with more or less accuracy what that person will do in a given situation.

Each time you smile and wave at your new friend, you are testing that model. You probably expect your friend to smile and wave back, and the expected response confirms your model. If your new friend grunts and makes an obscene gesture at you, your opinion of that person may need revision; but then again, upon some reflection, you may conclude that you have just cut your new friend out of a parking space.

The more you know about a person, the more sophisticated your model of that person becomes, but no mental model is ever more than an approximation of reality. People sometimes lose sight of this fact. They tend to confuse their models with reality itself. When it happens in other people, this phenomenon is called prejudice. When it happens in our own thinking, well—of course, it doesn't happen in our own thinking.

The history of science has been a continual revolution in which physical laws continue to be discovered and old models of the universe give way to new. From time to time it might seem that we have such a good understanding of the universe that no future revision of our theories is possible. When that happens, we tend to ignore inconsistencies, thinking of them as minor or accommodating them as special cases. Prejudice thus tends to creep into our thinking about physical, as well as social, matters. We become too fond of our mental models and confuse them with reality itself.

For example, Einstein's theory of relativity, first published in 1905, flew in the face of commonly accepted principles of the day. The accepted truth fell before the upstart theory only with great pain and suffering on the part of people who loved the old ways of looking at things. Einstein's Nobel prize citation in 1921 nowhere mentions the theory of relativity, as many people still hoped they could prove it wrong. Instead, the prize was for "his services to Theoretical Physics, and especially for his discovery of the law of the photoelectric effect."

Einstein's explanation for the photoelectric effect became one of the cornerstones of quantum mechanics, and yet Einstein did not believe in quantum mechanics.[4] Just as his own critics had hoped that the theory of relativity was wrong, Einstein always hoped this new way of thinking was wrong.

Figure 1-11
You smile and wave.

---

[4]Albert Einstein and Niels Bohr loved to argue, hours on end. One of their favorite topics was the Bohr model of the atom. Einstein liked it. Bohr didn't.

In Bohr's opinion the new quantum mechanics, developed by Bohr's own graduate students at his Copenhagen institute, made the beautiful Bohr model obsolete. The Copenhagen interpretation of quantum mechanics held that an electron in an atom is not really any place in particular but can be described only in terms of its probability of being found at some particular point.

This Einstein could not believe. "God does not play dice," he insisted. "Stop telling God what to do," Bohr answered on more than one occasion.

Nowadays, quantum mechanics tends to be accepted as truth in all of its glory. A little skepticism is a wise thing to keep, however, no matter how sophisticated or well accepted a model of reality seems to be. The trick is to understand the limitations of our models as well as their successes. That is not a trick to be lightly regarded in our study of physics since the whole thrust of that study will be to fit the exactitude of mathematical logic to the inexactitude of the real world.

All of geometry is based on mathematical contrivances more perfect than any real triangle, circle, or cube. Some of the mental models we will encounter are more algebraic than triangles, circles, and cubes and therefore more difficult to visualize. Although expressed in algebraic form, these mathematical relationships are also modeled after reality but are more perfect than the real thing. It is the business of science to invent such models and then test them to see how close they come.

It is said that there are no real physics problems that are simple enough to be solved. Reality is always too complex for us to take into account all of the variables that influence any real life situation. We can only solve some approximation of reality and then check to see if the results are good enough for a given purpose. Our judgment of what is good enough always depends, then, upon the purpose at hand.

One of the main virtues of science is that the ideas of how things should work are simple enough to test with numbers. In music and politics, mental models are far more complex and the tests involve far more complicated judgments than in science. Just as numbers simplify the application of scientific ideas to the real world, they also simplify the exploration of the point at which theory and practice must necessarily diverge.

Proportions

Thus far in this chapter, the crucial points have been an introduction to the metric system and an understanding of error. These have been discussed at some length because they are complex. There are two more things that you will need throughout this book, but they are much simpler ideas. They are the **conversion factor**, an algebraic tool for dealing with units, and the algebraic idea of the simple **proportion**. They are both concepts that we frequently use in our daily life, whether or not we think of them as being mathematical.

First, let us address the **direct proportion**. It is one of the most useful mathematical models and the simplest. There are plenty of things in daily life that are proportional to one another. For example, you have certainly noticed that when you get in your car and head down the freeway, the distance you go depends on the time. You also realize that this is only a first approximation. If the speed of traffic and the way you drive are constant, the model works better.

For predicting results and making decisions in daily life, we seldom need to bring the mathematical formalism of algebra to bear on this simple model. To make our point, however, we will express this proportionality as an equation. To say that the distance traveled, let's call it *d*, depends on the time spent on the freeway, *t*, is to say that *d* is *proportional* to *t*. We would write that like this:

$$d \propto t$$

where the symbol $\propto$ is read "is proportional to."

A feature of a proportionality is that it can always be written as an equality if we add a *proportionality constant*. Here, we can use the velocity *v* as a proportionality constant between distance *d* and time *t*:

$$d = vt$$

To the degree that *v* is constant, twice the *t* will bring about twice the *d*. Ten times the *t* will bring about ten times the *d*, and so on.

If you want to figure out how long it will take you to get from one place to the other, you can do a little algebra. Dividing both sides by the velocity, we can express this equation in terms of time.

$$t = \frac{d}{v}$$

Units and Conversion Factors

Suppose you want to take a drive from a really nice place like cool foggy San Francisco to find summer in some place like San Jose, a distance of 76 km.

Figure 1-12

A Sunday drive to find a summer day.

Now suppose that you hop into your antique English sports car. As long as we are supposing, let it be a Morgan with wind in your face, and it might as well be a beautiful day. You notice that Sunday traffic is moving along at 55 miles per hour, British Imperial measure. Putting the data into the above mathematical model, we have

$$t = \frac{(76 \text{ km})}{(55 \text{ mi/hr})}$$

Little difficulties with units like these in daily life are not really a problem. Since a mile is a little more than a kilometer and a half, we can see that its going to take us something like an hour to make the trip. For doing more difficult problems, however, let us build a more powerful tool for dealing with units.

The tool we need is called a **conversion factor**. We can build one that will get time in hours out of the denominator of the denominator and, in the same operation, change it into minutes.

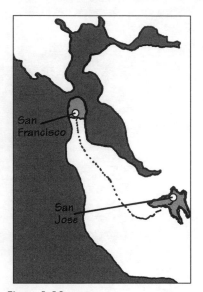

Figure 1-13

The driving distance from San Francisco to San Jose is 76 km.

The idea is to multiply by a fraction where the numerator is equal to the denominator. Such a fraction is equal to one, or *unity*. In this case, we would like a conversion factor that has hours in the denominator.

$$1 = \left(\frac{60 \text{ min}}{1 \text{ hr}} \right)$$

We know 60 minutes is exactly equal to an hour, so 60 minutes divided by an hour is exactly equal to one, in the sense that, by "one", we mean unity. Since one times anything is that thing itself, it won't disturb the equality in our expression for time if we multiply by it to get the following

$$t = \frac{(76 \text{ km})}{\left(\frac{55 \text{mi}}{\text{hr}} \right)} \cdot (1)$$

$$= \frac{(76 \text{ km})}{\left(\frac{55 \text{mi}}{\text{hr}} \right)} \cdot \left(\frac{60 \text{ min}}{1 \text{ hr}} \right)$$

Hours in the denominator of the conversion factor get rid of the hours in the denominator of the denominator, and we are left with minutes in the numerator.

In the same way, you can get rid of miles and kilometers if you know that a mile is 1.61 kilometer. The conversion factor

$$1 = \left(\frac{1 \text{ mi}}{1.61 \text{ km}} \right)$$

should have kilometers in the denominator since that is what we are trying to get rid of in the numerator.

$$t = \frac{(76 \text{ km})}{\left(\frac{55 \text{mi}}{\text{hr}} \right)} \left(\frac{60 \text{ min}}{1 \text{ hr}} \right) \left(\frac{1 \text{ mi}}{1.61 \text{ km}} \right)$$

$$= 51 \text{ min}$$

So sunny San Jose is only 51 minutes from foggy San Francisco. Start at eleven, and you're there by noon.

Suppose now that you do not know that a mile is 1.61 kilometers, or suppose that you need more than three significant figures of precision. We can start with the identity that a mile is equal to itself.

$$1 \text{mi} = 1 \text{mi}$$

Using the definition that a mile is exactly 5,280 feet and an inch is exactly 2.54 cm we can build such conversion factors as

$$1 = \left(\frac{5,280 \text{ ft}}{1 \text{ mi}}\right)$$

$$1 = \left(\frac{2.54 \text{ cm}}{1 \text{ in}}\right)$$

If we string them all together and multiply by our identity, we get:

$$1\,\text{mi} = 1\,\text{mi}\left(\frac{5,280 \text{ ft}}{1 \text{ mi}}\right)\left(\frac{12 \text{ in}}{1 \text{ ft}}\right)\left(\frac{2.54 \text{cm}}{1 \text{in}}\right)\left(\frac{1 \text{ m}}{100 \text{ cm}}\right)\left(\frac{1 \text{ km}}{1000 \text{ m}}\right)$$

$$= 1.609344 \text{ km}$$

Each of these conversion factors are based on a legal definition. It might seem that 2.54 cm is only good to three significant figures, but an inch was redefined by the standards people in 1959 to be exactly equal to 2.54 cm. Therefore, this equality is exactly true. It is digital.

In the second example problem in this chapter's problem set, we will see that the older definition of the inch made for a mile that was a tiny bit larger in the last significant figure. Using the old definition, the third conversion factor would still be correct, but it would have an error. It would be analog.

In 1959, they shrunk the mile! And hardly anyone noticed. The distance from San Francisco to San Jose changed by more than that in the last earthquake.

SUMMARY OF CONCEPTS

The **metric system,** where larger and smaller units for the same quantity are related to each other by powers of ten, has the virtue of using the same Latin prefixes to identify relative sizes of all the different quantities being measured. For example, the basic unit of length is the meter; the millimeter, centimeter, and kilometer are, respectively, a thousandth, a hundredth, and a thousand meters.

Millimeter	1 mm $= 10^{-3}$ m	0.001	meter
Centimeter	1 cm $= 10^{-2}$ m	0.01	meter
Kilometer	1 km $= 10^{3}$ m	1,000	meters

The same Latin prefixes can be used for all the other metric quantities except the centi-, which is generally only used in conjunction with the meter. A milliliter is a thousandth of a liter, where a liter was originally defined as the volume of a cube 10 cm on a side, so a milileter is the volume of a cube 1 centimeter on a side. That makes it smaller by a factor of ten in each of three dimensions. In the same way, a kilogram, originally defined as the mass of 1 liter of pure water, is one thousand times as big as the gram.

Data in a problem consist of both **numbers** and **units**. Units should always be inserted into the general solution of a problem along with the numbers, as a check on the algebra used to obtain the general solution.

Units may also serve as tools for reasoning. When you know the units you want are not the units you have, you can frequently reduce the problem to the need for a conversion factor or two. A **conversion factor** is a fraction equaling unity (1) by having a numerator equal to the denominator but expressed in different units. The units of the numerator and the denominator are chosen to cancel the units you are trying to get rid of and replace them with the ones you want.

While we are converting to the metric system from the English system, it will be useful to know a few common conversion factors by memory. Some good ones are:

1 inch = 2.54 cm (exactly)
1 mile = 1.61 km
1 gallon = 3.78 liter

One of the mathematical models that we will use frequently in this book is the direct **proportion**. The variable x is said to be directly proportional to y if equal increases in one variable are accompanied by equal increases in the other. This is expressed:

$$x \propto y$$

which says "x is proportional to y."

The nature of proportionalities is that they can always be expressed in terms of an equality sign with a proportionality constant:

$$x = Cy$$

The results of calculations should never be reported to more **significant figures** than justified by the data. Generally, an answer should not be reported to more than one significant figure greater than the least accurate data. Keeping insignificant figures is incorrect, because they imply that you know more than you actually do.

The number of significant figures is the number of digits needed to express data in **scientific notation**, or **power-of-ten** notation. Write the number as a single digit followed by a decimal and however many figures you know accurately multiplied by ten to a power. For example, if you know an airplane is traveling at 800 km/hr to the nearest 10 km/hr, you would write it 8.0×10^2 km/hr.

EXAMPLE PROBLEMS

Try to do the Examples yourself before looking at the Sample Solutions.

1-1 The legal definition of an inch was changed in 1959 to make a rule of thumb exact. Instead of saying how many inches there are in a meter, the number of centimeters in an inch was rounded off to an exact number. As a result, the United States statutory mile became a tiny bit smaller.

Given: Prior to 1959, a meter was equal to exactly 39.37 inches. Now, an inch is defined to equal exactly 2.54 centimeters. The definition for a mile, 1 mile = 5,280 feet (exactly), remained the same, as did the definition that a foot is 12 inches.

Find: (a) By the old definition of an inch, how long was a mile in meters?
(b) By the new definition of the inch, how long is a mile in meters?
(c) By how many millimeters did the statutory mile become shorter in 1959?

BACKGROUND

This problem is an exercise in conversion factors. We start with the identity 1mi = 1mi and multiply by one or more conversion factors, which are fractions equal to unity (1) and therefore do not disturb the equality.

1-2 Quadrille paper is excellent for doing physics homework. Both horizontal and vertical rulings, marked lightly in blue ink, form a grid to help keep your work straight and beautiful in both directions on the page. A ruler marked in English and metric scales is also handy to have on your desk, not only for measuring but mostly for drawing straight lines. Anyone who has a micrometer caliper on his or her desk, however, might be a little extreme…

Given: You are extreme. With your ruler you measure a piece of quadrille paper to be 27.94 centimeters long and 21.59 centimeters wide. With your micrometer caliper, you measure the paper to be 1.5×10^{-3} centimeters thick.

Find: (a) Calculate the area of the paper to the proper number of significant figures.
(b) Calculate the volume of the paper to the proper number of significant figures.

BACKGROUND

This problem illustrates the role of limited accuracy in dealing with numbers. It also points out the difference between precision and accuracy. The micrometer caliper is far more precise than an ordinary ruler in measuring tiny distances. It uses a fine pitched screw that makes one or two turns per millimeter. From the fraction of a turn, one can typically estimate distances to the thousandth of a millimeter, which is a micrometer. Even so, even with its great precision, we see that measurement with the micrometer is what limits the accuracy of the result because the thing being measured is also very tiny. Accuracy is relative to that which is being measured.

1-3 Billie Farmer is hauling good Kansas wheat at harvest time. At the grain elevator, the truck is unloaded and credit is given in bushels, the proper way to measure wheat in Kansas. The bushel is a measure of volume based on the traditional English doubling system.

Given: Billie's grain truck bed measures 6.65 m by 2.25 m and has sides that will hold wheat up to a level of 1.5 m deep. A bushel is defined as 8 gallons and a gallon is 231 in³. Remember that 1 inch is defined as 2.54 cm.

Find: (a) What is the volume of the wheat in m³?
(b) How many bushels of wheat are on the truck?

BACKGROUND

We should make the assumption that the volume is length times width times height—the familiar volume of a rectangular solid. Of course, the real world is never that simple, but our assumption will have to do. We have no better data to go on, and it seems like it ought to be a reasonable approximation. The sides of the truck may not be exactly square, so measuring to the closest centimeter is probably pushing it. The height, however, is less well defined. The grain will be piled up in the center, so the height is the limiting factor in the accuracy of our result.

1-4 "Time is money," or so the saying goes. When April Lake's swim team decided to sell candy bars as a fund-raiser for a swim meet, the relationship between time and money was obvious. Faced with a choice between selling or eating, she decided that athletic performance could depend on maintaining enthusiasm for selling. Enthusiasm being relatively constant, the money earned was roughly proportional to time on task.

Given: On Monday, Sue sold for 2 hours and earned $12.00. Tuesday, Sue sold for 1 hour and earned $6.50, and Wednesday, Sue had a burst of enthusiasm and sold for a 4-hour spree, earning $23.00. The goal was to earn $100.00.

Find: (a) Write the proportion between dollars D and time t, using a proportionality constant E, for enthusiasm.

(b) Solve for the constant E using each data point to see to what degree enthusiasm remained roughly constant. What are the units of E?

(c) Predict, using an average value for E, how many hours it will take to reach the goal.

BACKGROUND

This is a problem on proportionalities. If you have two things that are proportional to one another, you can experiment and take data to find a constant number that relates the two.

1-5 To illustrate the relationship between weight and gravity, a physics instructor climbs on a scale in class. Looking down at the scale in horror, he exclaims that gravity must have increased while he was on vacation.

Given: The physics instructor's weight, before vacation, was 931 newtons and his mass was 95 kilograms. After vacation his weight has mysteriously increased to 960 newtons.

Find: (a) Express the proportionality between weight w and mass m as an equality using a proportionality constant, which is nearly constant on the surface of the earth.

(b) What value of g is indicated by the instructor's mass and weight before he left for vacation?

(c) If the increase in weight had been due to a new and stronger gravity, what would be the value of g after vacation, assuming a constant mass?

(d) If, as is more likely, the increase in weight is due to an increase in mass, what is the new mass, assuming a constant g?

BACKGROUND

The idea that mass and weight are proportional to one another is not obvious. It is, however, an important idea which will be discussed at greater length in Chapter 5.

SAMPLE SOLUTION 1-1

Given:

Prior to 1959, a meter was equal to exactly 39.37 inches. Now, an inch is defined to equal exactly 2.54 centimeters. The definition for a mile, 1 mile = 5,280 feet (exactly), remained the same, as did the definition that a foot is 12 inches.

Find:

(a) By the old definition of an inch, how long was a mile in meters?

(b) By the new definition of the inch, how long is a mile in meters?

(c) By how many millimeters did the statutory mile become shorter in 1959?

> Part of your job is to learn proper use of your calculator, so you should get it out and check our results. You should also set the book aside and set up the problem yourself.

$$\text{(a)} \ 1 \ mi_{old} = 1mi\left(\frac{5,280 \ ft}{1 \ mi}\right)\left(\frac{12 \ in}{1 \ ft}\right)\left(\frac{1m}{39.37 \ in}\right)$$

$$= \boxed{1609.347219 \ m}$$

(Precision limited only by calculator)

> use exponent key on calculator

$$\text{(b)} \ 1mi_{new} = 1mi\left(\frac{5,280 \ ft}{1 \ mi}\right)\left(\frac{12 \ in}{1 \ ft}\right)\left(\frac{2.54 \times 10^{-2} \ m}{1 \ in}\right)$$

$$= \boxed{1609.344000 \ m}$$

> calculator won't show these zeros

$$\text{(c)} \ \Delta \ mile = (1 \ mi_{old}) - (1 \ mi_{new})$$

$$= 0.003219\left(\frac{mm}{10^{-3} \ m}\right)$$

$$= \boxed{3.219 \ mm}$$

DISCUSSION

The units in the conversion factors are selected to cancel the units you are trying to get rid of. If one conversion factor doesn't leave you with the units you want, just use another conversion factor. Don't do the arithmetic at each step. The arithmetic is not really the point. What is important is how you set the problem up. That is easier to see if you do not actually do the calculations a little at a time but instead show how the calculations are to be done. So string up a bunch of conversion factors and do the arithmetic all at once as shown in this example.

To find the change in part (c) we subtract answers for parts (a) and (b) and convert to more appropriate answers.

In this particular problem, the only limit on the known number of significant figures is your calculator. This is one of the rare problems where we can carry out the computations to as many significant figures as you please because all the data come from definitions, and are therefore exact.

(a) A = lw

= (27.94 cm)(21.59 cm)

= $\boxed{603.2 \ cm^2}$

calculator says 603.2246

(b) V = lwh

= $(27.94 \ cm)(21.59 \ cm)(1.5 \times 10^{-3} \ cm)\left(\dfrac{1ml}{1cm^3}\right)$

= $\boxed{0.90 \ ml}$

calculator says 0.9048369

Given:

You are extreme. With your ruler you measure a piece of quadrille paper to be 27.94 centimeters long and 21.59 centimeters wide. With your micrometer caliper, you measure the paper to be 1.5×10^{-3} centimeters thick.

Find:

(a) Calculate the area of the paper to the proper number of significant figures.

(b) Calculate the volume of the paper to the proper number of significant figures.

DISCUSSION

The area in part (a) should be computed to four significant figures, because both length and width are known to that accuracy. The calculator says 603.22460, but everything after 603.2 is garbage. Even the last significant figure is a little shaky. Add a little to the length, for example, so you multiply $27.944 \times 21.59 = 603.31096$, which would round to 603.3. While the last significant figure is not solid, however, it's not pure garbage. We should keep it because it has some significance, and throwing it away might introduce a round-off error.

The volume in part (b) is limited to two significant figures by the accuracy with which we know the thickness. The calculator says 0.9048369, but only the 0.9 is solid. Add a bunch of fours to the thickness, making it 1.54444×10^{-3}, which would still round off to the data given, and the calculator says 0.93164420. When we report the answer as 0.90, we lack confidence in the final zero but we report it because we know it can't be much larger than a three. We could report the answer as 0.90 ± 0.03 if we were all that worried about it.

You probably don't have a micrometer caliper on your desk, but if you did it would look like this.

← Workpiece being measured

← Spindle runs on a screw having a pitch of one or two turns per millimeter.

← Sleeve has a main scale reading directly in millimeters.

Thimble turns with spindle on same screw. Has a micrometer scale which divides each main scale division into 100 parts in terms of parts of a turn of the screw.

SAMPLE SOLUTION 1-3

Given:

Billie's grain truck bed measures 6.65 m by 2.25 m and has sides that will hold wheat up to a level of 1.5 m deep. A bushel is defined as 8 gallons and a gallon is 231 in³. Remember that 1 inch is defined as 2.54 cm.

Find:

(a) What is the volume of the wheat in m³?

(b) How many bushels of wheat are on the truck?

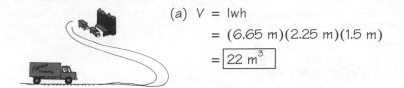

(a) $V = lwh$

$= (6.65 \text{ m})(2.25 \text{ m})(1.5 \text{ m})$

$= \boxed{22 \text{ m}^3}$

(b) $V = \left(22 \text{ m}^3\right)\left(\dfrac{1 \text{ in}}{2.54 \times 10^{-2} \text{ m}}\right)^3\left(\dfrac{1 \text{ gallon}}{231 \text{ in}^3}\right)\left(\dfrac{1 \text{ bushel}}{8 \text{ gallons}}\right)$

$= \boxed{726 \text{ bushels}}$

DISCUSSION

In part (a) the calculator says 22.44375, but we are limited to two significant figures by our height, so everything after the decimal place is garbage. If we were to use 1.54 for our height, which would still round to 1.5, our given data, the calculator would say 23.04225. Therefore, even the second significant figure has some uncertainty in it.

In part (b) we illustrate the trick of taking the cube of the conversion factor since we are trying to get rid of cubic meters and cubic inches, turning them into cubic inches. Then we use another conversion factor to get rid of gallons. Finally, we can get rid of gallons with a third conversion factor to arrive at bushels, the desired unit.

(a) $D = Et$

(b) $E = \dfrac{D}{t}$

$E_1 = \dfrac{\$23.00}{4\ hr}$

$\quad = \boxed{5.75 \dfrac{\$}{hr}}$

$E_2 = \dfrac{\$12.00}{2\ hr}$

$\quad = \boxed{6.00 \dfrac{\$}{hr}}$

$E_3 = \dfrac{\$6.50}{1\ hr}$

$\quad = \boxed{6.50 \dfrac{\$}{hr}}$

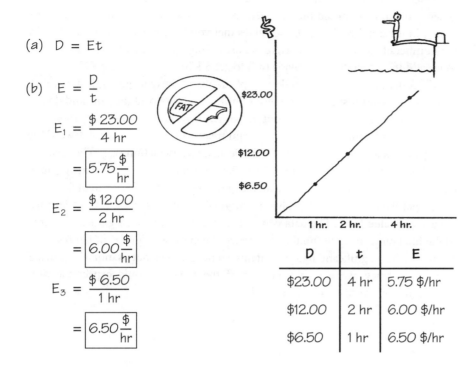

D	t	E
$23.00	4 hr	5.75 $/hr
$12.00	2 hr	6.00 $/hr
$6.50	1 hr	6.50 $/hr

$23.00

$12.00

$6.50

1 hr.   2 hr.   4 hr.

(c) $D_{Average} = \dfrac{\$23.00 + \$12.00 + \$6.50}{3} = \$100$

$E_{Average} = \dfrac{5.75\ \$/hr + 6.00\ \$/hr + 6.50\ \$/hr}{3} = 6\ \$/hr$

$t = \dfrac{D}{E_{Average}}$

$\quad = \dfrac{(\$100)}{(6\ \$/hr)}$

$\quad = \boxed{17\ hr}$

Given:

On Monday, Sue sold for 2 hours and earned $12.00. Tuesday , Sue sold for 1 hour, and earned $6.50, and on Wednesday, Sue had a burst of enthusiasm and sold for a 4-hour spree, earning $23.00. The goal was to earn $100.00.

Find:

(a) Write the proportion between dollars D and time t, using a proportionality constant E, for enthusiasm.

(b) Solve for the constant E using each data point to see to what degree enthusiasm remained roughly constant. What are the units of E?

(c) Predict, using an average value for E, how many hours it will take to reach the goal.

DISCUSSION

If you define the constant, E, in such a way that it increases with the amount of dollars earned, D, then the relationship is $D = Et$. It is equally correct to write the equation: $DC = t$ using a different proportionality constant C for candy consumption. If you did that, then your answers for C would be the inverse of enthusiasm.

The choice of which side of the equality to put your proportionality constant changes the meaning of the constant. Your choice depends on how you want to look at the situation.

For part (b), solve the equation in part (a) for E, and find a numerical value for each of the three sets of information. The units of the constant E are dollars per hour. These numbers look close enough to verify a rough proportional relationship between hours and dollars. Notice that a data table and a graph are handy for looking at a number of data points. When using a data table, a sample calculation will frequently suffice.

Since we do not have any one value for E that is better than any other, to get a general solution we should find an average value to work with. Average values are found, in general, by adding all the values that you have together, and dividing by the number of values you used. Here, we are averaging together three values for E, so we add them together and divide by 3 to get 6 $/hr.

Of course, this is only a crude way of taking an average, giving just as much weight to the 4-hour spree as the 1-hour outing. Adding up all the time and dollars is a little better, but the result is about the same.

Some people might notice that a careful treatment would use a fancy weighted average, called a mean, with a small sample standard deviation σ_{n-1}. The result of careful analysis is that you average 5.93 ± 0.28 dollars per hour. All that means is that sixty percent of the time you would expect to get within 25¢ or so of $6/hr.)

To get the answer for (c), solve the original equation for the time, t, and use the average value for E. The data is only accurate to about two significant figures, so the final value for time should be rounded down to two significant figures. Here again, the last significant figure contains some garbage but is not all garbage. (Using our fancy average, sixty percent of the time we could expect the time needed to be between 15.4 hr and 17.4 hr.)

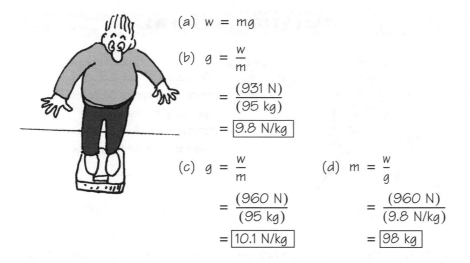

(a) $w = mg$

(b) $g = \dfrac{w}{m}$

$= \dfrac{(931 \text{ N})}{(95 \text{ kg})}$

$= \boxed{9.8 \text{ N/kg}}$

(c) $g = \dfrac{w}{m}$

$= \dfrac{(960 \text{ N})}{(95 \text{ kg})}$

$= \boxed{10.1 \text{ N/kg}}$

(d) $m = \dfrac{w}{g}$

$= \dfrac{(960 \text{ N})}{(9.8 \text{ N/kg})}$

$= \boxed{98 \text{ kg}}$

Given:

The physics instructor's weight, before vacation, was 931 newtons and his mass was 95 kilograms. After vacation his weight has mysteriously increased to 960 newtons.

Find:

(a) Express the proportionality between weight w and gravity g as an equality using a proportionality constant m, which we call mass.

(b) What value of g is indicated by the instructor's mass and weight before he left for vacation?

(c) If the increase in weight had been due to a new and stronger gravity, what would be the value of g after vacation, assuming a constant mass?

(d) If, as is more likely, the increase in weight is due to an increase in mass, what is the new mass, assuming a constant g?

DISCUSSION

It is good form to start a physics problem by explicitly stating the appropriate mathematical model. It is like saying, "Oh, this is one of those proportionality problems." This shows where you are starting from (which might well be good for part credit if you later make some stupid algebra mistake) at the same time that it reduces the chances that you will make an algebra error by trying to do too much in your head.

The model is then solved algebraically to get the general solution to the problem before inserting data, both numbers and units, to get the particular solution or answer. Recording your thoughts on paper in this way allows you or someone else to find your error should your answer be incorrect.

In part (b) the answer is properly rounded to two significant figures since mass is only known to that accuracy.

In part (c), however, the answer is properly rounded to three significant figures, one more than the number describing the mass. This is an example where following the rule of thumb *plus one* is a good idea. It is correct to keep an extra significant figure because error is really a relative thing. The leading digit in the answer is 1 whereas the leading digit in the least accurate data is 9. Rounding off to the same number of significant figures would therefore leave you with less accuracy than when you started. We can see that the answer in part (c) is only a little bigger than the answer in part (b), where the precision is to tenths of a N/kg. The unqualified rule of thumb would have us round the answer in part (c) to the closest N/kg, which is ten times the error in the slightly smaller number. Clearly, the rule of thumb needs to be qualified.

In part (d), we are back to two significant figures because of the limits with which we know the answer to part (b).

ESSENTIAL PROBLEMS

Be sure you can do the Example Problems (without peeking at the Sample Solutions) before working these Essential Problems.

1-6 "A pint is a pound the world round," or so the old saying goes. Actually, the amount in British Imperial measure depends somewhat upon what we are talking about. In England ale is commonly served in "Full Measure" pint sized glasses, and weighs a bit more than the same full measure of water. The difference, while considerable in nutritional value, is slight in weight.

Given: There are 8 pints to a gallon. One gallon of good English Bitter, ready to drink, has a mass of 3.831 kg. One gallon of pure clear water, also at drinking temperature, has a mass of 3.785 kg. A pound, on the other hand, is anything that has a mass of 0.4535 kg. There are, by definition, 16 ounces in a pound.

Find: (a) How heavy is a pint of English Bitter in pounds?
(b) How heavy is a pint of pure clear water in pounds?
(c) How much more massive is a pint of Bitter than a pint of water in ounces?

1-7 Grace is making a Ping-Pong table for her new apartment. She must get the lumber home and paint it to match her bed and two chairs. To buy paint she needs the area of the plywood. To estimate the weight of the table (she needs to be able to move it by herself), she needs to know the volume.

Given: The table top is 1.52 m wide, 2.74 m long and is 1.6×10^{-2} m thick.

Find: (a) Calculate the area of the top of the Ping-Pong table to the proper number of significant figures.
(b) Calculate the volume of the piece of wood to the proper number of significant figures.

1-8 Rat Man needs to save his sidekick, Sparrow, who has been nefariously locked in an air-tight safe for "safe-keeping." Knowing that Sparrow is prone to panicking, Rat Man should rescue his young friend before Sparrow has breathed all the air in the safe one time.

Given: The safe, a bank vault, is 2.4 meters by 1.5 meters on the floor and 2.2 meters high. Breathing normally, before panicking, Sparrow uses 0.92 liters of air each time he breathes. Remember, one liter is 1000 cm^3.

Find: (a) What is the volume of the air in the safe in m^3?

(b) How many breaths of air are in the safe?

1-9 Harry Homeowner discovers that he uses electricity faster in his home the more rooms he keeps lighted.

Given: He finds that he uses 2.9 kilowatt hours (kWh) per day when he is alone and only keeps one room at a time lit. He finds that it takes 9.3 kWh / day to keep 3 rooms at a time lit and 11.8 kWh / day when his family keeps 4 rooms at a time lit.

Find: (a) State a hypothetical proportionality between electricity used, E, and the number of rooms, R, using an equality sign and a proportionality constant C. Define C so that a larger value would reflect a larger amount of the electricity used, E.

(b) Solve this equality for the proportionality constant C and test its value on the data.

(c) Predict, using an average value for C, how many kilowatt hours it would take per day to keep all 8 rooms of Harry's house lit.

1-10 Carla Medina, taking an interplanetary vacation, is on a whirlwind tour of club locations in the solar system. Waking up one morning, after a heavy night doing party things in a singles bar, she can't quite remember where she is. "If it's Monday, it must be Saturn," she muses, "but maybe we're still on Uranus." To get a hint, she steps onto the bathroom scale. Like all interplanetary travelers, Carla realizes her weight w is proportional to gravity g, assuming her mass m is constant.

Given: Carla's weight, when she last checked it on Uranus, was 366 N (newtons). She figured that meant that her mass was 54 kg (kilograms). She finds that she now weighs 603 N.

Find: (a) Express the proportionality between w and g, assuming m is the proportionality constant.

(b) What is g on Uranus?

(c) In the unlikely event that Carla is still on Uranus, what is her new mass?

MORE INTERESTING PROBLEMS

1-11 A cubit an ancient unit of measure. It is the distance from the elbow to the fist plus one to four fingers, depending on the size of the person doing the measuring. Dudley, finding himself without tools, decides to measure the width of a tool cabinet at a garage sale in cubits.

Given: The tool chest is 2 ⅓ cubits long. An Egyptian Royal cubit is 52.3 cm. A furlong is an Old English measure, the length of a farmer's field. A furlong is exactly ⅛ of a mile, or 201.168 meters. A foot is twelve inches. An inch is exactly 2.54 cm.

Find: (a) How many meters long is the tool cabinet?
(b) How many furlongs long is the tool cabinet?
(c) How many feet long is the tool cabinet?

1-12 While scraping the mildew off the grout in his bathroom, Bob notices that he can just barely stick both his thumbnails into the space between tiles.

Given: The tiles are 2 thumbnail-thicknesses apart. A thumbnail has a thickness of about 0.90 mm. An inch is exactly 2.54 centimeters. One algae cell is 9.45 μm (micrometer = 10^{-6} m) across.

Find: (a) How wide is the gap between tiles in centimeters?
(b) How wide is the gap in inches?
(c) How many algae-thicknesses is the gap between tiles?

1-13 David Miler is running around a track. To amuse himself, he decides to figure out how fast he is running in furlongs per fortnight. A furlong is an Old English measure based on the length of a farmer's field. A fortnight is an Old English unit of time equal to two weeks.

Given: David is running at 3.528 m/s. A furlong is exactly 1/8 of a mile, or about 201.2 m. A fortnight is exactly two weeks.

Find: (a) How fast is David running in furlongs per fortnight?
(b) How fast is a furlong per fortnight in SI units?

1-14 Linnea Jonquil, the romance novelist, is on a writing retreat in a remote cabin for a week. She brought with her a supply of drinking water for the week, but finds that she has forgotten to pack a water glass with her cooking gear. Fortunately, she has packed a shot glass with her other supplies.

Given: Linnea brought 26 liters of drinking water for the week. A shot is 1.0 oz. An ounce of water is about 29.6 ml.

Find: (a) On average, how many shots of water will she drink per hour if she wants to finish drinking all her water in a week?

(b) What is a shot per hour in metric units (liter per second)?

1-15 The force necessary to drag a trunk across a floor is found by experiment to be directly proportional to the trunk's weight. The proportionality constant involved is called the coefficient of friction. The coefficient of friction is usually denoted with the Greek letter mu, written "μ," and is defined such that a larger value, say on a rougher floor surface, would reflect a larger force for the same weight.

Given: A 200-newton (N) trunk requires 90 N of force to make it slide. The same trunk, with more stuff in it, so that it weighs 300 N, requires 135 N of force to make it slide.

Find: (a) Write an equation relating force, F, weight, W, and the coefficient of friction μ.

(b) What is the coefficient of friction in this case?

(c) How much force would be necessary to pull the trunk if it contained lead, so that it weighed 550 N?

1-16 The force necessary to stretch a spring is found by experiment to always be directly proportional to how far the spring is stretched. The proportionality constant involved is called the spring constant, which is defined so that a larger spring constant describes a spring which requires a stronger force to stretch it the same distance.

Given: As an exercise, a woman pulls on both sides of a stiff spring to work on her chest muscles. The spring stretches 7.2 cm when she pulls on it with a force of 111 N. She then makes a greater effort and stretches the spring 17 cm.

Find: (a) Write an equation relating the force, F, with which the spring is pulled, the distance the spring is pulled, x, and the spring constant k.

(b) What is the spring constant in this case?

(c) What force would be necessary to stretch the spring 17 cm?

1-17 Billie Farmer's neighbors all take their good Kansas wheat to the same local grain elevator, a building where wheat is stored for market. The storage building is made of cylindrical sections.

Given: The base of each section of the grain elevator is a circle with a radius of 16 ft. The grain elevator is 180 ft high. A foot is a unit of measure still commonly used in Kansas and is equal to 30.48 cm. One truck load of grain has a volume of 22 m³. One Thursday, after this section of the grain elevator has been emptied, 52 trucks visit the grain elevator. (Remember: the area of a circle is πr^2, and the volume of a cylinder is the height times the area of the base.)

Find: (a) How much grain does the grain elevator hold, in ft³?
(b) How many truck loads of grain does the grain elevator hold?
(c) How much of the grain elevator was filled this particular Thursday?

1-18 Donald is canning good California peaches from his tree. He has cut the peaches into fourths and preserved them, and wonders how many cans he needs to can all his peaches. (Remember: the area of a circle is πr^2, and the volume of a cylinder is the height times the area of the base.)

Given: The base of a can has a 1 inch radius. The can is 4.50 inches high. An inch is a unit of measure still commonly used in California and is equal to 2.54 cm. Each peach, after cooking, takes up an average of 143 cm³ in the can. Donald started with 73 peaches.

Find: (a) What is the volume of the can in cubic inches?
(b) How many peaches, on average, does each can hold?
(c) How many cans will Donald need to can the peaches?

1-19 On a volleyball court there is a line 10 feet from the net called the 10-foot line. In international play it is called the 3-meter line.

Given: One inch is exactly 2.54 cm, one foot is exactly 12 inches and one meter is exactly 100 cm.

Find: (a) How many centimeters is the net from the 10-ft line?
(b) How many centimeters is the net from the 3-m line?
(c) By what percent is the American court different from the International court? (Hint: the percent difference is found by dividing the difference of the two numbers by the value for the international court and multiplying by 100.)

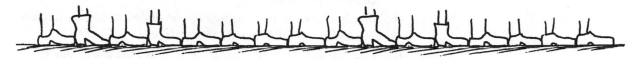

1-20 A royal warrant of 1522 had the town Sheriff stop sixteen men, tall and short, as they left church one Sunday and put their left feet one behind the other. That defined "a right and lawful rood to measure and survey the land with, and the sixteenth part of it shall be a right and lawful foot." People, usually, go to church fully clothed, so the Medieval foot was really an average measure of their shoes.

Given: The average length of the shoes in one man's closet is 12.17 inches from heel to toe. The naked foot that fits in those shoes measures 0.2731 m from heel to toe with freshly trimmed toenails.

Find: (a) What is the average length of the shoes in centimeters?
(b) What is the length of the naked foot in centimeters?
(c) By what percent is the naked foot different from the average shoe length? (Hint: the percent difference is found by dividing the difference of the two numbers by the average shoe length and multiplying by 100.)

1-21 Here is how the rich get richer and the poor get poorer. It takes money to make money, and the more money you have the faster you make it. It is the same with debt, but in reverse. The relationship looks proportional for the short haul, but not in a longer term.

Given: You buy a certificate of deposit that is worth $500 in 7 years. There is an option, however, to keep it longer so that it will be worth $673 in 10 years. The maximum value is $1,000 if you keep it for 14 years. "Someday," you say to yourself, "this investment could be worth $2,000."

Find: (a) Invent, as a first approximation, a proportionality between dollars D and time t, using a proportionality constant A, for accumulation of assets.

(b) Solve for the constant A using each set of data to see at what rate your assets ascend. What are the units of A?

(c) Predict, using an average value for A, how long it will take for "someday" to arrive if the proportional model were to hold.[5]

---

[5]Bank rates are not actually proportional. They are actually **exponential**. In dealing with exponential numbers, the proportionality is not constant, but the *doubling time* is. Here, this is the time needed to double your assets. Since it took 7 years to get to $500, and another 7 years to double to $1000, the doubling time here is 7 years. It will only take 7 years to double your money again to $2000, for a total time period of 21 years—substantially less than the answer you will get with the simple proportion. After 140 years, your investment would be worth $2.6 million. After 300 years it would be worth 2×10^{13}. To find doubling time, divide percentage growth rate into 70, which is approximately ln 2 (the natural logarithm of 2) expressed as a percent. (See Chapter 13.) In this case, a doubling time of 7 years corresponds to an interest rate of about 10%.

1-22 Harry Shark, the nice looking loan officer at the bank, has offered you a pre-approved credit card. Now you can buy that cute little CD player that you couldn't quite afford. If you let the payment ride for a few years, your debt looks proportional. But watch out, it's exponential! Plastic has many victims.

Given: Pay off your debt in 48 months and you owe $206. Pay in 60 months and you owe $247. By then, the CD player will have long since stopped working, but in 72 months, you will need to pay $296 for it. "Someday," you realize, "I may have to pay $400 for it."

Find: (a) Invent, as a first approximation, a proportionality between dollars D and time t, using a proportionality constant P, for paying for the broken player.

(b) Solve for the constant *P* using each set of data to see at what rate you are paying for past pleasure. What are the units of *P*?

(c) Predict, using an average value for *P,* how long it will take for "someday" to arrive if the proportional model were valid.[6]

[6]"Someday" turns out to be sooner than the proportional model predicts. See Chapter 13 for a discussion of the exponential function. For now, let us remember that you are paying interest on interest. This problem uses common credit card interest rates that will make you owe $400 in 92 months. In twenty years, you will owe $3,720 for that broken CD player. In 55 years, you will owe $2 million. Pass the debt down to your children, and in 199 years they will owe $424 × 10^{15}. Someday, at this rate, your family could owe the national debt. Of course, that won't happen, but Harry could easily own all of your assets. In Chapter 13 we will see how, if you divide the rate at which Harry is getting his hooks in you into about 70, you can figure out how long it takes for him to double his money.

ANSWERS

Solutions to **ESSENTIAL PROBLEMS**

1-6   (a) 1.056 lb; (b) 1.043 lb; (c) 0.2080 oz

1-7   (a) 4.16 m^2; (b) 6.7×10^{-2} m^3 or 67 liters

1-8   (a) 7.9 m^3; (b) 8.6×10^3 breaths of air

1-9   (a) $E = CR$; (b) $C_1 = 2.9$ kWh/room, $C_2 = 3.1$ kWh/room, $C_3 = 2.95$ kWh/room; (c) 23.87kWh = 24 kWh

1-10   (a) $w = mg$; (b) 6.8 N/kg; (c) 89 kg.

Solutions to **MORE INTERESTING PROBLEMS,** *odd answers*

1-11   (a) 1.22 m; (b) 6.07×10^{-3} furlongs; (c) 4.00 ft

1-13   (a) 2.121×10^4 furlongs/fortnight; (b) 1.663×10^{-4} m/s (or 1.000 cm/min)

1-15   (a) $F = w\mu$; (b) 0.45; (c) 248 N

1-17   (a) 1.4×10^5 ft^3; (b) 186 truck loads; (c) 0.28 of the elevator, or 28% was filled.

1-19   (a) 304.8 cm; (b) 300 cm; (c) 1.6%

1-21   (a) $D = At$; (b) $A_1 = 71$ dollars/year, $A_2 = 67$ dollars/year, $A_3 = 71$ dollars/year; (c) 28.55 years = 29 years

2 *Optics*

The Nature of Light

PEOPLE HAVE ALWAYS PONDERED the nature of sight. Ancient Greeks thought of sight as rays of light originating in the eye, reaching out in straight lines to sense objects, much as the fingers reach out from the hand. Today, you can still see this ray model of light portrayed in comic books. It is always used when a super hero possesses the power of x-ray vision.

By the seventeenth century, the Dutch astronomer Christian Huygens was thinking of light in terms of something that travels the other direction, from the object to the eye. He introduced the idea that light travels in the form of waves. With waves he was able to explain how lenses work by assuming that light travels more slowly in glass than air.

Huygens' wave theory of light gained many enthusiastic supporters. Among them in England was Robert Hooke, Isaac Newton's nemesis. Newton and Hooke hated each other. If Hooke liked waves, Newton naturally looked for another point of view.

Newton suggested that, while he saw evidence that light could be waves of some kind, he preferred to think it was made of particles because of the sharp edges that shadows cast. He was able to explain how lenses work with the assumption that light particles speed up when they enter glass. Newton's reputation and authority were so great that his preference for particles over waves was accepted as almost certain fact for over a hundred years.

Then, starting in about 1800, Thomas Young looked closely at the edges of shadows and saw evidence of wave behavior. He did a famous experiment with overlapping light from two parallel slits that revived Huygens' wave theory. By 1850, Jean Foucault had measured the speed of light under water and showed that it was in fact slower than in air. That seemed to put

Figure 2-1

Comic book artists use a ray coming from the eye to represent x-ray vision.

Figure 2-2

Young Albert Einstein working in the patent office came up with a new theory of light. He invented the concept we call the photon.

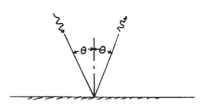

Figure 2-3

A perfectly elastic ball would bounce from the floor the same as a photon bounces from a shiny surface.

a lock on it. By 1860 the mathematical basis for the wave theory of light was well established, first by Augustin Fresnel in France and then by James Clerk Maxwell in England. For the next 45 years, light was certainly waves. No doubt about it.

Then, in 1905, a young patent examiner in the Swiss Patent Office published a number of papers on theoretical physics concerning subjects he had been thinking about in his spare time. The paper that gained him the most fame was the most controversial at the time. It was on the theory of relativity and, of course, his name was Albert Einstein.

Einstein's Nobel prize was not for the theory of relativity, however, it was for his research in light. He showed that light is made of particles. Wavy particles, but particles nonetheless.

Light, as it turns out, travels like a wave but interacts with matter like a particle. Its true nature lies somewhere between that of a wave and a particle. Our present day name for Einstein's wavy particles is in the common vernacular. We call them "photons." If this seems confusing, that is because it *is* confusing. We can visualize particles as very little marbles shooting around and waves like the waves in the ocean, but our everyday experience does not equip us with the tools we would need to visualize a photon which acts like both a particle and a wave.

Toward the end of this book, we will find that the photon model for light is essential for building our present understanding of atomic structure. In this chapter, however, we will be discussing the physical properties of light as we deal with them on a day-to-day basis. We will concentrate on the way lenses and mirrors work.

If we want to know what happens when we look at a mirror or through a magnifying glass, we do not need to wonder whether light is made of particles or waves. The photon model is far more sophisticated than we need. We can understand mirrors and lenses very well with a far more primitive model.

The model we will use to explore the behavior of light is the old ray model of ancient Greece, right out of the comic books. This model is easy to understand, and it is more than adequate for understanding images formed by lenses and mirrors.

In the ray model of light, we assume that light travels in straight lines until it is blocked by something. Then it does one of two things. If it bounces off the surface, we call that **reflection**. If it goes into the surface, perhaps into glass or water, we call that **refraction**. Often, it does both at the same time. There are simple rules that govern both these cases. First, we will discuss mirrors and the laws of reflection.

Reflection

When light bounces off a surface, at whatever angle it comes in, it is reflected at the same angle. We call this rule the **law of reflection**. The angle of incidence equals the angle of reflection:

$$\angle i = \angle r$$

Angle of incidence = Angle of reflection

The *angle of incidence* $\angle i$ is the angle at which the light hits the surface, and the *angle of reflection* $\angle r$ is the angle at which the light is reflected from the surface. These angles are usually measured in relation to the **normal**. The normal is an imaginary line that is perpendicular to the surface you are dealing with. That means that the normal always hits the surface at a 90 degree angle. We measure angles from the normal and not from the surface because, when we start dealing with curved surfaces, it is easier to find the normal than to deal with the curve of the surface.

The law of reflection illustrates the fact that light interacts with matter as if it were made up of particles. Light rays bounce off a flat surface in the same way balls bounce off surfaces. The next time you play miniature golf or pool notice how, when the ball hits a wall, it bounces off at the same angle that it hit. The same goes for bouncing a bouncy ball off the floor. Unless there are special circumstances, the ball leaves the floor at the same angle that it approached the floor.

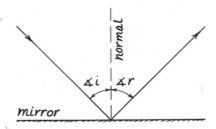

Figure 2-4

The law of reflection states that the angle of incidence equals the angle of reflection.

CHECK QUESTION

Light approaches a mirror along a line making an angle of 53° with the mirror's surface. (a) What is the angle of incidence? (b) At what angle to the normal will the light be reflected?

Answer: (a) 37°, (b) 37°

Images

Every morning, when you go to the bathroom mirror and look yourself in the eye, that nice-looking person peeking out from behind the mirror is properly called an **image**. The image in a flat mirror is located behind the mirror.

Flat mirror. Only one outcome.

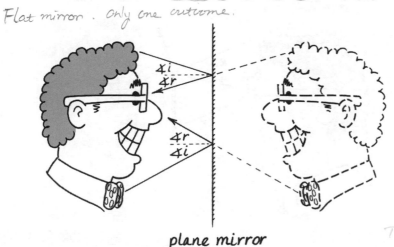

plane mirror

Figure 2-5

The virtual image in a plane mirror looks exactly like the object and is the same distance behind the mirror as the object is in front of the mirror.

The same size
The same distance.

mirror, as though there were someone in the next room. Of course, the light from the image does not really travel behind the mirror, but it acts as if it did. Light from your nose bounces to your eyes along a path similar to the path it would take had it come from behind the mirror.

Since the light does not actually pass through the image behind the mirror but behaves virtually as if it did, the image in your bathroom mirror is called a **virtual image**. The space behind the mirror from which the light seems to come, the Alice-in-the-Looking-Glass space, is called virtual space. The images of real objects formed by flat mirrors always lie in virtual space.

In day-to-day life, an image in a mirror is sometimes referred to as a "reflection," as in, "Carmen looked at her reflection in the mirror." When discussing the physics of reflection and images, however, the two terms are not interchangeable. We should be careful to use the two different words to distinguish the two related but different ideas. Reflection is something that happens at the surface of the mirror, when light is reflected off the smooth surface. When we see an image, what we see is not at the surface of the mirror—it is behind the mirror in virtual space.

In a mirror, as we shall see, size of the image and distance of the image from the mirror are related. In a flat mirror the image sits in virtual space behind the mirror exactly as far away from the mirror as the object and exactly the same size. In curved mirrors, the size and distance of the object and the image are still related to one another, but they are no longer equal. This gives curved mirrors their interesting properties of making images larger or smaller.

Spherical Mirrors

A **convex** mirror may be thought of as a plane mirror that has been bent so that it bulges out in the middle like the surface of a Christmas tree ball. Images formed by such a mirror are smaller than those formed by a plane

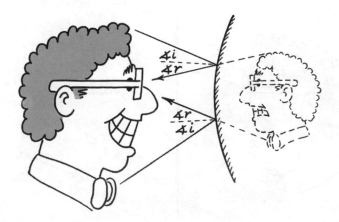

Figure 2-6

Images in a convex mirror are smaller and closer to the mirror than the object.

mirror. It might appear that the images are further away since they are small, but, actually, the image in a convex mirror is closer to the mirror than the object is. This is why wide-angle rear-view mirrors mounted on automobiles and trucks carry the warning: "Objects in this mirror are closer than they appear."

Convex mirrors are also frequently used in stores for their wide-angle view to spot shoplifters. When you next see one in a store, carefully observe the relative distances between images and objects as well as their relative sizes. A close examination of a store's security systems is always good for a little excitement.

A mirror that bulges in the other way—away from you in the middle like a *cave*—is called a **concave** mirror. At least for close objects, we will see that a concave mirror behaves the opposite way from a convex mirror. Virtual images in a concave mirror are larger and further away than the object. Concave mirrors are often used as magnifying mirrors.

Figure 2-7

The rear-view mirror on a truck is frequently a large plane mirror with a small convex mirror in one corner to give a wide-angle view.

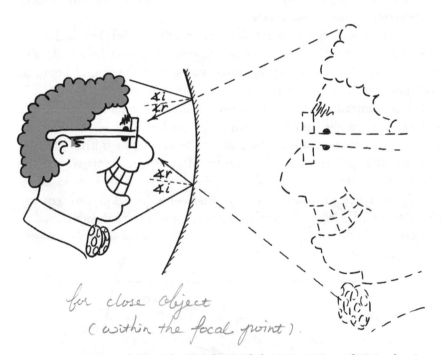

for close object
(within the focal point).

Figure 2-8

Images in a concave mirror are larger and farther away from the mirror than the object as long as the latter is within the focal point. (object)

We can gain a full understanding of the properties of curved mirrors using only one rule: the law of reflection.

We might now pause to reflect, so to speak, on our convention of measuring angles from the normal rather than from the surface itself. The normal, remember, is an imaginary line that is perpendicular to the surface of the mirror.

Each spherical mirror is a piece of a sphere. If you were to break a shiny, spherical Christmas tree ornament and pick up any one piece, even a small one, the inside surface would be a concave mirror and the outer surface would be a convex mirror. For a spherical mirror, then, the normal

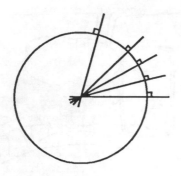

※ **Figure 2-9**

The normal lines to a circle pass through the center of the circle.

lines to the surface all pass through the center of the sphere. Every line drawn from the center of the Christmas tree ball will be perpendicular to the surface just as every radius of a circle is perpendicular to the tangent to the circle when the radius and tangent touch the circle at the same place.

In summary, if you draw lines perpendicular to any curved surface, these lines are called normals. The place where the normals meet is called the **center of curvature** of the surface. The distance from the center of curvature to the surface is called **radius of curvature**.

Focal Point

A concentrating solar collector makes use of the center of curvature of a curved surface to gather energy. Solar energy is widely diffused by the time it reaches the earth. That's a problem if you are trying to collect it. While spread out over a wide area, however, this energy is traveling in pretty much the same direction by the time it hits the earth. That sameness of direction can be used to advantage.

Light rays from any far distant object are nearly parallel. The further the object, the more nearly parallel. The best approximation we have to parallel light is starlight. The sun being the closest star, sun rays may not be quite as parallel as light from some of the other stars, but they are pretty close.

The concentrating solar collector uses reflection from a concave surface to bring these parallel sun rays together. To see how that works, consider first a ray of light that comes in right along the **axis** of the mirror. The axis is the line that passes through the center of curvature on its way to the mirror. Other rays come in parallel to this axis. Rays that come in above the axis are reflected downward. Rays that come in below this axis are reflected upward. After reflection, all the rays come toward the axis.

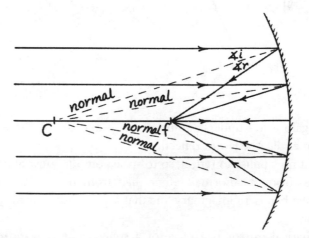

Figure 2-10

Parallel light is brought to a focus (f) *at a point that is halfway to the center of curvature* (C).

Parallel rays farther from the axis have larger angles of incidence and, hence, larger angles of reflection. The place where the various reflected rays come together is called the **focal point** of the mirror. The focal point is about half way to the center of curvature of the mirror. The **focal length** is the distance from the focal point to the surface of the mirror. We say the focal length for a mirror is half the radius of curvature

$$f = \tfrac{1}{2}c$$

where *f* stands for the focal length and *c* for the radius of curvature.[1] In solar collectors, the focal point is where the solar energy is most intense.

CHECK QUESTION

You accidentally drop and break a round glass Christmas ornament. While cleaning up the glass, you pick up one of the larger pieces and get interested in the concave inner surface. The ball was originally 8.0 centimeters in diameter. (a) What is the radius of curvature of the concave surface, neglecting the thickness of the glass? (b) What is the focal length of this concave surface?

Answer: (a) 4.0 cm, (b) 2.0 cm

Just as parallel light from a far distant object will be brought to focus at the focal point, light coming from the focal point will be reflected and go out parallel. The law of reflection works the same both ways. The angles of incidence and reflection are interchangeable, since the law of reflection says only that they are equal, without regard to which way the light is traveling. Thus, parallel light is associated with light traveling through the focal point whether the light is coming or going.

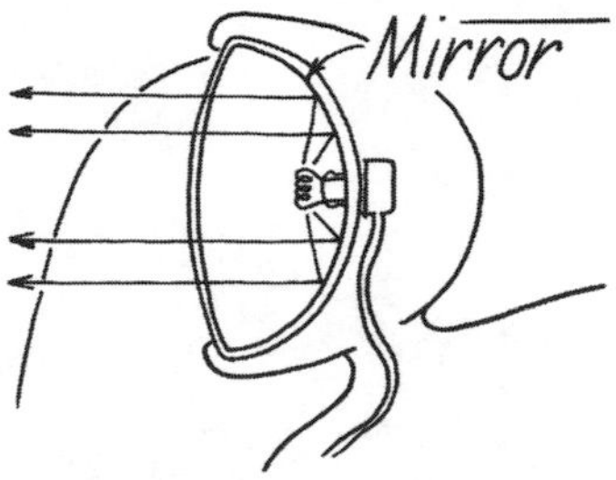

The ability of a concave mirror to form a parallel beam of light finds even more applications than its ability to bring light into focus. Search lights, flashlights, and automobile headlights all use concave mirrors to form beams of parallel light from light that starts out from a point. A white-hot wire filament, electric arc, or other bright source of light is placed at the focal point. Light coming from this source is reflected by the mirror into a parallel beam. This beam is so nearly parallel, in fact, that automobile head lamps have a lens with a rippled surface to diffuse the beam and spread it out in front of the car. Some car enthusiasts add driving lamps to their cars that are like headlights except that they have clear glass in place of the diffusing lens. Such driving lamps must be used judiciously, however, since they produce a "pencil beam," which is so nearly parallel that it can temporarily blind another driver miles away.

Figure 2-11

The driving light on a sports car has a concave mirror in the back to produce a parallel beam of light. Driving lights, unlike headlights, have no diffusing lens on their front surface to spread the light in front of the car.

[1]You can use this equation to solve problems, but keep in mind that it is an approximation that works for rays close to the axis. The closer to the axis, the better it works. It is called the **paraxial ray approximation**. Like many mental models, it works so amazingly well that we might be tempted to forget that it is not reality itself. To give it a name, we call the failure of this approximation *spherical aberration*. For solar collectors, spherical aberration is no problem at all. For the Hubbell Space Telescope in 1990, it was a billion dollar problem.

Ray Diagrams for Concave Mirrors

An image is, in general, a point in space from which light rays appear to come. In a mirror, an image is where the rays appear to come from after they have been reflected. We will first use geometry to find the image in a curved mirror. Our reasoning can also be turned around. Given an image and a mirror, we can use the same geometry to find the object.

We will then generalize our logic in the form of algebraic expressions. It would be a bad thing, however, to put too much faith in punching numbers into a calculator without remembering the geometry of the situation. A little mistake with the numbers, like reversing a minus sign, will lead to ridiculous results. A picture showing the paths of light rays is always needed to see if an answer is reasonable.

It is possible to apply the law of reflection to any number of rays to predict the location of the image. Using a protractor to measure the angle of incidence, we can always construct an equal angle of reflection to determine the path of a reflected light ray. Use a computer, and you can trace as many rays as you want.[2]

Some rays, however, are easier to construct than others. If we know either the center of curvature of the mirror or the location of the image, we can draw three lines without any special tools. These three rays can then be generalized to represent what all the other rays do.

Consider a candle located beyond the center of curvature of a concave mirror. Assume that the candle is right side up with its base resting on the mirror's axis, as shown in Figure 2-12. Of all the light rays hitting the

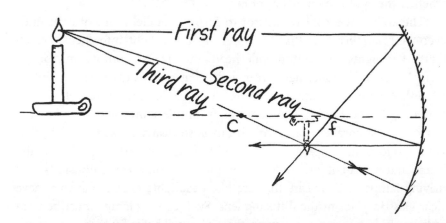

Figure 2-12

Three rays can be used to locate an image formed by a concave mirror.

surface of the mirror, the three rays easiest to locate are the following: The first is the one that comes from the tip of the candle and travels parallel to

[2]Optical designers do in fact use computers to trace many rays through an optical system. Spherical aberration is only one of the many aberrations they need to worry about. Using ray tracing, they can see the effect of modifications without actually making different lenses mirrors as they did in the days before digital computers. Ray tracing has thereby had a major impact on the cost of such things as fancy camera lenses. While we use ray tracing as a starting place in our thinking, these techniques are where the experts wind up.

the axis of the mirror. It will be reflected at such an angle as to pass through the focal point.

The first ray we might call the solar collector ray. That is how we defined the focal point. All light coming parallel to the mirror's axis is reflected to the focal point as in the solar collector. The ray does not stop at the focal point. It keeps going.

The second ray we might call the search light ray. It comes from the tip of the candle and passes through the focal point on its way to the mirror. This ray strikes the mirror at an angle as it would if it had originated from the focal point. The difference is that this ray doesn't start there but passes through on its way to the mirror. It comes in at an angle as if it had originated at the focal point, however, and is therefore reflected out parallel to the axis.

The third ray we might call the perpendicular ray. It passes through the center of curvature on its way to the mirror. This ray strikes the mirror perpendicular to the surface, traveling along a radius of curvature of the mirror, and is therefore reflected back on itself.

We find the image where these three rays cross. Any two rays could be used to locate the image, but we should draw all three. When making ray diagrams, not every place where two lines happen to cross is an image. What happens to three rays, however, starts to show a trend.

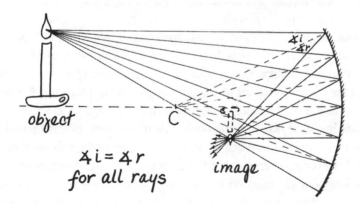

Figure 2-13

All the rays striking the mirror pass through the same image as the three rays that are easy to draw.

We could draw more than three rays to locate the image. All the light striking the mirror from a certain point on the object will be reflected to cross at a corresponding point on the image. The other rays would be more difficult to construct, but the mirror treats them the same as the three we find easy to draw.

The image you see in the mirror is really formed by all the rays that hit the mirror. They all contribute to forming the image. Even if several of the rays are obstructed, the image will be formed just the same by the light that does manage to get through. The mirror doesn't find the three that we find easy to draw as particularly special. A few specks of dirt on the surface of the mirror obstructing our special rays won't make much difference in the quality of the image.

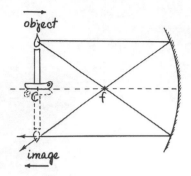

Figure 2-14

The object and image meet at the center of curvature.

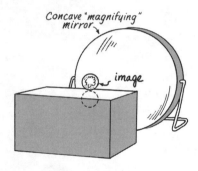

Figure 2-15

A real image is formed of a coin, hidden in a shoe box, by placing it at the center of curvature of a magnifying mirror.

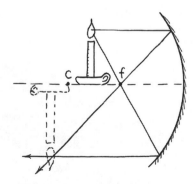

Figure 2-16

The image is beyond the center of curvature when the object is between the center of curvature and the focal point.

We could mask half the mirror and we would still get an image of the whole candle, and the image would be the same size as before we masked the mirror—it would just be dimmer, since not as much light would have gotten through. We can therefore construct a ray even if the mirror in our drawing is not large enough to reflect it. A ray that misses the mirror will still act as a tool to locate the image formed by all the rays that do hit the mirror.

We have only traced rays originating from the tip of the candle. Light actually comes from all points on the candle. For every point on the candle, there is a corresponding point on the image where reflected light rays meet. Light reflected from a drip on the candle forms rays that come in closer to the axis of the mirror and are reflected to cross at a drip on the image. The image looks just like the candle except that it is upside down, or inverted, and not necessarily the same size.

The farther the object from the mirror, the closer the image will be to the focal point. In fact, the focal point is the image of a far distant object. Light from a far distant object comes in nearly parallel and the place it crosses is the focal point. As the object approaches the mirror, starting from infinity, the image grows larger and moves away from the focal point toward the center of curvature. The image and object both reach the center of curvature at the same time. At that point, they are both the same distance from the mirror and they are both the same size.

In this example, we are discussing a particular kind of image which is formed by light actually crossing at a point. This kind of image is called a **real image**. Real images exist in real space, as opposed to virtual images which exist in the virtual space behind a mirror. A number of magic tricks are based on the lifelike appearance of a real image. One can create, for example, the magical illusion of a coin that you cannot feel. Use a magnifying mirror and a coin hidden in a shoe box. If the coin is suspended with tape inside the shoe box and the open side of the shoe box is turned away from view (toward the mirror), all that can be seen is the reflected image of the coin. If the coin is arranged at the center of curvature of the mirror, the image will appear to stand on the top edge of the shoe box. It will be the same distance from the mirror as the actual coin and the same size, as shown in Figure 2-15.

When the object moves closer to the mirror than the center of curvature, but not as far as the focal point, as shown in Figure 2-16, the image moves to a point beyond the center of curvature. The image is still inverted, but is larger and further from the mirror.

As the object approaches the focal point, its image moves still further away, always increasing in size. When the object finally reaches the focal point, the reflected light goes out in a parallel beam, as in a search light. Since parallel light rays never cross, we say the image has receded to infinity. In becoming infinitely far away, the image also becomes infinitely large.

Strange as it might seem to say, the image can go even *beyond infinity.* The image approaches infinity as the object approaches the focal point, and it clearly must go somewhere when the object crosses over the focal point. If you are wondering where this strange place beyond infinity is physically located, the answer is that just beyond infinity is negative infinity. After the real image has receded to an infinitely large distance on the same side of the mirror as the object, a virtual image comes in from infinity from the opposite direction. Positive distances are measured on the same side of the mirror as the object; negative distances are behind the mirror.

While real images are upside down in the real space in front of the mirror, virtual images are right side up in the negative space behind the mirror. Although light never really passes behind the mirror, it does come away from the mirror at the same angle as though it did. For a concave mirror, the virtual image is always larger and farther away from the mirror

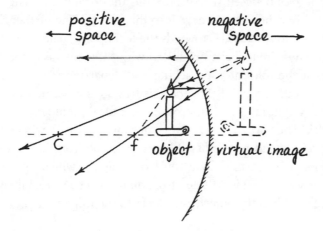

Figure 2-17

A magnified virtual image is formed in negative space, behind the mirror, when the object is within the focal distance of a concave mirror.

than the object. This is how a concave mirror magnifies your nose. You move your nose to a point within the focal length of the mirror and observe the "reflection" or virtual image behind the mirror. The image of your nose is magnified and is further from the mirror than your real nose (see Figure 2-17).

Plane Mirrors

The mirrors you are probably most familiar with are flat mirrors, which are called **plane mirrors**. A plane mirror can be thought of as a concave mirror having an infinite radius of curvature. The longer the radius of curvature, the flatter the surface. People used to think that the surface of the earth was flat because its radius of curvature was 4,000 miles. A still longer radius of curvature would mean a still flatter surface. A plane mirror would therefore have an infinite radius of curvature.

Since half of infinity is still infinity, a plane mirror also has an infinite focal length. All real objects are therefore within the focal length of a plane mirror and produce virtual images that are right side up and behind the mirror (in negative space).

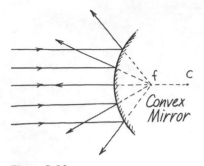

Figure 2-18

Parallel light is reflected by a convex mirror so that is goes away from the focal point.

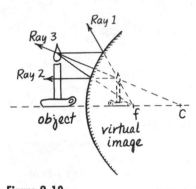

Figure 2-19

The image formed by the convex mirror is virtual, erect, and smaller than the object.

Convex Mirrors

A convex mirror may be thought of as a spherical mirror having its center of curvature located on the negative side of the mirror. Just as for a concave mirror, the focal point is half the radius of curvature.

$$f = \tfrac{1}{2}c$$

Parallel light incident on a convex mirror is reflected away from, instead of toward, the focal point. (Physicists refer to light hitting a mirror as being "incident on" the mirror.)

Just as with concave mirrors, the focal point may be regarded as the image of a far distant object. Objects viewed in a Christmas tree ball appear as virtual images between the surface and the focal point. The closer the object, the closer the virtual image is to the surface. The further the object, the further the image recedes towards the focal point, halfway to the center of the ball. These images can be located with three rays as before, but keep in mind that the roles of approaching and departing rays are reversed. A light ray approaching parallel to the axis (ray 1) is reflected away from the focal point. A light ray which approaches the focal point (ray 2) is reflected out parallel. The virtual image is located at the place where these two departing rays would have crossed had they come from inside the mirror. This position is confirmed by a third ray (ray 3), which approaches the center of curvature and is reflected back on itself. The virtual image is right side up, smaller than the object, and closer to the mirror.

The Object-Image Formula

Ray diagrams are valuable tools for understanding an optical system. It is also helpful to have an algebraic formula relating object and image distances so that the accuracy of results will not depend on the precision of drawing tools. We shall derive such a formula from the ray diagram of a concave mirror. One common convention is to call the height of the object a, the height of the image b, the object distance p, and the image distance q.

Figure 2-20

One set of similar triangles formed by a ray diagram relates the object distance, p, and focal length, f, to the object and image heights, a and b.

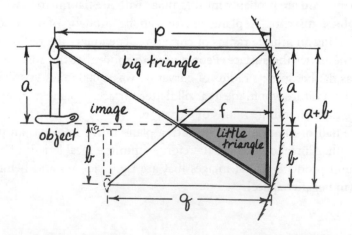

When we draw a ray diagram it forms two sets of similar triangles. The first set, shown in Figure 2-20, is made up of a little triangle having the focal length f and the image height b as sides and a big triangle having the object distance p and the sum of object and image heights $a + b$ as sides. These triangles are geometrically similar because they have parallel sides and therefore the same angles. The similar sides of similar triangles are proportional.

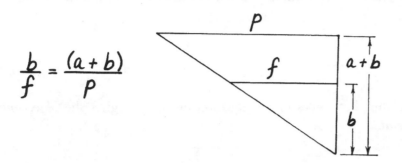

$$\frac{b}{f} = \frac{(a + b)}{P}$$

Figure 2-21

The ratio of the sides of the little triangle is equal to the ratio of the sides of the similar big triangle. The equality of these ratios reduces part of the geometry of a ray diagram to an algebraic expression.

The second set of similar triangles, shown in Figure 2-22, is made up of a little triangle having the focal length f and object height a as two sides

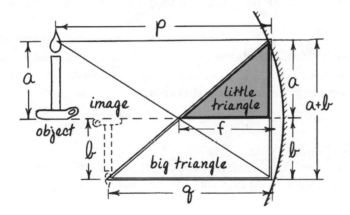

Figure 2-22

Another set of similar triangles is formed by a ray diagram relating the image distance, q, and focal length, f, to the object and image heights, a and b.

and a big triangle having the image distance q and the sum of the object and image heights $a + b$ as two sides. Having parallel sides, these two triangles are also geometrically similar. We can again set up a proportionality between similar sides of similar triangles.

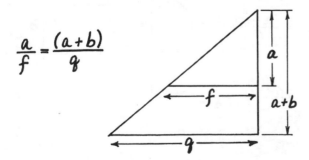

$$\frac{a}{f} = \frac{(a + b)}{q}$$

Figure 2-23

The ratio of the sides of the little triangle is also equal to the ratio of the sides of the big triangle for the two similar triangles that relate the focal length to the image distance.

There is an easy way to eliminate *a* and *b* from the two proportionalities in Figure 2-21 and 2-23. We simply add the two equations together:

$$\frac{a}{f} + \frac{b}{f} = \frac{a+b}{p} + \frac{a+b}{q}$$

We can factor 1/*f* from the left side of this expression and (*a* + *b*) from the right side.

$$\frac{1}{f}(a+b) = (a+b)\left(\frac{1}{p} + \frac{1}{q}\right)$$

Dividing both sides of the equation by (*a* + *b*) gives the **object-image formula**:

$$\boxed{\frac{1}{f} = \frac{1}{p} + \frac{1}{q}}$$

This equation is useful in many problems. It says that the reciprocal of the focal length is equal to the sum of the reciprocals of the image and object distances.

The focal length, *f*, can be eliminated from the same two geometrical proportions to yield a relationship between the sizes and distances of the object and image. These two proportions, restated, can each be cross-multiplied to get everything out of the denominator:

$$bp = f(a+b)$$
$$aq = f(a+b)$$

Since the right sides are equal, so are the left sides.

$$bp = aq$$

Figure 2-24

The equality of ratios based on both sets of similar triangles can be combined to eliminate either (a+b) or the focal length f.

$$\frac{b}{f} = \frac{(a+b)}{p}$$

$$\frac{a}{f} = \frac{(a+b)}{q}$$

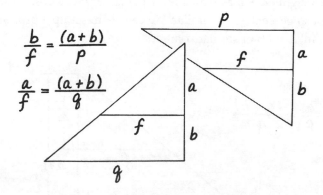

Dividing both sides of this equation by a and p,

$$\frac{b}{a} = \frac{q}{p}$$

We have shown that the ratio of the image and object heights b/a is equal to the ratio of the image and object distances q/p. The ratio of the image height b to object height a, defined as the magnification M, tells us how much bigger the image is than the object. The preceding equation says that, if you look at it just right, the image is as much bigger as it is farther away. In other words, the size is proportional to the distance.

$$\boxed{M \equiv \frac{b}{a} = \frac{q}{p}}$$

('The three-lined equal sign $\equiv$ means, as you recall, "is defined as.") This is the **magnification formula**. It says that the magnification, which is defined as b over a (the ratio of the image height to the object height), is equal to q over p (the ratio of the image distance to the object distance).

Applications of the Object-Image Formula

Just to see how the object-image formula works, let us use it to find the position of the image formed by a concave mirror when the object is at the center of curvature. Of course, we already know from our discussion of ray diagrams that the image will also be at the center of curvature and inverted. That was the situation in the magic trick we discussed where the image of the coin in the shoe box was made to stand on the top of the shoe box. It is instructive, however, to apply the object-image formula as a test. This situation was pictured in Figures 2-14 and 2-15.

The image distance p in this case is equal to the distance to the center of curvature c, which is twice the focal length.

$$p = c = 2f$$

Solving the object-image formula for $1/q$,

$$\frac{1}{q} = \frac{1}{f} - \frac{1}{p}$$

Inserting our given value for the object position,

$$\frac{1}{q} = \frac{1}{f} - \frac{1}{2f}$$

Reducing both fractions to the same common denominator, 2*f*,

$$\frac{1}{q} = \frac{2}{2f} - \frac{1}{2f} = \frac{1}{2f}$$

Taking the reciprocal of both sides, we see that the image is also at twice the focal length—which is the center of curvature:

$$q = 2f$$

We can also see that the magnification formula tells us that the image and object are the same size:

$$M \equiv \frac{b}{a} = \frac{q}{p}$$
$$= \frac{2f}{2f}$$
$$= 1$$

The magnification *M* is equal to unity in this case. Thus, the object and image are the same size.

CHECK QUESTIONS

1. An object is located at three times the focal length of a concave mirror, thus p = 3f. (a) Where will the image be located in terms of the focal length? (Hint: Replace 2f in above calculation with 3f.) (b) What will the magnification be in this case?
    Answer: (a) $\frac{3}{2}f$, (b) $\frac{1}{2}$

2. An object is located 15 cm from a mirror. The mirror forms a real image of this object at a point 30 cm from the mirror. (a) What is the focal length of the mirror? (b) What is the magnification in this case?
    Answer: (a) 10 cm, (b) 2

As another example of the object-image formula, let us consider the special case of the plane mirror. We already know that the image formed in a plane mirror is located the same distance behind the mirror as the object is in front, but now we can show that this is the case using the image-object formula:

$$\frac{1}{f} = \frac{1}{p} + \frac{1}{q}$$

Remembering that the focal length of a plane mirror, as well as its radius of curvature, is infinite, the reciprocal of the focal length is zero.[3]

$$\frac{1}{\infty} = \frac{1}{p} + \frac{1}{q}$$

$$0 = \frac{1}{p} + \frac{1}{q}$$

Solving for the image distance q in terms of the object distance p,

$$-\frac{1}{q} = \frac{1}{p}$$

We find that the image is as far behind the mirror as the object is in front:

$$q = -p$$

This agrees with our common experience as we stand before the bathroom mirror each morning.

We can also find from the magnification formula that the image and object are the same size:

$$M \equiv \frac{b}{a} = \frac{q}{p}$$

$$= \frac{-p}{p}$$

$$= -1$$

The magnification in this case is negative unity. Since from the definition of magnification,

$$b = Ma$$

$$= (-1)a$$

$$= -a$$

the image height is minus the object height. The minus means that the image is right side up. The derivation of the formula used the implied assumption that the image was both real and upside down. A negative image distance places the image in negative space behind the mirror, making it virtual; a negative image height measures the image above the axis instead of below it.

---

[3]The reciprocal of larger and larger numbers is closer and closer to zero. One one-millionth is smaller than one one-thousandth. One one-billionth is smaller than one one-millionth, and so on.

Overview of Reflection

We have used the law of reflection to understand the formation of images by flat and curved mirrors. In so doing, we have treated light as though it consisted of rays, in much the same way as the ancient Greeks thought of it.

We first described the positions of images formed by mirrors under various conditions, and then we derived the image positions by means of ray diagrams. Three rays were used to locate the crossing place of all the other rays in the system. Two of the three rays were associated with parallel light and with the focal point. The third ray was associated with the center of curvature and was found to be perpendicular to the mirror.

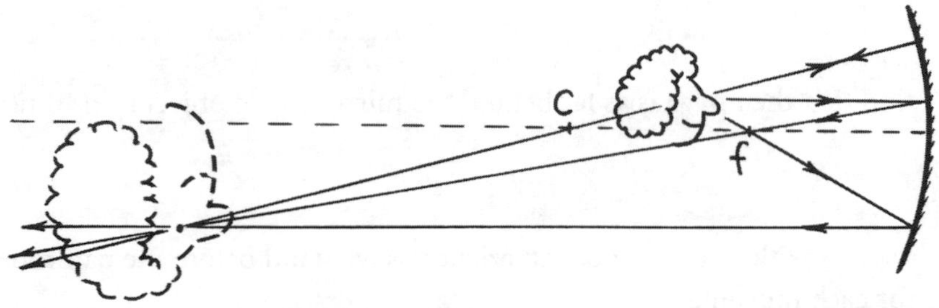

Figure 2-25

Two of the three rays that can be used to locate an image are associated with the focal point. The third is associated with the center of curvature.

We found that real images are formed by light actually crossing at a point after being reflected from the mirror. Such images can be captured on a screen. Virtual images are formed by light reflected at such an angle that it appears to come from a point behind the mirror. Such images are said to be virtual because they exist in negative space behind the mirror, where light never really travels.

Although we derived the object-image and magnification formulas from the geometry of a concave mirror ray diagram, these formulas can be made to work for convex and plane mirrors as well. All you need to do is let the focal length be negative in the case of a convex mirror and infinite in the case of a plane mirror. These modifications allow the formulas to apply to all three kinds of mirrors.

Snell's Law

When a light ray interacts with an object, it can bounce off by reflection or it can penetrate the surface and pass into the object by **refraction**. Reflection is the property that governs the interaction of light with mirrors and

Figure 2-26

Light is bent away from the normal when it passes from water into air, allowing the bottom of a fountain to be seen when full, although it could not be seen when empty.

refraction is the property that governs the interaction of light with transparent surfaces like glass, water or lenses. Just as there is a relationship, the law of reflection, that allows us to predict the angle of the reflected light, there is a relationship that allows us to predict the path refracted light will take inside a translucent object. The law of refraction is a little more complicated than the law of reflection. The ancient Greeks knew about both reflection and refraction, and they could predict angles for reflection; but they did not understand the relationship between the angles for refracted light. They did know, however, that light is bent toward the surface (or away from the normal) when it passes from water into air. They reported that they could see the bottom of a fountain when it was full of water although they could not see it from the same angle when the fountain was empty.

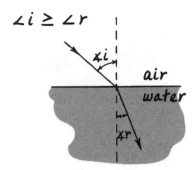

Figure 2-27

The angle of incidence is greater than the angle of refraction for light passing from air into water. Both the angle of incidence and the angle of refraction are measured to the normal.

They also knew that light going the other way, incident on the water from air, would be refracted away from the surface by the same amount. In terms of angles measured to a line drawn perpendicular to the surface (the normal), they knew that the angle of incidence for light coming from air and going into water would be greater than the angle of refraction, but that was the extent of their knowledge. They may have tried a proportionality between the two angles, but if they did, they found that it did not work. It was not until 1621 that a Dutch astronomer by the name of Willebrod Snell noticed that while the angles themselves are not proportional, their sines are.

$$\sin \angle i \propto \sin \angle r$$

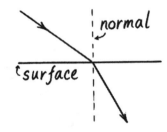

Figure 2-28

The "normal" is defined as a line that is drawn perpendicular to the surface at the point where a ray of light is refracted.

Snell did not understand why the sines of the angles are proportional; he just observed it as an experimental fact. He noticed that the proportionality constant is different for different materials, and he called it the **index of refraction**, *n*, of the material. In terms of this proportionality constant, **Snell's law** can be written

$$\sin \angle i = n \sin \angle r$$

No one knows why Snell chose the letter *n* for his proportionality constant (perhaps it was because *n* is the second letter of the word *index* and is also the second letter of Snell's name), but this letter can be taken as a reminder that the angles are measured from the normal rather than from the surface itself. (If Snell had measured to the surface, his law would be expressed as a proportionality between the cosines of the angles.)

We now understand that light bends as it passes into more dense material because it slows down. Imagine how a person on roller skates would change direction as he came across a rug at an angle, one foot slowing down before the other. Light rays are deflected toward the normal in the same way when they come across a surface that slows them down.

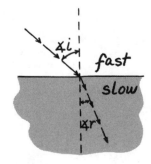

Figure 2-29

Light is deflected toward the normal when it is slowed down.

The index of refraction turns out to be inversely proportional to the velocity of light in the material. The proof of Snell's law[4] in terms of the velocity of light is easy, but we will leave that for a later course. For our purposes, we can think of Snell's law in the same way it was discovered, as a proportionality that just seems to work.

CHECK QUESTION

Light is incident on water from air. The angle of incidence, 23.6°, has a sine of 0.400. The angle of refraction in water is found to be 17.5°, which has a sine of 0.301. What is the index of refraction of water?
    Answer: $n = 1.33$

Convex Lenses

A light ray bends as it enters glass and bends again as it leaves. If a piece of glass has just the right shape, it can bend parallel rays of light so that they all cross to form a focal point just as with the spherical mirror. These curved pieces of glass, plastic or crystal are called *lenses*. It turns out that a spherical convex lens provides the same kind of positive focal point as does a concave mirror.

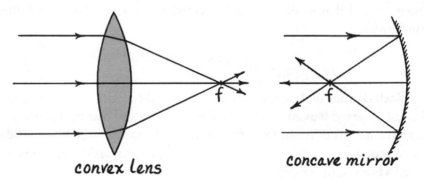

Figure 2-30

A convex lens has a positive focal length in the same sense that a concave mirror has a focal length.

    Light that comes in parallel to the axis and strikes the top of the lens is bent toward the normal to the first surface. Since this first normal is directed downward, the light is bent downward at the first surface. After the light passes through the lens, it strikes the second surface and is bent away from the normal as it emerges into air. The normal to the second surface is directed upward, however, and the light is bent downward again. Light coming in parallel below the axis is also deflected toward the first normal and away from the second, but the directions of the normals are

---

[4]This version of Snell's law assumes that light is passing from air or some other medium having an index of refraction close to that of a vacuum, namely $n = 1.00$, into a medium having a greater index of refraction. These are the sort of conditions under which Snell originally made his observations. A more general form of Snell's law, good for light passing from any medium having an index of refraction n_1 into any other medium having an index of refraction n_2, is stated: $n_1 \sin \angle i = n_2 \sin \angle r$

reversed, and the light is bent upward each time. All rays that make reasonably small angles with the surface pass through the same point after being deflected by the lens. This is called the focal point of the lens.

Lenses have two focal points. Light that comes in parallel to the lens is brought to a focus on the opposite side of the lens. Light that comes in parallel from the other side is brought to a focus on the other side of the lens. For lenses that are symmetrical, the focal points are at equal distances on opposite sides of the lens.

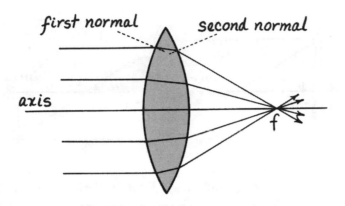

Figure 2-31

In a convex lens, parallel light is brought into focus by being bent toward the normal to the first surface of the lens and away from the normal to the second surface.

Lens Images

A real image is formed by a lens in much the same way as by a mirror. The light from the object is directed by the lens to cross at the image position. Any number of rays can be traced through the lens by use of Snell's law. However, as was the case for mirrors, there are three rays that are particularly easy to trace. Light coming from the tip of the object parallel to the

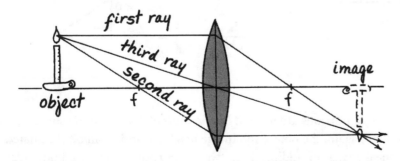

Figure 2-32

Three light rays may be constructed to locate the image.

lens axis is necessarily deflected so as to pass through the focal point on the far side of the lens. Light passing through the focal point on the near side of the lens is deflected and goes out parallel, since the angle of incidence is the same as if the light had *originated* at the focal point. The third ray passes through the center of the lens and is not deflected at all. This ray enters the lens at the same angle to the normal as it leaves; the first surface is curved up and the second surface is curved down. The amount that this ray is deflected toward the first normal is equal to the amount that it is deflected away from the second normal. The net effect is to displace the ray slightly to one side but to leave its direction unchanged. As long as the lens is thin compared to the focal length, we can ignore this tiny sideways displacement within the lens.

The first two rays described are also displaced within the lens. The rays are actually deflected at both surfaces, but only a small error is involved in pretending that the entire deflection takes place at the center of the lens. We use this thin lens approximation when drawing thin lens diagrams, making just one bend in the ray at the center of the lens.

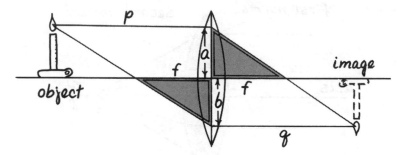

Figure 2-33

Two sets of similar triangles are formed by a ray diagram of a convex lens.

The formula for object and image distances derived for a mirror also applies to lenses. The object and image distances are again called p and q; and the object and image heights are labeled a and b. In a similar manner two sets of similar triangles are formed by the parallel rays and the rays passing through the focal points, together with a line having the same length as the object and image heights. The little triangle is cut from the similar big triangle by the **lens axis**. In this case, however, the two sets of similar triangles are on opposite sides of the lens.

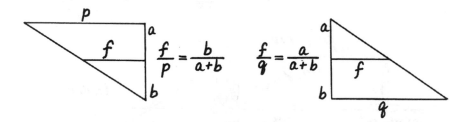

Figure 2-34

The ratios of the sides of the big and little triangles are equal for two sets of similar triangles that relate the image and object distances to the focal length.

$$\frac{f}{p} = \frac{b}{a+b} \qquad \frac{f}{q} = \frac{a}{a+b}$$

Two proportions are formed by the ratio of sides of these similar triangles. Once again, the object and image heights a and b can be eliminated by the sneaky trick of adding these two proportions. Dividing both sides of this equation by f, the focal length, produces the same relationship as for a mirror. That is, the reciprocal of the focal length is equal to the sum of the reciprocals of the object and image distances:

$$\frac{1}{p} + \frac{1}{q} = \frac{1}{f}$$

This relationship is called the **lens equation**, although we found it convenient to derive it first for the mirror. With this equation we can predict the position of an image produced by either a lens or a mirror.

Notice that the image is larger in one of the lens diagrams (Figure 2-35) and a little bit smaller in the other (Figure 2-33). Just as we did with the mirror, we define the magnification as the ratio of the image height to the object height. It turns out to be equal to the ratio of the image to the object distance.

$$M \equiv \frac{b}{a} = \frac{q}{p}$$

When the object is far away, the image is real, inverted, small, and close to the focal point. As the object approaches the focal point, the image remains real and inverted but grows larger and moves away from the focal point. When they are both two focal lengths distant from the lens, the image is the same size as the object at that point. The object is on one side and the image is on the other. As the object gets increasingly closer to the focal point, the image grows ever larger and goes increasingly further away from the lens. The image becomes infinitely large and is located at infinity when the object is at the focal point. This is how a penlight works. Light from the object is deflected outward in a parallel beam by a lens built into the penlight bulb.

When the object passes over the focal point, the image goes beyond infinity, coming in at a negative infinity on the same side of the lens as the object. The image is now erect, larger, and further away from the lens than the object, but virtual. When the object is within the focal distance of the lens, the light rays diverge so greatly that the lens cannot converge them enough to make them cross. The rays of light leave the lens still diverging.

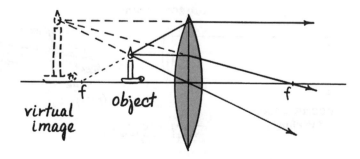

virtual
image

object

Figure 2-35

A magnifying glass is a convex lens used to form a virtual image of an object within its focal distance.

They diverge as though they had come from a point in space behind the location of the object. Since this point in space, the virtual image, is always further from the lens than the object, the image is always larger than the object. This is how a convex lens is used as a magnifying glass.

You can only see things with your naked eye if they are a certain distance away. When things get too close, they look blurred. Look at a friend's nose and then get closer and closer. The point at which the nose becomes blurred is called your **near point**. A jeweler can examine a diamond very close to his eye with the aid of a little magnifying glass mounted in a cylinder, called a loupe (pronounced loop). The loupe not only makes an enlarged image, but also moves it back beyond the jeweler's near point so he can see it.

Optical Instruments

Your eyes and a camera work on the same principle. The lens in the front of your eye forms a real image of whatever you're looking at on the retina in the back of your eye. In a camera the image is projected on film instead of retinal tissue. The image is upside down in both cases but that fact is compensated for in both cases. You simply turn camera film around to look at it, and your brain has learned to turn around the images it receives from your retina.

A telescope also uses a lens to form a real image of distant objects. The real image is not caught on a light-sensitive screen, such as a retina or photographic film, but is projected in space to be examined by another lens used as a magnifying glass. The second lens, called the eyepiece, is positioned so that the image produced by the first lens is within its focal point. The eyepiece forms an enlarged virtual image of the real image. When you look through a telescope, you are looking at an image of an image.

A compound microscope forms a real image of a close object. Since the image is further from the lens than the object, it is larger. A second lens, called the eyepiece, forms a virtual image of the first image, further enlarged. It is called a compound microscope because it enlarges an already enlarged image.

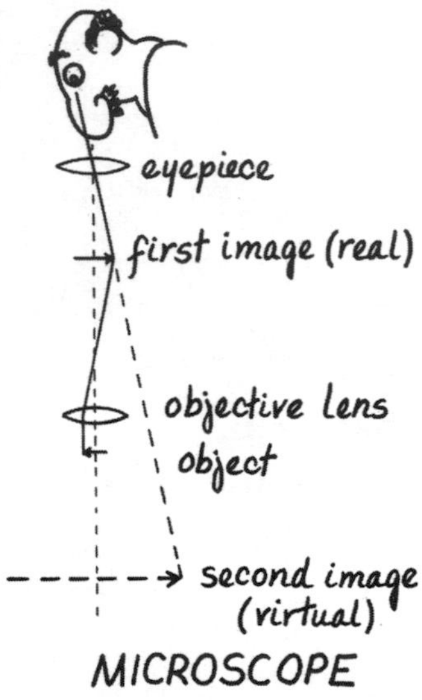

Figure 2-36

A compound microscope uses one lens to form an enlarged real image and another to magnify it again.

Figure 2-37

A telescope also uses one lens to form a real image and another as a magnifying glass to examine the real image.

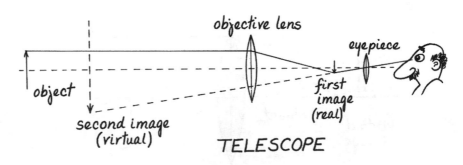

SUMMARY OF CONCEPTS

The problems in this chapter are on the optics of reflection and refraction. The law of reflection states that the angle of incidence equals the angle of reflection:

$$\angle i = \angle r$$

The convention of abbreviations that we use in this chapter are:

focal length = f
radius of curvature = c
object height = a
image height = b
object distance = p
image distance = q

A concave mirror brings light to a focus at a point halfway to the radius of curvature.

$$f = \tfrac{1}{2}c$$

The focal point and center of curvature of a mirror can be used to trace three rays to locate an image. Two of these rays are associated with parallel light and the focal point. Incident parallel light is reflected so that it passes through the focal point:

$$\| \rightarrow f$$

Incident light that passes through the focal point is reflected so that it goes out parallel:

$$f \rightarrow \|$$

The third ray is associated with the center of curvature of the mirror. Incident light that passes through the center of curvature is reflected back on itself:

$$c \rightarrow c$$

An image may be found numerically from the object-image formula, which states that the reciprocal of the focal length equals the sum of the reciprocals of the object distance, p, and the image distance, q.

$$\frac{1}{f} = \frac{1}{p} + \frac{1}{q}$$

The object-image formula assumes the following sign convention:

The focal length is positive for a converging mirror or lens and negative for a diverging mirror or lens.

The image height, b, is positive for an inverted image and negative for an erect image.

The image distance, q, is positive for a real image and negative for a virtual image.

Magnification, defined as the ratio of the image height, b, to the object height, a, is equal to the ratio of the image distance to the object distance:

$$M \equiv \frac{b}{a} = \frac{q}{p}$$

The law of refraction is **Snell's law,** which states that the ratio of the sine of the angle of incidence to the sine of the angle of refraction is a constant, called the index of refraction of the material:

$$\frac{\sin \angle i}{\sin \angle r} = n$$

Three rays may also be used to locate the image formed by a lens. Two are associated with parallel light and the focal points, while the third passes through the center of the lens and is not deflected. The image may also be found from the object-image formula, the same as for a mirror.

EXAMPLE PROBLEMS

Try to do the Examples yourself before looking at the Sample Solutions.

2-1 To produce the illusion of a ghostly candle burning under water, place the candle in front of a window pane and a glass of water an equal distance on the other side. A dark background and a screen to hide the candle from direct view helps to perfect the illusion.

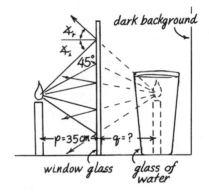

Given: The candle is placed 35 cm in front of the window glass. In understanding the position of the virtual image, you decide to trace rays that make an angle of 30° and 45° with the surface of the window glass.

Find: (a) What will be the angle of reflection for these rays?
(b) Draw several reflected rays showing the position of the virtual image.
(c) How far behind the window pane should the glass of water be placed?

BACKGROUND

In a famous old magic book entitled Stage Magic and Scientific Diversions, *Albert Hopkins describes several illusions based on superimposing a virtual image on a real object. This book is probably responsible for the commonly held notion that many magic tricks are done with mirrors. Actually, very few illusions are done with mirrors, but the ones that are can be very striking.*

2-2 In a commercial movie projector, the light from the lamp is concentrated by a concave mirror before being projected onto the screen by a lens.

Given: The lamp is 20 cm from the mirror, while the image formed by the light source in the lamp, a glowing filament, must be near the projection lens some 80 cm from the mirror. The image must be about 5.0 cm in diameter to illuminate each frame on the 35-mm movie film.

Find: (a) What focal length should the mirror in this projector have?
(b) How large should the light source be to produce the proper size image?

BACKGROUND

Light from the lamp is imaged on the projection lens so that it will be totally out of focus on the screen. You want a nice crisp image of the film on the screen, but you do not want to see any hint of the structure of the light source on the screen. The image produced by the mirror is an object for the projection lens, but making this object distance zero causes the object-image formula to blow up, since the reciprocal of zero is infinite. The result is that light from the light source fills the screen evenly (in the absence of film).

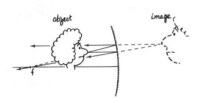

2-3 A woman is looking at the virtual image of her nose in a concave shaving mirror.

Given: The woman is holding her nose 13.3 cm away from the mirror, which produces an image at 20.0 cm behind the mirror. She has a 3.5-cm nose.

Find: (a) What is the focal length of the shaving mirror?
(b) What is the magnification?
(c) How large will the image of her nose be?

BACKGROUND

This problem illustrates the use of a concave mirror as a magnifying mirror. Shaving mirrors are sometimes concave so that you can see defects in your face on an enlarged scale.

In order to use a concave mirror to produce a magnified virtual image, the object must be placed inside the focal point of the mirror. Light from the object will be converged by the mirror but not enough to actually cross and produce a real image. Light from the object will be so strongly diverging that the converging effect of the mirror can only make the reflected light appear to come from a point further behind the mirror than the object is in front.

2-4 Convex mirrors are sometimes used at blind intersections to provide a wide-angle view of oncoming traffic.

Given: A convex mirror having a focal length of −1.0 m is placed at a blind intersection. An oncoming car is 3.0 m from the mirror.

Find: (a) What is the position of the virtual image of the oncoming car produced by the convex mirror?

(b) What is the magnification?

(c) Draw a ray diagram showing the three rays associated with parallel light, the focal point, and the center of curvature of the mirror.

BACKGROUND

This problem illustrates the use of a convex mirror to produce a wide-angle view. Since the images of real objects are smaller in the mirror, more images fit within the field of view than would fit within the field of view of a plane mirror.

Just as is the case for a concave mirror, the focal point is the image of a far distant object. But instead of concentrating parallel light to cross at the focal point, a convex mirror diverges parallel light so that it is reflected away from the focal point. Reflected light "crosses" at the focal point because that is where it appears to come from.

2-5 A merman is looking at a handsome young woman. He is underwater while she is standing on a rock at the edge of the sea.

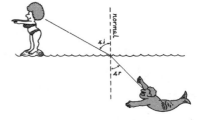

Given: Light from her nose on the way to his eye is incident at an angle of 60° to the normal. Light from his chin on the way to her eye makes an angle of 38° to the normal. The index of refraction of water is $n = 1.33$.

Find: (a) What is the angle of refraction for light from her nose on the way to the merman's eye?

(b) At what angle does the light from his chin leave the water's surface on the way to the handsome young woman's eye?

BACKGROUND

This problem illustrates the use of Snell's law both for light going from air into water and for light going from water into air. Light can travel both ways, although the form of Snell's law that we have been considering is only set up for light going from air into a medium having a higher index of refraction.

2-6 A college student is amazing her little brother with a demonstration of real images. She uses a strong magnifying glass (about seven power) to project an image of a candle on a piece of paper several centimeters away.

Given: The lens has a focal length of 3 cm. The candle flame is about 1 cm tall and is held 5 cm from the lens.

Find: (a) Predict the position of the image and approximate size by drawing a ray diagram showing three rays that a little brother could understand.
(b) What is the image distance from the object-image formula?
(c) How large will the image be, from the magnification formula?

BACKGROUND

A good test of your understanding of a subject is to try to explain it to a child. If you don't have any children around the house, go out and try your neighborhood playground. It takes some planning, however, to get even a very smart child through an explanation involving a ray diagram. Start with the image of a bright nearby object and show how the image distance changes with the object distance. Then shift to a discussion of the focal point by using the magnifying glass to burn your initials in a piece of wood. Shift back to a discussion of real images by showing how things look upside down when you hold the magnifying glass far away from your eye. Socratic teaching is difficult, but it is fun.

2-7 When a physics instructor focuses on an exam paper, the lens in his or her eye forms a real image of the paper on the retina in the back of the eye. The light-sensitive retina then translates the patterns of light and dark areas to the brain, which in turn translates them into a grade.

Given: An exam paper is 20 cm long and is located 28 cm from a physics instructor's eye. The eyeball is 3.0 cm in diameter.

Find: (a) What is the focal length of the lens in the instructor's eye?
(b) How large will the image of the exam paper be when projected by the lens on the back surface of the eyeball?

BACKGROUND

The eye works very much like a camera in that a real image is formed on a light-sensitive film. This image, like all real images, is upside down. The brain is used to that, however, and only becomes confused if the image is right side up.

One of the classical experiments in psychology is based upon wearing special glasses that invert the image so that it is in fact right side up on the retina. It takes the brain between one and two days to learn to flip this image. After that, a person wearing the glasses can function normally. Everything is fine until the person takes off the glasses.

2-8 A detective is examining evidence with a magnifying glass. He holds the glass, a double convex lens, close enough to the evidence so that it produces an enlarged virtual image. He adjusts the position of the magnifying glass so that the image occurs at his distance of most distinct vision.

Given: The magnifying glass has a focal length of +15 cm. The detective adjusts its position to produce an image at −22 cm from the lens (on the same side of the lens as the object).

Find: (a) How far from the evidence should the detective hold the lens?
(b) What is the magnification?

BACKGROUND

This problem illustrates the use of a converging lens to produce a virtual image. The object must be located within the focal length if the image is to be virtual.

SAMPLE SOLUTION 2-1

Given:

The candle is placed 35 cm in front of the window glass. In understanding the position of the virtual image, you decide to trace rays that make an angle of 30° and 45° with the surface of the glass.

Find:

(a) What will be the angle of reflection for these rays?

(b) Draw several reflected rays showing the position of the virtual image.

(c) How far behind the window pane should the glass of water be placed?

(a) $\angle i = \angle r$

$\angle i_1 = \angle r_1$
$= (90° - 30°)$
$= \boxed{60°}$

$\angle i_2 = \angle r_2$
$= (90° - 45°)$
$= \boxed{45°}$

(b)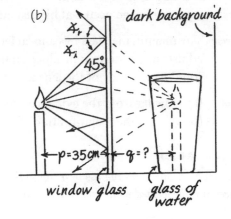

(c) $\dfrac{1}{f} = \dfrac{1}{p} + \dfrac{1}{q}$

$f = \infty$ for a plane mirror

$\dfrac{1}{q} = \dfrac{1}{f} - \dfrac{1}{p}$

$= \dfrac{1}{\infty} - \dfrac{1}{p}$

$= 0 - \dfrac{1}{p}$

$q = -p$

$= \boxed{-35\text{cm}}$

DISCUSSION

The solution to part (a) is based on the law of reflection. The angle of incidence and the angle of reflection are, however, usually measured to the normal rather than to the surface itself. Since the normal is defined as a line that is perpendicular to the surface, the angle of incidence is found by subtracting the given angle from 90°.

The drawing in part (b) shows that the position of the virtual image is independent of the angle of incidence.

Part (c) is based on the object-image formula applied to a plane mirror, where the focal length is infinite. The image and object distances are equal in this case.

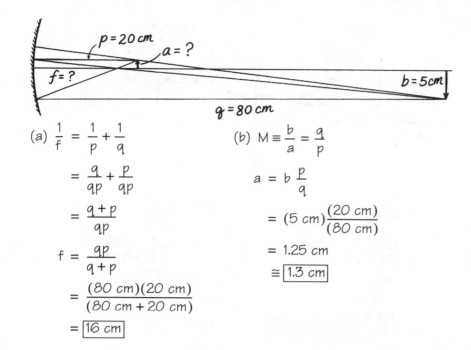

(a) $\dfrac{1}{f} = \dfrac{1}{p} + \dfrac{1}{q}$

$\quad = \dfrac{q}{qp} + \dfrac{p}{qp}$

$\quad = \dfrac{q+p}{qp}$

$f = \dfrac{qp}{q+p}$

$\quad = \dfrac{(80\ cm)(20\ cm)}{(80\ cm + 20\ cm)}$

$\quad = \boxed{16\ cm}$

(b) $M \equiv \dfrac{b}{a} = \dfrac{q}{p}$

$a = b\,\dfrac{p}{q}$

$\quad = (5\ cm)\dfrac{(20\ cm)}{(80\ cm)}$

$\quad = 1.25\ cm$

$\quad \cong \boxed{1.3\ cm}$

Given:
The lamp is 20 cm from the mirror, while the image formed by the light source in the lamp, a glowing filament, must be near the projection lens some 80 cm from the mirror. The image must be about 5.0 cm in diameter to illuminate each frame on the 35-mm movie film.

Find:
(a) What focal length should the mirror in this projector have?

(b) How large should the light source be to produce the proper size image?

DISCUSSION

The ray diagram is not called for in the problem and might not be part of your solution. While not required, however, ray diagrams are always helpful for understanding what is going on and for checking the numerical results to see if they look reasonable. Arrows are frequently used in such diagrams to represent the object and the image.

Part (a) can be solved by solving the object-image formula for f. Part (b) can be solved using the definition of magnification.

The result of part (b) must be rounded off because the image height is known to only one significant figure. The rule of thumb of significant figures would have us round the answer off to 1 cm, but that would imply an error of ±0.5 cm, or ±50%. That is less accurate than we know the height ($b = 5$ cm ±0.5 cm, which is an error of ±10%), so we stretch the rule a bit and report the answer to two significant figures. Three significant figures would, however, be stretching the rule too far.

SAMPLE SOLUTION 2-3

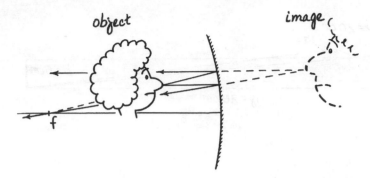

Given:

The woman is holding her nose 13.3 cm away from the mirror, which produces an image at 20.0 cm behind the mirror. She has a 3.5-cm nose.

Find:

(a) What is the focal length of the shaving mirror?

(b) What is the magnification?

(c) How large will the image of her nose be?

(a) $\dfrac{1}{f} = \dfrac{1}{p} + \dfrac{1}{q}$

$\quad = \dfrac{q+p}{pq}$

$f = \dfrac{pq}{p+q}$

$\quad = \dfrac{(13.3 \text{ cm})(-20.0 \text{ cm})}{(13.3 \text{ cm}) + (-20.0 \text{ cm})}$

$\quad = \boxed{40 \text{ cm}}$

(b) $M = \dfrac{q}{p}$

$\quad = \dfrac{(-20 \text{ cm})}{(13.3 \text{ cm})}$

$\quad = \boxed{-1.5}$

(c) $M \equiv \dfrac{b}{a}$

$b = Ma$

$\quad = (-1.5)(3.5)$

$\quad = \boxed{-5.3 \text{ cm}}$

DISCUSSION

Here again, while a ray diagram is not required by the problem, it helps to show how the image is formed and allows you to see if the result is reasonable.

Part (a) can be worked by solving the object-image formula for f and then inserting the data. Parts (b) and (c) use the magnification formula. The results in parts (b) and (c) are negative because the image is virtual and therefore right side up. The formulas were derived from the geometry of a real image, which is upside down. Negative upside down is right side up.

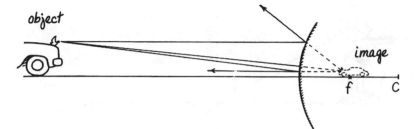

object

image

f   C

(a) Given $\begin{cases} f = -1.0 \text{ m} \\ p = 3.0 \text{ m} \end{cases}$

$$\frac{1}{f} = \frac{1}{p} + \frac{1}{q}$$

$$\frac{1}{q} = \frac{1}{f} - \frac{1}{p}$$

$$= \frac{p - f}{pf}$$

$$q = \frac{pf}{p - f}$$

$$= \frac{(3.0 \text{ m})(-1.0 \text{ m})}{(3.0 \text{ m}) - (-1.0 \text{ m})}$$

$$= \boxed{-0.75 \text{ m}}$$

(b) $M \equiv \dfrac{b}{a} = \dfrac{q}{p}$

$$= \frac{(-0.75 \text{ m})}{(3.0 \text{ m})}$$

$$= \boxed{-0.25}$$

Given:

A convex mirror having a focal length of −1.0 m is placed at a blind intersection. An oncoming car is 3.0 m from the mirror.

Find:

(a) What is the position of the virtual image of the oncoming car produced by the convex mirror?

(b) What is the magnification?

(c) Draw a ray diagram showing the three rays associated with parallel light, the focal point, and the center of curvature of the mirror.

DISCUSSION

The negative image distance in part (a) and negative value for the magnification in part (b) mean that the image is in virtual space (behind the mirror) and right side up. The formulas were derived from a ray diagram for a real image, which assumes the image is on the same side of the mirror as the object and that the image is upside down. Negative image distances are just measured in the other direction, on the opposite side of the mirror as the object. Negative magnification means that the image height will be negative if the object height is positive. Negative upside down is right side up. Virtual images are in general right side up.

SAMPLE SOLUTION 2-5

Given:

Light from her nose on the way to his eye is incident at an angle of 60° to the normal. Light from his chin on the way to her eye makes an angle of 38° to the normal. The index of refraction of water is n = 1.33.

Find:

(a) What is the angle of refraction for light from her nose on the way to the merman's eye?

(b) At what angle does the light from his chin leave the water's surface on the way to the handsome young woman's eye?

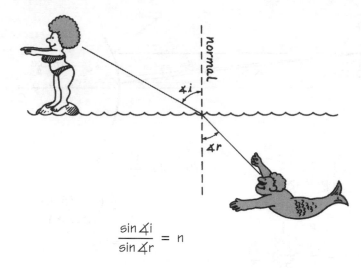

$$\frac{\sin \angle i}{\sin \angle r} = n$$

(a) $\sin \angle r = \dfrac{\sin \angle i}{n}$

$ = \dfrac{\sin(60°)}{1.33}$

$ = 0.651$

$\angle r = \boxed{41°}$

(b) $\sin \angle i = n \sin \angle r$

$ = (1.33)\sin(38°)$

$ = 0.819$

$\angle i = \boxed{55°}$

DISCUSSION

Part (a) is worked by solving Snell's law for the sine of the angle of refraction. The sine of 60° is found from your calculator. After dividing by the index of refraction to get the sine of the angle we are looking for, the angle itself can be found by using the inverse sine function on your calculator. If no calculator is handy, the angles can also be found by looking at a table of trigonometric functions.

Part (b) is found by solving Snell's law for the angle of incidence, although the light from the merman's chin is actually going the other way in this case. If you can figure out the path that light will take going one way, the same path will work for light going in the reverse direction. This rule even works for the most complicated optical system. It is called the **principle of reversibility** of optics.

The reason we must reverse the direction of the light ray in part (b) is that we have only concerned ourselves with the simplest form of Snell's law, which assumes that light is passing from air or some other medium having an index of refraction close to that of a vacuum into a medium having a greater index of refraction.

(a)

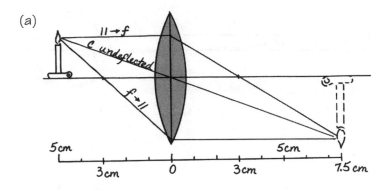

(b) $\dfrac{1}{f} = \dfrac{1}{p} + \dfrac{1}{q}$

$\dfrac{1}{q} = \dfrac{1}{f} - \dfrac{1}{p}$

$\quad = \dfrac{p - f}{pf}$

$q = \dfrac{pf}{p - f}$

$\quad = \dfrac{(5\ cm)(3\ cm)}{(5\ cm) - (3\ cm)}$

$\quad = \boxed{7.5\ cm}$

(c) $M \equiv \dfrac{b}{a} = \dfrac{q}{p}$

$b = a\,\dfrac{q}{p}$

$\quad = (1\ cm)\dfrac{(7.5\ cm)}{(5\ cm)}$

$\quad = \boxed{1.5\ cm}$

Given:
The lens has a focal length of 3 cm. The candle flame is about 1 cm tall and is held 5 cm from the lens.

Find:
(a) Predict the position of the image and approximate size by drawing a ray diagram showing three rays that a little brother could understand.

(b) What is the image distance from the object-image formula?

(c) How large will the image be, from the magnification formula?

DISCUSSION

Part (b) is worked by solving the object-image formula for the image distance. This algebra was done in an earlier example, but it is best to repeat it for each problem. You could memorize the general solution of every physics problem, but it is far better for you to know how to obtain it.

Part (c) is solved by the magnification formula and the definition of magnification. Magnification is defined as *b* over *a*, but it is equal to *q* over *p*. Solving for the image height and substituting the numerical data gives the answer.

SAMPLE SOLUTION 2-7

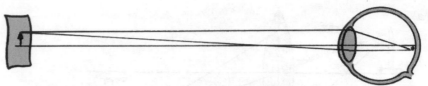

Given:

An exam paper is 20 cm long and is located 28 cm from a physics instructor's eye. The eyeball is 3.0 cm in diameter.

Find:

(a) What is the focal length of the lens in the instructor's eye?

(b) How large will the image of the exam paper be when projected by the lens on the back surface of the eyeball?

(a)
$$\frac{1}{f} = \frac{1}{p} + \frac{1}{q}$$
$$= \frac{q + p}{qp}$$
$$f = \frac{qp}{q + p}$$
$$= \frac{(28 \text{ cm})(3.0 \text{ cm})}{(28 \text{ cm}) + (3.0 \text{ cm})}$$
$$= \boxed{2.7 \text{ cm}}$$

(b) $M \equiv \dfrac{b}{a} = \dfrac{q}{p}$

$$b = a\frac{q}{p}$$
$$= (20 \text{ cm})\frac{(3.0 \text{ cm})}{(28 \text{ cm})}$$
$$= \boxed{2.1 \text{ cm}}$$

DISCUSSION

The focal length of the lens is found from the object-image formula. The trick is to recognize that the diameter of the eyeball is the image distance.

This problem is useful on examinations to discriminate between people who do the example problems and those who do not. People who see this problem for the first time tend to confuse it with a mirror situation and take the focal length as one-half the radius of curvature. Those who have thought about it some realize that the back side of the eyeball is not a reflecting surface but is the location of the image.

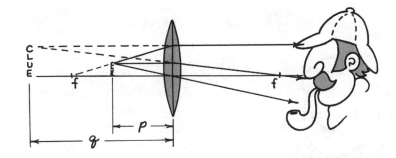

Given:
The magnifying glass has a focal length of +15 cm. The detective adjusts its position to produce an image at −22 cm from the lens (on the same side of the lens as the object).

Find:
(a) How far from the evidence should the detective hold the lens?

(b) What is the magnification?

(a) $\dfrac{1}{f} = \dfrac{1}{p} + \dfrac{1}{q}$

$\dfrac{1}{p} = \dfrac{1}{f} - \dfrac{1}{q}$

$\quad = \dfrac{q - f}{qf}$

$p = \dfrac{qf}{q - f}$

$\quad = \dfrac{(-22 \text{ cm})(15 \text{ cm})}{(-22 \text{ cm}) - (15 \text{ cm})}$

$\quad = \boxed{8.9 \text{ cm}}$

(b) $M \equiv \dfrac{b}{a} = \dfrac{q}{p}$

$\quad = \dfrac{(-22 \text{ cm})}{(8.9 \text{ cm})}$

$\quad = \boxed{-2.5}$

DISCUSSION

Parts (a) and (b) use the object-image formula and the magnification formula. Just as was the case with mirrors, a negative magnification means the image is right side up, since the formulas are derived from the geometry of a real-image ray diagram. The only difference between the sign conventions for a lens and for a mirror is the positive direction for image location. Since a real image is produced by a mirror on the same side of the mirror as the object, that is considered to be the direction of positive image distances. A lens produces a real image on the opposite side from the object, so positive image distances are measured in that direction for a lens.

ESSENTIAL PROBLEMS

Be sure you can do the Example Problems (without peeking at the Sample Solutions) before working these Essential Problems.

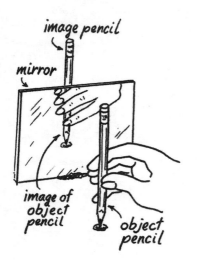

image pencil

mirror

image of object pencil

object pencil

2-9 In attempting to understand the nature of virtual images, a student sets up a little experiment on the kitchen table. A lump of clay is used to stand a small pocket mirror up vertically in the middle of the table. Two pencils are stuck into two smaller lumps of clay, one to act as the object and the other to locate the image. The object pencil is stood up a short distance in front of the mirror, and the image pencil is moved around behind the mirror until it lines up with the image even when viewed from different angles.

Given: The object pencil is placed 20 cm from the mirror. Viewed from one angle, the light reflected from the mirror makes an angle of 75° with the mirror. Viewed from a different angle, the reflected light makes an angle of 65° with the normal.

Find: (a) What is the angle of incidence for these rays?
(b) Draw a top-view diagram showing the incident and reflected rays. Indicate the position of the image by showing where the reflected rays seem to come from after they leave the mirror.
(c) How far behind the mirror should the image pencil be placed to coincide with the image formed by the mirror?

2-10 Just for fun, a physics student takes a magnifying mirror off the dressing table and uses it to form a real image of a light bulb on the wall.

Given: The student finds it necessary to hold the mirror 102 cm from the light bulb to produce a nice sharp image on the wall 204 cm away. The image of the light bulb is 20 cm tall.

Find: (a) What is the focal length of the mirror?
(b) How tall is the light bulb?

2-11 In examining the inside surface of his eyeglasses one day, a young man notices a magnified reflection of his eye. He starts wondering what the focal length might be for this partially reflecting concave surface.

Given: When he holds his eye 2.0 cm from the inside surface of the lens, the reflected image is 3.5 cm on the other side of the surface (in negative space). His eye, measured vertically eyelash to eyelash, is 1.2 cm.

Find: (a) What is the focal length of the reflecting surface?
(b) What is the magnification?
(c) How tall is the image of the young man's eye?

2-12 A shoplifter is shopping for convex mirrors one day. He admires one that is mounted in the corner of a store.

Given: The mirror has a focal length of 2.0 m on the negative side, and he is standing 2.0 m away from the mirror.

Find: (a) How far is the virtual image on the other side of the mirror?
(b) What is the magnification?
(c) Draw a ray diagram showing the three rays associated with parallel light, the focal point and the center of curvature of the mirror.

2-13 A wicked magician has trapped an Arabian princess in a crystal cube. Light from the magician's evil eye is deflected toward the normal as it enters the crystal due to the increased index of refraction.

Given: Light from the evil eye is incident at an angle of 45° from the normal and is refracted at an angle of 25°. Light from the evil chin is incident at an angle of 75° from the normal.

Find: (a) What is the index of refraction of the crystal?
(b) What is the angle of refraction for light from the magician's chin?

2-14 A photographic enlarger is basically a high-quality projector used to expose photosensitive print paper with a real image of the negative. The size of this image can be adjusted by changing the relative distances between the negative, or object, and the lens and the print paper, or image.

Given: A normal lens for enlarging 35-mm film has a focal length of 5.0 cm. A photographer who wishes to enlarge a feature 2.5 cm tall on the negative places the negative 8.3 cm from the lens.

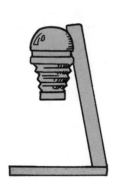

Find: (a) Predict the position of the image and approximate size by drawing a ray diagram showing three rays.
(b) What should be the distance from the lens to the print paper from the object-image formula?
(c) How large will the image of the desired feature be from the magnification formula?

2-15 The heat lamp in the bathroom has been <u>absentmindedly</u> left on. 心 不在焉 The light shines across the room, hitting a slender perfume bottle with convex sides which focuses the light into a small dot which is slowly but surely burning a hole in the wall behind the bottle…

Given: The heat lamp has a <u>diameter</u> of 14 cm. The heat lamp is 1.9 meters 直径 from the perfume bottle, which 9.8 cm away from the wall with the smoldering hole.

Find: (a) What is the focal length of the bottle of perfume when it acts as a lens?

(b) What is the diameter of the image of the heat lamp?

2-16 A jeweler's loupe is a small magnifying glass of short focal length that is held up close to the eye so that the jeweler can take a very close look at a gem or a watch part, holding the object much too close to be seen with the naked eye. The lens forms a virtual image of the object far enough away for the jeweler to see it.

Given: The magnifying glass has a focal length of 2.2 cm. The jeweler is looking at a gem that he positions just far enough away from the lens so that the image distance is −22 cm from the lens (on the same side of the lens as the gem).

Find: (a) How far from the lens should the jeweler hold the gem?

(b) What will be the magnification?

2-17 A physics instructor is playing with a magnifying mirror while watching television. He notices that he can use the mirror to project a real image of the TV screen on the surface of a beer can that happens to be close at hand.

Given: The beer can is 25 cm from the mirror, and the TV set is 3 m across the room. The TV screen is 20 cm tall.

Find: (a) What is the focal length of the mirror?
(b) How tall is the image projected on the beer can?
(c) Is the image on the beer can right-side up or upside-down?

2-18 A girl is looking into her magnifying makeup mirror. Behind her is a sunny window. She notices that, if she holds a piece of paper just right in front of the mirror, she can "catch" an image of the window on it.

Given: The mirror is 2.79 m from the window which is 1.37 m tall. For the image to be in focus, the girl holds the paper 49.0 cm from the mirror.

Find: (a) What is the focal length of the mirror?
(b) How tall is the image?
(c) Is the image right-side up or upside-down?

2-19 When light enters glass it changes direction. We now know that this is because the light slows down. You can easily see this change in direction with a laser light. If you shine the laser light on a thick piece of glass, you can see the angle change while the light is in the glass.

Given: You shine a laser at a piece of glass at an angle of 27° measured with respect to the normal. You can see the light deflect when it goes into the glass and measure the angle with a protractor as 17° with respect to the normal. You then change the angle of the laser to hit the glass at 11° with respect to the normal.

Find: (a) What is the index of refraction of the piece of glass?
(b) When you change the angle of incidence to 11°, what is the new angle of refraction?

2-20 A sunbeam shining on a clear pond changes direction when it enters the water.

Given: In the morning, the sunbeam hits the surface of the pond at an angle of 41° measured with respect to the normal. It is deflected 屈折 when it hits the surface of the pond to an angle of 29°, to hit a rock below. An hour later the sunbeam is incident on the pond at an angle of 28°.

Find: (a) What is the index of refraction of the water in this pond?
(b) At the later time, what is the angle of refraction of the sunbeam?

2-21 A woman is admiring herself in the convex surface of a fun-house mirror. She notices that her image seems far away when she looks at it with one eye closed because the image is so small. When she looks at the image with both eyes, however, her binocular vision tells her that the image is really closer behind the surface of the mirror than she is in front.

Given: The woman is 2.5 m from the mirror. The mirror has a focal length of −0.75 m. The woman is 1.78 m tall.

Find: (a) Draw a ray diagram showing the position and approximate size of the image using three reflected rays.
(b) Find the position of the woman's image using the object-image formula.
(c) How tall is the woman's image?

2-22 Eleanor drives a pick-up truck that has a convex mirror stuck onto the side mirror to help her maneuver. She finds it especially useful for parking. One time while pulling out of a driveway she watches the image in the mirror of a tree that she wishes to avoid hitting.

Given: The mirror has a focal length of −6.5 cm. The tree is 3.2 m tall and is 4.6 m from the mirror.

Find: (a) Draw a ray diagram showing the position and approximate size of the image using three reflected rays.
(b) What is the position of the image with respect to the surface of the mirror?
(c) How tall is the image of the tree in the mirror?

2-23 A movie theater projectionist must know how to select the proper projection lens to fill the screen with an image of the movie film.

Given: The picture on the movie film is 3.4 cm wide. In a particular theater the screen is 10 m wide and is located 30 m from the projector.

Find: (a) What magnification is needed to fill the projection screen?
(b) What object distance is needed to produce this magnification?
(c) What should be the focal length of the lens?

2-24 The type of camera most frequently used by professional photographers to take portraits in the studio uses a large film format. (The old standard was 4 inches by 5 inches.) The image can therefore be much larger compared to the grain of the film than is possible for a small camera.

Given: The object being photographed is 97 cm tall. The camera is set so that the film is 12.8 centimeters from the lens of a large-format view camera. The photographer wishes the image to be 9.0 cm tall on the film and needs to select a lens of the proper focal length to produce this result.

Find: (a) What is the magnification of the object?
(b) How far should the photographer position the object being photographed from the lens of the camera?
(c) What is the focal length of the lens?

2-25 One way to find the focal length of a lens is to form a real image of a far-distant object. Unfortunately, you find yourself wishing to know the focal length of a double convex lens that you are holding in your hand while standing in a storeroom where there are no far-distant objects. The farthest thing away is an overhead light fixture. You form an image of this light fixture on the floor and decide to approximate the distance from the lens to the floor as the focal length even though you know that the light from the fixture is not really parallel.

Given: The distance from the lens to the light fixture is about 2 m. The distance from the lens to the floor is 35 cm.

Find: (a) What is the focal length of the lens from the object-image formula?

(b) How much error is involved in just assuming the object distance to be infinite?

(c) What is the percentage error? (How great is the error compared to the best value?)

2-26 Horace is playing with a convex lens while watching TV. While waiting for the next exciting commercial break, Horace makes an image of the TV screen on the opposite wall. He figures that the focus of the lens must be pretty close to the image distance.

Given: When the image of the TV is in focus, he is holding the lens 2.9 m away from the screen and 31 cm away from the wall.

Find: (a) What is the focal length of the lens from the object-image formula?

(b) How much error is involved in forgetting about the lens equation and just assuming the object distance to be infinite?

(c) What is the percentage error? (How great is the error compared to the best value?)

2-27 A common type of Christmas tree ornament is a glass ball having a mirrored surface. You can use such an ornament as a convex mirror to reinforce your understanding of virtual images. The radius of curvature is, naturally, half the diameter of the ball.

Given: You are looking at a Christmas tree ball having a diameter of 10 cm. You compare the image position of your nose to the image position of the fireplace and decide that the fireplace image is probably near the focal point. Your nose is 30 cm from the surface of the ball, and the fireplace is 5.00 m across the room.

Find: (a) What is the focal length of the mirrored surface?

(b) What is the image position of your nose?

(c) How much error is involved in assuming the fireplace image to be the focal point of the mirror?

(d) Draw a diagram showing parallel light from a far-distant object being reflected in accordance with the law of reflection and show that it comes away from the focal point.

2-28 Ali has a set of measuring spoons that are shaped like half-spheres and polished until they reflect images. While cooking one day, he leaves the tablespoon measure bowl-side up on the counter and becomes interested in the images.

Given: The tablespoon measure has a diameter of 4.1 cm. Ali can see the image of a bell pepper which is on the counter 8.0 cm from the measuring spoon. He can also see an image of the coffee maker on the other end of the counter, 2.3 m away. On inspection, Ali figures that the image of the coffee maker is close to the focal point of this spherical mirror.

Find: (a) What is the focal length of the measuring spoon?
(b) What is the image position of the bell pepper?
(c) How much error is involved in assuming the coffee machine image to be the focal point of the spoon?
(d) Draw a diagram showing parallel light from a far-distant object being reflected in accordance with the law of reflection and show that it comes away from the focal point.

2-29 A man is admiring himself in the shiny concave surface of a half-spherical stainless steel mixing bowl. He notices that his image is upside down and smaller than his object (himself) when he holds the bowl at arm's length.

Given: The bowl has a diameter of 30 cm. The distance from the bowl to the object is 75 cm, and the object size is 6 cm.

Find: (a) What is the focal length of the mirrored surface of the bowl?
(b) What is the position of the image?
(c) What is the size of the image?

2-30 While decking a Christmas tree, an ornament breaks. The ornament was round with a mirrored surface. When it breaks, you notice that the inside of the ornament is mirrored and you can see your face, small and up-side down, in its surface.

Given: The ornament originally had a diameter of 8.0 cm. You are holding the ornament piece 38 cm away from yourself to study it. Assume that you have a 19-cm face.

Find: (a) What is the focal length of the ornament fragment?
(b) What is the position of the image of your face?
(c) How tall is the image of your face?

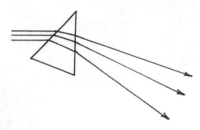

2-31 Prisms are nice. They spread white light into a rainbow of colors. This is because the index of refraction of the glass depends on the frequency of light. White light is a mixture of frequencies, each frequency a different color. When a beam of white light is shone on the glass each frequency has a different angle of refraction within the glass. The result is that the colors come out of the prism separated from one another.

Given: Heavy flint glass has an index of refraction $n = 1.675$ for blue light and $n = 1.638$ for red light. A beam of white light containing these two colors shines on the glass at an angle of $36°$ (with respect to the normal).

Find: (a) What is the angle of refraction for the blue light?
(b) What is the angle of refraction for the red light?
(c) Both colors of light travel at the same speed outside of air. Which one do you think is slowed down more when it enters the glass? What is your reasoning?

2-32 Rainbows are formed when white light from the sun hits raindrops in the atmosphere. Water has different indexes of refraction for different colors of light, so water drops separate the different colors from light.

Given: Water has an index of refraction of $n = 1.3311$ for red light and $n = 1.3330$ for yellow light. A sunbeam shines on a raindrop at an angle of $41°$ (with respect to the normal).

Find: (a) What is the angle of refraction for the red light?
(b) What is the angle of refraction for the yellow light?
(c) Both colors of light enter the raindrop at about the same speed. Which one do you think is slowed down more while inside the water droplet?

2-33 A magician is practicing before a mirror. He needs to be able to see both his left hand, in which he has palmed a coin, and his right hand, which is holding an imaginary coin. His left arm is vertical while his right arm is horizontal.

Given: The distance from his eye to the tip of the fingers of his open left hand is 1.03 m. The horizontal distance from his right fist to his left hand is 1.15 m. He is standing 2.0 m from the mirror.

Find: (a) What is the angle of incidence for light reflected from the magician's open left hand to his eye?

(b) What is the minimum size of the mirror needed to see both hands at once? (Hint: you may want draw several rays between the magician's eye and his body.)

(c) How far is it from the magician to his image?

2-34 In the story *The Gift of the Magi*, O. Henry describes the thin mirror between the windows of a room. "A very thin and agile person may, by observing his reflection in a rapid sequence of longitudinal strips, obtain a fairly accurate conception of his looks. Della, being slender, had mastered the art."

Given: When Della looks in the rectangular mirror she can see her right eye, her right shoulder and the hem of her dress at the same time. It is 0.21 m from Della's eye to her shoulder and 1.05 m from her eye to the hem of her dress. She is standing 1.2 m from the mirror.

Find: (a) What is the angle of incidence for light reflected from Della's eye to the hem of her dress?

(b) How big is the mirror? (You may want to draw some sample rays from Della's eye to her shoulder to get a conceptual idea for this problem.)

(c) How far is Della from her image?

ANSWERS

Solutions to **ESSENTIAL PROBLEMS**   A11

2-9   (a) 15° and 25°; (c) 20 cm

2-10   (a) 68 cm; (b) 10 cm

2-11   (a) 4.7 cm; (b) −1.8; (c) −2.1 cm

2-12   (a) 1.0 m; (b) −0.5

2-13   (a) 1.7; (b) 35°

2-14   (b) 13 cm; (c) 3.8 cm

2-15   (a) 9.3 cm; (b) 0.72 cm

2-16   (a) 2.0 cm; (b) −11

Even to Hand in (½?)

Solutions to **MORE INTERESTING PROBLEMS**, *odd answers*
Odd to Read

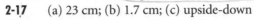

16

2-17   (a) 23 cm; (b) 1.7 cm; (c) upside-down

2-19   (a) 1.6; (b) 7.1°

2-21   (b) −0.58 m; (c) 41 cm

2-23   (a) $294 \cong 2.9 \times 10^2$; (b) 10.2 cm; (c) 10.2 cm or 10.1 cm

2-25   (a) 30 cm; (b) 5 cm; (c) 17%

2-27   (a) −2.5 cm; (b) −2.3 cm; (c) 0.4% or 0.5%

2-29   (a) 7.5 cm; (b) 8.3 cm; (c) 7 mm

2-31   (a) 20.54°; (b) 21.03°; (c) The red slows down more, since the path is deflected farther

2-33   (a) 14°; (b) 51.5 cm by 57.5 cm; (c) 4.0 m

CHAPTER 3 *Force*

So Christopher Robin took hold of Pooh's front paws and Rabbit took hold of Christopher Robin, and all Rabbit's friends and relations took hold of Rabbit, and they all pulled together.... —WINNIE-THE-POOH[1]

Figure 3-1

The concept of force is as sophisticated as it is simple.

IF YOU WERE TRYING TO MOVE SOMEONE who was stuck in a tight place, you would certainly apply a force. If you found that the force was not great enough, you would get help. The force is greater, according to our common sense, when everyone pulls together than when one person pulls alone.

If some object near you suddenly started to move, you would look to find the cause. Finding none, you might become alarmed. Inanimate objects are just not supposed to move by themselves in the absence of demons and black magic. We don't believe much in supernatural or

---

[1]From *Winnie-the-Pooh* by A. A. Milne, illustrated by Ernest H. Shephard. Copyright, 1926, by E. P. Dutton & Co.; renewal ©1954 by A. A. Milne. Reprinted by permission of the publishers, E. P. Dutton & Co., Inc.

magical powers these days because we have acquired so much confidence in certain kinds of force that we feel are natural. The success of science to date is based on explaining the vast diversity of almost all physical phenomena in terms of a very few well-behaved forces of nature.

Hooke's Law, A Way of Conceptualizing Force

There are certain underlying similarities in the way things behave. As an example, consider the relationship between the stretch (or the strain) in a solid object and the force applied to it. It turns out that the stretch (or the strain) is always nearly proportional to the force, as long as you stay within certain limits. This fact was first noticed by Sir Robert Hooke, an acquaintance of Isaac Newton, and is known as **Hooke's law**.

Picture yourself holding one end of a fishing line and a fish holding the other end. The harder you two pull against each other, the more the fishing line stretches. If you pull twice as hard, the fishing line stretches twice as much up to a point. If you get too near the breaking point of the fishing line, it will stretch to its limit and will not spring back after the force is released. Snap! But as long as you remain below what is called the **elastic limit** of the fishing line, you will find that the stretch is very nearly proportional to the force. This will be true whether the fishing line is made out of cotton, nylon, or even steel. The reason that the same proportional rule works for all materials is that all solids are held together by the same kind of electrical forces acting between atoms; the main point of our discussion is that the proportion exists as an experimental fact.

Figure 3-2a

There are certain similarities in the way things behave.

Figure 3-2b

The harder the fish pulls, the more the fishing line stretches.

You could do some experiments to confirm Hooke's law for yourself. You might use rubber bands hanging from a drawer pull or a long string hung from the ceiling. A bunch of coat hangers all made out of the same kind of wire make good weights. You could use a ruler to measure the stretch of the rubber band or string and see if it isn't proportional to the number of coat hangers used to cause the stretch. On the other hand, you might prefer to go fishing.

We can use the stretch of the fishing line as a way of visualizing the concept of force. You can think of the force acting on your hand in terms of the stretch in the line. The amount of force is something proportional to the amount of stretch, but the direction of the force is important too. The direction of the force must be along the fishing line, since any sideward force would make the line bend. Of course, the fishing line is actually pulling in two directions. It is pulling one way on your hand; it is pulling the other way on the fish. When you use the concept of force, you need to direct your attention to one object at a time. If you are thinking of your hand, the force is along the fishing line in the direction of the fish. If you are thinking of the fish, the force is along the fishing line in the other direction.

Suppose you start thinking of the force acting in the fishing line. The fishing line is pulling in two opposite directions, holding you and the fish together. The word **tension** is frequently used to refer to this pair of balanced forces. If you found yourself grasping a tree branch to keep from being pulled into the water, as a matter of fact, it would be proper to think of the balanced forces acting on your body in terms of tension. This concept of tension turns out to be useful in many force problems.

Hooke's law works as well for bending, twisting, and compression of solids as it does for stretching. A coil spring, for example, can be made to stretch over great distances by small forces applied to the ends, but most of the actual give in the spring comes from the twisting of the wire in the coils of the spring. Very little of the stretch in the spring comes from stretching the wire.

Coil springs, which are simple springs made of a coiled piece of wire, are often used in physics labs to study Hooke's law because they can be stretched over such long distances and still spring back to their original shape. These coil springs are frequently provided with hooks at either end for ease of applying forces. Many people have come away from a physics course believing that Hooke's law got its name from this convenient structure on the ends of a coil spring.

The proportionality between the force F applied to the bottom hook of the spring and the distance x by which the spring stretches may be written with a proportionality sign:

$$F \propto x$$

It looks more official, however, with an equality sign and a proportionality constant:

$$F = kx$$

This proportionality constant k is called the **spring constant** of the spring.

The SI unit for force is the newton, named after Sir Isaac Newton. It is abbreviated N. The units of the spring constant are therefore in newtons divided by meters.

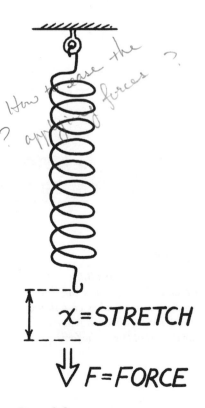

x=STRETCH

F=FORCE

Figure 3-3

The stretch in a spring is proportional to the force applied.

CHECK QUESTIONS

1. *A force of 3.0 N stretches a spring by 6.0 cm. What is the spring constant?*
   Answer: $k = 0.50$ N/cm

2. *How far would the same spring stretch if a force of 2.3 N were applied to it?*
   Answer: $x = 4.6$ cm

Hooke's law for small displacements is universally applicable to all solid objects. It works for everything from building beams to eyelashes. An ant crawling across the floor makes the floor sag by an amount proportional to the ant's body weight. An ordinary-sized ant might be so light that the sag would be too small to measure, but the sag is still there, at least in principle. The small force you can apply to a 10-in steel I-beam like those used in the construction of large buildings will make the beam compress, bend, and twist by an amount that can easily be measured with an optical instrument called an interferometer.

Of course, Hooke's law is only an approximation of any real situation. Such things as frictional losses in a spring, "creep" in a rubber band, and thermal expansion enter into any real situation to limit the applicability of our simple model. In many cases, however, Hooke's law works so well that it is used as a measure of force. Fishermen frequently keep a spring scale in their tackle box to weigh their catch. A simple spring scale is nothing more than a spring within a housing and a pointer to show how much the spring has stretched.[2]

A spring scale is also a useful device for conceptualizing force. If you think of something as being pulled by a rope with a spring scale in between, the scale gives the strength of the force while the rope gives the direction of the force.

Figure 3-4

Forces have both magnitude and direction. The magnitude can be visualized as the reading on a spring scale, while the direction is along a rope that would exert the force.

---

[2]A spring scale is not the only way to measure force. A more accurate way to measure force is to balance it with a known force in the opposite direction. Meat market scales marked "no springs" are of the balance type.

Force is a quantity that has both magnitude and direction. Both the strength and direction of a force must be considered when combining it with another force. This kind of quantity is called a **vector**. Other quantities, which are fully described by magnitude alone, are called **scalars**. A scalar quantity is measured by "how much," while a vector must be described by "how much" *and* "which way."

The distinction becomes important when adding things together. If you add 1 liter of beer to a half-full 1-liter stein, obviously it runs over. We know from experience that an amount of beer is a scalar quantity. One liter added to half a liter always results in one and a half liters, even if part of it is in someone's lap.

Force is a different story. Suppose you and a friend of yours are both applying a force to a cart. You are applying a force of 500 N. Your friend, being somewhat stronger, is applying a force of 1,000 N. The resultant, or vector sum, of these two forces is 500 N + 1,000 N = 1,500 N only if you agree with your friend on the direction you are pushing. If you and your friend disagree and decide to push in opposite directions, the resultant is 500 N − 1,000 N = −500 N in the direction you are pushing (or +500 N in the direction your friend is pushing).[3]

Stationary objects can only remain stationary when all of the forces acting on them are balanced. A spring scale supporting a 50-N weight can only remain stationary if the force acting on the bottom hook is balanced by an equal force pulling upward on its top ring. Most people would correctly guess that the reading on the scale in this situation should be 50 N. A few people might guess that the reading should be 100 N, reasoning that that would be the total force acting on the scale. The 50 N acting upward, however, does not add to the 50 N downward. The forces subtract because they are in opposite directions. A 620-N woman standing on a bathroom scale would be startled if the scale were to read 1,240 N. Fortunately, bathroom scales are not constructed to read the *numerical sum* of the forces acting on them. Bathroom scales are also not constructed to read the *vector*

Figure 3-5

Forces in the same direction add.
Forces in opposite directions subtract.

---

[3]Vectors add when they are in the same direction and subtract when they are in opposite directions. It is fortunate that forces are vectors and not scalars. Otherwise, nothing could ever be stationary. You are able to remain in your seat only because the upward and downward forces acting upon your body are balanced. The forces exerted on you toward the right are balanced by forces acting upon you to the left. If the forces in one direction added to, rather than subtracted from, the forces in the other direction, you would find yourself in a state of accelerated motion.

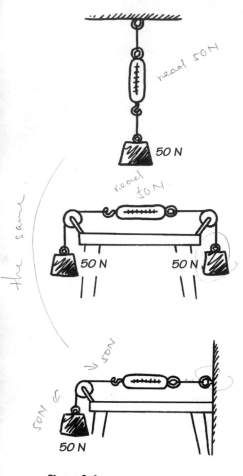

read 50 N

50 N

read 50 N

the same

50 N   50 N

50 N ← → 50 N

50 N

Figure 3-6

Three similar force situations, all having the same tension in the cord.

sum of the forces acting on them. The same person would be nearly as startled if the scale were to read 0 N. Scales are constructed to measure only the force applied to one side; they assume that an equal but opposite force is applied to the other side to hold the whole scale stationary.

Many people who have no difficulty understanding a simple vertically supported spring scale have considerably more difficulty with the same scale supported horizontally with a 50-N weight at each end by means of strings and pulleys, as shown in the center picture of Figure 3-6. They tend to think that the reading should now be 100 N because that is the total weight being held up, but that is not the case. The 50 N on one side cannot possibly add to the 50 N on the other, as their weights are pulling in opposite directions. The vector sum of these two forces is zero, just as the force pulling up on the vertical scale exactly balances the force pulling down on the hook. The reading on the horizontal scale is 50 N, just as it was on the vertical scale.

Let us now consider a similar situation, in which the pulley and weight on one side of the horizontal scale is replaced with a hook in the wall. A force applied to the top of the scale by a string attached to a wall has the same effect as attaching to the weight hung over a pulley. In both cases, the force is only as strong as necessary to hold the scale stationary. The top of the scale does not care what is at the other end of the string. The only thing that effects the scale is the tension in the string that is holding it in place, which is the same in all three of these cases. The scale between the weight and the hook in the wall looks a great deal like the first force situation with the scale between the weight and a hook in the ceiling. The only difference is that we are using a pulley to pull horizontally. We know that the tension in the string, however, is the same on both sides of the pulley. Otherwise the pulley would turn.

CHECK QUESTIONS

Two forces, one 500 N (newtons) and the other 700 N, are acting on the same object. What is the vector sum of these two forces (a) if they are acting in the same direction, and (b) if they are acting in opposite directions?

Answer: (a) 1,200 N, (b) 200 N

Resultant

Figure 3-7

The resultant of two forces has a direction that is a combination of the two original directions.

We have only considered examples so far in which the forces have all been along the same line, either in the same or opposite directions. It is easy to find cases, however, in which one force pulls off at some angle relative to the other. You might be pulling, for example, on the cart in a direction perpendicular to the direction your friend is pulling. Maybe the cart is sloshing around in the mud, and you want it to go off to the side as well as forward. The resulting force, called the **resultant**, is a force equivalent to the combined efforts of both you and your friend. It would be somewhat in your direction and somewhat in his. We shall see that the resultant of 500 N in your direction and 1,000 N in your friend's direction would, for example, be a force of 1,118 N at an angle of about 63° off to your side from your friend's direction. Since your friend is pulling harder than you, you would certainly expect the resultant to be more in your friend's direction. Since you are both pulling, you would expect the resultant to be larger in this case than either of your forces pulling separately. It is clear that we need a general rule for figuring out the magnitude as well as the direction of the resultant of any two forces.

Vector Spaces

One useful convention for conceptualizing the addition of vectors is to draw a picture in which all the vectors are represented by arrows. We let the length of the arrows be proportional to the magnitude of the vectors and their direction be the same as that of the vectors.

One way to look at the representation of vectors by arrows is to think of yourself as operating in a special mathematical space where direction is the same as in real space but where the length of the arrows corresponds to the magnitude of whatever quantity you wish to represent. Such mathematical spaces are called vector spaces. The force arrows, for example, allow length to stand for strength.

When working problems, the usual convention is to write vector quantities in bold face and scalar quantities normally, so that **A** is a vector and *A* is a scalar quantity. The vector **A**, as discussed in the last section, has both a length and a direction. The corresponding scalar *A* is exactly as long as the vector **A**, but has no direction associated with it. Since they are the same length, however, *A* is said to have the same **magnitude** as **A**.

The general rule for finding the resultant is to move the tail of one vector up to the head of the other. This is demonstrated in Figure 3-8. The resultant then starts at the tail of the vector you did not move and ends up at the head of the one you did move. The maximum length of a resultant is achieved when the two original vectors are in the same direction, and the minimum length occurs when the original vectors are in opposite directions.

Figure 3-8

A resultant is formed by moving one vector to the head of the other.

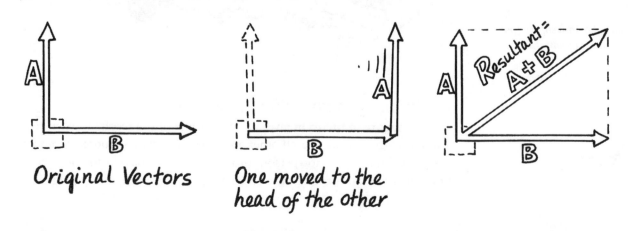

Original Vectors   One moved to the head of the other

' Any other angle produces a resultant whose magnitude is between these two extremes and whose direction is somewhat in the direction of one vector and somewhat in the direction of the other. It doesn't matter which vector is moved and which is left stationary. If we draw a parallelogram, the sides of which are the original vectors, the resultant will have the same length and be along the direction of the diagonal of the parallelogram.

This parallelogram rule works for adding forces as well as any other vectors. One vector quantity for which the rule obviously works is displacement. If you walk 150 paces east and 100 paces north, you get to the same place as if you had walked 100 paces north first and then 150 paces east. Displacement is thus a vector quantity in that it has both magnitude and direction.

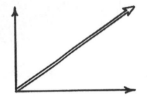  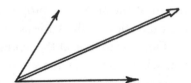

Figure 3-9

A resultant is somewhat in the direction of each of the original vectors and is smaller when they tend to be in the opposite directions but larger when they tend to be in the same direction.

Sometimes it is convenient to draw arrows representing force on a picture depicting the real thing. Force-space diagrams are shown superimposed on real space. Such diagrams are useful. They demonstrate what force is acting on what object. Sometimes people lose the distinction, however, between the real objects and the force space diagram and make the mistake of thinking that the strength of the force has something to do with the length of the rope, or whatever is applying the force. This is a failure to distinguish the model from reality itself.

Figure 3-10

A force space diagram superimposed on a real space diagram.

Addition of Perpendicular Vectors

The **Pythagorean theorem** gives the magnitude of the resultant when two vectors at right angles to each other are added together. Suppose you and a friend are building a snowperson. In the process of making the body, you find yourselves both rolling the same snowball across the snowy field. The snowball is still too small for you both to get on the same side and push in the same direction, so you push north while your friend pushes east. If the snowball rolls more east than north, it is evident that your friend is pushing harder than you are.

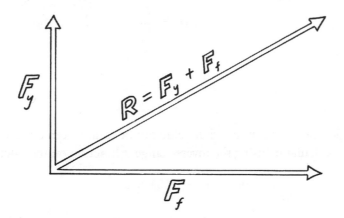

Figure 3-11

The resultant of the forces is a force that would be equivalent to your force F_y and your friend's force F_f combined.

Suppose you push with a force of 60 N while your friend pushes with 120 N of force. Call these forces $\mathbf{F}_y$ for "your force" and $\mathbf{F}_f$ for "friend's force." The snowball will roll twice as far in your friend's direction as it does in yours. The path of the snowball will be the direction of the resultant, which is defined as a force equivalent to you and your friend's forces combined. The resultant in this case is the hypotenuse of a right triangle whose sides are equal to the individual forces exerted by you and your friend. The Pythagorean theorem tells us that the square of the hypotenuse of a right triangle is equal to the sum of the squares of the sides:

$$R^2 = F_y{}^2 + F_f{}^2$$

where R stands for the magnitude of the vector $\mathbf{R}$. F_y and F_f are equal to 60 N and 120 N, the magnitudes of F_y and F_f. (It bears repeating that it is always a good idea to write a vector, such as $\mathbf{R}$, different from a scalar, such as R, so that you can keep track of which rules are appropriate for addition.) Taking the square root of both sides, we find that the magnitude of it is this.

$$
\begin{aligned}
R &= \sqrt{F_y{}^2 + F_f{}^2} \\
&= \sqrt{(60\ \text{N})^2 + (120\ \text{N})^2} \\
&= \sqrt{3{,}600\ \text{N}^2 + 14{,}400\ \text{N}^2} \\
&= 134\ \text{N}
\end{aligned}
$$

The direction part of **R** can be found from the definition of the tangent of an angle (the ratio of the side that is opposite the angle to the side adjacent to the angle). If we measure the angle θ (theta) from your friend's direction to **R**, the definition gives us

$$\tan\theta = \frac{F_f}{F_y}$$

In our case,

$$\tan\theta = \frac{60 \text{ N}}{120 \text{ N}} = 0.50$$

The angle whose tangent has that value can be found from a trigonometry table or a calculator (using the inverse tangent function on the calculator):

$$\theta = \tan^{-1}(0.50) = 26.6°$$

Thus, the resultant *R*, which is the vector sum $F_y + F_f$, is a force of 134 N in the direction east by 26.6° north (26.6° north of east).

CHECK QUESTIONS

A force of 100 N toward the north is added to a force of 100 N toward the east. What is the (a) magnitude and what is the (b) direction of the resultant?
    Answer: (a) 141 N, (b) 45° east of north

Resolving a Force into Components

The equivalence between the resultant **R** and the combination of **F**$_y$ and **F**$_f$ works both ways. When we already know the resultant, we can find the two perpendicular forces that will add together to produce it. This is called **resolving** the initial force into **components**. Suppose that we were trying to find out how hard you and your friend were pushing from the observation that the snowball rolls 26.6° north of east, the same way that it would if there were an applied force of 134 N in the direction it is moving. We want to know how much of this 134 N is in the northward direction and how much is in the eastward direction.

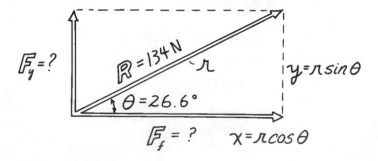

Figure 3-12

The resultant can be resolved back into components using trigonometry.

The northward force **F**$_y$ is found from the sine of the angle θ (since the definition of the sine is the ratio of the side opposite the angle, y, to the hypotenuse, r), and the magnitude of force F_y, is the same as the length of the side of the triangle opposite θ:

$$F_y = R \sin \theta$$

The eastward force F_f is found from the cosine of the angle θ (since the definition of the cosine is the ratio of the adjacent side, x, to the hypotenuse, r), and your friend's force is the side of the triangle adjacent to θ:

$$F_f = R \cos \theta$$

Given the magnitude and direction of the force acting on the snowball as 134 N at 26.6°, we can calculate your force as

$$
\begin{aligned}
F_y &= R \sin \theta \\
&= (134 \, \text{N})(\sin 26.6°) \\
&= (134 \, \text{N})(0.448) \\
&= 60 \text{N (north)}
\end{aligned}
$$

while your friend's force is

$$
\begin{aligned}
F_f &= R \cos \theta \\
&= (134 \, \text{N})(\cos 26.6°) \\
&= (134 \, \text{N})(0.894) \\
&= 120 \, \text{N (east)}
\end{aligned}
$$

We therefore have a way of getting back to our starting place.

Sailboats

> *One ship drives east and another drives west*
> *With the self-same winds that blow,*
> *'Tis the set of the sails and not the gales*
> *Which tells us the way to go.*

ELLA WHEELER WILCOX, *WINDS OF FATE*

To illustrate the great power of the reasoning by which a force is broken into components, let us investigate the ability of a sailboat to sail upwind. Christopher Columbus would have given a great deal for this secret. He had a square rigger and had to wait for the wind to blow from east to west before he could make any progress from Spain toward America. Since the prevailing winds are from the other direction, he had to spend most of his time with his sails furled and his sea anchor out. It was well known that the wind filling the sails exerts a force that is directly perpendicular to the sail, and thus it seemed that only when the wind was behind him could he make any progress.

Figure 3-13

Square-rigged vessels were best going in the direction of the wind.

It seemed to make little sense for him to change the direction of his sail. Not only would the sail catch less wind but the force of the wind, acting perpendicular to the sail, would not be in the direction he wanted the boat to go. Only the component in the forward direction would make the boat go. The other component would act against the keel and tend to tip the boat over. Thus, such an arrangement seemed to be full of disadvantages.

However, had Columbus seen this diagram (with one slight modification) it would have blown his mind. Just reverse the roles of the incident and reflected winds, and the forces stay the same. The force on the sail resulting from the wind bouncing off is still perpendicular to the sail, and there is still a forward component.

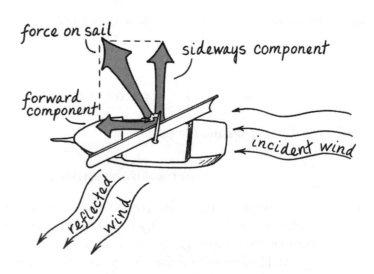

Figure 3-14

One component of the force on the sail is in the forward direction.

A sailboat cannot go directly into the wind, but it can angle back and forth to get upwind.[4] It can actually go faster tacking upwind than when going downwind. The maximum velocity when going downwind, neglecting friction, would be the velocity of the wind. But the faster the boat goes upwind, the harder the wind seems to blow. It is common for the speed of an ice boat, a type of sailboat on ice skates, to be four or five times that of the wind.

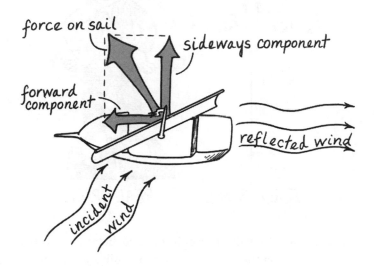

Figure 3-15

One component of the force is still in the forward direction even though the direction of the wind is reversed.

Equilibrium Problems

There are many instances in modern society where forces must be balanced to equal zero. Some engineer would be very embarrassed if the forces acting on a building or bridge did not balance. An imbalanced force always causes acceleration, and acceleration is usually not so good for buildings and bridges. If you want something to stay put, you have to arrange things so that the sum of all the forces acting on it is zero. When the forces acting on an object are all balanced, the object is said to be in equilibrium. We will be considering situations in this chapter involving **static equilibrium**. We will assume that an object is stationary to start with, and we will examine the conditions necessary to keep it that way.

The simplest equilibrium problems are those in which all the forces act along the same line in the same or opposite directions. Consider, for example, a friend of yours who steps on a bathroom scale one morning to discover that the pull of gravity on his body is too great. Not satisfied, he

---

[4]Only an experienced helmsperson would be ordered to sail a close-hauled vessel "full and by," as close to the wind as possible. An inexperienced person would be given the order "by and large," to sail slightly off the wind since this tack would leave the sailor in less danger of being "taken aback." Hence today a person speaking "by and large" about a subject is not speaking directly to the point. William Morris, *Dictionary of Word and Phrase Origins* (New York: Harper and Row, 1962).

purchases another scale. This scale also produces too high a reading, 100 kg. His solution is to stand on both scales at the same time. If he stands evenly on both feet, what will be the reading on each scale?

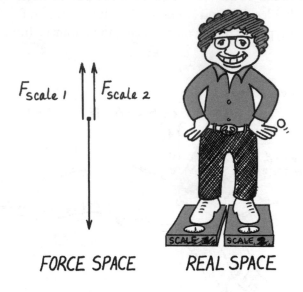

Figure 3-16

The force of each leg is only half the weight.

Your friend's weight will be evenly divided between the two scales. When he is just standing there, he has to be in equilibrium. The sum of all the forces acting on him has to be equal to zero, since he is not moving.

$$\text{Sum of the forces} = 0$$

If there were a greater force in the upward direction than in the downward direction, for example, he would go up. Since he is going neither up nor down, the forces in these two opposite directions must be equal. We can express this condition of equilibrium in vector notation, using the Greek letter sigma "Σ" to stand for "sum of":

$$\Sigma \mathbf{F} = 0$$

In equilibrium problems it is important to focus attention on one object and consider all the forces acting on that object alone. The object in this case is your friend. The sum of the forces $\Sigma \mathbf{F}$ is therefore the two forces $\mathbf{F}_1$ and $\mathbf{F}_2$ by which the scales are pushing up, and the weight of your friend, $\mathbf{w}$, which is pulling down:

$$\mathbf{F}_1 + \mathbf{F}_2 + \mathbf{w} = 0$$

Since, as we initially stated, your friend is standing evenly on both scales, these two forces are equal and in the same direction:

$$\mathbf{F}_1 + \mathbf{F}_2 = 2\mathbf{F}$$

Including this fact in our condition of equilibrium,

$$2\mathbf{F} + \mathbf{w} = 0$$

and solving for the force

$$\mathbf{F} = -\frac{\mathbf{w}}{2}$$

we see that the reading in each scale is equal to half the weight. (The minus sign comes from the fact that we are dealing with vector quantities and the force provided by each scale is in the opposite direction from the weight. If we call the weight positive, we are assigning downward as the positive direction. The upward force on your friend would therefore be negative.)

Having 100 kg divided between two legs is one thing, but this result would be more important to your friend if he were being hung by his thumbs. Would it be kinder to hang him by one thumb or by both thumbs? The force upward must equal the force of his entire weight downward. The force acting on one thumb would be twice as great as if the force were divided between the two thumbs. Now consider the relative merits of hanging with one's arms straight up as against hanging with one's arms separated by some angle. The former turns out to be easier on the thumbs. When the arms are stretched out at some angle, it is the upward component of their force that is equal to half the weight. But now there is also a horizontal component because of the angle involved. The condition of

Figure 3-17

The force on each thumb is only half the weight.

Figure 3-18

The force on each thumb is greater than half the weight, since it also has a horizontal component.

equilibrium applies to both the horizontal and vertical components of force independently. This means the horizontal forces by themselves must equal zero and vertical forces by themselves must equal zero.

$$\Sigma\mathbf{F} = 0$$

The only way this condition can be satisfied is for the sum of the force components in the upward direction to equal the sum of the forces in the downward direction:

$$\Sigma F_\uparrow = \Sigma F_\downarrow$$

(Note that the vector notation is dropped because we are now talking about components along only one direction.) If the upward and downward forces weren't equal, no amount of sideward force could bring the forces into balance. In the same way, the sum of the forces to the left, $\Sigma F_\leftarrow$, must equal the sum of the forces to the right, $\Sigma F_\rightarrow$:

$$\Sigma F_\leftarrow = \Sigma F_\rightarrow$$

We can now apply these equilibrium conditions to the situation at hand. The sum of the downward forces, $\Sigma F_\downarrow$, is nothing more than the weight of your friend, w, since that is the only force acting in the downward direction. All the other forces are components of the forces in the arms.

Drawing a force diagram, and letting θ be the angle between the arms and the vertical, the vertical components are the sides adjacent to the angle. The vertical equilibrium condition becomes

$$F_{\Sigma y} = 0$$
$$F_1 \cos\theta - F_2 \cos\theta - w = 0$$
$$F_1 \cos\theta - F_2 \cos\theta = w$$

The horizontal part of the equilibrium condition says that the pull to the right equals the pull to the left.

$$F_{\Sigma x} = 0$$
$$F_1 \sin\theta - F_2 \sin\theta = 0$$
$$F_1 \sin\theta = F_2 \sin\theta$$

Assuming the angles are the same on both sides, the horizontal part shows that the forces in the two arms are the same.

$$F_1 = F_2$$

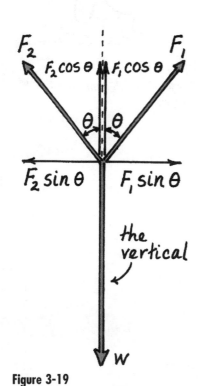

Figure 3-19

The horizontal components to the two upward forces balance each other and the two upward components balance the weight.

We let F stand for both F_1 and F_2 in the above equation:

$$2F\cos\theta = w$$

$$F = \frac{w}{2\cos\theta}$$

This answer applies as well to the situation in which your friend's arms were straight up. In that case, the angle θ is zero and the cosine of zero degrees is equal to 1. The result is the same as when we were considering the force applied to your friend by the two scales placed side by side:

$$F = \frac{w}{2(1)}$$

Any larger angle produces a larger force in the two arms since any angle larger than 0° has a cosine smaller than unity (1). The cosine of the angle is in the denominator of our general solution. Making it smaller thus has the effect of making the result larger:

$$F = \frac{w}{2\cos\theta}$$

We might see what happens, for example, when the angle θ is equal to 60°. The arithmetic is very easy in this case because the cosine of 60° equals 0.5. The result is reasonable, because the total angle between the two upward forces adds up to the same 120° as remains between either of the two upward forces and the weight. Everything is therefore symmetrical for this particular angle, and you expect all three forces to be the same.

CHECK QUESTION

An object weighing 180 N is supported by two equal forces, each making an angle of 25.8° with the upward vertical. The cosine of 25.8° is 0.900. How great are the two upward forces?
Answer: 100 N

Picture Frame Problems

We have been considering a class of problems in which something is held up by two equal forces that pull upward and outward. The force situation is nearly the same in another class of problems, in which two equal forces pull upward and *inward* to support the weight of an object. A common object supported in this way is an ordinary picture frame. In looking at this class of problems, we will see how the same force space diagram can apply to two different real space situations.

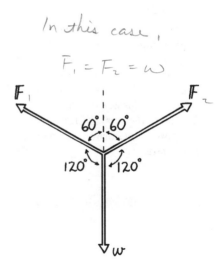

In this case,

$F_1 = F_2 = w$

Figure 3-20

When the angle to the vertical is 60°, all three forces are equal, as you might expect from the fact that all three angles are equal.

Before we proceed with this discussion, let us go back to make certain that you understand why each of the upward forces must be greater than half the weight as long as the angle to the vertical is greater than 0°. We simplify the situation by looking at only half the picture frame problem. Let's begin by having only one upward and outward force. Assume we have a monkey in the zoo holding on to a rope with one hand and the side of the cage with the other. If the second hand exerts only a horizontal force to pull him over to the side of the cage, none of this force helps to support his weight. The first hand, which is holding on to the rope, is therefore supporting all of the monkey's weight as well as balancing the sideward force exerted by the second hand. The vector sum of these two perpendicular forces is by necessity greater than either of the two forces taken alone, just as the hypotenuse of a right triangle is always greater than either of the two legs.

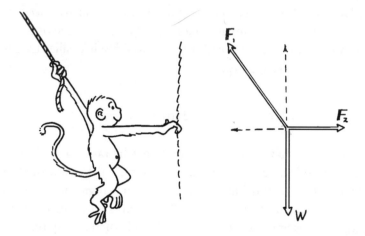

Figure 3-21

The force with which the monkey holds on to the rope <u>must be greater than his weight</u> since the force in that arm must balance the horizontal force with which he holds on to the zoo cage as well as to lift his weight. The force diagram in the second picture depicts the upward and sideward components of this force.

The monkey hanging from the rope represents half of the force problem discussed in the previous section where a weight is supported by two equal forces pulling upward and outward. The horizontal force exerted by the monkey's second hand is replaced by the horizontal component of the other upward and outward force in the situation where we had your friend hanging by his thumbs. The monkey also represents half of the force situation involved in supporting a picture frame.

A picture frame is generally supported by a single wire attached to the two sides of the frame and stretched over a nail in the wall. The frame is usually balanced so that the tension in the wire will be the same on both sides of the nail, since too great an imbalance in this tension would cause the wire to slip over the nail.

One way to look at a picture frame is to say that each end of the wire is holding up one side of the picture frame. That would be to look upon it as two monkey problems. The wire on the left of the nail is lifting up half the weight of the frame and pulling it toward the right. The wire on the right is lifting the other half of the frame and pulling toward the left. The sideward

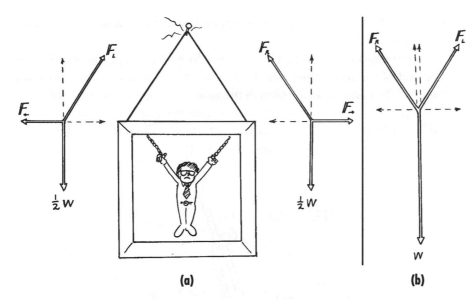

(a)   **(b)**

Figure 3-22

(a) One way to look at a picture frame problem is to think of it as two problems like the monkey hanging from the rope. (b) Another equally valid way is to use a single vector diagram that turns out to be nearly the same as we used for the friend hanging by the thumbs.

force of one balances the sideward force of the other, and together they lift the whole picture frame.

Another equally valid way of treating the picture frame problem uses a single force space diagram. Looking upon the frame as a single object, we see one weight pulling down and two forces pulling upward and sideward. The upward components work together to balance the weight; the sideward components work in opposite directions to balance each other. The force space diagram looks very much like the diagram we used for your friend being hung by his thumbs. The only difference is that the wire that is physically on the left of the nail applies a force to the frame that is depicted on the right in the force space diagram. The wire on the right of the nail, on the other hand, pulls up and to the left. It is therefore properly shown as a vector in force space which is upward and toward the left.

This reversal of position between force space and real space results from the picture frame (extended over real space) being represented as a point in force space. While the two ends of the wire are attached to the frame at two points physically located some distance apart, they both are acting on the same object. When making force diagrams, it is essential to focus your attention on one object at a time and consider the forces acting on that object alone.

Real Space vs. Force Space

Let us illustrate this trick of representing something that is extended over real space with a point in force space by considering the forces acting on the various parts of a rope that is tied in a leverage knot. Teamsters use this kind of knot to tie things down on flat-bed trucks. You are likely to see it on trucks driving down the highway. The operation of the knot is not hard to understand if you are willing to ignore the fact that it is extended over real space.

A leverage knot is formed by making a loop in the rope and passing the free end of the rope, called the "bitter end" of the line, around a hook on the truck bed and back up through the loop. The rope in the loop then acts like a pulley; the truck driver can get nearly three times as much force in the part of the rope going up and over the load, called the "standing part" of the line, as he must apply to the bitter end.

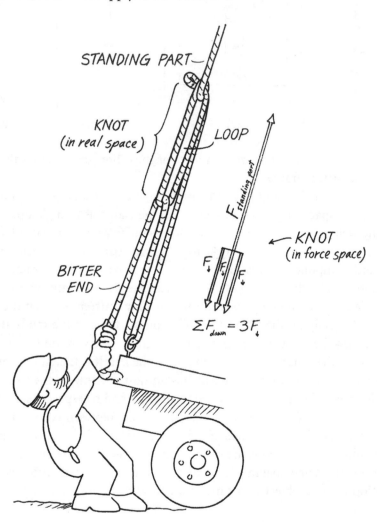

Figure 3-23

A leverage knot is used by teamsters to hold down loads on the back of flat-bed trucks. A greater force will be exerted on the standing part of the line than is exerted by the truck driver on the bitter end.

If we look only at the parts of the rope going into the knot, we see one part going up and three parts going down. The one part going up, the standing part, must therefore exert a force on the knot which is equal in magnitude but opposite in direction to the three forces that are pulling down. To the extent that friction can be ignored, the three parts of the rope pulling down on the knot all exert equal forces. As the force on the bitter end becomes much greater than the part of the rope passing from the loop and going around the hook on the truck bed, the latter part of the rope will slip through the loop. As this part of the rope becomes tighter than the other part of the rope going to the hook, the rope will slip around the hook. In practice there is considerable friction as the rope slides through the loop and around the hook in the truck bed, but the truck driver can largely compensate for this friction by pulling on the bitter end with one hand and helping the rope slide around the hook with the other.

This is a useful knot for tying down camping gear to the luggage rack on your station wagon, or whatever. It can be tied off by making a simple half-hitch on the other two downward parts of the line with the bitter end just below the loop. To be more secure, you can put an extra half-hitch on the bend forming the loop with the standing part. Made with this extra half-hitch (see Figure 3-24), this knot will hold the load on a truck from coast to coast without slipping.

After tying the knot you only have two parts of the rope pulling down and one part pulling up. The force situation gets simpler, even though the twists and turns of the rope making up the knot get more complicated. No matter how gnarled the knot becomes, however, you can figure out what is happening outside the knot by looking only at the ropes going into it and representing the whole knot as a point in force space, disregarding how much it might be extended over real space.

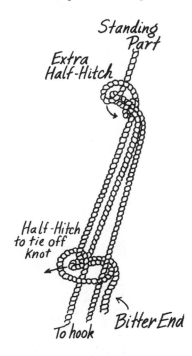

Figure 3-24

The leverage knot shown in Figure 3-23 can be tied by using the bitter end to make a half-hitch around the two parts of the line going to the hook just below where it passes through the loop. The knot can be made more secure by using the standing part to put an extra half-hitch on the bend which forms the loop.

PROBLEM SET 3

SUMMARY OF CONCEPTS

The problems in this chapter deal with forces and how they combine. **Hooke's law** provides a way of conceptualizing force. Hooke's law states that the force in a spring or almost any other object is proportional to the amount of stretch, twist, bend, or other displacement. The proportionality constant k is called the **spring constant** of the object.

$$F = kx$$

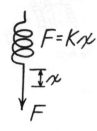

Force has both magnitude and direction, which must be taken into account when two or more forces are added together. Quantities that have both magnitude and direction are called **vectors**. The rules for adding vectors are as follows: (1) Vectors in the same direction add, while vectors in the opposite direction subtract. The resultant is in the direction of the

larger vector. (2) Perpendicular vectors are added by Pythagorean theorem. The resultant has a magnitude which is the square root of the sum of the squares of the two original vectors, and a direction given by the angle whose tangent equals the ratio of the two sides.

A vector is resolved into rectangular components by dropping perpendiculars to the coordinate axes and finding two perpendicular vectors whose resultant would be the original vector. The component adjacent to the angle is the vector times the cosine of that angle, and the other component is the vector times the sine of the angle.

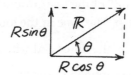

A body is said to be in equilibrium if the sum of all the forces acting on it is equal to zero. This condition implies the sum of all the vertical components is equal to zero, and the sum of all the horizontal components is separately equal to zero.

$$\Sigma F_x = 0 \qquad \Sigma F_y = 0$$

EXAMPLE PROBLEMS

Try to do the Examples yourself before looking at the Sample Solutions.

3-1 A nylon fishing line is stretched when a fish pulls on it and returns to its original length when the force is removed.

Given: The line stretches from 100 m to 103 m when the fish pulls with a force of 120 N.

Find: (a) What is the spring constant of the fishing line?
(b) How much will this line stretch if the fish pulls with only 50 N?

BACKGROUND

This problem illustrates Hooke's law, a useful tool for conceptualizing the idea of force. The direction of the force is, of course, along the fishing line and the magnitude of the force with which the fish pulls is assumed to be directly proportional to the stretch of the fishing line. The spring constant is the proportionality constant between the force and the amount of stretch.

3-2 Teamsters use a leverage knot to secure a load to a flat-bed truck. The knot is formed by making a loop in the rope and passing the free end of the rope, called the "bitter end" of the line, around a hook on the truck bed and back up through the loop. The rope in the loop acts like a pulley.

Given: A truck driver pulls on the bitter end with a force of 880 N.

Find: (a) How much force is applied to the load?

(b) How hard does the rope pull on the truck-bed hook?

BACKGROUND

This is an equilibrium problem like problems involving pulleys. It illustrates the addition of forces that all act in the same or opposite directions. We can assume that the tension in all parts of the rope is the same, because any difference in force would cause the ropes to slip. In doing equilibrium problems, remember to focus your attention on all the forces acting on a single object at one time.

3-3 An elephant is pulling a log through the jungle, one end of a rope being tied to the elephant and the other end being tied to the log.

Given: The rope makes an angle of 30° with the horizontal and is under a tension of 4,000 N.

Find: (a) What is the horizontal component of the force?

(b) What is the vertical component of the force?

BACKGROUND

This is an equilibrium problem in which force is doing two things. Part of the force is pulling up and part is pulling sidewards. You are given the magnitude and direction of the force and must find the components. That is like changing from polar coordinates to rectangular coordinates in force space.

3-4 Paul Hew is rolling the lawn. The force he applies to the lawn roller handle performs two functions. It helps to hold the roller on the ground so it doesn't slip, and it pushes the lawn roller forward at a constant velocity.

Given: Paul is pushing with a force of 250 N at a direction of 45° into the ground. The lawn roller weighs 100 N.

Find: (a) What is the horizontal force making the lawn roller go forward?
(b) What is the total vertical force holding the roller against the ground?

BACKGROUND

One of the forces in this problem must be resolved into components, as in the previous problem. The force in the handle is pushing both sidewards and down. The downward component, however, is being aided by a second force, the weight of the lawn roller.

3-5 Two men are carrying a trunk. They each support half the weight, but they also pull sidewards against each other to hold the trunk away from their legs.

Given: Each man pulls with a force of 200 N at a direction 12° from the vertical.

Find: (a) How heavy is the trunk?
(b) What is the horizontal force with which each man pulls against the trunk?

BACKGROUND

There are three forces in this problem: two upward forces and the weight. The problem is simpler than it might otherwise be because the two upward forces are equal and make the same angle with the vertical when pulling away from each other.

3-6 A picture frame is supported by a wire hooked over a nail in the wall. The picture is hung squarely on the wall, so the tension in the wire is the same on both sides of the nail.

Given: The tension in the wire is 500 N, and the wire makes an angle of 35° with the horizontal top part of the picture frame.

Find: (a) What is the sideward (horizontal) component of the force exerted by one half of the wire?
(b) What is the upward component of the force exerted on the frame by one half of the wire?
(c) What is the weight of the picture frame?

BACKGROUND

In the previous problem, two equal forces pulled upward and outward to hold up an object. In this problem the forces pull inward but the force diagram will be the same for this problem as the last one, since there are the same number of forces: two upward, symmetrical forces and one downward force.

3-7 A concrete bucket on low friction wheels is being held stationary on a ramp. The cable holding the bucket is parallel to the ramp. The force exerted by the ramp on the wheels is perpendicular to the ramp.

Given: The mass of the bucket and contents is 1,000 N, and the ramp makes an angle of 25° to the horizontal.

Find: (a) What is the force with which the ramp pushes on the bucket?
(b) What is the tension in the cable?

BACKGROUND

Many balanced force problems are best worked out by resolving the forces into horizontal and vertical components. Situations involving inclined planes, however, are best solved by resolving forces into components that are parallel and perpendicular to the plane. Using a tilted frame of reference frequently reduces the number of forces that need to be resolved into components.

Solutions to **EXAMPLE PROBLEMS**

$x_1 = \text{STRETCH} = \Delta d$

$\quad = 103 \text{ m} - 100 \text{ m}$

$\quad = 3 \text{ m}$

$F_1 = 120 \text{ N}, \ k = ?$

$F_2 = 50 \text{ N}, \ x_2 = ?$

(a) $F = kx$

$\quad k = \dfrac{F}{x}$

$\qquad = \dfrac{(120 \text{ N})}{(3 \text{ m})}$

$\qquad \cong \boxed{4 \times 10^1 \text{ N/m}}$

(b) $F_2 = kx_2$

$\quad x_2 = \dfrac{F_2}{k}$

$\qquad = \dfrac{(50 \text{ N})}{(4 \times 10^1 \text{ N/m})}$

$\qquad \cong \boxed{1 \text{ m}}$

SAMPLE SOLUTION 3-1

Given:
The line stretches from 100 m to 103 m when the fish pulls with a force of 120 N.

Find:
(a) What is the spring constant of the fishing line?

(b) How much will this line stretch if the fish pulls with only 50 N?

DISCUSSION

Hooke's law $F = kx$ is solved for the proportionality constant k in part (a) and for the stretch x in part (b). The subscripts on F and x are used to show that Hooke's law is being applied in two different situations. There is no subscript on the spring constant k, because it is the same in both cases.

The answer in part (a) is only good to one significant figure, since that is the accuracy with which we know the stretch x_1. This result also limits the accuracy of the answer in part (b) to one significant figure.

Given:

A truck driver pulls on the bitter end with a force of 880 N.

Find:

(a) How much force is applied to the load?

(b) How hard does the rope pull on the truck-bed hook?

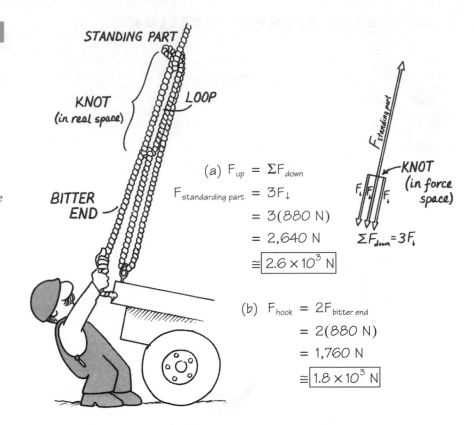

(a) $F_{up} = \Sigma F_{down}$

$F_{standarding\,part} = 3F_{\downarrow}$

$= 3(880\ N)$

$= 2,640\ N$

$\cong \boxed{2.6 \times 10^3\ N}$

(b) $F_{hook} = 2F_{bitter\,end}$

$= 2(880\ N)$

$= 1,760\ N$

$\cong \boxed{1.8 \times 10^3\ N}$

DISCUSSION

This is a useful knot to know if you are ever going to want to tie camping gear or the like to the top of your car. You may wish to find a piece of rope and tie this knot over the back of a chair to better understand how it works. (The figure shows the loop held to the standing part by a simple half hitch. For long distance runs, teamsters double this hitch to hold the loop more securely, but you can tie a loop in the standing part any old way and the knot will function as a leverage knot. The important features are that the bitter end, the end you pull on, passes through the loop, back around a hook of some kind, and then back to the knot.)

The trick to understanding this and many other force situations is to ignore the fact that the knot (or whatever has forces applied to it) is spread out over real space. A force space diagram for part (a), in which the object of interest is the knot, shows three forces pulling in one direction and only one pulling in the other. The three forces in the one direction are assumed to be equal, neglecting the frictional forces with which the rope slides through the loop and around the hook, so that the one force in the other direction is three times as great. Part (b), in which the object of interest is the hook, shows a similar situation, but there are only two equal forces pulling in one direction against the one force in the other direction.

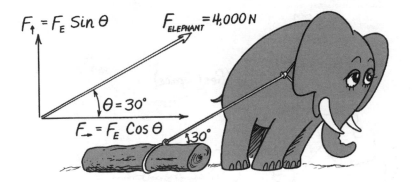

Given:
The rope makes an angle of 30° with the horizontal and is under a tension of 4,000 N.

Find:
(a) What is the horizontal component of the force?

(b) What is the vertical component of the force?

(a) $F_\rightarrow = F_E \cos\theta$

$= (4{,}000 \text{ N})(\cos 30°) = 3{,}464 \text{ N}$

$\cong \boxed{3{,}500 \text{ N}}$

(b) $F_\uparrow = F_E \sin\theta$

$= (4{,}000 \text{ N})(\sin 30°)$

$= \boxed{2{,}000 \text{ N}}$

DISCUSSION

The force in the force space diagram has the same direction as the rope in real space, but the length of the force space arrow has nothing to do with the length of the rope.

The side of the force space triangle adjacent to the angle is clearly the hypotenuse times the cosine of the angle. The other side, which is not adjacent to the angle, is the hypotenuse times the sine, since it is the same length as a side of the right triangle that would be opposite the angle.

SAMPLE SOLUTION 3-4

Given:

Paul is pushing with a force of 250 N at a direction of 45° into the ground. The lawn roller weighs 100 N.

Find:

(a) What is the horizontal force making the lawn roller go forward?

(b) What is the total vertical force holding the roller against the ground?

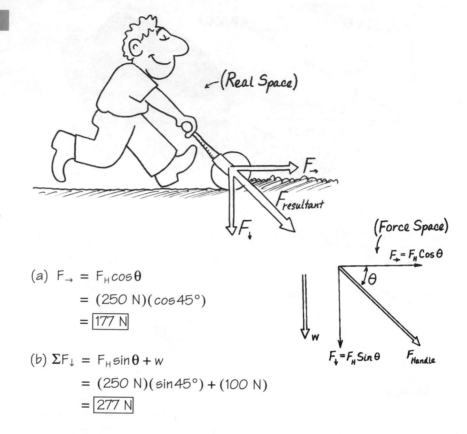

(a) $F_\rightarrow = F_H \cos\theta$

$\quad = (250\ N)(\cos 45°)$

$\quad = \boxed{177\ N}$

(b) $\Sigma F_\downarrow = F_H \sin\theta + w$

$\quad = (250\ N)(\sin 45°) + (100\ N)$

$\quad = \boxed{277\ N}$

DISCUSSION

The two actual forces acting on the lawn roller, the force in the handle and the weight, are shown as double-line arrows in the force space diagram. The components of the force in the handle are shown as single-line arrows. The forward component found in part (a) is the side adjacent the angle and is therefore given by the hypotenuse times the cosine of the angle.

The total downward force acting on the roller found in part (b) is the downward component of the handle force added to the weight of the roller, since both act in the same direction.

Given:
Each man pulls with a force of 200 N at a direction 12° from the vertical.

Find:
(a) How heavy is the trunk?

(b) What is the horizontal force with which each man pulls against the trunk?

(a) $F_\downarrow = \Sigma F_\uparrow$

$w = 2(F\cos\theta)$

$w = 2(200\text{ N})\cos 12°$

$= \boxed{391\text{ N}}$

(b) $F_\leftarrow = F_\rightarrow$

$F\sin\theta = F\sin\theta$

$F\sin\theta = (200\text{ N})\sin(12°)$

$= (200)(.208)$

$= \boxed{42\text{ N}}$

DISCUSSION

All the forces are assumed to be balanced in both parts of this problem such that the upward forces are balanced by downward forces and the sideward forces in one direction are balanced by sideward forces in the other direction. Part (a) deals with forces in the upward and downward direction and is solved from the assumption that the one downward force, the weight of the trunk, is equal to the sum of the two upward forces, the upward components of the forces in the two men's arms. Since these are the components that are adjacent to the known angle, they are found by multiplying the known forces in the arms by the cosine of the angle.

The sideward components are found in part (b) by multiplying the known forces in the arms by the sine of the angle, since these components are opposite the known angle. (Remember that the side adjacent is associated with the cosine and the side opposite is associated with the sine of the angle.) The sideward components of the force are equal in magnitude but opposite in direction so that they balance each other.

SAMPLE SOLUTION 3-6

Given:

The tension in the wire is 500 N, and the wire makes an angle of 35° with the horizontal top part of the picture frame.

Find:

(a) What is the sideward (horizontal) component of the force exerted by one half of the wire?

(b) What is the upward component of the force exerted on the frame by one half of the wire?

(c) What is the weight of the picture frame?

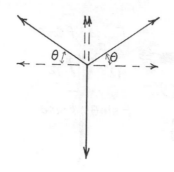

(a) $T_x = T\cos\theta$

$\quad = (500\ N)\cos(35°) = \boxed{410\ N}$

(b) $T_y = T\sin\theta$

$\quad = (500\ N)\sin(35°) = \boxed{287\ N}$

(c) $\Sigma F_{DOWN} = \Sigma F_{UP}$

$\quad w = 2T_y$

$\quad = 2(287\ N) = \boxed{574\ N}$

DISCUSSION

The tension of the wire is the force acting against gravity to hold up the picture frame. The tension in the two halves of the wire are forces that both act in upward directions, but in different sidewards directions from one another. The angle the wires make with the horizontal was given. Forces like tension and weight can be treated as interchangeable forces in computation, as long as the units are comparable.

Parts (a) and (b) deal with resolving the force of one wire into the upward and sideward directions. You can choose either wire. They will give the same results since the strength of the forces on both sides are the same, even if they pull in different directions. Part (a) asks for the sideward component of the angle. The sideward component of the force is adjacent to the angle that we know, so the sideward force of one side of the wire will be cosine of the angle multiplied by the tension in the wire. Similarly, part (b) asks for the upward component, which is the force opposite the known angle. The upward component of this half of the wire will be the sine of the angle times the tension in the wire.

Part (c) asks for the weight of the frame, which is the downward force in the force diagram. As long as the picture is stationary, the forces are equal. The horizontal forces balance each other as do the vertical forces. The weight, therefore, must be balanced by the upward force components of the two wires, so it will be twice the force of one, or twice the answer from part (b).

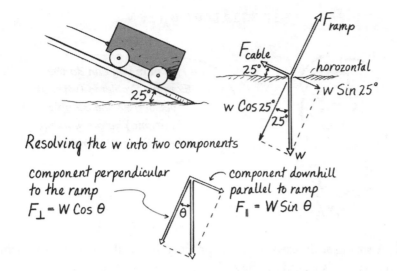

Resolving the w into two components

component perpendicular
to the ramp
$F_\perp = W \cos\theta$

component downhill
parallel to ramp
$F_\parallel = W \sin\theta$

SAMPLE SOLUTION 3-7

Given:
The weight of the bucket and contents is 1,000 N, and the ramp makes an angle of 25° to the horizontal.

Find:
(a) What is the force with which the ramp pushes on the bucket?

(b) What is the tension in the cable?

(a) $F_{ramp} = F_\perp$

$\qquad = w \cos\theta$

$\qquad = (1{,}000 \text{ N})(\cos 25°)$

$\qquad = \boxed{906 \text{ N}}$

(b) $F_{cable} = F_\parallel$

$\qquad = w \sin\theta$

$\qquad = (1{,}000 \text{ N})(\sin 25°)$

$\qquad = \boxed{423 \text{ N}}$

DISCUSSION

In (a) the force exerted by the ramp is perpendicular to the cable force, assuming that there is no friction in the wheels. We can therefore choose a tilted frame of reference in which these forces are already resolved into components, and then we have only to work out the weight if we want to find the conditions where everything balances. (We will see in Chapter 6 that the forces are balanced even if the object is moving, so long as the velocity is constant.)

In (b) the force in the cable is in the direction of the ramp, namely 25° to the horizontal.

The angle between the ramp and the horizontal is the same as the angle between the weight and the perpendicular to the ramp. One way to see this is to consider that both angles are complements of the same angle.

ESSENTIAL PROBLEMS

Be sure you can do the Example Problems (without peeking at the Sample Solutions) before working these Essential Problems.

3-8 A spring scale used to weigh produce is a coil spring in a housing. When a weight is placed on the scale, the spring inside is pulled longer. The stretch of the spring is proportional to the force applied to it.

Given: The coil spring inside a spring scale in a produce store stretches form 15 cm to 17.5 cm when 11 N of apples are weighed.

Find: (a) What is the spring constant of the coil spring?
(b) How far will the spring stretch if only one apple weighing 2 N is weighed?

3-9 Henry Hiker is sitting in a tree. In an attempt to lift his backpack out of harm's way he has attached a cord to his backpack. One end if the cord is attached to the backpack. It then goes over the tree limb, loops back through the shoulder strap on the backpack, than back up to Henry, who is pulling on the end.

Given: Henry is pulling on the free end of the rope with a force of 179 N.

Find: (a) What is the force on the backpack, due to Henry's exertions?
(b) How hard does the rope pull on the shoulder strap of the backpack?

3-10 A cabin cruiser has a rowboat in tow. The force on the rowboat due to the tow line is somewhat upward, tending to lift the rowboat as well as pull it horizontally.

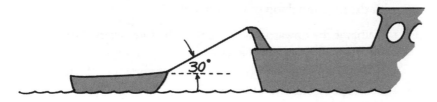

Given: The tow line makes an angle of 30° with the horizontal, and the tension in the line is 240 N.

Find: (a) What is the horizontal force?
(b) What is the vertical force acting on the rowboat?

3-11 A weighted floor polishing mop is used to polish waxed floors. The weighted head of the mop helps to increase the downward force against the floor. The rest of the downward force comes from the force applied to the mop handle.

Given: The head of the floor-polishing mop weighs 11 N. It is being pushed with a force of 35 N in the direction of the handle, which is held at 30° from the vertical.

Find: (a) What is the horizontal force making the mop head go forward?
(b) What is the net downward force holding the mop head against the floor?

3-12 Two monkeys are holding hands in the zoo. They are hanging from equal lengths of rope, which are attached to the top of the monkey cage some distance apart.

Given: The lengths of rope each make an angle of 25° with the vertical, and each monkey exerts a force of 194 N in the direction of the rope.

Find: (a) What is the combined weight of the two monkeys?
(b) With what horizontal force does each monkey pull on the other monkey?

3-13 A mirror is hung with a wire passing over a hook attached to the wall. The tension in the wire is the same on both sides of the hook.

Given: The tension in the wire is 500 N, and the wire makes an angle of 15° with the horizontal top of the mirror.

Find: (a) What is the upward component of the force exerted on the frame by one end of the wire?
(b) What is the sideward (horizontal) component of the force exerted by one end of the wire?
(c) What is the weight of the mirror?

3-14 In one of the nicest known forms of public transportation, people ride in cars drawn up steep hills by cables running through a slot between the tracks. The gripperson who operates such a car activates a cable-gripping mechanism that fits down inside the cable slot.

Given: The weight of a Cable Car on the Powell Street Line in San Francisco fully loaded with passengers is 85,000 N. At one point on the Hyde Street hill, the street makes an angle of 8.5° with the horizontal.

Find: (a) What is the force with which the cable pulls the car up the hill?
(b) Neglecting friction and assuming the brakes are not set, what is the force with which the track on Powell Street pushes against the wheels?

MORE INTERESTING PROBLEMS

3-15 A mountain climber finds herself dangling from the end of her nylon safety line one morning after she has slipped over the edge of a cliff.

Given: The length of rope between the climber and the edge of the cliff has a spring constant of 8.3 N/cm. The climber's weight is 780 N. Before it was stretched, this piece of rope was 3.25 m long.

Find: (a) What is the amount of stretch in the rope due to the climber's weight?
(b) What is the stretched length of the rope?

3-16 Rose, the cat, decides to attack the cord that pulls the curtains shut. As the curtains are already closed, the cord stays stationary but stretches due to the force applied by the cat.

Given: Before the cat attacked it, the cord was 2.05 m. The spring constant of the cord is 6.5 N/cm. Rose pulls on the cord with 18 N of force, not quite her entire weight.

Find: (a) What is the amount of stretch in the curtain cord due to the force of the cat?
(b) During the attack, what is the stretched length of the cord?

3-17 Canal boats are pulled along the canal by a team of horses who walk along a tow path on one bank of the canal. The canal boat resists being pulled into the bank, setting its angle away from the bank by using its rudder.

Given: A canal boat is being towed by a rope 10 m long. When the boat is 3.0 m from the tow path, a tension in the rope of 240 N keeps the boat moving at a uniform constant velocity.

Find: (a) What is the angle the rope makes with the canal bank?
(b) How much of the tow rope force is in the forward direction?
(c) How much of the tow rope force is sidewards?

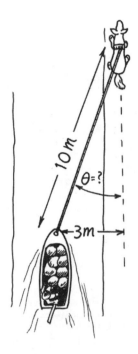

3-18 A water-skier plans to leap from a ramp. To get on the ramp, the skier pulls to one side of the path of the boat, so that the boat can pass on one side of the ramp and the skier can be pulled onto the ramp and fly thru the air to the applause of on-lookers.

Given: A water-skier is pulled behind a motorboat with a rope 8 meters long. As she is approaching a ramp, she pulls herself 2 meters to the side of the path of the boat.

Find: (a) What is the angle the rope makes with the path of the boat?
(b) How much of the tow rope force is in the forward direction?
(c) How much of the tow rope force is sidewards?

3-19 A kind daddy is pushing his little girl on a swing in the park. One fun thing to do is to hold her in equilibrium for a moment with a horizontal force before releasing her so she can swing back. The vector sum of daddy's force, the little girl's weight, and the force of the two chains of the swing is then zero.

Given: The little girl has a weight of 200 N, and the chains of the swing make an angle of 25° to the vertical.

Find: (a) What is the combined tension in the two chains?
(b) With what force must daddy push horizontally to hold his little girl in equilibrium?

3-20 The daring young man on a flying trapeze is preparing for a special trick. He is hanging from his trapeze but is holding himself stationary by holding onto the jumping platform so that he can let go at just the right time to coordinate the swing of his trapeze with the swing of his partner's trapeze.

Given: The trapeze artist has a weight of 512 N, and as he is holding his trapeze stationary, the two ropes of the trapeze are at an angle of 40° to the vertical.

Find: (a) What is the combined tension in the two ropes?
(b) With what force must he hold onto the jumping platform to hold himself in equilibrium, presuming this force is purely horizontal?

3-21 A janitor's broom has a heavy head which helps increase the downward pressure against the floor. The rest of the downward pressure comes from the force applied to the broom handle.

Given: The head of the broom weighs 18 N. When the broom is being pushed the components of the force in the handle are 21 N sidewards and 29 N downwards.

Find: (a) What is the net downward force acting on the broom head?
(b) What is the magnitude of the resultant force acting on the mop head due both to its weight and the force of the handle?
(c) What is the direction of the resultant force?

3-22 The force applied to the handle of a reel-type manual lawn mower serves two functions. The most obvious is that it drives the lawn mower forward, but it also helps the weight of the lawn mower to hold the wheels against the ground so that they do not slip. That is important because the lawn mower mechanism gets its power from the wheels.

Given: The weight of the lawn mower is 120 N, and the components of the force in the handle are 125 N along the lawn and 215 N downwards.

Find: (a) What is the net downward force acting on the lawn mower?
(b) What is the magnitude of the resultant force acting on the lawn mower due both to its weight and the force applied in pushing it?
(c) What is the direction of the resultant force (sum of weight and handle force)?

3-23 A sailboat is tacking upwind. Although the wind comes across the bow, or the front of the boat, the force of the wind against the sail has a forward component that drives the sailboat forward

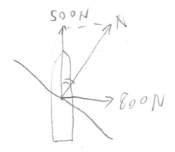

Given: The forward component of the force is 500 N while the perpendicular component of the force, which tends to tip the sailboat over, is 800 N.

Find: (a) What it the magnitude of the force exerted by the wind against the sail?
(b) In what direction is this force, measuring from the forward direction that the sailboat is traveling?

3-24 One day on the slopes, a skier decides to get a ride back to the lodge on a toboggan. One member of the ski patrol pulls the toboggan toward the medical station while another exerts a force perpendicular to the direction of travel to keep the toboggan from sliding downhill.

Given: The forward force is 30 N while the perpendicular force is 40 N.

Find: (a) What single force would do the job if only one member of the ski patrol were available to pull the toboggan?
(b) In what direction would this single force need to be exerted to keep the toboggan moving toward the medical station?

3-25 Harry, the sign painter, has rigged up a boatswain's chair with which to pull himself up the side of a tall building with a block and tackle. He usually ties the free end of the rope back onto the chair, but one day he decides to tie it onto a flagpole as shown.

Given: Harry and his chair weigh 850 N. The rope is secured at the roof of the building, passes through a pulley attached to the chair, back up through a pulley attached to the roof, and then down to Harry again.

Find: What is the tension in the rope if
(a) Harry ties the rope onto the chair?
(b) Harry ties the rope onto the flagpole?

3-26 A traveling prince comes upon a princess who has been imprisoned at the top of a tall tower by an evil stepparent. The only access to the tall tower is the window at which the princess is standing, but there seems to be an easy way up. A bucket has been provided with a rope that goes up and over a pulley attached to a beam projecting from the top of the tower, and a cleat has been conveniently provided beside the window for securing the rope. Not recognizing the wicked stepparent's trap, the prince steps into the bucket and, pulling on the free end of the rope, hauls himself up to the window.

Given: The prince, dressed in the traditional armor of the day, has a weight of 1,100 N.

Find: (a) What is the tension in the rope as long as the free end is in the prince's hands?
(b) What is the tension in the rope after the prince secures the end to the cleat and lets go of it, so that he can have both hands free?

3-27 A strong man is trying to pull a chain out horizontally when the chain has a physics book attached at its midpoint. He finds that the chain is never straight, no matter how hard he pulls on its ends.

Given: The physics book weighs 24 N. The strong man first pulls on both sides with a force of 100 N and then, summoning all of his strength, he pulls with 400 N.

Find: What is the angle the chain makes with the horizontal with
(a) a force of 100 N on each end of the chain?
(b) a force of 400 N on each end?

3-28 Some children are playing tug-of-war, a game where two teams of children pull on opposite ends of a rope to try and pull down the other team. To make the game more interesting, they have placed a weight on the middle of their rope and are trying to pull hard enough to make the rope straight.

Given: The two teams of children are well-matched, and they pull on the rope with exactly the same force. They have placed a 65 N weight at the center of the rope. The teams first pull with 750 N of force each, then, in an heroic effort to straighten the rope, pull with 1,500 N of force.

Find: What is the angle the rope makes with the horizontal with
(a) a force of 750 N on each end of the rope?
(b) a force of 1,500 N on each end?

3-29 A picture frame is supported by a wire having breaking tension, called the "test strength" of the wire, of somewhat more than half the weight of the picture.

Given: The test strength of the wire is 300 N, and the picture, together with the frame, weighs 450 N. The support points on the frame to which the wire is attached are 80.0 cm apart.

Find: (a) What is the minimum angle the wire can make with the horizontal before it breaks?
(b) What is the minimum length of wire necessary to support the picture frame?

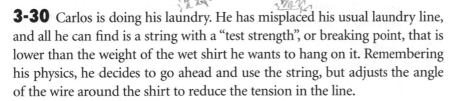

3-30 Carlos is doing his laundry. He has misplaced his usual laundry line, and all he can find is a string with a "test strength", or breaking point, that is lower than the weight of the wet shirt he wants to hang on it. Remembering his physics, he decides to go ahead and use the string, but adjusts the angle of the wire around the shirt to reduce the tension in the line.

Given: The test strength of the string is 15 N. The wet shirt that Carlos wants to hang out to dry weighs 22 N. Carlos hangs the shirt on the middle of the string. The walls that the laundry cable is attached to are 2.45 meters apart.

Find: (a) What is the minimum angle the string can make with the horizontal before it breaks?
(b) What is the minimum length of string necessary to support the wet shirt?

ANSWERS

Solutions to **ESSENTIAL PROBLEMS**

3-8 (a) 440 N/m $\cong$ 4.4 × 10^2 N/m; (b) 0.45 cm

3-9 (a) 537 N; (b) 358 N

3-10 (a) 208 N $\cong$ 2.1 × 10^2 N; (b) 120 N

3-11 (a) 18 N; (b) 41 N

3-12 (a) 352 N; (b) 82 N

3-13 (a) 129 N; (b) 483 N; (c) 259 N

3-14 (a) 1.3 × 10^4 N; (b) 8.4 × 10^4 N

Solutions to **MORE INTERESTING PROBLEMS,** *odd answers*

3-15 (a) 94 cm; (b) 4.2 m

3-17 (a) 17.5°; (b) 229 N; (c) 72 N

3-19 (a) 221 N; (b) 93 N

3-21 (a) 47 N; (b) 51 N; (c) 24° to the vertical

3-23 (a) 943 N; (b) 58°

3-25 (a) 283 N; (b) 425 N

3-27 (a) 6.9°; (b) 1.7°

And so no force, however great,
Can stretch a cord, however fine,
Into a horizontal line
That shall be absolutely straight.[5]

3-29 (a) 49°; (b) 1.2 m

[5]Quoted in the *Oxford Dictionary of Quotations* as an example of accidental meter and rhyme. (2nd ed.; New York: Oxford University Press, 1955. p. 566.) Printed in prose in Whewell's *Elementary Treatise on Mechanics,* 1819.

CHAPTER 4

Velocity and Acceleration

AS SOPHISTICATED AS THEY WERE, the ancient Greeks did not fully develop the concept of velocity. They thought in terms of distance and in terms of time sequence, but they did not combine these two quantities.

In the seventeenth century, Galileo Galilei helped develop the more complete modern view. Motion through space was described in terms of **displacement**, with both magnitude and direction, and time as something that moved onward forever independent of space.

Today, when we want to describe motion over a distance, it is customary to use displacement, **d**, a vector having both magnitude and direction. Average velocity is defined as a ratio of displacement over time, or the direction and distance traveled divided by the time it took to travel that distance.

$$\mathbf{v} \equiv \frac{\mathbf{d}}{t}$$

In everyday usage, people often discuss velocity as a scalar quantity. The word *speed* is frequently used to refer to the magnitude, or scalar part, of velocity. Thus, the speedometer in your car is properly named. It tells you how fast you are going, but not which way. To know your velocity, however, you need both a speedometer and a compass. A vector quantity, as you recall, is specified by both "how much" and "which way."

Galileo understood the full vector nature of velocity. These rules stood for centuries, only to be revised by Albert Einstein. It was Einstein's genius to recognize that Galileo's rules had to be revised at velocities near the speed of light.

Most people in our present society are intuitively comfortable with Galileo's concept of velocity but find Einstein's interpretation difficult to comprehend. Galileo's level is, however, quite sophisticated. The ancient Greeks would have had as much difficulty with it as most people now have with Einstein. It is interesting to speculate how people 200 years from now will understand the concept. The spirit of the time does indeed "teach" us speed, if not in exactly the same sense that Shakespeare's bastard intended.

This chapter will deal mostly with the scalar components of velocity and acceleration because they have many applications in our society. The vector natures will be mentioned so the ideas are familiar when acceleration and velocity vectors are developed in Chapter 5.

Velocity

Velocity, the vector quantity **v**, is defined as the ratio of displacement **d** over time t, where displacement has both a magnitude and a direction.

$$\mathbf{v} \equiv \frac{\mathbf{d}}{t}$$

Velocity can be treated as a scalar quantity if we deal only with the magnitude of the displacement, which is the distance traveled, d.

$$v \equiv \frac{d}{t}$$

If you walk at a velocity of 6 kilometers per hour, that means that if you walk at that speed for one hour, you will have covered 6 kilometers. Likewise, when you are driving at 15 meters per second, if you have moved 15 meters, it took you one second to do it.

Instantaneous vs. Average Velocity

Unless velocity is constant, we must distinguish between the concept of instantaneous and average velocity. By instantaneous velocity we mean, as before, the ratio of displacement over time,

$$v \equiv \frac{d}{t}$$

except we specify that t be only an instant of time. An instant of time refers to an infinitesimally small bit of time, smaller than anything that we can imagine. **Average velocity**, on the other hand, refers to the same definition of displacement over time, except that *t* is some finite interval of time. The distinction only becomes important, as we shall see, when the velocity is changing. An automobile speeding down the highway at 100 kilometers per hour will travel a distance of 100 kilometers if it can maintain this average velocity for a whole hour. It is, however, impossible to maintain this constant velocity exactly for an hour, particularly if there is traffic. The driver would probably find that his instantaneous velocity is sometimes a bit over and sometimes a bit under the average. Converting from kilometers to meters and from hours to seconds, we find that 100 kilometers per hour equals about 28 m/s.

$$100 \, \frac{\text{kilometers}}{\text{hour}} \left(\frac{1000 \text{ meters}}{1 \text{ kilometer}} \right) \left(\frac{1 \text{ hour}}{3,600 \text{ seconds}} \right) = 28 \, \frac{\text{meters}}{\text{second}}$$

It is much easier to keep the velocity of a car constant for a second than for an hour. We might think that 28 m/s is our instantaneous velocity. And it might be for many practical purposes, but that is not what is really meant by instantaneous. At an average velocity of 28 m/s an automobile will travel a distance of 0.028 meters in a millisecond. This, however, is still an average velocity. No matter how small we make the time interval, the displacement will be correspondingly small. The ratio of displacement to time will approach a constant value no matter how small the time interval becomes, as long as it does not actually become zero (in the language of calculus, this is called taking the first derivative of distance with respect to time.) As long as the time interval remains finite, no matter how small it is, the ratio of displacement to time is technically an average velocity. Only when the time is an instant, by definition something smaller than anything we can imagine, is the ratio of displacement to time truly **instantaneous velocity**. It is important to make a distinction between average and instantaneous velocities because, when you are not dealing with an instantaneous velocity, the velocity changes over time and you must use an average value. This affects the way that you solve problems.

From a practical point of view, the distinction between instantaneous and average velocities can be disregarded as long as the velocity remains constant over the time interval in question. The size of the time interval that can be considered an instant depends on how fast the velocity is changing and how particular you might be in your measurements. If a car

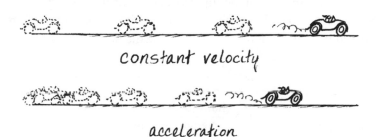

constant velocity

acceleration

Figure 4-1

The difference between constant velocity and constant acceleration is sometimes elusive.

is traveling along at 28 m/s, a second might well be regarded as an instant for all practical purposes. If the car accelerates from 0 to 28 m/s in 11 s, a second would be too long for judging instantaneous velocity. The velocity at the end of any one second would be 28 m/s greater than at the start of that second. At that acceleration a millisecond might be a better time interval to use as an instant.

Acceleration

A concept closely related to velocity is the change of velocity. The two concepts are so closely related, in fact, that people frequently confuse the two. Velocity is already a change in something, a change in distance. A change in velocity is something else again. We call it **acceleration**, and abbreviate it *a*.

$$\mathbf{a} \equiv \frac{\text{change of } \mathbf{v}}{\text{change of } t}$$

Acceleration is defined as the **rate of change of velocity**. Just like velocity, acceleration is a vector quantity. It has magnitude and direction. The accelerator in an automobile is a pedal that produces an acceleration in the same direction as the initial velocity. The brake is also an accelerator in the sense that it produces an acceleration. The direction of the acceleration is, however, opposite to the direction in which you were originally going. An acceleration in this direction is frequently called a deceleration, but it is an acceleration nonetheless. Even the steering wheel is an accelerator of sorts. The direction of the acceleration it produces is perpendicular to the original direction of motion.

We will worry about the true vector nature of acceleration in Chapter 6, when we deal with motion along a curved path. For now, we can treat acceleration as a scalar quantity by concerning ourselves only with accelerations that are in the same direction as the original velocity, or in the opposite direction by using a negative sign to deal with deceleration.

Just as with velocity, we may distinguish between average and instantaneous acceleration. Instantaneous acceleration is the preceding definition with the added qualification that the time interval be smaller than anything you can imagine. This distinction, however, may be disregarded as long as the acceleration is constant over the time interval considered.

Concepts of velocity and acceleration are similar in several ways. The displacement we spoke of in defining velocity is really a change in position, just as the time interval is really a change in time. Thus, the definition of velocity can be made to look a great deal like the definition of acceleration.

$$a \equiv \frac{\text{change of velocity}}{\text{change of time}}$$

$$v \equiv \frac{\text{change of position}}{\text{change of time}}$$

Velocity is rate of change of position with respect to time, while acceleration is rate of change of velocity with respect to time. Acceleration, however, is the rate of change of a rate of change.

$$a \equiv \frac{\text{change of } \left(\dfrac{\text{change of position}}{\text{change of time}} \right)}{\text{change of time}}$$

This business of the rate of change of one thing with respect to something else is very common in mathematics at the level of calculus. While we don't need such sophisticated mathematics for this course, we might borrow a symbol to make our notation look a little simpler. The Greek letter delta, Δ, is frequently used to stand for "the change of." That is a shorthand way of saying the final value of something minus the initial value. Using this delta notation, the definition of velocity can be written

$$v \equiv \frac{\Delta d}{\Delta t}$$

People who are familiar only with algebra sometimes want to try to cancel the Δ in the numerator with the Δ in the denominator. That is not possible, because Δd, as it is used here, means the final distance minus the initial distance, $d_f - d_i$. It does not mean Δ times d. In the same way, Δt means the difference between the final and initial times, $t_f - t_i$.

This delta notation is particularly useful in showing the similarity between the definitions of velocity and acceleration. Using it, the definition of acceleration can be written

$$a \equiv \frac{\Delta v}{\Delta t}$$

Here again, Δv means $v_f - v_i$ and Δt stands for $t_f - t_i$.

The distinction between velocity and acceleration must be clearly understood; these two concepts are frequently confused. The units of velocity and acceleration look a great deal alike. They both have units that involve the ratio of distance and time. A sports car accelerating from 0 to 100 kilometers per hour in 11 s, for example, has an acceleration that is expressed in units having distance in the numerator and time in the denominator:

$$a \equiv \frac{\text{change of velocity}}{\text{change of time}}$$

$$= \frac{100 \text{ kilometers/hour}}{11 \text{ seconds}}$$

$$= 9.1 \frac{\text{kilometers}}{\text{hour} \cdot \text{second}}$$

Figure 4-2

Some people think that acceleration is fun.

The difference is that acceleration has time in the denominator twice. The automobile is increasing its velocity at the rate of 9.1 kilometers per hour per second. At the end of each second, the car is going 9.1 kilometers per hour faster than it was before. This distinction is clear, because hours and seconds are different units for time. The confusion comes when the same units of time are used for both velocity and acceleration. Since 100 kilometers per hour equals 28 meters per second, acceleration may be expressed in meters per second per second, or m/s^2.

$$a = \frac{28 \text{ m/s}}{11 \text{ s}}$$

$$= 2.5 \text{ m/s}^2$$

The acceleration of the car is such that it gains 2.5 m/s of velocity for each second that it accelerates. At the end of 1 s, starting from rest, it is going 2.5 m/s. At the end of 2 s, it is going 5.0 m/s. At the end of 3 s, it is going at a velocity of 7.5 m/s, and so on.

The units "meters per second" and "meters per second squared" sound alike, and people sometimes forget that they are different. But acceleration is as different from velocity as velocity is different from distance.

You might ask yourself how far the car in the preceding example will go in 1 s, starting from rest. While you are thinking about that, we might point out that the instantaneous velocity of the automobile in our example is constantly changing. It starts out at zero, since the car is initially at rest. It then increases as time progresses, reaching a velocity of 2.5 m/s at the instant that falls at the very end of the first second.

If you said to yourself that the car went a distance of 2.5 meters in the first second, you are absolutely... wrong. The car would go that far if it had kept up a constant velocity of 2.5 m/s for the whole second. The car in our example, however, has an instantaneous value that is always smaller than 2.5 m/s until the very last instant of that first second. In the following sections we will see how to solve problems like this one. In this case, the actual distance traveled in the first second is only 1.25 meters.

Velocity-Time Space

We can draw a graph of the velocity of some object, plotting the velocity on one axis and time on the other. Such a graph may be thought of as a representation in a special mathematical space, which we may call **velocity-time space**. Velocity is one dimension of this space, and time is the other.

It turns out that distance is represented by area in velocity-time space. Distance is, after all, equal to the product of velocity and time—just by algebraically rearranging the definition of velocity.

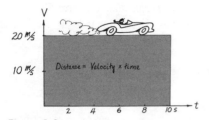

Figure 4-3

The distance traveled at constant velocity is the area of a rectangle on a velocity-time graph.

$$v \equiv \frac{d}{t}$$

$$d = v \cdot t$$

Since velocity is one dimension in our special mathematical space, and time is the other, the product of velocity and time would be area. The total distance traveled by an automobile at a rate of 20 m/s for 10 s is 200 m, since

$$d = vt$$

$$= (20 \text{ m/s})(10 \text{ s})$$

$$= 200 \text{ m}\left(\frac{\text{s}}{\text{s}}\right)$$

$$= 200 \text{ m}$$

The distance traveled by an automobile accelerating is still represented by area in velocity-time space. On a graph in velocity-time space the velocity at any instant of time increases along a more or less straight line from an initial to a final value. The initial velocity v_i is zero in this case, and the final velocity v_f is the instantaneous velocity at the time t_f. We can see from Figure 4-4 that the area under the curve (which is a straight line as long as the velocity increases at a constant rate) is less than it was for the constant velocity case. Instead of being the area of a rectangle, the distance traveled under constant acceleration is the area of a triangle in our mathematical space. Since the distance is less, the *average* velocity is less. By average velocity, we mean the ratio of distance to time where the time interval is larger than an instant.

$$v_{\text{ave}} = \frac{d}{t}$$

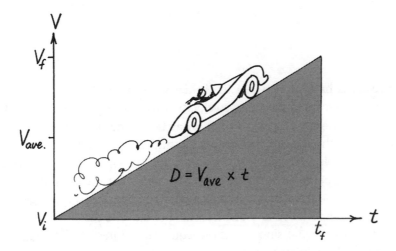

Figure 4-4

The distance traveled from rest at constant acceleration is the area of a triangle on a velocity-time graph.

As you recall, instantaneous velocities, such as v_i and v_f are the values of this ratio of distance to time when the time interval is very small. The average velocity is this same ratio when the time interval is large, such as all the way from zero to t_f.

We can show that the average velocity, defined in this way, is the same thing as the average of the velocities, as long as the acceleration is constant. This would mean that the average velocity v_{ave} as shown on a graph would have a value halfway between v_i and v_f. Let us go ahead, however, and include a more general case in which the initial velocity is not necessarily zero. Let us work from a numerical example involving two constant velocities. Suppose a car travels 10 m/s for 5.0 s and then somehow instantaneously speeds up to 20 m/s for the next 5.0 s. (The time intervals must be equal for the argument to work.) The distance traveled in the first 5.0 seconds would be

$$d = vt$$
$$= (10 \text{ m/s})(5 \text{ s})$$
$$= 50 \text{ m}$$

The distance traveled in the next 5.0 seconds would be twice as great, 100 meters, since the velocity would be twice as great. The total distance traveled would be the sum of the two rectangular areas, as shown in Figure 4-5:

$$d_{total} = 50 \text{ m} + 100 \text{ m}$$
$$= 150 \text{ m}$$

By definition, average velocity would be the total distance traveled over the total time:

$$v \equiv \frac{d}{t}$$
$$= \frac{150 \text{ m}}{10 \text{ s}}$$
$$= 15 \text{ m/s}$$

Our result is halfway between the initial velocity of 10 m/s and the final velocity of 20 m/s. So the average velocity turns out to be the *average* of the velocities. Although the car did not go as far in the first 5.0 s as it would have gone at the average velocity (the distance it did *not* go being represented by the blank rectangle below the v_{ave} line in Figure 4-5), this distance is exactly compensated for by the extra distance that the car did go in the next 5.0 s. This extra distance is represented by the shaded rectangle above the v_{ave} line. The shaded and blank rectangles and the distances they represent are exactly equal since we chose equal time intervals. The v_{ave} line is therefore halfway between v_i and v_f.

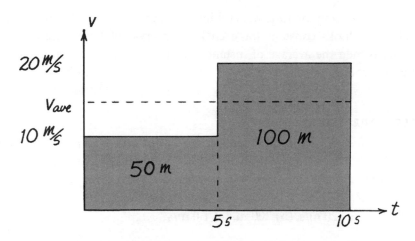

Figure 4-5

The average velocity is the constant velocity that would produce the same displacement as the two different velocities.

The preceding example involving two constant velocities is unrealistic, because it is impossible for a car to jump from one velocity to another with a short but infinite acceleration. The same reasoning, however, can be applied, with minor modification, to a constant acceleration situation. Constant acceleration would be represented by a straight line from initial velocity v_i at initial time to final velocity v_f at final time t_f. The distance traveled, represented by the whole shaded area, would be the same as if the car had traveled for the whole time at average velocity v_{ave}. Since they have parallel sides, the shaded triangle above the v_{ave} line is similar to the blank triangle below the line. Moreover, these triangles are congruent, and so have equal areas, providing that v_{ave} is halfway between v_i and v_f, since that would give the triangles one equal side. The distance not traveled in the first half of the time is exactly made up by the extra distance traveled in the second half.

This discussion assumes constant acceleration and will not work otherwise. Average velocity is not the average of initial and final velocities unless the area above the average velocity line is equal to the area below. We can, however, use constant acceleration to illustrate physical principles in this course, and the arguments can be extended to non-constant acceleration

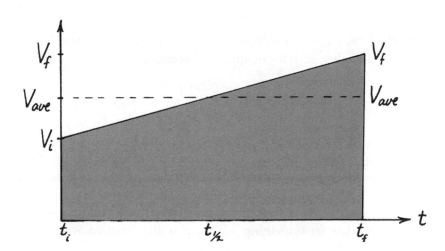

Figure 4-6

If the acceleration is constant, the average velocity is halfway between the initial and final velocities.

by simply breaking the time interval into enough equal chunks so that the acceleration looks constant for each little interval of time. The average velocity is then the average of instantaneous velocities at the end of each little time chunk.

CHECK QUESTION

An automobile enters a freeway on-ramp at 15 m/s and accelerates uniformly up to 25 m/s in a time of 10 s. (a) What is the automobile's average velocity? (b) How far does the automobile travel in this amount of time? (c) What is the acceleration?

Answer: (a) 20 m/s, (b) 200 m, (c) 1.0 m/s²

Motion Under Constant Acceleration

We now have enough theory to solve any problem involving constant acceleration. We can find time, distance, or average velocity, assuming we know the other two quantities.

$$v_{ave} \equiv \frac{d}{t} \tag{1}$$

If we don't know the average velocity, we can find it from the initial and final velocities:

$$v_{ave} = \frac{v_f + v_i}{2} \tag{2}$$

As long as the acceleration is constant, we have seen that the average velocity is the same as the average of the velocities. We can therefore combine definition (1) and equation (2) to obtain distance and time directly in terms of the initial and final velocities:

$$\frac{v_f + v_i}{2} = \frac{d}{t} \tag{1 \& 2}$$

We also have another handle on the initial and final velocities. We can find one or the other from the definition of acceleration:

$$a \equiv \frac{v_f - v_i}{t} \tag{3}$$

It now appears that we have two equations with five unknown quantities. That might seem pretty bad, but there is a ray of hope. We can always reduce the problem to one equation with four unknowns. This turns out to be better than it sounds, because it means that we can solve for either the distance or the final velocity in terms of the initial conditions. If we want to know how far something will travel under constant acceleration, for example, we could find the final velocity from the definition of acceleration, which is equation (3), and use it to find distance from the combined equation (1) & (2). Problems of this nature are so common, however, that many people prefer to work the problem out in general and remember the result. The same is true, as well, for the general solution of final velocity in terms of initial velocity, acceleration, and distance.

The algebra necessary to get these two general solutions looks a little simpler if we express the combined equations (1) & (2) as well as the definition (3) in terms of the sum and difference of the initial and final velocity:

$$v_f + v_i = \frac{2d}{t} \qquad\qquad (1) \text{ \& } (2)$$

$$v_f - v_i = at \qquad\qquad (3)$$

It is now clear that we can eliminate the final velocity v_f by subtracting these two expressions. (We could instead get rid of the initial velocity v_i by adding them, but most problems are given in terms of initial conditions.)

$$0 - 2v_i = at - \frac{2d}{t}$$

This relationship can be expressed in terms of distance d. Multiply both sides by $t/2$ and rearrange:

$$\boxed{d = \tfrac{1}{2}at^2 + v_i t}$$

This formula is worth remembering. Before going on to derive the other formula, let us consider a couple of special cases as examples. If the acceleration happens to be equal to zero, this formula reduces to rate times time.

$$d = 0 + v_i t$$

This is the result we would get directly from the definition of velocity. If initial velocity, on the other hand, is equal to zero, the formula tells us how far something will move from a position at rest:

$$d = \tfrac{1}{2}at^2 + 0$$

We can see, for example, that a car accelerating with a constant acceleration of 4.0 m/s^2 and starting from rest will travel 2 meters in the first second:

$$d = \tfrac{1}{2}(4 \text{ m/s}^2)(1 \text{ s})^2 = 2 \text{ m}$$

We could have figured out the same thing without the formula by using the idea of an average velocity. At the instant the car starts to move, it is going at 0 m/s; after 1 s it has accelerated up to an instantaneous velocity of 4 m/s. The average of 0 and 4 m/s is 2 m/s. After having traveled 1 s with this average velocity, the car will therefore have traveled 2 m.

We have just seen how the formula for distance under constant velocity can be applied to situations in which either the acceleration or initial velocity is zero. We have also seen how we could get the same result using the idea of an average velocity instead of the formula. This should come as no surprise, since that is how we got the formula in the first place. It is only a general solution of the problem as to the distance traveled under a constant velocity. The only reason for remembering it is to save time. If you are going to use it, however, you should know where it came from and be sure you know how to solve the same problem with the idea of an average velocity.

The other formula worthy of being remembered is the general solution for the final velocity in terms of the initial conditions. This formula may be found by eliminating time, t, between equations (1) & (2) and (3). The easy way to do that is to multiply them together in the form in which they were last expressed on the previous page (in terms of the sum and difference of the velocities). This eliminates time, t, since it is in the numerator of the right-hand side of one and the denominator of the right-hand side of the other:

$$(v_f + v_i)(v_f - v_i) = \left(\frac{2d}{t}\right)(at)$$

The product on the left-hand side of this expression is the difference of the squares of the velocities, since, as you remember from algebra, $(a + b)(a - b) = a^2 - b^2$ (see Appendix A).

$$v_f^2 - v_i^2 = 2ad$$

Subtracting the initial velocity squared from both sides gives the general solution of the final velocity, or at least its square, in terms of the initial velocity, acceleration, and the distance:

$$\boxed{v_f^2 = v_i^2 + 2ad}$$

The square of the final velocity is therefore equal to the square of the initial velocity plus two times the acceleration times the distance. This formula is also worthy of remembering because it will save a great deal of time in solving problems.

As an example using the second formula, suppose we want to know how fast our car will be moving after it goes 20 meters at an acceleration of 4 m/s^2, again assuming it starts from rest, so that $v_i = 0$,

$$v_f{}^2 = 0 + 2(4 \text{ m/s}^2)(20 \text{ m})$$

$$= 160 \text{ m}^2/\text{s}^2$$

Or taking the square root of both sides,

$$v_f{}^2 = \sqrt{160 \text{ m}^2/\text{s}^2}$$

$$= 12.6 \text{ m/s}$$

CHECK QUESTIONS

1. An automobile enters a freeway on-ramp with an initial velocity of 15 m/s and accelerates at 1.5 m/s^2 for 10 s. How far does the automobile travel in this amount of time?

Answer: 225 m

2. A sports car accelerates at 2.0 m/s^2 for a distance of 50 m, starting with an initial velocity of 6.0 m/s. What is the sports car's final velocity at the end of this distance?

Answer: 15 m/s

Acceleration Due to Gravity

The way objects fall in Earth's gravity is one particular case of constant acceleration. All objects in free fall have the same constant acceleration— 9.8 m/s^2. This acceleration is the result of Earth's gravitational force, which pulls uniformly on all objects. The value of acceleration due to gravity on earth is only good to two significant figures in the general case, because, as we will discuss in Chapter 5, the force of gravity is different in different places on the earth, depending on how high you are above sea level and where you are with respect to the poles.

The fact that all objects on earth accelerate at the same rate when dropped was discovered for the first time by Galileo in the seventeenth century. Before Galileo, scholars believed that light objects fall more slowly than heavy objects. Aristotle taught that an object ten times as heavy as another object would fall ten times as fast. His authority was so strong that nobody ever bothered to experiment and see if he was right until Galileo's day. It is generally believed that Galileo disproved Aristotle's theory by dropping a light and a heavy stone at the same time from the Leaning Tower of Pisa in Italy to show that they landed at the same time. No one knows if he actually did this experiment, but he did write a book in which two characters did and then argued afterward about the meaning of the results. The story pointed out that experience is more valuable than relying on authority.

You have enough experience with falling objects to know that an object in free fall does not start falling suddenly at a constant velocity. You immediately sense the exaggeration in a cartoon character who walks off a ledge, looks around, and then suddenly plummets downward. Your sense of reality tells you that he should not acquire speed suddenly. He should speed up gradually.

Figure 4-7

A cartoon may exaggerate reality by showing a change in velocity as nearly instantaneous.

You would probably also agree that a coin dropped from your hand picks up more and more speed as it falls. You know you hit the ground harder when you fall further because you are going faster. Many other examples of ever-increasing velocity of falling objects are common to our experience. There still remains a lingering belief in some people's minds, however, that everything falls with the same constant speed. Again we should point out that this mistaken belief usually results from a failure to distinguish between velocity and acceleration.

There is, however, a grain of truth in the belief in a constant velocity for freely falling objects. Objects in free fall do stop accelerating as they reach a terminal velocity when the force of air friction balances their weight. This constant velocity is, however, far from being the same constant for all objects or even for the same object under different circumstances. A piece of paper floating downward at one terminal velocity can be made to fall nearly as fast as a book by crumpling it into a ball. We will discuss this in greater detail at the end of Chapter 5.

For now, we should note that objects do fall with the same acceleration as long as air friction is negligible. If two objects of different weights are dropped at the same time in a vacuum, their velocities will constantly increase together. At the surface of the earth the rate of change of velocity of a freely falling object is

$$g = 9.8 \text{ m/s}^2$$

The letter g is used to denote this particular value of acceleration.

The value of *g* has been experimentally measured. Various objects have been dropped and their velocities measured after certain time intervals. The value of *g* is found to vary (in the third significant figure) at different places on the earth's surface, but it is always found to be the same at any one place for any two different objects. The reason this works so consistently will be discussed in the next chapter.

Once the value of *g* is determined, it may be used as a perfectly good unit of acceleration. A sports car accelerating at 2.45 m/s^2 might be said to have an acceleration of 0.25 *g*:

$$a = 2.45 \text{ m/s}^2\left(\frac{1\ g}{9.8 \text{ m/s}^2}\right) = 0.25\ g$$

That is, the car is accelerating at one-quarter the rate that it would if it were dropped. An airplane pilot can take 10 *g* of acceleration if she is wearing a special tight-fitting *g*-suit to keep her blood from rushing from her head.

Figure 4-8 is a full-scale diagram showing the position of a falling penny at intervals of 0.01 s after it is released from rest. This drawing, or a copy of it, can be used to test someone's reaction time. Hold the drawing vertically upside down and have the other person hold his fingers at the zero-second mark. Drop the drawing between his fingers and let him catch the first image he can. A good reaction time is close to 0.15 s. A fair reaction time is closer to 0.20 s.

Because objects fall at the same rate due to gravity you will get the same results whether you drop a cutout of the picture, a copy of the whole page, or if you drop the whole book.

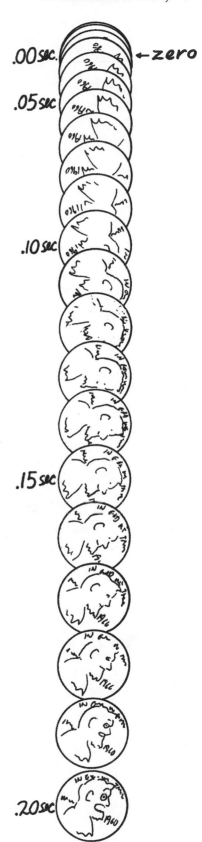

Figure 4-8

A full-scale drawing of a coin at equal time intervals after falling may be used to test reaction time.

SUMMARY OF CONCEPTS

The problems in this chapter are concerned with motion under constant velocity or under constant acceleration in a straight line. The definitions of **velocity** and **acceleration** are

$$v \equiv \frac{\Delta d}{\Delta t} = \frac{d_f - d_i}{t_f - t_i}$$

$$a \equiv \frac{\Delta v}{\Delta t} = \frac{v_f - v_i}{t_f - t_i}$$

If the time interval Δt is very small, an instant of time, the preceding definitions refer to instantaneous velocity and instantaneous acceleration. Otherwise they refer to average values of velocity and acceleration. (A time interval is short enough to be considered an instant if it is small compared to time over which velocity or acceleration fluctuates or changes.)

If velocity is changing, but at a constant rate of acceleration, the average velocity v_{ave} turns out to be the average of the initial and final velocities over a given time interval:

$$v_{ave} \equiv \frac{d}{t} = \frac{v_f + v_i}{2}$$

These three relationships can be combined in various ways. The following two shortcuts are frequently used. You should know how to derive them so that they do not seem mysterious.

The distance traveled under constant acceleration is:

$$d = \tfrac{1}{2}at^2 + v_i t$$

The final velocity may also be found as a function of acceleration and the distance over which the acceleration was constant:

$$v_f{}^2 = v_i{}^2 + 2ad$$

One common constant acceleration situation is a body in free fall. The acceleration due to gravity, *g*, is a constant near the surface of the earth:

$$g = 9.8 \text{ m/s}^2$$

This value is only constant to two significant figures for a few kilometers from the earth's surface and in the absence of air friction. Under those conditions the distance traveled in free fall is

$$d = \tfrac{1}{2}gt^2 + v_i t$$

assuming the positive direction to be downward.

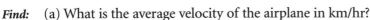

EXAMPLE PROBLEMS

Try to do the Examples yourself before looking at the Sample Solutions.

4-1 An airplane trip from San Francisco to New York has a layover in Chicago. You are on board and wish to estimate the time it will take you to get from Chicago to New York. The speed of an airplane varies depending on the wind it encounters and how high it flies, but you can get a fair estimate anyway.

Given: It is 3,400 km from San Francisco to Chicago and 1,300 km from Chicago to New York. It took you 4.2 hours of air time to get from San Francisco to Chicago. You look at your watch 35 minutes after leaving Chicago.

Find: (a) What is the average velocity of the airplane in km/hr?
(b) How long will it take to reach New York from Chicago?
(c) How many miles have you traveled since leaving Chicago?

BACKGROUND

Velocity is a function of distance and time. It is therefore possible to find any one of these things when the other two are known.

4-2 One measure of a sports car's acceleration is the time necessary for it to get from 0 to 60 mph. Another common measure is the amount of time necessary for it to go a quarter of a mile.

Given   One particular sports car in good tune can accelerate from 0 to 26.8 m/s[1] in 11.0 s. (Assume that the acceleration remains constant.)

Find: (a) What is the acceleration in meters per second squared?
(b) What is the distance traveled in 11.0 seconds?
(c) What is the time necessary to travel 400.0 m?

BACKGROUND

This problem is an example of motion under constant acceleration. The assumption that the sports car accelerates at a constant rate is really only a first-order approximation, however, since much greater acceleration is available in lower gears.

4-3 A 150-kg box falls from a flat-bed truck into the path of your car. You are wondering how far it will slide before it comes to rest.

Given: The initial velocity of the box is 5.0 m/s the instant it falls. It slides to a stop in 3.0 s under constant negative acceleration.

Find: (a) What is the acceleration of the box?
(b) What is the average velocity of the box?
(c) How far does the box slide?

BACKGROUND

This problem is an example of motion under constant acceleration. The box is assumed to decelerate at a constant rate until it stops. Please note the distance traveled by the box is not as great as if it continued moving at a constant velocity.

[1]Is this the same as 60 miles per hour? Check it on your own.

4-4 It's hard to stop a freight train. After the engineer sees an amber warning signal and applies the brakes, neither the engineer nor the brakeman can do much but watch and take note. When, after a few moments, they see the reason for stopping, they note the time and distance elapsed. But other than that, all they can do is watch and hope for the best.

Given: The brakes are applied when the train has an initial velocity of 30 m/s. After 64 s, when note is taken, the train has traveled a distance of 1.6 km.

Find: (a) Between setting the breaks and the noting of time and distance, what is the average velocity of the train?

(b) Assuming that initial, average, and final velocities plotted on a velocity-time graph all lie on a straight line, what is the velocity of the train when note is taken? (Hint: The average velocity is the average of velocities if acceleration is constant. Draw the graph.)

(c) What is the deceleration (negative acceleration) of the train?

(d) How much more time will the train take to make a complete stop?

BACKGROUND

Accident reconstruction is serious business these days. Notes taken by a train crew are frequently the best evidence for establishing liability in the inevitable law suits that result from the inevitable accidents that happen in the normal operation of a railroad. Train crews are therefore taught to observe and take good notes.

We can think of this problem in two parts, before and after note is taken. The velocity at that instant can be thought of as the final velocity for the first part of the problem and the initial velocity for the second part of the problem. The final velocity for the second part of the problem is zero. Parts (a), (b), and (c) all have to do with the first part of the problem, and part (c) uses the acceleration calculated to predict what will happen assuming the acceleration remains constant.

4-5 A man jumps off a bridge to see how high it is. He measures his time of descent. At this point in his life, he decides to ignore the effects of air friction.

Given: He is in free fall for 3.0 s before splashing into the water below.

Find: (a) How high is the bridge above the water?
(b) How fast is the man going as he hits the water?

BACKGROUND

This is an example of motion under constant acceleration in that special case where the acceleration is the acceleration due to gravity.

4-6 A foolish archer is one who shoots an arrow directly upward and then waits for the arrow to return without moving. Neglecting air resistance, the arrow will return to earth with the same velocity as that with which it left the bow.

Given: An arrow is shot directly upward with an initial velocity of 50 m/s.

Find: (a) How long does the arrow travel upward before it stops?
(b) How high does it go?
(c) When will it return?

BACKGROUND

When something, like an arrow, is propelled upwards against gravity it slows down, or decelerates, and then stops for an instant before it begins to accelerate back downwards. This means that the highest point of any arc will be the place where the object is traveling at zero velocity.

This is another example of motion under constant acceleration. The initial and final velocity ($v_f = 0$) are given, and you are asked to find the time. From the time it takes to go up and stop and the initial velocity, you should be able to find how high the arrow goes before returning. You should also be able to figure out how long it takes the arrow to make the total trip, going up and coming back down.

Solutions to EXAMPLE PROBLEMS

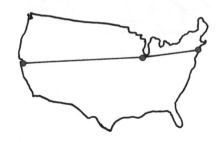

(a) $\quad v \equiv \dfrac{\Delta d}{\Delta t}$

$\qquad = \dfrac{(3,400 \text{ km})}{4.2 \text{ hr}}$

$\qquad = \boxed{810 \text{ km/hr}}$

(b) $\quad v \equiv \dfrac{\Delta d}{\Delta t}$

$\qquad \Delta t = \dfrac{\Delta d}{v}$

$\qquad = \dfrac{(1,300 \text{ km})}{810 \text{ km/hr}}$

$\qquad = \boxed{1.6 \text{ hr}}$

(c) $\quad v \equiv \dfrac{\Delta d}{\Delta t}$

$\quad \Delta d = v \Delta t$

$\qquad = (810 \text{ km/hr})(35 \text{ min}) \left(\dfrac{1 \text{ hr}}{60 \text{ min}} \right)$

$\qquad = \boxed{473 \text{ km}}$

SAMPLE SOLUTION 4-1

Given:
It is 3,400 km from San Francisco to Chicago and 1,300 km from Chicago to New York. It took you 4.2 hours of air time to get from San Francisco to Chicago. You look at your watch 35 minutes after leaving Chicago.

Find:
(a) What is the average velocity of the airplane in km/hr?

(b) How long will it take to reach New York from Chicago?

(c) How many miles have you traveled since leaving Chicago?

DISCUSSION

All the parts of this question are found from the definition of velocity. A three-line equality sign is used to mean "is defined as." Velocity is more than equal to the change of distance over the change in time, it is *defined* as the change of distance over the change in time.

The average velocity of the airplane is found directly from this definition. It is found from the known quantities: the time from San Francisco to Chicago and the distance from San Francisco to Chicago. The equation is given for changing quantities. In this case, the "change of" distance is actually the distance from zero to the traveled distance, here 3,400 km. Therefore, the final distance is 3,400 km, the initial distance is 0 km, so the change is 3,400 km. In later problems it will be critical to note that we are dealing with the change of distance, even though the distance and the "change of" the distance are the same here.

Since we have found the average velocity of the airplane, that value can be used to find the time in part (b) and the distance in part (c), by simple manipulations of the original equation.

SAMPLE SOLUTION 4-2

Given

One particular sports car in good tune can accelerate from 0 to 26.8 m/s in 11.0 s. (Assume that the acceleration remains constant.)

Find:

(a) What is the acceleration in meters per second squared?

(b) What is the distance traveled in 11.0 seconds?

(c) What is the time necessary to travel 400.0 m?

(a) $\quad a \equiv \dfrac{\Delta v}{\Delta t} = \dfrac{v_f - v_i}{t}$

$\qquad = \dfrac{26.8 \text{ m/s} - 0 \text{ m/s}}{11.0 \text{ s}}$

$\qquad = \boxed{2.44 \text{ m/s}^2}$

(b) $\quad d = v_{ave}t$

$\qquad = \left(\dfrac{v_f + v_i}{2}\right)t$

$\qquad = \left(\dfrac{26.8 \text{ m/s} + 0}{2}\right)(11.0 \text{ s}) = \boxed{147 \text{ m}}$

$(\text{or } d = \tfrac{1}{2}at^2 + v_i t = \tfrac{1}{2}(2.44 \text{ m/s}^2)(11 \text{ s})^2 + 0)$

(c) from $d = \tfrac{1}{2}at^2$

$\qquad t = \sqrt{\dfrac{2d}{a}}$

$\qquad t = \sqrt{\dfrac{2(400.0 \text{ m})}{(2.44 \text{ m/s}^2)}} = \sqrt{327 \text{ s}^2} = \boxed{18.1 \text{ s}}$

DISCUSSION

The answer in part (a) comes directly form the definition of acceleration: the change of velocity divided by the change in time.

The distance in part (b) can be found either from the average velocity or from the equation for distance traveled under constant acceleration, with the understanding that the initial velocity is zero, since the car begins from rest.

The time in part (c) can be found by solving this equation for time. The same average velocity cannot be used, because the final velocity turns out to be 44 m/s or about 90 mph, which is not the same as in part (b).

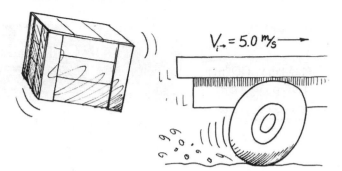

Given:
The initial velocity of the box is 5.0 m/s the instant it falls. It slides to a stop in 3.0 s under constant negative acceleration.

Find:
(a) What is the acceleration of the box?

(b) What is the average velocity of the box?

(c) How far does the box slide?

(a) $a \equiv \dfrac{\Delta v}{\Delta t}$

$= \dfrac{(5.0 \text{ m/s})}{(3.0 \text{ s})} = \boxed{1.67 \text{ m/s}^2}$

(b) $v \equiv \dfrac{d}{t} = \dfrac{v_f - v_i}{2}$

$= \dfrac{0 - (5.0 \text{ m/s})}{2} = \boxed{-2.5 \text{ m/s}}$

(c) $v_{ave} \equiv \dfrac{d}{t} = \dfrac{v_f - v_i}{2}$

$d = v_{ave}t$

$= (-2.5 \text{ m/s})(3.0 \text{ s}) = \boxed{-7.5 \text{ m}}$

DISCUSSION

The acceleration can be found from the definition of acceleration, since the change in velocity and the time are given. We make the answer negative since the acceleration is in the opposite direction from that in which the box is traveling.

Average velocity is the same thing as instantaneous velocity, as long as the acceleration is constant. Distance can either be found from this average velocity or from the equation giving distance traveled under constant acceleration. The answers in parts (b) and (c) are negative because of our choice of direction in terms of the velocity of the truck.

SAMPLE SOLUTION 4-4

Given:

The brakes are applied when the train has an initial velocity of 30 m/s. After 64 s, when note is taken, the train has traveled a distance of 1.6 km.

Find:

(a) Between setting the breaks and the noting time and distance, what is the average velocity of the train?

(b) Assuming that initial, average, and final velocities plotted on a velocity-time graph all lie on a straight line, what is the velocity of the train when note is taken? (Hint: The average velocity is the average of velocities if acceleration is constant. Draw the graph.)

(c) What is the deceleration (negative acceleration) of the train?

(d) How much more time will the train take to make a complete stop?

(a)
$$v_{ave} \equiv \frac{d}{t} = \frac{v_f + v_i}{2}$$

$$= \frac{(1.6 \text{ km})}{(64 \text{ s})} \cdot \frac{(10^3 \text{m})}{(\text{km})}$$

$$= \boxed{25 \text{m/s}}$$

(b)
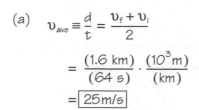

$$v_{ave} = \frac{d}{t} = \frac{v_f + v_i}{2}$$

$$2v_{ave} = v_f + v_i$$

$$v_f = 2v_{ave} - v_i$$

$$= 2(25 \text{m/s}) - (30 \text{m/s})$$

$$= \boxed{20 \text{m/s}}$$

(c)
$$a \equiv \frac{\Delta v}{\Delta t}$$

$$= \frac{(20 \text{m/s}) - (30 \text{m/s})}{(64 \text{s})}$$

$$= \boxed{-0.156 \text{m/s}^2}$$

(d)
$$a \equiv \frac{\Delta v}{\Delta t}$$

$$\Delta t = \frac{\Delta v}{a}$$

$$= \frac{0 - (20 \text{m/s})}{(-0.156 \text{m/s}^2)}$$

$$= \boxed{128 \text{s}}$$

DISCUSSION

The average velocity asked for in part (a) is just distance over time. Although we know the initial velocity v_i, we have yet to find the final velocity v_f at the instant that note is taken. While it is always good to write down the whole relationship, true for constant acceleration, we only need the definition part for part (a):

$$v_{ave} \frac{d}{t}$$

When you draw the velocity-time graph for part (b), you know that the average velocity is half-way between the initial and final velocities, so that if the initial velocity is 5 m/s more than the average, the final velocity will be 5 m/s less than the average velocity. This value can be checked by the second part of the relationship between average and instantaneous velocities when acceleration is constant:

$$v_{ave} = \frac{v_i + v_f}{2}$$

Part (c) is found from the definition of acceleration.

Part (d) can also be found from the definition of acceleration. Since the train is undergoing constant deceleration, the acceleration for the first 64 seconds, found in part (c), is the same acceleration for the remaining time. The answer can be found either by using the initial velocity as the velocity the train is going at the end of 64 seconds – $v_i = 20$ m/s, or you could find the time for the train to make a stop from the time the breaks were applied, using $v_i = 30$ m/s, and then subtracting 64 seconds away from the total time.

SAMPLE SOLUTION 4-5

Given:

He is in free fall for 3.0 s before splashing into the water below.

Find:

(a) How high is the bridge above the water?

(b) How fast is the man going as he hits the water?

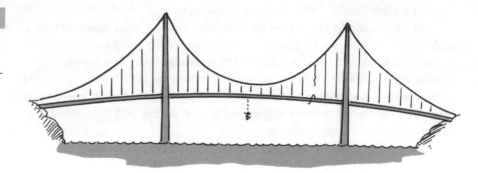

(a) $d = \frac{1}{2}at^2 + v_1 t$

$= \frac{1}{2}(9.8 \text{ m/s}^2)(3.0 \text{s})^2 + 0$

$= \boxed{44\text{m}}$

(b) $a \equiv \dfrac{\Delta v}{\Delta t}$

$\Delta v = a\Delta t$

$= (9.8 \text{ m/s}^2)(3.0 \text{s})$

$= \boxed{29 \text{ m/s}}$

(which is about 100 km/hr
or 60 mph, California speed)

DISCUSSION

The distance is found from the formula for distance under constant acceleration, where the acceleration is *g* and the initial velocity is zero. The final velocity can be found from the definition of acceleration, since we know the time.

(a) $\quad a = \dfrac{v_f - v_i}{t}$

$\quad t = \dfrac{v_f - v_i}{a} = \dfrac{0 - (v_\uparrow)}{(-g)}$

$\qquad = \dfrac{-(50 \text{ m/s})}{(-9.8 \text{ m/s}^2)}$

$\qquad = \boxed{5.1 \text{ s}}$

(b) $\quad v_{ave} = \dfrac{d}{t}$

$\quad d = (v_{ave})(t)$

$\qquad = \left(\dfrac{v_f + v_i}{2}\right)(t)$

$\qquad = \left(\dfrac{0 \text{ m/s} + 50 \text{ m/s}}{2}\right)(5.1 \text{ s}) = \boxed{128 \text{ m}}$

(c) $\quad v_{ave} = \dfrac{d}{t}$

$\quad t = \dfrac{d}{v_{ave}} = \dfrac{d}{\frac{1}{2}(v_i + v_f)}$

$\quad t_\downarrow = t_\uparrow$ since $(v_i + v_f) =$ same thing in both
$\qquad\qquad\qquad\qquad\qquad\quad$ up and down cases

$\quad t_{total} = t_\uparrow + t_\downarrow$

$\qquad = 2t_\uparrow = 2(51 \text{ s}) = \boxed{10.2 \text{ s}}$

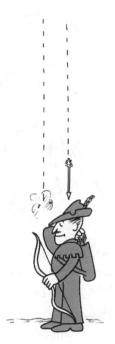

Given:
An arrow is shot directly upward with an initial velocity of 50 m/s.

Find:
(a) How long does the arrow travel upward before it stops?

(b) How high does it go?

(c) When will it return?

DISCUSSION

The time necessary to reach the top of the arc can be found from the acceleration and the change of velocity. The acceleration is just the acceleration due to gravity. The negative sign for acceleration, −g, is used, because the acceleration is in the direction opposite to that of the initial velocity. It is a deceleration. While the arrow is going up, it is falling to a stop.

The distance in part (b) can either be found from the average velocity or from the formula for velocity under constant acceleration.

$$v_f{}^2 = v_i{}^2 + 2ad$$

You might find it interesting to solve this equation for *d* and try it.

The total time can be found in part (c) by considering that the time necessary for it to fall up is equal to the time necessary for it to fall the same distance down. An alternative approach would be to set the distance equal to zero in the equation for distance traveled under constant acceleration. This approach will be discussed in Chapter 6.

Be sure you can do the Example Problems (without peeking at the Sample Solutions) before working these Essential Problems.

4-7 An ocean liner going from San Francisco to Tokyo stops for a few days in Honolulu. You are on board and wish to estimate the time it will take you to get to Tokyo.

Given: It is 3,700 km from San Francisco to Honolulu and 6,200 km from Honolulu to Tokyo. It took you 5.1 days to reach Honolulu from San Francisco.

Find: (a) What is the average velocity of the ocean liner in km/day?
(b) How long will it take to reach Tokyo from Honolulu?
(c) 2.6 days after leaving Honolulu, how many miles have you traveled?

4-8 The brakes on a parked car fail and it accelerates at a constant rate, rolling into a lamp post down the hill.

Given: The car reaches a velocity of 4.3 m/s by the time it smashes into the lamp post. The time between the brakes failing and the smashing of the car is 3.7 s.

Find: (a) What is the car's acceleration?
(b) How far is the lamp post from where the car was parked?
(c) How long did it take for the car to roll 6.0 m?

4-9 The deceleration, or negative acceleration, of an automobile in a full skid on dry pavement is a little greater than one-half "*g*," or half the acceleration you would experience in free fall. Suppose you are going down the highway and you suddenly decide you need to stop your car.

Given: Before braking, you were traveling at 26.7 m/s. It took you 4.9 s to make a complete stop.

Find: (a) What is the car's deceleration?
      (b) What was the average velocity of the car while stopping?
      (c) How many meters did it take you to stop?

4-10 A car slams on the brakes before entering an intersection. After that, the driver and passenger are witnesses to what happens next. They agree on the following facts.

Given: The car had an initial velocity of 25 m/s before the brakes were applied. Before entering the intersection, the car traveled 30 m in 2.0 s.

Find: (a) After the brakes are applied but before entering the intersection, what is the average velocity of the car?
      (b) Assuming that initial, average, and final velocities plotted on a velocity-time graph all lie on a straight line, what is the velocity of the car upon entering the intersection? (Hint: The average velocity is the average of velocities if acceleration is constant. Draw the graph.)
      (c) What is the deceleration (negative acceleration) of the car?
      (d) If nothing were to change this deceleration, how much more time would it take the car to stop moving?

4-11 A little girl throws a coin into a wishing well and makes a wish. Knowing that it is never too soon to begin learning about physics, her mother times the descent of the coin. Not wishing to complicate matters, she ignores air friction.

Given: Mother and daughter see the splash of the coin in the water at the bottom of the well 1.7 seconds after the little girl drops it.

Find: (a) How far is it from the top of the well to the water?
      (b) How fast was the coin traveling when it hit the water in the well?

4-12 A juggler tosses a ball upward and waits for its return. During the first half of the time that the ball is in the air, it is falling to a stop. After reaching its highest point, the ball returns to the juggler's hand.

Given: The initial upward velocity of the ball is 4.8 m/s.

Find: (a) How long does it take the ball to reach its highest point?
(b) How high does it go?
(c) When will it return?

MORE INTERESTING PROBLEMS

4-13 Windows are placed at the top of the Washington Monument to give tourists a view of their nation's capital and a gasp of fresh air. The Park Department has had to put guard screens over these windows to keep people from throwing objects out and injuring other tourists below.

Given: The Washington Monument is 167 m high. A tourist manages to squeeze a nickel through the guard screen to see if it will dent the sidewalk below.

Find: (a) How much time will the nickel take to fall?
(b) How fast will it be traveling when it strikes?

4-14 People sometimes throw rocks off the Golden Gate Bridge in San Francisco to see how long it is until they see a splash at the bottom.

Given: The Golden Gate Bridge rises 45 meters above the water. A kid drops a rock off the edge and watches its descent.

Find: (a) How long will it take for the rock to hit the water?
(b) How fast will it be traveling when it hits the surface?

4-15 BART trains (Bay Area Rapid Transit—modern mass transportation in the San Francisco Bay area) are computer-controlled to accelerate up to a given velocity, determined by a signal received from a particular track section.

Given: A BART train accelerates at the rate of 4.8 (km/hr)/s.

Find: (a) What is the acceleration in m/s^2?
(b) How long will it take for a BART train to move 50 m while pulling out of the station?
(c) What will be the average velocity over the first 50 m?

4-16 Roller coasters pull a long car to the top of a peak, and then release the car to let gravity and inertia take care of the rest. When the car is released it accelerates at a constant rate until the track curves upward again.

Given: A car on a roller coaster is released from rest at the top of a peak. The distance from the top to the bottom of the first peak is 30 m. The car has a constant acceleration of 4.3 m/s².

Find: (a) How long will it take the train to get to the bottom of the first peak?

(b) What will the average velocity be from the top to the bottom of the peak?

4-17 The planet Mars is much smaller than earth. A rock takes longer to fall from a certain height above the surface of the planet Mars than it does to fall from the same distance above the surface of the earth.

Given: The acceleration due to gravity on Mars is 0.38 of the acceleration due to gravity on the earth. On the earth, it takes a rock 1.5 s to fall from a certain height.

Find: (a) How long would it take the same rock to fall this distance on Mars?

(b) How high would that be above the surface of Mars?

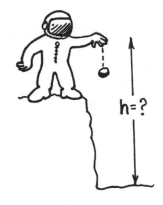

4-18 There are film clips from the moon landing in 1969 of the astronauts jumping around on the moon, playing with the unfamiliar gravity. They can jump much higher than on the earth, and they fall much more slowly than expected.

Given: An astronaut on the moon jumps into a crater. The same jump on would have taken 0.15 seconds on Earth. The acceleration due to gravity on the moon is one-sixth of that on the earth.

Find: (a) How long does this jump take the astronaut on the moon?

(b) How deep was the crater?

4-19 A blunderbuss is an obsolete short firearm with a flaring muzzle for use at close quarters. It was usually loaded with a number of balls, like a sawed-off shotgun.

Given: A blunderbuss ball is fired directly upward from the ground and returns to its starting point in 8.8 s.

Find: (a) How much time did the ball take to reach its maximum height?

(b) What was the muzzle velocity of the ball?

(c) How high does the ball go?

4-20 A basketball player is throwing the ball in the air.

Given: The basketball is thrown directly upward from the ground and returns to its starting point in 0.80 s.

Find: (a) How long did the ball take to reach its maximum height?
(b) What was the initial velocity of the basketball?
(c) How high did the ball go?

4-21 A proposed ion propulsion rocket would drive a spacecraft forward by ejecting electrically charged particles out the back. The problem with this rocket would be that it would not give large accelerations under existing technology, but the little acceleration produced would last for long periods of time.

Given: An ion propulsion unit accelerates a spaceship at 0.6 cm/s^2. The distance from the earth to the moon is 390,000 km.

Find: (a) How long would it take to go a distance equal to that from the earth to the moon, starting from rest and traveling at this constant acceleration?
(b) How fast would you be going when you got there?

4-22 City buses spend most of their time either accelerating or decelerating. When a driver puts the acceleration pedal all the way down, the bus accelerates at a constant rate.

Given: A bus driver puts her foot on the accelerator when the light at the intersection turns green and keeps the bus constantly accelerating until it reaches the other side of the intersection. The bus accelerates at 4.9 m/s^2 and the distance from one end of the intersection to the other is 5.6 meters.

Find: (a) How long did it take the bus to cross the intersection?
(b) How fast was it going when it got to the other end?

4-23 Relativistic effects do not begin to become noticeable below about one-tenth the velocity of light. Above that velocity, people on a spacecraft will experience a time dilation that will shorten their journey. The rules we have learned apply only below that velocity.

Given: A spacecraft accelerates up to 3.0×10^7 m/s (which is about a tenth the velocity of light) with a constant acceleration of 9.8 m/s^2.

Find: (a) How many days will it take to reach this velocity?
   (b) How far will the spacecraft have traveled in the meanwhile?

4-24 Earl is using the programmable treadmill in his health club. When he starts the treadmill from rest, he programs in the speed at which he would like to walk and presses the start button. He notices that the treadmill makes a slow constant acceleration until it reaches the speed he chooses.

Given: Earl wants to walk at 1.78 m/s. The treadmill has a constant acceleration of 0.290 m/s^2.

Find: (a) How long does it take the treadmill to get up to speed?
   (b) How far will Earl have walked before he reaches the speed he selected?

4-25 Drivers' manuals published by the highway departments of most states point out the dangers of fast driving by showing, usually graphically, the disproportionately large stopping distances for high velocities.

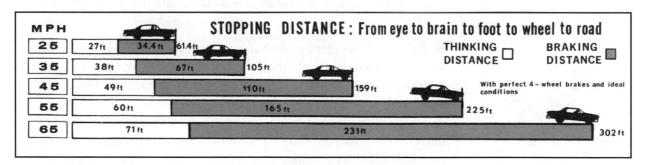

Given: The 1976 California Driver's Handbook showed the above chart. There are 5,280 feet in a mile.

Find: (a) What is the reaction time associated with the "thinking distance"?
   (b) What is the deceleration associated with the "braking distance" used in calculating these stopping distances?

4-26 Bunny Bernard, snowed in at a mountain resort, decides to learn how to ski. The first lesson, on how to stop, turns out to be most valuable. As Bunny learns to go faster, he discovers that he travels further both before and after executing the maneuver known as the "snowplow." Over steaming hot chocolate, Bunny reports his findings to new friends in the form of a chart drawn on a paper napkin.

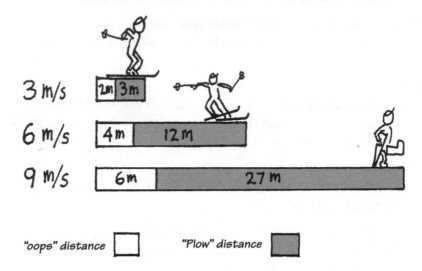

"oops" distance ☐   **"Plow" distance** ▧

Given: "Oops" to stand for distance traveled at constant velocity before getting into snowplow position. "Plow" to stand for distance traveled while decelerating.

Find: (a) What is the reaction time associated with the "Oops" distance?
(b) What is the deceleration associated with the "Plow" distance?

4-27 People frequently say that you can estimate the distance between yourself and a flash of lightning by counting the seconds between a flash of lightning and the thunder. This is true, because it takes almost no time for the light to get to you, whereas the sound travels relatively slowly.

In the story of the race between a rabbit and a turtle, little is said of the officials who covered the start and finish lines. Mother Nature started the race with a flash of lightning near the starting line and a bear watched the finish. The bear determined the distance from the start to the finish by counting the time between the lightning and the thunder.

Given: The rabbit covers the first 1.61 km (1.00 mile) in 4.00 min. The distance from the start to the finish line is 2.90 km. The speed of light is 3.00×10^8 m/s. The speed of sound is 350 m/s.

Find: (a) What is the velocity of the rabbit in meters per second?
(b) How long will it take the rabbit from start to finish (no stops)?
(c) How great is the time lapse between the lightning and thunder as observed by the bear at the end of the race?

4-28 The next time you feel an earthquake, pay attention. You may notice a delay between the bouncing and the shaking, depending on how far you are from the epicenter where the earthquake energy is released. Pressure waves, called P waves, travel faster than transverse (sideways) waves, called S waves. Then come the surface waves, called Raliegh waves, which wiggle the telephone poles and make the wires dance.

Given: The Rayleigh waves cover the first 1.61 km (1.00 mile) in 2.0 s. The distance from you to the epicenter is 290 km. P waves travel at 6.1 km/s. S waves travel at 3.5 km/s.

Find: (a) What is the velocity of the Rayleigh waves in meters per second?
(b) How long does it take for you to feel the Rayliegh waves?
(c) How great is the time lapse between the start of the P waves and the start of the S waves?

ANSWERS

(The answers to parts (b) and (c) may depend on how you round off in previous parts. Your best bet is to keep one extra digit in an intermediate result to avoid a round-off error in the final answer.)

Solutions to **ESSENTIAL PROBLEMS**

4-7   (a) 725 km/day $\cong 7.3 \times 10^2$ km/day; (b) 205 hr $\cong$ 8.5 days;
(c) 1.9×10^3 km

4-8   (a) 1.16 m/s^2; (b) 8.0 m; (c) 3.2 s

4-9   (a) 5.4 m/s^2; (b) −13.4 m/s; (c) 65.4 m

4-10   (a) 15 m/s; (b) 5 m/s; (c) −10 m/s^2; (d) 0.5 s

4-11   (a) 14.2 m; (b) 17 m/s

4-12   (a) 0.49 s; (b) 1.2 m; (c) 0.98 seconds later

Solutions to **MORE INTERESTING PROBLEMS,** *odd answers*

4-13   (a) 5.84 s; (b) 57.2 m/s

4-15   (a) 1.33 m/s^2; (b) 8.66 s or 8.67 s; (c) 5.8 m/s

4-17   (a) 2.4 s; (b) 11 m

4-19   (a) 4.4 s; (b) 43 m/s; (c) 95 m

4-21   (a) 3.6×10^5 s; (b) 2.2×10^3 m/s

4-23   (a) 35 days; (b) 4.6×10^{10} km

4-25   (a) 0.74 s; (b) 20 ft/s$^2 \cong 0.63$ g

4-27   (a) 6.71 m/s; (b) 432 s; (c) 8.29 s

CHAPTER 5 The Laws of Motion

The First and Second Laws of Motion

IN A SCIENCE FICTION MOVIE of the early l950s, *Destination Moon,* a crew of six climbed aboard a single-stage rocket ship and shot off into outer space. Once the large initial acceleration was over, all the crew put on their space suits and magnetic boots to go out and clomp around on the outside of the space ship.

The audience gasped as one member of the crew seemed to get too close to the rocket engines. The script had emphasized the fact that the ship was going thousands of miles per hour, and the audience just assumed that the engines had to be running to keep the ship going that fast. This audience had grown up reading Flash Gordon, who always had fire coming out of his rocket ship as he zipped through space. As the crew member in this new movie leaned over and looked into one of the engines, however, *nothing happened.* The crew member was not blown away, because the engines were not in operation. The new movie was more sophisticated than Flash Gordon. An object in motion does not need a force to keep it in motion. The rocket ship was coasting toward the moon, and it did not need power.

This idea that no force is needed to keep an object moving has been with us since Galileo. Before him people had assumed that a force was necessary to keep an object in motion. They couldn't see the frictional forces, so they naturally assumed that the force they had to apply was the only force acting. They had a hard time, however, explaining the fact that a ball, once thrown, continues moving. They thought that the air parted in front of the ball and rushed around to push it from the rear. Galileo was the first to point out that if constant velocity was considered the natural state of

Figure 5-1

Galileo was thinking of a frictionless surface

things, complicated mechanisms required to maintain motion were unnecessary. An external force was only needed, he said, to balance the invisible forces of friction. If friction could be reduced to nothing, then there would be no need of an external force to keep an object moving.

His closest approximation to frictionless motion was a ball rolling on a very smooth surface. If he had been able to reduce friction still further, he would have found that his theory worked even better. An air track would have done beautifully. It is a hollow track with little air holes spaced along its surface. Air from the pressure side of a vacuum cleaner is forced through these holes, making a glider placed on the track float on a thin layer of air in nearly frictionless contact. Galileo would have loved owning an air track. His theories of motion might not have been improved, but he would have had an easier time convincing his contemporaries that his ideas were correct.

Galileo's contemporaries were not ready to accept his theories of motion, which were contrary to the theology of his day. The idea that an object in motion tends to remain in motion lent credence to the Copernican theory, which said that the same thing is going on in the heavens. The idea that the same sort of motion occurs in the heavens as on earth was just too much. The authorities showed Galileo the instruments of torture and asked that he take back his teachings. He was no fool. He took them back.

Culture is always evolving. Movie audiences in the early 1950s still believed that Flash Gordon's rocket ship on its way to the moon had to have its engines operating all the way to maintain its speed. Most school children today know better than that. They can tell you, from what they have seen on television, that a space ship will coast along under its own inertia until ground control orders a "burn."

A generation or so after Galileo, Newton said nearly the same thing and got away with it. People were willing to accept the heresy of Galileo as the gospel of Newton. With little change, therefore, Galileo's idea is now called Newton's first law of motion.

Newton's First Law: The Law of Inertia

Every object in the universe has the property of resisting change in motion in the absence of a force acting upon it. This fact is called Newton's first law. Galileo said it first, but Newton improved it a little. He gave the name **inertia** to this property of resisting change in motion, and he defined what is meant by a change in motion. A change in the magnitude of velocity is clearly a change in motion, but so is a change in direction. As stated by Newton, the first law says this: **An object at rest will, in the absence of a force, remain at rest, and an object in motion will remain in motion at constant velocity along a straight line.** Since we have already discussed velocity as a vector quantity, we can summarize Newton's law by saying that the velocity of an object will remain constant (in both magnitude and direction) as long as the force applied to it is zero. Since a change in velocity is defined as acceleration, we could also say that *if the force is zero, the acceleration will be zero.*

$$\text{If } \boldsymbol{F} = 0, \text{ then } \boldsymbol{a} = 0 \qquad \text{(Newton's first law)}$$

You can perform several experiments that will confirm the first law. One of the most dramatic is to jerk the tablecloth out from under a full service of china, crystal wine goblets, and dinnerware. This experiment works best when the plates are filled with food and the goblets with wine. The more stuff there is, the more inertia it has. Before the experiment, tell your guests, "An object at rest will remain at rest." One small variation of the experiment is to pull the tablecloth more slowly and demonstrate that objects in motion will remain in motion.

Like all physical concepts, the first law is only a model of reality. There is no real object in the universe for which the sum of all the forces acting on it is exactly zero. This condition is never exactly satisfied. There is therefore no real object in the universe for which the first law really works. It *nearly* works, however, when the force is very small. Also, the acceleration is hardly noticeable when the force acts only for a very short time. The smaller the force or the shorter the time, the better the first law works. We say that the first law holds true *in the limit where the force is zero*, since it works better the closer we get to that limit. The force acting on the place settings in the tablecloth trick, for example, is not really zero; but if the cloth is whipped out quickly enough, the small force exerted will not last long enough to disturb things—much.

Another dramatic experiment demonstrating inertia is done by putting an anvil on someone's stomach and striking it with a sledge hammer. The anvil has enough mass that it starts to move slowly, so slowly that the blow from the sledge hammer does not harm the stomach. This has been an effective, although dangerous, lecture demonstration. One excellent physics teacher successfully performed this demonstration in four out of five of his classes one semester, using student volunteers to strike the anvil. He found that some students are more accurate than others and has "war wounds" to show where students have missed the anvil. Amazingly, he still performs this dramatic demonstration every semester.

Figure 5-2

The anvil's inertia protects the instructor's stomach.

Newton's Second Law of Motion

Both the anvil and the tablecloth experiments involve unbalanced forces, but the effects of these forces are small because they are applied for such a short period of time that only a small change in motion is produced. The anvil rebounds a little from the sledge hammer blow, but its motion is small enough to be absorbed by even a skinny person's stomach. The wine goblets in the tablecloth trick do have some velocity after the tablecloth is whipped out from underneath them. For this reason, it is best to use a waxed table, so that the goblets slide gently to a stop without tipping over.

Thus, experience tells us that the converse of the first law of motion is also true. If the force is not zero, the acceleration is not zero:

$$\text{If } \boldsymbol{F} \neq 0, \text{ then } \boldsymbol{a} \neq 0$$

Further, the amount of acceleration will be proportional to, and in the same direction as, the force applied:

$$F \propto a$$

This proportionality is Newton's second law of motion. The proportionality depends upon something called the object's "inertia." An anvil has more inertia than a wine glass; it requires greater force to give it the same acceleration.

The second law can be written with an equality sign when using a proportionality constant, the value of the constant being a measure of the object's inertia.

$$\boxed{\mathbf{F} = m\mathbf{a}}$$

The proportionality constant m *is* the **mass** of the object. In words, Newton's second law says: **Force equals mass times acceleration**.

Mass is the best measure of how much of an object you have. The same force will accelerate a particular object the same amount anywhere in the universe. In the old British system, things were measured by weight. When you bought 10 lb of sugar, 10 lb was the gravitational force between the earth and the bag of sugar. On the moon, the same bag of sugar would weigh one-sixth as much. If you buy 10 *kilograms* of sugar, however, it will be 10 kilograms anywhere in the universe, including the moon. This is one of the many advantages of the metric system.

Mass and acceleration are defined in the metric system in terms of fundamental quantities. The mass of an object is compared to the mass of a standard kilogram reference block kept in France. Acceleration is measured in meters and seconds, both of which are defined in terms of absolute standards.

Figure 5-3

Newton was thinking of an unbalanced force.

Force in the metric system is defined in terms of mass and acceleration. One newton of force is that amount which will accelerate a 1 kg mass at the rate of 1 m/s². Thus, a newton is defined in terms of Newton's second law

$$\mathbf{F} = m\mathbf{a}$$

where the mass and acceleration are unity (1) as:

$$1 \text{ N} \equiv (1 \text{ kg})(1 \text{ m/s}^2)$$

Figure 5-4

The same force produces the same acceleration anywhere in the universe.

This same force will accelerate the same object at the same rate anywhere in the universe.

The force referred to in Newton's second law is the **net**, or unbalanced, force. Many forces may be acting upon an object, some canceling the effect of others. It is the vector sum of all the forces that produces acceleration. If a large weight is suspended by a single rope from the ceiling, the rope exerts whatever force is necessary to balance the weight. As long as the net force on it is zero, a stationary weight will remain at rest. If you push horizontally on it while it is in its equilibrium position, and there is nothing to counter this force, the weight will accelerate. If you repeated this experiment on the moon, it would take only one-sixth as much force in the rope balancing one-sixth as much weight; but the same horizontal force would result in the same acceleration. Since gravity is balanced in both cases, it should have no effect on our thought experiment. If the vector sum of all the forces is the same, the same acceleration results anywhere you happen to be.

Of course, once the weight swings any distance at all from its equilibrium position, the situation becomes altogether different. The force in the rope would then have a vertical component to balance its weight and a horizontal component to oppose the force you applied. Any force applied to the weight would effect its acceleration. Your hand, the weight, components of the rope force, air friction, and everything else would have to be taken into account. The force in

$$\Sigma F = m\mathbf{a}$$

is the resultant, or vector sum, of *all* of the forces acting. The upper case Greek letter sigma, Σ, is used to denote "the sum of."

CHECK QUESTION

An object of mass 5.0 kg is found to accelerate at 2.0 m/s² when a certain unbalanced force acts upon it. (a) How large is this unbalanced force? (b) How much will the same object accelerate if an unbalanced force of 20 N acts upon it?

Answer: (a) 10 N, (b) 4.0 m/s²

Weight

If a force of 1.0 N will produce an acceleration of 1.0 m/s^2 when applied to a 1.0-kg mass, a force of 2.0 N will accelerate the same mass at 2.0 m/s^2. That is what is meant by a proportionality.

This proportionality holds the other way around as well. If a 1.0-kg mass is observed to accelerate at a rate of 4.0 m/s^2, we may assume that an unbalanced force of 4.0 N is being applied to it. If a complex set of forces causes this object to accelerate at 8.0 m/s^2, you know that the resultant of all those forces is 8.0 N and that its direction is the same as that of the acceleration.

Drop the 1.0-kg object on Earth and you will observe that gravity accelerates it downward at the rate of 9.8 m/s^2. Since its mass is 1.0 kg, the force of gravity necessary to cause this acceleration must be 9.8 N.

The same reasoning can be applied to any object, large or small. A 10.0-kg stone will require 10.0 N to accelerate it at 1.0 m/s^2. If you observe a 10-kg stone accelerating at 10 m/s^2, you may assume that the resultant force acting on it must be 100 N. If it accelerates downward at 9.8 m/s^2, the gravitational force acting on it, its weight must be 98 N.

$$
\begin{aligned}
F &= ma \\
&= (10.0 \text{ kg})(9.8 \text{ m/s}^2) \\
&= 98 \text{ N} \qquad \text{since } 1 \text{ N} = (1 \text{ kg})(1 \text{ m/s}^2)
\end{aligned}
$$

The situation in which the force of gravity produces acceleration amounts to a special case of Newton's second law. Remember from the last chapter that we use the letter "g" to express the acceleration of objects due to the Earth's gravity, and g is always equal to 9.8 m/s^2.

$$
\begin{aligned}
F_{\text{gravity}} &= ma_{\text{gravity}} \\
w &= mg
\end{aligned}
$$

Weight is defined as the force due to gravity. We can use this relationship to determine mass or weight when one or the other is known.

$$
\boxed{w = mg}
$$

The Principle of Equivalence

We have been assuming that the mass of something is what gives it weight. We could draw a distinction between the inertial mass of an object, which is a measure of its resistance to acceleration, and the gravitational mass, which would be the thing that gives it weight. There is really no reason why the inertial mass has to be the same as the gravitational mass, when you stop to think about it. There is no reason, that is, except for the experimental fact that we have never found anything that had even a bit more of the one without having exactly the same amount more of the other.

What would we do if we found something that had more inertial mass than gravitational mass? We would use it to make hammers. Carpenters need to carry around several different sizes of hammers for different sizes of nails. A 20-oz hammer, which is good for framing a house, is too large for small work. It has so much mass that it tends to knock cabinets apart rather than knock them together. A 16-oz hammer, which is good for general work, does not have enough mass to efficiently drive the big nails used in structural lumber. Carpenters would welcome a set of light hammers that would be easy to carry around but still had the required mass. Unfortunately for carpenters, there is no such material. They don't make hammers out of aluminum, for example, because aluminum lacks mass in exactly the same proportion as it lacks weight.

The misfortune of the carpenter turns out to be good fortune for theoretical physicists. The fact that the inertial mass of an object is always equivalent to the gravitational mass is called the **principle of equivalence**, and the principle of equivalence is the foundation of the general theory of relativity. That is heavy physics, so to speak. The principle of equivalence, however, is well within most people's understanding of reality. Hand a penny and a dime to some unsuspecting person and ask which is heavier. Explain that the dime is a tiny bit smaller but is made from a different metal, which might be more dense. (A real silver dime would be great for this experiment if you could lay your hands on one.) The result is not important. The point is to watch what that person does in trying to decide which is heavier. People almost always toss small objects in their hand when trying to resolve small weight differences. When they hold the coin stationary in their hand, they are testing for gravitational mass. When they toss the coin gently about, they are testing for inertial mass. They just naturally assume that if it has more of one, it will have more of the other.

Newton's Third Law

Newton's first and second laws describe the relationship between force and acceleration. Newton's third law states something about the very nature of force. It says that a force is only half of an interaction. It says that: **for every force of action there is an equal but opposite force of reaction**. It says that there is always the other half of the interaction to consider.

We will need to discuss the third law when we develop the ideas of energy and momentum. That will be in a few chapters from now.

The Law of Universal Gravitation

Here on Earth, we tend to think of gravity as the force that holds us to the ground and makes objects fall downwards. That is true, but it is only part of the whole picture. It turns out that the force of gravitation affects all objects equally proportionate to their mass and how far apart they are from one another. The earth is the biggest mass that is very near to us, so it is the gravitational force we feel the strongest. It turns out, however, that not only do you have a gravitational attraction to the Earth, the Earth has a gravitational attraction to you. What is more, all the objects in the room in which you are sitting have a gravitational attraction to each other, and you to them. You probably do not notice the gravitational attraction of people and objects around you, however, since gravitational force is a very weak force compared to the other forces like electric fields or nuclear forces. But any force that keeps you on the earth, with the earth going around the sun is consequential enough to study.

Newton, in addition to his laws of motion, came up with a relationship that describes the way that objects are affected by gravitational force. It is called **Newton's law of universal gravitation**. This law states that: **every object in the universe pulls on every other object with a force proportional to the product of their masses and inversely proportional to the square of the distance between their centers**. We will look at each of these relationships separately and then put them back into a single equation.

First, let us assume that some kind of attraction between any two objects exists. We might expect the strength of this pull to be directly pro-

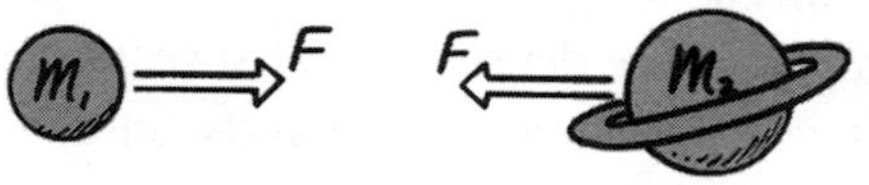

Figure 5-5

There is a gravitational force of attraction between any two masses.

portional to the mass of the object since, if we were to replace one of the objects with two identical objects of the same size, we would expect the force to double. Tripling the objects would triple the force, and so on.

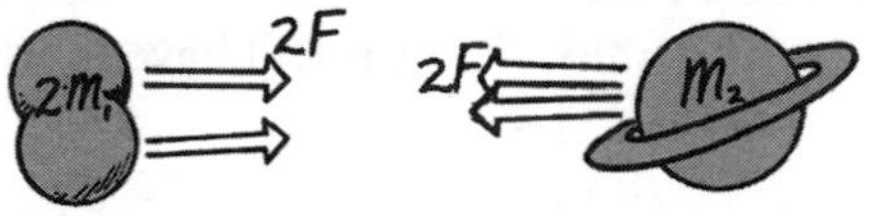

Figure 5-6

The gravitational force is directly proportional to the mass.

Therefore, the force might reasonably be proportional to the mass of object number one.

$$F \propto m_1$$

Similarly, the force would reasonably be proportional to the mass of object number two.

$$F \propto m_2$$

Any time a single quantity is independently proportional to two independent quantities, it is also proportional to their product. If either m_1 or m_2 is doubled, F will double. Therefore, the force between any two objects may be expected to be proportional to the product of their masses.

$$F \propto m_1 \cdot m_2$$

The force between any two objects gets weaker as the objects get further apart. The fact that the force is *inversely* proportional to the *square* of the distance has been observed experimentally. This relationship,

$$F \propto \frac{1}{d^2}$$

just seems to work, but it works in a reasonable fashion. When the objects are twice as far apart, the force is only one-fourth as great as before. When they are three times as far apart, the force is only one-ninth as great, and so on. Combining the inverse square law with the proportionality for the product of the masses:

$$F \propto \frac{m_1 m_2}{d^2}$$

We note that doubling either of the masses doubles the force. If we double both of the masses simultaneously, the force is doubled squared, or four times as great. Doubling the distance then puts you right back where you started.

$$F \propto \frac{(2m_1)(2m_2)}{(2d)^2} = \frac{m_1 \cdot m_2}{d^2}$$

Newton's law of universal gravitation can be expressed in terms of an equality sign and a proportionality constant:

$$\boxed{F = G\frac{m_1 \cdot m_2}{d^2}}$$

CHECK QUESTION

A small object a certain distance from the center of Planet X is found to experience a gravitational attraction of 4.0 N toward the center of that planet. (a) How much force of attraction would the same object experience if it were the same distance from the center of Planet Y, assuming Planet Y has 10 times the mass of Planet X? (b) How much force of attraction would the object experience toward the center of Planet X if it were 2.0 times as far from the center of that planet?

Answer: (a) 40 N, (b) 1.0 N

The value of the proportionality constant G is determined by experiment. Just hold two objects a certain distance apart and see how hard they pull on each other.

One way of measuring this force, for example, is to hang a pair of masses, say bottles of milk, on the end of a stick, supported at the middle by a thread. Pulling two other masses, say boxes of sand, up close to the two masses hanging from the stick will cause tiny gravitational forces to act, tending to turn the stick and twist the thread. It is just a question of using a weak enough string and making it long enough so that this tiny force makes a measurable difference. You can then measure the amount of force it takes to twist a shorter piece of the same string, and you will have a way of getting an experimental value for the force of attraction, F, that exists between room-sized objects. If you measure the masses of the milk bottles, m_1, and the boxes of sand, m_2, as well as the distance d between the centers of the bottles and sand boxes, you will have measured all the parameters in Newton's law of universal gravitation. The only unknown is therefore the value of the proportionality constant G.

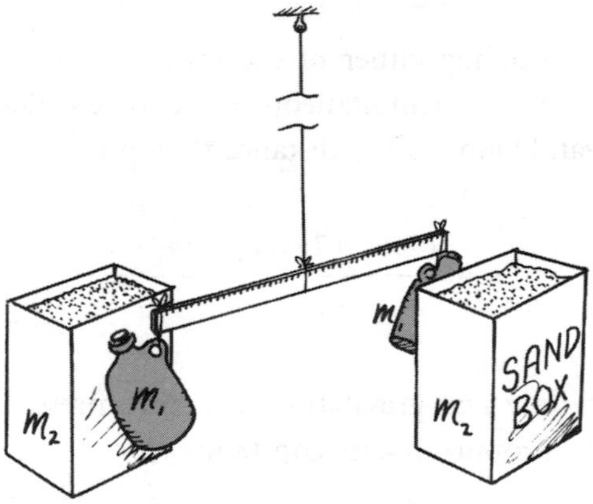

Figure 5-7

A torsion balance may be used for measuring gravitational forces between room-sized objects.

$$F = G\frac{m_1 \cdot m_2}{d^2}$$

Solving the gravitational law for this constant, we get

$$G = \frac{Fd^2}{m_1 \cdot m_2}$$

It turns out that you always get the same value for G whenever you do the experiment or wherever you do it, even on other planets. Whenever you plug in the experimental values for everything on the right-hand side of this equation, you get the same result:

$$G = 6.67 \times 10^{-11} \text{N} \cdot \text{m}^2/\text{kg}^2$$

People sometimes confuse this gravitational constant, capital G, with the g used for the acceleration due to gravity. People sometimes use the word *gravity* to refer to both constants. They might say that the universal law of gravitation states, "The force between any two objects in the universe is equal to gravity times the product of their masses divided by the square of the distance between them." They might then say that the relationship between mass and weight

$$w = mg$$

is that "weight equals mass times gravity." This question of whether *gravity* best refers to G or g *is* further complicated by the fact that most people would say that gravity is the thing that holds us down on the earth's surface. In this latter case, they clearly mean the *force* of gravity, or weight, w. When discussing physics it is therefore wise always to specify the phenomenon you are thinking of when you say gravity. Speak either of the gravitational constant G, the acceleration due to gravity g, or the force of gravity w (the weight), but never of "gravity" alone.

"Weighing" the Earth

Experiments to measure the universal gravitational constant, such as the one we have just described, are sometimes described as "weighing the earth." This is because there is a close relationship between G and *g*. If one of the masses in the law of universal gravitation is the mass of the earth,

$$m_1 = m_e$$

and the distance is the radius of the earth,

$$d = r_e$$

the force given by the law of universal gravitation is simply the weight on the earth's surface:

$$w = G\frac{m_e m_2}{r_e^2}$$

This weight is the same thing as is given by the relationship between weight and mass:

$$w = m_2 g$$

We can get the relationship between G and g by equating these two expressions for weight:

$$m_2 g = G\frac{m_e m_2}{r_e^2}$$

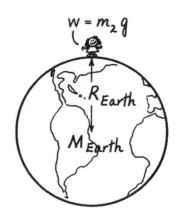

Figure 5-8

Weight is force due to gravity at the earth's surface.

Dividing both sides by the mass m_2, which is the same quantity in both expressions, we see that g *is* related to G by the mass of the earth and the radius of the earth.

$$g = G\frac{m_e}{r_e^2}$$

All we need now to find the mass of the earth is to measure g by dropping something and to measure the radius of the earth, r_e, by taking a long walk. Once we know the value of G, g, and r_e, we can solve this equation for the mass of the earth:

$$m_e = \frac{gr_e^2}{G} = \frac{(9.8 \text{ m/s}^2)(6 \times 10^6 \text{ m})^2}{6.7 \times 10^{-11}\text{N} \cdot \text{m}^2/\text{kg}^2}$$

People who do such things always seem to come out with the same result:

$$m_e = 6 \times 10^{24}\text{kg}$$

The Acceleration Due to Gravity

We have said that everything on Earth falls with the same acceleration. This experimental fact is directly related to the principle of equivalence. We can prove this fact mathematically by using the preceding argument on weighing the earth. Assuming that the only force acting on an object is the gravitational force of attraction exerted on the object by the earth, we can show that the resulting acceleration is the same without regard to the mass of the object. First, we can express this gravitational force in terms of the mass of the earth, m_e, the radius of the earth, r_e, (which is also the distance between the center of the object and the center of the earth), and the mass m of the object.

$$F = G\frac{m_e m}{r_e^2} \qquad \textbf{Gravitational}$$

Still assuming that this is the only force acting on the object, the resulting acceleration a is given by Newton's second law of motion, as usual:

$$(F = ma) \qquad \textbf{Inertial}$$

The principle of equivalence assumes that the gravitational mass in the law of universal gravitation is the same quantity as the inertial mass in the second law of motion. The same symbol, m, represents mass in both equations. If the gravitational force in the first equation is the force used to overcome inertia of the object in the second equation, the right-hand sides of the equations are also equal:

$$ma = G\frac{m_e}{r_e^2} \cdot m$$

Assuming that the masses are the same on both sides of the equation, in agreement with the principle of equivalence, we can simplify this expression by dividing both sides by m.

$$a = G\frac{m_e}{r_e^2}$$

The acceleration therefore does not depend on the mass of the object but only upon the universal gravitational constant, G, the mass of the earth, m_e, and the radius of the earth, r_e. These quantities are all the same for any object on the earth's surface. Plug in these numbers and we find:

$$a = g = 9.8 \ \text{m/s}^2$$

You might argue that r_e would be different on the top of a mountain. That is actually true. Values for g at different locations can be found in handbooks. A few examples are listed in Table 5-1.

This change of values of g is, however, negligible compared to the 4,000 mile interior radius of the earth. All life as we know it is confined to a bubble-thin shell, called the biosphere. Every point within this shell is so nearly the same distance from the center of the earth that almost all objects within our direct experience fall with the same acceleration.

Location	m/s²
North Pole	9.832
Equator	9.780
New York City	9.802
San Francisco	9.799
Pikes Peak	9.789

Table 5-1

The force of gravity at different locations on the earth.

$g = 9.8$ m/s² is only correct to two significant figures.

Air Friction and Terminal Velocity

Objects on earth all fall with the same acceleration, but only when there are no other forces acting on the objects. Frequently, however, there are other frictional forces at work when objects fall, such as air friction. Eliminate air friction, and even a feather will fall like a stone. This can be shown by placing both a feather and a stone together in a glass tube from which the air can be removed. With air in the tube, the stone will fall like a stone and the feather will float down like a feather. Evacuate the tube of air, however, and the feather will fall from one end of the tube to the other with the same acceleration as the stone.

The feather and stone experiment, which has been a classic demonstration for hundreds of years, can be done almost as well by dropping your physics book at the same time as you drop a sheet of paper. The paper will accelerate for only a tiny distance before it reaches its **terminal velocity,**

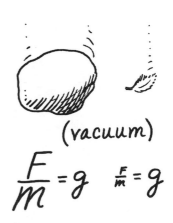

Figure 5-9

All objects fall with the same acceleration in the absence of air friction.

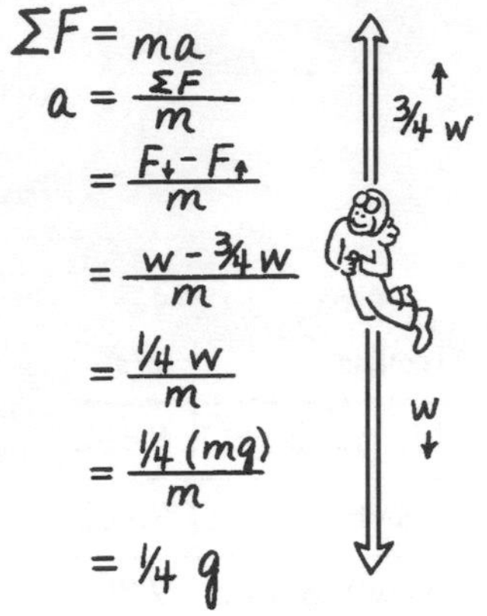

$$\Sigma F = ma$$
$$a = \frac{\Sigma F}{m}$$
$$= \frac{F_\downarrow - F_\uparrow}{m}$$
$$= \frac{w - \tfrac{3}{4}w}{m}$$
$$= \frac{\tfrac{1}{4}w}{m}$$
$$= \frac{\tfrac{1}{4}(mg)}{m}$$
$$= \tfrac{1}{4}g$$

$\tfrac{3}{4}w\uparrow$

$w\downarrow$

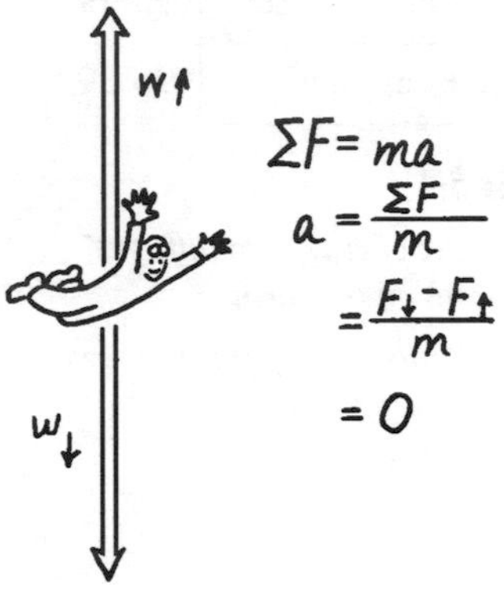

$w\uparrow$

$$\Sigma F = ma$$
$$a = \frac{\Sigma F}{m}$$
$$= \frac{F_\downarrow - F_\uparrow}{m}$$
$$= 0$$

$w\downarrow$

Figure 5-10

The skydiver falls with terminal velocity only after air friction becomes large enough to balance her weight. If she pulls her arms in, air friction will decrease and she will accelerate downwards.

the velocity at which air friction balances downward force. With the paper crumpled into a ball to minimize air friction, however, the paper will be found to strike the floor at about the same time as the book. But if you repeat the experiment once more, throwing both objects out of a high window, you can observe that the force of air friction will again become important. The book will eventually reach a higher terminal velocity than the paper.

Skydivers make use of air friction. They do not open their parachutes for a long time after jumping out of the airplane. Much of the time a skydiver is falling, however, she is not accelerating much at all. Acceleration occurs only from the time the skydiver jumps out the airplane to the time she gets up to a constant speed of about 100 mph. Air friction then balances the weight and the skydiver falls with a constant velocity. This is the same as saying she falls with zero acceleration. This is not really free fall, because there is no unbalanced force to cause an acceleration. The skydiver can control this terminal velocity, however, by holding her hands and arms in different positions. If she wishes to join another skydiver below, she has only to pull her arms in close to her body to reduce air friction. Suppose, because of this, that the force of air friction drops to only three-quarters of her weight; the unbalanced force acting on her will be one-fourth of her weight, and she will accelerate downward at 0.25 g. As she reaches a new terminal velocity, the air friction increases to balance her weight and acceleration drops to zero.

If Galileo could only have jumped out of an airplane, he could have once more shown that an object in motion tends to stay in motion at a constant velocity if all the forces are balanced to zero. Then imagine Newton tucking his arms in and saying, "But if the forces are unbalanced…."

SUMMARY OF CONCEPTS

The problems in this chapter deal with situations in which an unbalanced force acting on an object produces acceleration. The amount of acceleration, according to Newton's second law of motion, is proportional to the vector sum of all the forces acting on the object:

$$\Sigma F = ma$$

The proportionality constant m is called the mass of the object. Force in the metric system of units is defined in terms of mass and acceleration. One **newton** of force is defined as the amount of force that will produce 1 m/s^2 of acceleration when acting on a mass of 1 kg:

$$(1 \text{ N}) \equiv (1 \text{ kg})(1 \text{ m/s}^2)$$

A special case of Newton's second law is the relationship between mass and weight. If the only force acting on an object is its weight (the apparent gravitational force between it and the earth), the same acceleration will always result. This particular acceleration is called the **acceleration due to gravity**, g. The relationship between mass and weight is that weight equals mass times the acceleration due to gravity:

$$w = mg$$

The acceleration due to gravity at the earth's surface has a value of 9.8 m/s^2.

$$g = 9.8 \text{ m/s}^2$$

The fact that all objects fall with the same acceleration shows that the same property of an object that gives it inertia, that is, mass in Newton's second law, also gives it weight, or gravitational force. Newton's **law of universal gravitation** states that every body in the universe pulls on every other body with a force proportional to the product of their masses and inversely proportional to the square of the distance between them:

$$F = G \frac{m_1 m_2}{d^2}$$

The proportionality constant G is called the **universal gravitational constant** and has a value of:

$$G = 6.67 \times 10^{-11} \text{ N} \cdot \text{m}^2/\text{kg}^2$$

EXAMPLE PROBLEMS

Try to do the Examples yourself before looking at the Sample Solutions.

5-1 A bullet in a rifle barrel is propelled by expanding gases. Gun owners select gunpowders with different burning rates depending on the length of the gun barrel and the mass of the bullet.

Given: A 30-caliber bullet typical for sporting use has a mass of 11.7 g. It is accelerated up to a velocity of 620 m/s over a distance of 0.51 m.

Find: (a) What is the acceleration of the bullet?
(b) How much force on the average is exerted on the bullet by the expanding gases?
(c) How long, in seconds, is the bullet in the rifle barrel?

BACKGROUND

This problem illustrates Newton's second law of motion. The greater the unbalanced force acting on an object, the greater will be its acceleration. The unbalanced force used to accelerate a bullet can be calculated by figuring out its acceleration and its mass.

5-2 A nice mother is taking her baby for a stroll on level ground. To give the baby a little thrill, she gives the baby carriage a shove with a force greater than that necessary to overcome friction. She then lets go and allows the carriage to roll to a stop.

Given: The carriage and baby have a mass of 45 kg. She shoves with a force of 90 N and finds that the carriage accelerates at 1.8 m/s^2.

Find: (a) How much unbalanced force is necessary to produce this acceleration?

(b) How much of the mother's shove is used in overcoming friction?

(c) What will be the deceleration of the carriage when friction alone acts?

BACKGROUND

This is an example where more than one force is acting on an object. The force of friction is in the opposite direction from the driving force and partly diminishes its effect. If you know the driving force, you should be able to figure out the friction by its effect.

5-3 All pizzas, small, medium and large, accelerate downward at the same rate when dropped. In New York City, when air drag can be neglected, the acceleration is 9.802 m/s^2.

Given: A pizza cook throws a 1.5-kg pizza upward with a force of 67 N.

Find: (a) How much does this pizza weigh?

(b) What is the unbalanced force on the pizza as it is thrown?

(c) What is the resulting upward acceleration of the pizza?

BACKGROUND

This problem is similar to the previous one, except that one of the forces is the force of gravity, or weight mg.

5-4 A skydiver falling at a constant velocity with her arms out-stretched wishes to fall faster to execute a kiss pass to catch up with another skydiver below. She pulls her arms in close to her body to reduce air friction.

Given: Her mass is 60 kg. With her arms out, she was falling at a constant velocity. With her arms pulled in, she accelerates downward at 2 m/s^2.

Find: (a) What was the force of air friction while her arms were still outstretched?
(b) What is the force of air friction after she pulls her arms in?

BACKGROUND

As in the previous problem, there are two forces acting on the object, the skydiver. In this case, air friction is the other important force.

5-5 One way to measure the acceleration of your car is to see how far a pendulum swings from the vertical. That is why so many highly intelligent drivers have objects of art hanging from their rear-view mirrors.

Given: When the car is accelerating, an object of art having a mass of 0.2 kg is found to swing 15° from the vertical.

Find: (a) What is the weight of the object?
(b) What is the force of acceleration necessary to make it swing that much?
(c) What is the acceleration of the car?

BACKGROUND

This is another example of an acceleration produced by the combined effect of two forces acting on an object. The forces are neither in the same nor opposite direction, however, and must be combined by the rules of vector addition.

5-6 A bricklayer inadvertently rediscovers a device known in the physics community as the Atwood machine, sometimes used in the undergraduate laboratory to measure acceleration due to gravity. This arrangement is composed of two unequal masses supported by a cord passing over a pulley. Assume the cord does not stretch.

Given: The bricklayer has a mass of 75 kg, and the bricks together with the barrel have a mass of 110 kg, which, as he notices, is heavier. The bricks start out 12 m from the ground. He decides to hold on.

Find: (a) What is the differnece between the weights of the bricks and the bricklayer?
(b) What is the total mass being accelerated?
(c) What is the acceleration of the bricklayer?

BACKGROUND

The acceleration and velocity of the bricklayer are the same magnitude but opposite direction from the bricks. It is the difference in the weights that cause this acceleration.

5-7 Both the sun and the moon cause tides in the earth's oceans by virtue of their gravitational forces acting on water on the earth's surface.

Given: The sun has a mass of 2.0×10^{30} kg and is at an average distance of 1.5×10^{11} m from the earth. The moon has a mass of 7.5×10^{22} kg and is at an average distance of 3.8×10^{8} m from the earth. The universal gravitational constant has a value of 6.67×10^{-11} N · m²/kg².

Find: (a) What is the gravitational force of attraction between the sun and 1.0 kg of water on the earth's surface?
(b) What is the force of gravity between the moon and 1.0 kg of water on the surface of the earth?

BACKGROUND

Most people think of the moon alone as causing the tides. That is because the tides caused by the moon are about twice as large as those caused by the sun. When these tides get together, however, the result is an extra large tide, called a spring tide.

This problem is an application of Newton's law of universal gravitation. It will give you a chance to compare the gravitational force on earthly objects caused by the sun and the moon.

5-8 An object on Mars weighs less than the same object on earth, because Mars has much less mass. The smaller mass, however, is somewhat compensated for by the shorter distance from the surface of that planet to its center as compared to the earth.

Given: The mass of Mars is 0.11 times the mass of the earth, and the radius of Mars is 0.53 times that of the earth. A certain physics instructor weighs 900 N on the earth.

Find: (a) How much would he weigh at a distance equal to the earth's radius from the center of Mars?
(b) How much would he weigh on the surface of Mars?

BACKGROUND

This problem is an example of Newton's law of universal gravitation. This law combines the effects of mass and distance to produce a complex proportionality. One way to see the combined effect is to look at a change in mass and distance at separate steps. To see what your weight would be on Mars, for example, imagine that someone were to sneak in at night when you were not looking and replace the earth with Mars. Your weight would decrease because of the change in mass of the planet attracting you, but you would find yourself considerably above the surface of the smaller planet. Your weight would then start to increase somewhat as you fell to the surface of Mars.

(a) $v_f{}^2 = v_i{}^2 + 2ad$

$$a = \frac{v_f{}^2 - v_i{}^2}{2d}$$

$$= \frac{(620 \text{ m/s})^2 - 0}{2(0.51 \text{ m})}$$

$$= \boxed{3.77 \times 10^5 \text{ m/s}^2}$$

(b) $F = ma$

$$= (11.7 \text{ g})(3.77 \times 10^5 \text{ m/s}^2)\left(\frac{1 \text{ kg}}{10^3 \text{ g}}\right)\left(\frac{1 \text{ N}}{\text{kg} \cdot \text{m/s}^2}\right)$$

$$= \boxed{4.4 \text{ kN}}$$

(c) $v_{ave} \equiv \dfrac{d}{t} = \dfrac{v_f + v_i}{2}$

$$t = \frac{2d}{v_f + v_i}$$

$$= \frac{2(0.51 \text{ m})}{(620 \text{ m/s}) + 0}$$

$$= \boxed{1.65 \times 10^{-3} \text{ s}}$$

SAMPLE SOLUTION 5-1

Given:
A 30-caliber bullet typical for sporting use has a mass of 11.7 g. It is accelerated up to a velocity of 620 m/s over a distance of 0.51 m.

Find:
(a) What is the acceleration of the bullet?

(b) How much force is exerted on the bullet by the expanding gases?

(c) How long, in seconds, is the bullet in the rifle barrel?

DISCUSSION

The acceleration can be calculated from the general solution of the velocities in terms of acceleration and distance, introduced in Chapter 4, and solved for the acceleration.

The force is calculated directly from Newton's second law. The combination of units, a kilogram times a meter divided by a second squared, is called a newton, since a newton of force is defined as the amount of force that will accelerate 1 kg of mass at the rate of 1 m/s².

In Part (c), we write the two definitions for average velocity and solve them for time.

SAMPLE SOLUTION 5-2

Given:

The carriage and baby have a mass of 45 kg. She shoves with a force of 90 N and finds that the carriage accelerates at 1.8 m/s².

Find:

(a) How much unbalanced force is necessary to produce this acceleration?

(b) How much of the mother's shove is used in overcoming friction?

(c) What will be the deceleration of the carriage when friction alone acts?

(a) $\Sigma F = ma$

$$= (45 \text{ kg})(1.8 \text{ m/s}^2)\left(\frac{1 \text{ N}}{1 \text{ kg} \cdot \text{m/s}^2}\right)$$

$$= \boxed{81 \text{ N}}$$

(b) $\Sigma F = F_{shove} - F_{friction}$

$F_{friction} = F_{shove} - \Sigma F$

$$= (90 \text{ N}) - (81 \text{ N}) = \boxed{9 \text{ N}}$$

(c) $F = ma$

$a = \dfrac{F}{m}$

$$= \frac{(9 \text{ N})}{(45 \text{ kg})} \cdot \left(\frac{1 \text{ kg} \cdot \text{m/s}^2}{1 \text{ N}}\right)$$

$$= \boxed{0.2 \text{ m/s}^2}$$

DISCUSSION

The total force can be found by a direct application of Newton's second law. Since this total force is the difference (or vector sum) of the force with which the mother shoves and the force of friction, the frictional force is the amount by which the total force is smaller than the mother's shove. If we know the size of the frictional force, Newton's second law tells us the rate at which deceleration of the baby carriage occurs when friction alone acts.

(a) $w = mg$

$\quad = (1.5 \text{ kg})(9.8 \text{ m/s}^2)\dfrac{1 \text{ N}}{1 \text{ kg} \cdot \text{m/s}^2}$

$\quad \cong \boxed{15 \text{ N}}$

(b) $\Sigma F = F_{up} - w$

$\quad = (67 \text{ N}) - (14.7 \text{ N})$

$\quad = \boxed{52 \text{ N}}$

(c) $F = ma$

$\quad a = \dfrac{F}{m}$

$\quad = \dfrac{(52 \text{ N})}{1.5 \text{ kg}}\left(\dfrac{1 \text{ kg} \cdot \text{m/s}^2}{1 \text{ N}}\right)$

$\quad = \boxed{35 \text{ m/s}^2}$

Given:

A pizza cook throws a 1.5-kg pizza upward with a force of 67 N.

Find:

(a) How much does this pizza weigh?

(b) What is the unbalanced force on the pizza as it is thrown?

(c) What is the resulting upward acceleration of the pizza?

F_{up}

w

DISCUSSION

The relationship between mass and weight can be looked upon as a special case of Newton's second law:

$$F = ma$$

If the unbalanced force is the weight, $F = w$, the resulting acceleration will be the acceleration due to gravity, $a = g$:

$$w = mg$$

You can figure out the unbalanced force acting on the pizza when it is tossed into the air from its mass and acceleration. Since you know that this unbalanced force is the difference between the upward force applied by the pizza cook and the weight (or vector sum, since they are in opposite directions), you can find the upward force by solving this equality.

After you know the unbalanced force, the acceleration can be found by a direct application of Newton's second law.

SAMPLE SOLUTION 5-4

Given:
Her mass is 60 kg. With her arms out, she was falling at a constant velocity. With her arms pulled in, she accelerates downward at 2 m/s².

Find:
(a)What was the force of air friction while her arms were still outstretched?

(b) What is the force of air friction after she pulls her arms in?

(a) $\Sigma F = ma$

$w = mg$

$\Sigma F = w - F_{friction}$

$F_{friction} = w - \Sigma F$

$= mg - ma$

$= m(g - a)$

$= (60 \text{ kg})(9.8 \text{ m/s}^2 - 0)\left(\dfrac{1 \text{ N}}{1 \text{ kg} \cdot \text{ m/s}^2}\right)$

$= \boxed{588 \text{ N}}$ or just her weight

(b) $F = m(g - a)$

$= (60 \text{ kg})[(9.8 \text{ m/s}^2) - (2 \text{ m/s}^2)]$

$= \boxed{468 \text{ N}}$

DISCUSSION

With her arms outstretched, the skydiver has reached *terminal velocity.* That is, she has reached the point where she has stopped accelerating downwards because the force of air friction is equal to her weight. In terms of the equation, when the velocity is constant, the acceleration is equal to zero.

When she pulls her arms in there is less air friction, so she is again acted on by an unbalanced force. We know that this unbalanced force is equal to the difference between the air friction and the skydiver's weight. This force is also equal to the mass times the acceleration. These two equations can be combined to solve for the air friction.

Given:
When the car is accelerating, an object of art having a mass of 0.2 kg is found to swing 15° from the vertical.

Find:
(a) What is the weight of the object?

(b) What is the force of acceleration necessary to make it swing that much?

(c) What is the acceleration of the car?

(a) $w = mg$

$$= (0.2 \text{ kg})(9.8 \text{ m/s}^2)\left(\frac{1 \text{ N}}{1 \text{ kg} \cdot \text{m/s}^2}\right)$$

$$= 1.96 \text{ N} \cong \boxed{2 \text{ N}}$$

(b) $F_\uparrow = w$ and $F_\rightarrow = ma_\rightarrow$

also

$$F_\rightarrow = F_{string} \sin\theta$$

$$F_\uparrow = F_{string} \cos\theta$$

$$\frac{F_\rightarrow}{F_\uparrow} = \frac{F_{string} \sin\theta}{F_{string} \cos\theta} = \tan\theta$$

$$F_\rightarrow = F_\uparrow \tan\theta = (2 \text{ N})\tan(15°) = \boxed{0.5 \text{ N}}$$

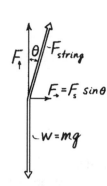

(c) $\dfrac{F_\rightarrow}{F_\uparrow} = \tan\theta$

$$\frac{ma}{mg} = \tan\theta$$

$$a = g\tan\theta = (9.8 \text{ m/s}^2)\tan(15°)$$

$$= \boxed{2.6 \text{ m/s}^2}$$

DISCUSSION

The force acting in the string has two components. The upward component balances the weight. By figuring out the weight in part (a), we can use this component to figure out the force in the string. We could then find the sideward component and the acceleration it produces to get the answers to parts (b) and (c). The solution shown here, however, shows a slightly different approach. The upward and sideward components of the force in the string are expressed in terms of the sine and cosine of the angle to the vertical, and these components are divided, one by the other, to produce the tangent of that angle. One component can then be solved in terms of the other. A similar approach is shown for the accelerations. The answers in parts (a) and (b) are limited to one significant figure by the accuracy to which we know the mass. The answer to part (c), however, can be reported to two significant figures, since the mass divides out of the result.

SAMPLE SOLUTION 5-6

Given:
The bricklayer has a mass of 75 kg, and the bricks together with the barrel have a mass of 110 kg, which, as he notices, is heavier. The bricks start out 12 m from the ground. He decides to hold on.

Find:
(a) What is the difference between the weights of the bricks and the bricklayer?

(b) What is the total mass being accelerated?

(c) What is the acceleration of the bricklayer?

(a) $\Sigma F = w_{bricks} - w_{man}$

$\quad = m_{bricks}g - m_{man}g$

$\quad = [(110 \text{ kg}) - (75 \text{ kg})](9.8 \text{ m/s}^2)\left(\dfrac{1 \text{ N}}{1 \text{ kg} \cdot \text{m/s}^2}\right)$

$\quad = \boxed{343 \text{ N}}$

(b) $\Sigma m = m_{bricks} + m_{man}$

$\quad = (110 \text{ kg}) + (75 \text{ kg})$

$\quad = \boxed{185 \text{ kg}}$

(c) $F = ma$

$\quad a = \dfrac{F}{m}$

$\quad = \dfrac{(343 \text{ N})}{(185 \text{ kg})}\left(\dfrac{1 \text{ kg} \cdot \text{m/s}^2}{1 \text{ N}}\right)$

$\quad = \boxed{1.85 \text{ m/s}^2}$

DISCUSSION

Previously, we have seen that the force is the sum of the unbalanced forces in a situation. In this problem, we see that when the acceleration is the same, we can find the force from the sum of the masses whether or not they are traveling in the same direction.

The acceleration can again be found with Newton's second law. The mass in this case is the combined masses of the objects and the unbalanced force is the difference of the weights. The velocity can be found using the formula for velocity in terms of acceleration and distance, as discussed in Chapter 4.

A more detailed analysis of this problem would look at the tension in the rope. You can draw separate force diagrams for each mass independently, using the difference between tension in the rope and each weight as the force that accelerates each mass.

That approach leads to two equations in two unknowns, which can be solved either for the tension or the acceleration. Although the level of mathematics required is not great, we choose to focus on the physics for present and not get involved with simultaneous equations.

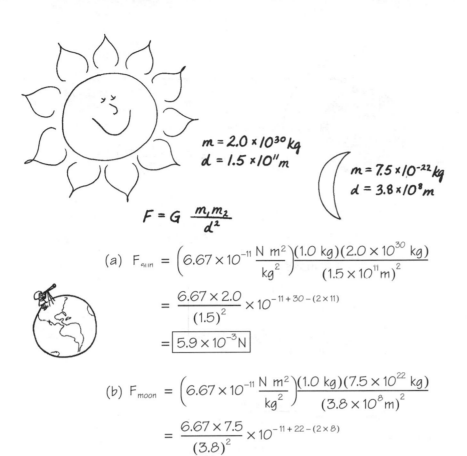

$$m = 2.0 \times 10^{30} \, kg$$
$$d = 1.5 \times 10^{11} \, m$$

$$m = 7.5 \times 10^{-22} \, kg$$
$$d = 3.8 \times 10^{8} \, m$$

$$F = G \, \frac{m_1 m_2}{d^2}$$

(a) $F_{sun} = \left(6.67 \times 10^{-11} \, \dfrac{N \, m^2}{kg^2} \right) \dfrac{(1.0 \, kg)(2.0 \times 10^{30} \, kg)}{(1.5 \times 10^{11} m)^2}$

$$= \frac{6.67 \times 2.0}{(1.5)^2} \times 10^{-11 + 30 - (2 \times 11)}$$

$$= \boxed{5.9 \times 10^{-3} \, N}$$

(b) $F_{moon} = \left(6.67 \times 10^{-11} \, \dfrac{N \, m^2}{kg^2} \right) \dfrac{(1.0 \, kg)(7.5 \times 10^{22} \, kg)}{(3.8 \times 10^{8} m)^2}$

$$= \frac{6.67 \times 7.5}{(3.8)^2} \times 10^{-11 + 22 - (2 \times 8)}$$

$$= \boxed{3.5 \times 10^{-5} \, N}$$

Given:

The sun has a mass of 2.0×10^{30} kg and is at an average distance of 1.5×10^{11} m from the earth. The moon has a mass of 7.5×10^{22} kg and is at an average distance of 3.8×10^{8} m from the earth. The universal gravitational constant has a value of $6.67 \times 10^{-11} \, N \cdot m^2/kg^2$.

Find:

(a) What is the gravitational force of attraction between the sun and 1.0 kg of water on the earth's surface?

(b) What is the force of gravity between the moon and 1.0 kg of water on the surface of the earth?

DISCUSSION

Both parts (a) and (b) are solved by inserting the given values into Newton's law of universal gravitation. The results, as you can see, are quite small. Some people feel, however, that these tiny forces are responsible for all sorts of human events. Astrologers make lengthy calculations based, some say, upon these forces.

The forces exerted by the sun on our 1-kg sample of water turn out to be a hundred and seventy-seven times greater than the force exerted by the moon. Yet the tides caused by the sun are smaller. That is because one cause of the tides is the differences in the forces acting on one side of the earth as compared to the other. Tides are formed when water on the side of the earth closest to the moon is pulled toward the moon harder than the water on the side of the earth that is farther from the moon. Roughly what happens is that the world's water is pulled into a bulge on the side of the earth with the greater gravitational pull. There is another bulge on the other side of the earth due to the fact that both the earth and moon revolve about their common center of mass. That sloshes out another tidal bulge. Both these causes of tides depend on the differences of the distances between the sun or the moon and the two sides of the earth. Since the sun is so far away, the force on the closest side of the earth is pretty similar to the force on the side away from the sun. Since the moon is relatively close to the earth, however, there is proportionately a big difference in the distances to the closer and further sides of the earth.

SAMPLE SOLUTION 5-8

Given:
The mass of Mars is 0.11 times the mass of the earth, and the radius of Mars is 0.53 times that of the earth. A certain physics instructor weighs 900 N on the earth.

Find:
(a) How much would he weigh at a distance equal to the earth's radius from the center of Mars?

(b) How much would he weigh on the surface of Mars?

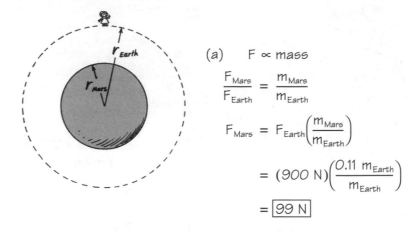

(a) $\quad F \propto mass$

$$\frac{F_{Mars}}{F_{Earth}} = \frac{m_{Mars}}{m_{Earth}}$$

$$F_{Mars} = F_{Earth}\left(\frac{m_{Mars}}{m_{Earth}}\right)$$

$$= (900\ N)\left(\frac{0.11\ m_{Earth}}{m_{Earth}}\right)$$

$$= \boxed{99\ N}$$

(b) $\quad F \propto \dfrac{mass}{(distance)^2}$

$$\frac{F_{Mars}}{F_{Earth}} = \frac{\left(\dfrac{m_{Mars}}{r_{Mars}^2}\right)}{\left(\dfrac{m_{Earth}}{r_{Earth}^2}\right)} = \frac{m_{Mars}}{r_{Mars}^2}\ \frac{r_{Earth}^2}{m_{Earth}}$$

$$F_{Mars} = F_{Earth}\ \frac{m_{Mars}\,r_{Earth}^2}{m_{Earth}\,r_{Mars}^2}$$

$$= (900\ N)\ \frac{(0.11\ m_{Earth})\,r_{Earth}^2}{(m_{Earth})(0.53\ r_{Earth}^2)}$$

$$= \boxed{352\ N}$$

DISCUSSION

The law of universal gravitation says that the force is directly proportional to the product of masses and inversely proportional to the distance squared:

$$F = G\frac{m_1 m_2}{d^2}$$

Part (a) just looks at the proportionality between force and mass. A direct proportion can be expressed as a ratio, which can then be solved for one of the unknowns. Part (b) uses the same approach but includes the inverse proportion with the square of the distance.

Be sure you can do the Example Problems (without peeking at the Sample Solutions) before working these Essential Problems.

5-9 An arrow is propelled by the string of a bow.

Given: An arrow has a mass of 8.6 g. It is accelerated up to a velocity of 35 m/s over a distance of 0.66 m.

Find: (a) What is the acceleration of the arrow?
(b) How much force is exerted on the arrow by the bow string?
(c) How long is the arrow in contact with the string?

5-10 Herman is rollerblading in New York's Central Park. He is on level ground.

Given: Herman and his equipment have a mass of 95 kg. Starting from rest, he kicks off with a force of 280 N, which gives him an acceleration of 2.9 m/s². He then lets himself roll to a stop without pushing again.

Find: (a) How much unbalanced force is necessary to produce this acceleration?
(b) How much of Herman's original kick is used to overcome friction?
(c) If Herman lets himself roll to a stop, what is his deceleration?

5-11 An elephant gets on an elevator. He presses a floor button and notices that the force with which the elevator presses up on his feet changes.

Given: The elephant has a mass of 2,000 kg. Just after the button is pressed, the force against his feet is 15,000 N.

Find: (a) What is the elephant's weight?
(b) How much unbalanced force acts on the elephant?
(c) What is the elephant's acceleration? Up or down?

5-12 The boss is making coffee for the office staff. A paper coffee filter has fluted edges and floats well, even with ground coffee in it. The boss has a filter in which he has put one scoop of ground coffee. By accident he knocks it off the counter top, and, fascinated by how well it floats, wads it up and drops it again for comparison.

Given: One scoop of coffee weighs 8.0 *g*. Open, the coffee filter falls at a constant velocity. Closed, it accelerates at 9.7 m/s^2.

Find: (a) What was the force of air friction with the filter open?
(b) What was the force of air friction on the filter after it had been wadded up?

5-13 Esmeralda is at an amusement park with her science class. As part of the class she has built a simple accelerometer, a device with which to measure acceleration, made of a cardboard protractor with a piece of lead as a weight hanging from a string in the center. When she wants to know how fast she is accelerating, she can hold the accelerometer upright and note the angle to which the lead weight swings.

Given: Esmeralda's accelerometer has a piece of lead with a mass of 0.25 kg. On one acceleration ride, Esmeralda notes that the lead swings 7.0° from the vertical at the start of the ride.

Find: (a) How much does the lead weigh in newtons?
(b) What was the forward force necessary to make it swing that much?
(c) What is the initial acceleration of the ride?

5-14 Donald is painting the mast of his sailboat. He has rigged up a clever way to keep his paint can nearby when he is high on the mast. He discovered that he could counter-balance his open can of paint with his radio by attaching them to either end of a rope on a pulley. This works well for a while, but as he uses more paint, the can ascends.

Given: The radio weighs 3.1 kg, as does the paint can when full. Donald notices he has a problem after he has used 0.2 kg of paint. The paint can travels 2.6 m before it hits the pulley at the top.

Find: (a) What is the unbalanced force acting on the system?
(b) What is the total mass being accelerated?
(c) What is the acceleration of the radio?

5-15 Some people believe that the position of the planet Venus has a significant influence on a child at birth.

Given: The mass of the planet Venus is 4.88×10^{24} kg. It is a distance of 2.1×10^{11} m from the child. The child's father, standing in the delivery room, has a mass of 105 kg and is standing 0.501 m from the child. The child has a mass of 3.21 kg. Remember, $G = 6.67 \times 10^{-11}$ N · m²/kg²

Find: (a) What was the gravitational force of attraction between Venus and the child?

(b) What is the gravitational force between the child and its father?

5-16 In the 1969 moon landing, the astronauts left behind a Hasselblad camera. The camera, still on the moon to this day, weighs much less than an identical camera on earth.

Given: The mass of the moon is 0.0123 times the mass of the earth. The radius of the moon is 0.273 times that of the earth. An identical camera weighs 1.36 N on the earth.

Find: (a) How much would the camera weigh at a distance equal to the earth's radius from the center of the moon?

(b) How much does the camera weigh on the surface of the moon?

MORE INTERESTING PROBLEMS

5-17 One measure of a sports car's performance is its ability to accelerate from rest up to a given speed in some amount of time. The better the engine is running, the shorter will be the time.

Given: A 1960 Triumph TR-3 with driver has a mass of 1,000 kg. In good tune, it is able to accelerate from 0 to 26.8 m/s in 11.0 s.

Find: (a) What is the acceleration in meters per second squared?
(b) What is the unbalanced force acting on the car, in newtons?

5-18 A space shuttle ignites its rockets for a burn to correct its course in accordance with calculations made by ground control.

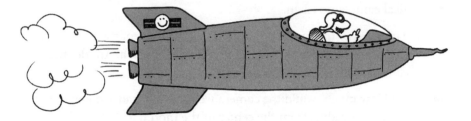

Given: The space shuttle has a mass of 17,000 kg. Ground control orders a burn of 4.5 s to change the velocity by 150 m/s.

Find: (a) What is the average acceleration in meters per second squared?
(b) What is the thrust of the rockets necessary to provide this acceleration?

5-19 Space travelers of the future will conduct commercial transactions in terms of the mass of a commodity rather than its weight, mass being independent of gravitational field.

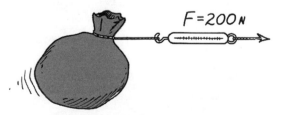

Given: A sack of dried leaves has a mass of 1.2 kg and is acted upon by an unbalanced force of 200 N for 1.5 s.

Find: (a) What is the acceleration of the sack?
(b) How far will the sack have gone at the end of the time period?

5-20 While building a space station, a massive beam is being maneuvered in outer space. In the absence of friction, a very small force can move a large mass if applied over time.

Given: A ton of metal is acted upon by an unbalanced force of 15 N for 25.0 s. A metric ton is defined as 1000 kg.

Find: (a) What is the acceleration of the metal?
(b) How far did it go while the force was applied?

5-21 One of the problems encountered in the Skylab program, an early venture into space, was measuring without the aid of gravity the mass of waste products. The technique actually used involved applying a given force and measuring the acceleration.

Given: An unknown mass of waste product experiences an acceleration of 8.3 m/s^2 when a force of 5.0 N is applied to it. This amount of force is applied for 0.25 s.

Find: (a) What is the mass of the waste product?
(b) How far did it go while the force was applied?

5-22 A kid is trying roller skates for the first time. She is having trouble getting started so her big sister gives her a little push.

Given: The kid, initially at rest, is given a push of 60 N for 0.4 seconds and experiences an acceleration of 1.9 m/s^2.

Find: (a) What is the mass of the kid?
(b) How far did the kid roll during the push?

5-23 A jet aircraft takes off after accelerating down the runway under full throttle.

Given: The mass of the plane is 65,000 kg. It reaches a takeoff velocity of 50 m/s in 35 s, using an engine thrust of 100,000 N.

Find: (a) What is the actual acceleration (assumed constant) of the aircraft?
(b) How much unbalanced force produces this acceleration?
(c) What was the force of friction that opposed the acceleration?

5-24 Bert is on the Tidal Wave, an amusement park ride in Great America in Northern California. He is accelerated from rest to enough speed to remain seated even as the roller coaster train goes upside down through a big loop.

Given: The train has a mass of 3,200 kg. The train reaches a velocity of 25 m/s in 1.2 s and is pushed with a force of 73,000 N.

Find: (a) What is the actual acceleration (assumed constant) of the roller coaster train?
(b) How much unbalanced force produces this acceleration?
(c) What was the force of friction that opposed the acceleration?

5-25 A dropped piece of paper that has not been crumpled quickly reaches its terminal velocity. A crumpled piece of paper continues to accelerate until it hits the floor.

Given: A piece of paper has a mass of 1.8 g. When dropped broad side down to the floor, it quickly reaches a constant velocity. When it is crumpled, it falls at a rate of 8.2 m/s^2.

Find: (a) What is the force of air friction on the paper after it reaches a constant velocity?
(b) What is the force of air friction on the crumpled paper?

5-26 The famous feather and coin demonstration can be done by putting a feather in an evacuated tube. With normal air pressure in the tube, the feather floats at terminal velocity. When all the air is removed from the tube, the feather falls as fast as the coin, or at *g*. With only some of the air removed the feather will fall at a rate between these two accelerations.

Given: A feather in a vacuum tube has a mass of 0.21 *g*. First, it is allowed to fall with normal air pressure and quickly reaches its terminal velocity. Next, some air is removed from the tube and the feather is dropped again. This time it accelerates at a rate of 5.6 m/s^2.

Find: (a) What is the force of air friction on the feather when it is falling at terminal velocity?
(b) What is the force of air friction when the feather is dropped in the partially evacuated tube?

5-27 Your body seems heavier than usual when you are in a jet airliner that is accelerating down the runway for a takeoff. You could actually measure the increased weight of a mass with a spring fish scale. Of course, the best mass to use for this experiment, to avoid suspicious looks from fellow passengers, would be a fish.

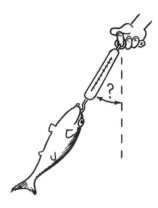

Given: The fish normally weighs 40 N. During takeoff the weight of the fish seems to increase to 47 N as it swings from the vertical.

Find: (a) To what angle does the fish swing from the vertical?
(b) What is the acceleration of the airliner?

5-28 A flyfisher is casting a light fishing line attached to a weight. When casting the line she gives it a swing sidewards.

Given: The fishing weight weighs 1.47 N. At the moment the weight is being swung to the side, the tension in the fishing line is 14.0 N.

Find: (a) At this moment, what angle does the fishing line make with the vertical?
(b) What is the acceleration of the weight through space?

5-29 A block of ice slides down a wooden plank with very little friction. The acceleration of the ice is less than *g*, because only the component of the weight parallel to the plank is unbalanced.

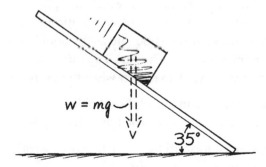

Given: The plank makes an angle of 35° with the horizontal. The block of ice has a mass of 11.5 kg.

Find: (a) How much force is acting on the block in the direction parallel to the plank?
(b) What is the acceleration of the block?
(c) What would be the acceleration of the ice block if it had twice as much mass?

5-30 A car rolls down a hill in neutral. There is very little friction resisting this action.

Given: The hill makes an angle of 2.6° with the horizontal. The car has a mass of 1,670 kg.

Find: (a) How much force is acting on the car in the direction parallel to the hill?
(b) What is the acceleration of the car?
(c) What would be the acceleration of the car if it had twice as much mass?

5-31 A person standing on a bus during rush hour finds it necessary to lean forward when the bus starts moving from the bus stop into traffic.

Given: The bus accelerated forward with an acceleration of 1.5 m/s². The person has a mass of 55 kg.

Find: (a) How much forward force must act on the person?
(b) What is the person's weight?
(c) At what angle to the vertical must the person lean to remain balanced, if it is impossible to find a polite place to hold on?

5-32 There is a commercially made cup holder designed to keep the contents of a coffee cup from sloshing when you have it in your car. It consists of a base and a movable platform on which you place your cup. This platform freely responds to the forces of acceleration of the car. The holder allows the cup to move at the same rate as the liquid, so the coffee doesn't slosh over the sides of the cup. Thus, the cup tilts toward the back of the car when the car moves forward, to the left when the car turns left, and so on.

Given: You have your cup in one of these holders. Leaving a stop sign, your car accelerates at 1.8 m/s². The full coffee cup has a mass of 0.69 kg.

Find: (a) How much forward force is acting on the coffee cup?
(b) How much does it weigh?
(c) At what angle to the vertical is the cup while it is being accelerated?

5-33 One common experiment used to illustrate Newton's second law of motion involves a cart on a horizontal track. The cart is accelerated by a weight, which is attached to a string running over a pulley. Just as in the Atwood machine, the unbalanced force accelerates both masses.

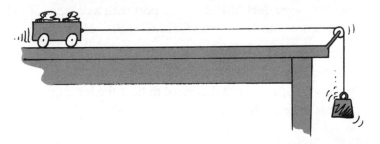

Given: The cart has a mass of 1.00 kg. The weight has a mass of 0.25 kg in one experiment and 2.50 kg in a second experiment.

Find: (a) What is the total mass being accelerated in the first experiment?
(b) What is the acceleration of the cart in the first experiment?
(c) What is the acceleration in the second experiment?

5-34 Clever Coyote has been tricked again. In trying to trick his arch enemy, he has accidentally ended up sitting in a cart attached by a rope to a boulder which has just rolled over a cliff....

Given: The cart and Clever Coyote combined have a combined mass of 78 kg. The boulder is 450 kg.

Find: (a) What is the total mass being accelerated?
(b) What is Clever Coyote's acceleration over the cliff?
(c) If Clever Coyote tries to be especially clever and set his trap a second time using a boulder twice as massive, what would be his acceleration if he accidentally ended up doing the experiment again?

5-35 The *Queen Elizabeth* and the *Queen Mary* were two famous ocean liners once operated by the same shipping company. They were sister ships and both had about the same mass.

Given: Both ocean liners had a mass of about 75,000 tonnes (7.5×10^7 kg). Once they were docked at the same port with a distance of only 306 m between their centers. ($G = 6.67 \times 10^{-11}$ N · m²/kg²)

Find: (a) What was the gravitational force of attraction between these two sister ships?

(b) How far apart would they need to be if the force of attraction were to be 1.0 N?

5-36 The huge stone blocks making up the ancient Stonehenge observatory, at Salisbury, England, are a sight to behold. They must have required hundreds of prehistoric people to drag each into place. A pilgrim to Stonehenge imagines he can sense the gravitational pull between two of the massive stones.

Given: The two stones weigh 45 tonnes each (4.5×10^4 kg). They stand with a distance of 2.7 m between their centers. ($G = 6.67 \times 10^{-11}$ N · m²/kg²)

Find: (a) What is the gravitational force of attraction between these two stones?

(b) How far apart would they need to be if the force of attraction were to be 0.10 N?

5-37 A spaceship would be much heavier on the surface of Jupiter than on the surface of Earth.

Given: A 25,000-kg spaceship has a weight of 2.45×10^5 N on the surface of the earth. Jupiter has a mass of 1.9×10^{27} kg and a radius of 7.0×10^4 km, and the earth has a mass of 6.0×10^{24} kg and a radius of 6.4×10^3 km. $G = 6.67 \times 10^{-11}$ N · m²/kg².

Find: (a) How much would the spaceship weigh if it were to move away from the center of the earth to a distance equal to Jupiter's radius?

(b) How much would the spaceship weigh if someone were to replace the earth with Jupiter when no one else was looking? (This would be the same as landing the spaceship on the surface of Jupiter.)

(c) What is the acceleration due to gravity on Jupiter?

5-38 The largest moon in our solar system is Ganymede, one of Jupiter's moons. Ganymede is bigger than the planet Mercury.

Given: A 2.57-kg American flag has a weight of 4.20 N on the surface of the moon. Ganymede has a mass of 1.54×10^{23} kg and a radius of 2,529 km. Our moon has a mass of 7.36×10^{22} kg and a radius of 1,738 km. $G = 6.67 \times 10^{-11}$ N · m²/kg².

Find: (a) How much would the flag weigh if it were to move away from the center of our moon to a distance equal to Ganymede's radius?

(b) How much would the flag weigh if someone were to replace our moon with Ganymede when no one else was looking? (This would be the same result as moving the flag to the surface of Ganymede).

(c) What is the acceleration due to gravity on Ganymede?

ANSWERS

Solutions to **ESSENTIAL PROBLEMS**

5-9   (a) 928 m/s$^2 \cong 9.3 \times 10^2$ m/s^2; (b) 8.0 N; (c) 3.8×10^{-2} s

5-10   (a) 276 N $\cong 2.8 \times 10^2$ N; (b) 4.5 N; (c) 4.7×10^{-2} m/s^2

5-11   (a) 1.96×10^4 N; (b) 4.6×10^3 N down; (c) 2.3 m/s^2 down

5-12   (a) 7.8×10^{-2} N; (b) 8×10^{-4} N

5-13   (a) 2.5 N; (b) 0.30 N; (c) 1.2 m/s^2

5-14   (a) 1.96 N $\cong$ 2.0 N; (b) 6.0 kg; (c) 0.33 m/s^2

5-15   (a) 2.37×10^{-8} N; (b) 8.96×10^{-8} N

5-16   (a) 1.67×10^{-2} N; (b) 0.224 N

Solutions to **MORE INTERESTING PROBLEMS,** *odd answers*

5-17   (a) 2.44 m/s^2; (b) 2,440 N

5-19   (a) 167 m/s$^2 \cong 1.7 \times 10^2$ m/s^2; (b) 188 m $\cong 1.9 \times 10^2$ m

5-21   (a) 0.60 kg; (b) 0.26 m

5-23   (a) 1.4 m/s^2; (b) 9.3×10^4 N; (c) 7.1×10^3 N

5-25   (a) 1.8×10^{-2} N; (b) 2.9×10^{-3} N

5-27   (a) 32°; (b) 6.0 m/s$^2 \cong 0.62$ g

5-29   (a) 65 N; (b) 5.6 m/s^2 = 0.57 g; (c) 5.6 m/s^2 = 0.57 g (same as part (b)).

5-31   (a) 83 N; (b) 539 N $\cong 5.4 \times 10^2$ N; (c) 8.7°

5-33   (a) 1.25 kg; (b) 1.96 m/s$^2 \cong$ 2.0 m/s^2; (c) 7.00 m/s^2

5-35   (a) 4.0 N; (b) 613 m

5-37   (a) 2.0×10^3 N; (b) 6.5×10^5 N; (c) 26 m/s$^2 \cong 2.6$ g

6 Motion Along a Curved Path

Projectile Motion

IN CHAPTER 5 WE DISCUSSED ACCELERATION produced by a force that was either acting in the same or opposite direction from that which the object was originally moving. The resulting motion was always confined to a straight line. For an object to vary from a straight-line path, it must experience a force that has a component perpendicular to its original direction. The result is a curved path—our subject for this chapter.

Suppose you are being served coffee in an airliner. Your coffee cup is traveling at 800 kilometers per hour in a horizontal path. As your coffee is being poured, it accelerates under its own weight downward towards the cup.

$$F = ma$$
$$w = mg$$

The acceleration of the coffee depends only upon its mass and weight. Neither one of these is affected by the fact that the coffee is traveling horizontally. You would be surprised, in fact, if the steward were to compensate for the velocity of the plane by pouring the coffee ahead of your cup.

The idea that motion in one direction is independent of motion in another is the **principle of independence of motion.** This principle is quite general and useful for understanding any motion along a curved path. Let us consider the motion of a bullet fired horizontally from an automatic pistol. As the bullet leaves horizontally the spent shell casing drops vertically. You might think that the bullet would hit the ground first, since it is going faster. Or you might argue that the shell casing would hit the ground first, since it does not go as far.

Neither of these contentions is correct. Let us neglect air friction and assume that the ground is level. The bullet and the shell casing fall the same vertical distance in the same amount of time, just as coffee pours the same in a jet plane traveling 800 kilometers per hour as it does when the plane is standing on the runway. Both the bullet and casing hit the ground at the same time, because they both start out with zero vertical velocity. The horizontal velocity of the bullet does not affect its motion in the vertical direction.

Let us study an example with the automatic pistol, assuming a convenient height and a reasonable distance for the bullet to go before hitting the ground, and we will see how to calculate the muzzle velocity of the bullet as it leaves the gun. We will assume that the bullet and shell casing both leave the gun without any vertical velocity. The large horizontal velocity of the bullet carries the bullet a considerable horizontal distance, say, 200 meters, while the shell casing falls right beside the shoe or boot of the person firing the gun.

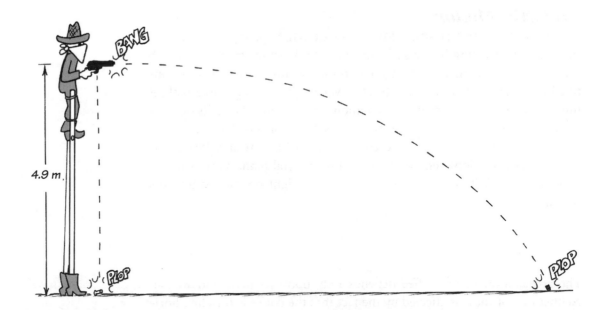

Figure 6-1

The shell casing ejected and the bullet fired horizontally from a 4.9-meter height will both require 1 second to hit the ground.

A nice height to deal with would be 4.9 meters. Although we would need a tall Texan to hold the gun, we already know that anything dropped from rest falls 4.9 meters in the first second. We therefore know that the shell casing takes exactly 1.0 second to hit the ground.

If you didn't know that, you could figure it out from the equation for distance traveled under constant acceleration, which we developed in Chapter 4:

$$d = \tfrac{1}{2}at^2 + v_i t$$

In this case, the distance is downward, the acceleration, g, is downward and the initial velocity is zero—at least in the downward direction.

$$d_\downarrow = \tfrac{1}{2}gt^2 + 0$$

This equation can be solved for time by multiplying both sides of the equation by 2, dividing both sides by g,

$$\frac{2d_\downarrow}{g} = t^2$$

and taking the square root of both sides,

$$t = \sqrt{\frac{2d_\downarrow}{g}}$$

Inserting 4.9 meters for the height $d_\downarrow$ and the known value for g, we get the time that it takes the shell casing to fall, which you probably already knew:

$$t = \sqrt{\frac{2(4.9 \text{ m})}{9.8 \text{ m/s}^2}}$$
$$= \boxed{1.0 \text{ s}}$$

The only new part of this discussion is that the same argument works for the bullet as works for the shell casing. Neither is being acted upon by any force other than its weight after it leaves the gun. Both start with zero initial vertical velocity, and so both have the same vertical height at all times. The bullet takes 1.0 second to fall 4.9 meters, just as the shell casing does.

In the meantime, the bullet travels a large horizontal distance, 200 meters, while it is falling. This is motion at a constant velocity, since there is no force in the horizontal direction, neglecting air friction. We can therefore use the definition of velocity to calculate the horizontal velocity with which the bullet left the gun:

$$v_\rightarrow = \frac{d_\rightarrow}{t}$$

$$= \frac{200 \text{ m}}{1.0 \text{ s}}$$

$$= \boxed{200 \text{ m/s}}$$

This is also the horizontal component of velocity that the bullet has as it strikes the ground. The only thing that happens to its velocity is that it picks up a vertical component as it falls.

Trajectories

The path of a projectile is called the projectile's **trajectory**. People frequently anticipate the trajectory of a projectile in order to predict its range (how far it will go), whether the projectile happens to be a wad of waste paper they are tossing into a waste basket or a package they are dropping from an airplane.

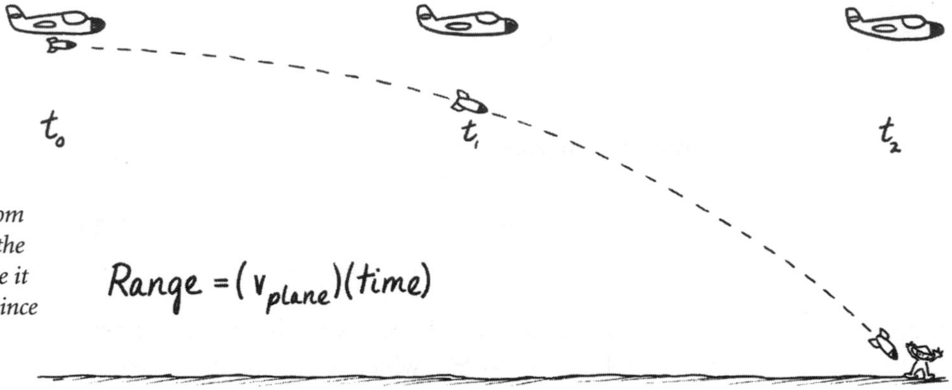

$$\text{Range} = (v_{plane})(time)$$

Figure 6-2

The range of an object dropped from an airplane flying horizontally is the velocity of the plane times the time it takes to get over the landing site, since the object always stays under the plane.

Perhaps you recall old newsreel films showing packages being released from airplanes. The packages were often long and cylindrical, with fins on the back so that they would line themselves up with wind to reduce air friction. The movies showed the packages dropping away from the plane but staying underneath the plane all the way down to the ground. The physics demonstrated by these old newsreels is clear, whether we agree with the application or not. The airplane is always directly over the packages, because of independence of motion. The plane stays over the packages, because they are both going at the same horizontal velocity. The packages are falling, but they are only falling down. They are coasting in the sideward direction, since their shape makes the sideward force of air friction small.

Suppose, if you like, that these packages contain food and medical supplies; and let us calculate the range or horizontal distance from the drop point to the place where the packages hit the ground. Knowing that the plane is overhead when our package hits the ground (since they are both traveling at the same horizontal velocity, $v_{\rightarrow}$), we can compute the range, $d_{\rightarrow}$, from the definition of velocity:

$$v_{\rightarrow} \equiv \frac{d_{\rightarrow}}{t}$$

We can get the time, t, in this expression from the vertical motion of the package, since it is moving in both directions at once. The time required for an object to drop a vertical height, $d_\downarrow$, from the plane is given by the formula for the distance traveled under constant acceleration:

$$d_\downarrow = \tfrac{1}{2}gt^2$$

As in the example with the automatic pistol, we can solve this equation for the time of flight in terms of this downward distance and downward acceleration g. First multiply both sides by 2, to get rid of the $\frac{1}{2}$, then solve for time and take the square root:

$$2d_\downarrow = gt^2$$

$$t = \sqrt{\frac{2d_\downarrow}{g}}$$

The time thus calculated from the accelerated motion in the downward direction can then be used to determine $d_\rightarrow$, the horizontal distance, from the constant horizontal velocity $v_\rightarrow$, since both motions are going on at the same time:

$$d_\rightarrow = v_\rightarrow t$$

$$= v_\rightarrow \sqrt{\frac{2d_\rightarrow}{g}}$$

We can calculate the range to the target area for a plane flying at 800 kilometers per hour (222 m/s) at a height of 3,000 meters by first calculating the time necessary for the package to fall:

$$t = \sqrt{\frac{2d_\downarrow}{g}}$$

$$= \sqrt{\frac{2(3{,}000 \text{ m})}{9.8 \text{ m/s}^2}}$$

$$= 25 \text{ s}$$

The distance to the target area can be found from the horizontal velocity of the plane, assuming that components of the package retain the horizontal component of velocity (222 m/s) while falling:

$$d_\rightarrow = v_\rightarrow t$$

$$= (222 \text{ m/s})(25 \text{ s})$$

$$= 5{,}550 \text{ m}$$

Thus the package must be dropped a horizontal distance of 5,550 meters (more than three miles) before the target.

A frequent error in reasoning is to try to put the acceleration due to gravity and the horizontal velocity $v_\rightarrow$ into the same equation for distance under a constant acceleration.

$$d_\nearrow = \tfrac{1}{2}g_\downarrow t^2 + v_\rightarrow t \qquad \text{WRONG!}$$

The resulting distance $d_\nearrow$ is neither in the horizontal nor vertical direction. This equation is wrong. Don't do it. The whole idea of independence of motion is that you have to treat velocity and distance in one direction independently from velocity and distance in the other. Only time links the two equations of motion for the two different directions.

CHECK QUESTION

An object is thrown horizontally with a velocity of 10 m/s from the top of a building that is 19.6 meters tall. (a) How long does it take the object to fall the vertical distance of 19.6 meters? (b) How far does the object travel horizontally while it is in free fall?
Answer: (a) 2.0 s, (b) 20 m

Vertical Components of Velocity

We have so far only considered cases where the initial motion of a projectile is entirely horizontal, so that there has been no initial vertical component to worry about. Under these conditions we found the formula for vertical displacement is

$$d_\uparrow = \tfrac{1}{2}\, gt^2$$

If there is an initial vertical component of velocity $v_\uparrow$ as well, the vertical displacement is

$$d_\uparrow = \tfrac{1}{2}\, gt^2 + v_\uparrow t$$

If, for example, a juggler throws a ball upward with a velocity of 19.6 m/s, the height to which the ball rises at time t is given by

$$d_\uparrow = \tfrac{1}{2}(-9.8 \text{ m/s}^2)\, t^2 + (19.6 \text{ m/s})\, t$$

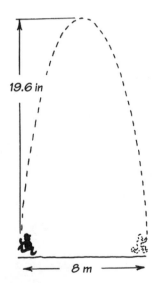

Figure 6-3

A juggler sees the path of an object moving along with him as motion in a vertical path alone. From the vantage point of the audience it is seen to travel a curved path called a parabola.

19.6 in

8 m

(If we consider up the positive direction, acceleration due to gravity is negative.) At the end of the first second $t = 1$ s and the height is given by

$$d_\uparrow = \tfrac{1}{2}(-9.8 \text{ m/s}^2)(1^2 s^2) + (19.6 \text{ m/s})(1 \text{ s})$$
$$= -4.9 \text{ m} + 19.6 \text{ m}$$
$$= 14.7 \text{ m}$$

At the end of the second second, $t = 2$ s and the height is given by

$$d_\uparrow = \tfrac{1}{2}(-9.8 \text{ m/s}^2)(4 \text{ s}^2) + (19.6 \text{ m/s})(2 \text{ s})$$
$$= -19.6 \text{ m} + 39.2 \text{ m}$$
$$= 19.6 \text{ m}$$

At the end of the third second $t = 3$ s. The acceleration term is beginning to gain on the velocity term, and the height is back to

$$d_\uparrow = \tfrac{1}{2}(-9.8 \text{ m/s}^2)(9 \text{ s}^2) + (19.6 \text{ m/s})(3 \text{ s})$$
$$= -44.1 \text{ m} + 58.8 \text{ m}$$
$$= 14.7 \text{ m}$$

At the end of the fourth second, however, we find that $d_\uparrow = 0$ ft. The acceleration term is equal to the velocity term, and the ball is back in the juggler's hand.

$$d_\uparrow = \tfrac{1}{2}(-9.8 \text{ m/s}^2)(16 \text{ s}^2) + (19.6 \text{ m/s})(4 \text{ s})$$
$$= -78.4 \text{ m} + 78.4 \text{ m}$$
$$= 0 \text{ m}$$

At any time after 4 seconds have elapsed, the negative acceleration term is even larger than the positive velocity term. The resulting negative distance just means that the ball has gone below the original starting position.

Thus, we see that the equation for distance under a constant acceleration can be used to predict the height of the ball, or any other object for that matter, at the end of any given time. We just insert the initial velocity and time into the equation

$$d_\uparrow = \tfrac{1}{2}(-g)t^2 + v_\uparrow t$$

We could also turn this equation around to find the time at which the ball will be at any particular height. It is just a quadratic equation using time as the variable, and you probably learned how to solve it when you took algebra. We could always haul out the old quadratic equation and solve it to find the time for any height we like.

There is one particular height of frequent interest, however, for which we don't even need the quadratic equation. That is the height zero. We often want to know the time that something thrown up in the air will come down again. As we saw from our numerical example, $d_\uparrow = 0$ when the ball returns to the juggler's hand after 4 seconds. We could find the time necessary for the ball to return to this height for any initial velocity if we solve the equation for values of time with the height equal to zero:

$$0 = -\tfrac{1}{2}gt^2 + v_\uparrow t$$

One way to find the values for time that will produce this result is to factor out time from the two terms in the preceding equation. We will then have the product of two things that are equal to zero.

$$0 = (-\tfrac{1}{2}gt + v_\uparrow) \cdot t$$

This expression is of the form $a \cdot b = 0$. We know that if that is true, either a must be equal to zero or b must be equal to zero (or they may both be equal to zero). Then we see that either

$$t = 0$$

or

$$-\tfrac{1}{2}gt + v_\uparrow = 0$$

will satisfy the equation. The first solution, $t = 0$, we knew already. That was when the ball was first thrown. The other solution gives the condition that must be satisfied when the ball returns. The minus term in the parentheses must equal the plus term.

$$\tfrac{1}{2}gt = v_\uparrow$$

These two terms are equal when the distance is equal to zero. If we solve this equation for time, we have a general solution to the question of how long it takes something thrown upward to return to the hand:

$$\boxed{t = \frac{2v_\uparrow}{g}}$$

For the juggler's ball in our example, the initial velocity $v_\uparrow$ was 19.6 m/s. The acceleration due to gravity, g, is just 9.8 m/s^2

$$t = \frac{2(19.6 \text{ m/s})}{9.8 \text{ m/s}^2} = 4 \text{ s}$$

We already found this result by the trial-and-error method. We started at time zero and put in larger and larger values for time until the height of the ball returned to zero.

The trial-and-error method could be used to find the time of return for any initial upward velocity. This is called calculation by successive approximation. It is useful if one has either a programmable computer or lots of time. The algebraic method is more powerful, however, because it brings the force of mathematical logic to bear on the problem and produces a general solution that is truc for a similar situation.

Suppose that the juggler in the preceding example was walking across the stage. He has a certain horizontal velocity, say, 2 m/s. The juggler would see the ball going up and right back down into his hand. The audience, however, would see the ball going up and back down in a curved path. The ball would meet the juggler after he had walked 8 meters.

$$v \equiv \frac{d}{t}$$

$$d_\rightarrow = v_\rightarrow t$$
$$= (2 \text{ m/s})(4 \text{ s})$$
$$= 8 \text{ m}$$

CHECK QUESTION

An object is thrown with a velocity having an upward component of 9.8 m/s and a sideward component of 30 m/s. (a) How much time does it take the object to go up and come back down to the same level? (b) What horizontal distance does the object travel in this amount of time?

Answer: (a) 2.0 s, (b) 60 m

Parabolic Motion

The shape of the path taken by the ball, in the last example, or any other projectile for that matter, is a parabola.

Let us examine the parabolic path of a projectile to see how we can control the range by changing the direction of the initial velocity of the projectile, usually called muzzle velocity. Assume that, given a constant magnitude for the muzzle velocity, we can point the muzzle whichever direction we like in order to hit any target within some maximum range. As we shall see, this maximum range happens when we point the muzzle upward at 45°.

It should come as no surprise to find that you can hit a closer target by pointing the muzzle upwards at a sharper angle. Increasing the angle with the horizontal decreases the horizontal component of the muzzle velocity while increasing the vertical component, as shown in Figure 6-4 and Figure 6-5. A smaller horizontal component of velocity should make for a shorter range, since the horizontal distance, $d_\rightarrow$, that something travels while in the air a certain amount of time, t, can be found from the definition of velocity (as we just saw in our discussion of the juggler):

$$d_\rightarrow = v_\rightarrow t$$

Thus, making $v_\rightarrow$ smaller also makes the range smaller.

However, you can do the same thing by decreasing the angle, thereby increasing $v_\rightarrow$. The reason is that a smaller angle to the horizontal has the effect of decreasing the time, t, that whatever we shot up in the air takes to go up and come down.

Figure 6-4

A parabolic trajectory is calculated by resolving the muzzle velocity into horizontal and vertical components.

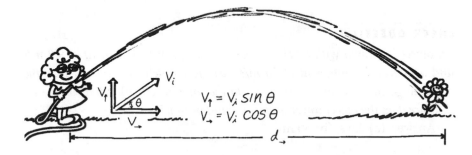

$$V_\uparrow = V_i \sin \theta$$
$$V_\rightarrow = V_i \cos \theta$$

As in our discussion of the juggler, we can use algebra to figure out the time it takes something to go up and come back down to the same level. We just use the formula for distance traveled under constant acceleration,

$$d_\uparrow = \tfrac{1}{2}(-g)\,t^2 + v_\uparrow t$$

and solve it for when the object is back down on the ground again, that is, when $d_\uparrow = 0$. As before, the two solutions for time that satisfy this condition are $t = 0$ when the projectile left the ground, and t equals twice the velocity divided by the acceleration due to gravity, that is, when the object got back:

$$t = \frac{2v_\uparrow}{g}$$

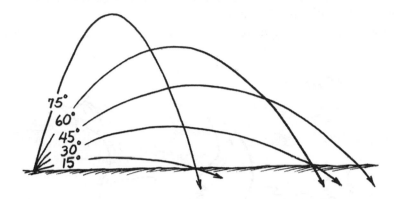

Figure 6-5

Any target within the maximum range can be hit by shooting high or shooting low.

We therefore have an algebraic expression for the range $d_\rightarrow$ in terms of the horizontal component of the velocity and the time, and another expression for the time in terms of the vertical component of velocity and the acceleration due to gravity. Putting these two formulas together, we have a general solution for range:

$$d_\rightarrow = v_\rightarrow \left(\frac{2v_\uparrow}{g}\right)$$

Expressing these components in terms of the angle θ that the initial velocity makes with the horizon, we must use trigonometry to determine where the projectile will land:

$$d_\rightarrow = v_i \cos\theta \frac{2(v_i \sin\theta)}{g}$$

Rearranging, we get

$$d_\rightarrow = \frac{2v_i^2 \sin\theta \cos\theta}{g}$$

This expression is symmetrical with respect to sine and cosine. Since the sine of an angle is the cosine of the complement of that angle, we can replace the angle in the expression with its complement and get the same range. This means that we will hit the same target aiming at θ above the horizontal as we will aiming at $(90° - \theta)$. The maximum range for any given velocity is achieved by shooting at 45° above the horizontal. At that angle, the sine and the cosine are equal.

A stone is thrown from a slingshot with a velocity of 9.8 m/s with a certain upward angle, starting from ground level. (a) How far does the stone go before hitting the ground if the certain angle is 45°? (Remember, sin 45° = cos 45° = $\sqrt{2}/2$.) (b) How far does the stone go if the certain angle is 37°? The sine of 37° is 0.60; the cosine of 37° is 0.80.

Answer: (a) 9.8 m, (b) 9.4 m

Circular Motion

The man in the moon
Came tumbling down,
And ask'd his way to Norwich.
He went by the south,
And burnt his mouth
With supping hot pease porridge.

RITSON'S GAMMER GURTON'S GARLAND *(1784)*

Have you ever wondered what holds the moon up? The gravitational force of the earth certainly extends as far as the moon. In fact the moon's gravitational force acting in the other direction is what causes our ocean tides. If this force of attraction exists, why is it that the moon does not come tumbling down?

The answer is that nothing holds the moon up. The moon is falling. The gravitational attraction of the earth causes the moon to accelerate continually toward the earth. Fortunately, it misses. Instead of falling into, it falls around, the earth. Newton's first law of motion tells us that the moon would travel at constant velocity along a straight line if there were no force acting upon it. The gravitational force acting on the moon causes it to fall *away from a straight line* and into a circle.

We started this chapter talking about the motion of coffee as it is poured in a moving airplane. In all our discussions of the motion of different objects in a gravitational field, we assumed that we were close to the surface of the earth and that the force on the object was always in the same direction, namely down. If, for some reason, you were to throw that cup of coffee up into the air, its path would be a parabola. The force of gravity acting on it would be constant in magnitude and constant in direction in the sense that the force at any one point would be parallel to the force at any other point. If an object is acted upon by a force that is constant in magnitude but directed always perpendicular to its direction of motion, the object will travel in a circle. Both situations are shown in Figure 6-6. The path of the coffee cup, shown in dotted lines, is a parabola. The path of the moon, which travels in a nearly circular path, is shown with the direction of its acceleration always pointing toward the center of the earth and always perpendicular to its direction of motion.

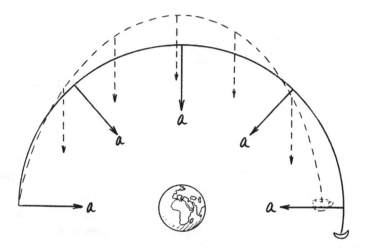

Figure 6-6

Acceleration in circular motion is always perpendicular to the path of the object and toward the center, whereas the acceleration in parabolic motion, shown in dotted lines for comparison, is always in the same (downward) direction.

This perpendicular acceleration produces a change in the moon's velocity—in direction, not magnitude. The moon is 380,000 km from earth today, and it is traveling at 3,700 km/hr. Tomorrow it will still be 380,000 km away and (assuming a circular path) will still be traveling at 3,700 km/hr. After having fallen for a whole day, the moon's velocity will have changed, but only in direction. This change in velocity may be represented as a velocity vector $\Delta\mathbf{v}$, which must be added to the first velocity, $\mathbf{v}_1$, in order to form the new velocity, $\mathbf{v}_2$, as a resultant.[1] This change in velocity can be most easily visualized by drawing a diagram in velocity space, a mathematical space having the same directions as real space but in which length represents speed.

An isosceles triangle (a triangle with two equal sides) is formed in real space by the two radii drawn from the center of the earth and by the distance traveled by the moon. (Actually the base of this "triangle" is the curved part of a circle, but we can think of it as straight if we choose a

---

[1] Remember from Chapter 4 that "Δ," the Greek letter delta, is a mathematical operator that stands for "the change in." In this case, we use vector notation:

$$\Delta v \equiv v_2 - v_1$$

$\Delta\mathbf{v}$ is the vector difference between the velocities.

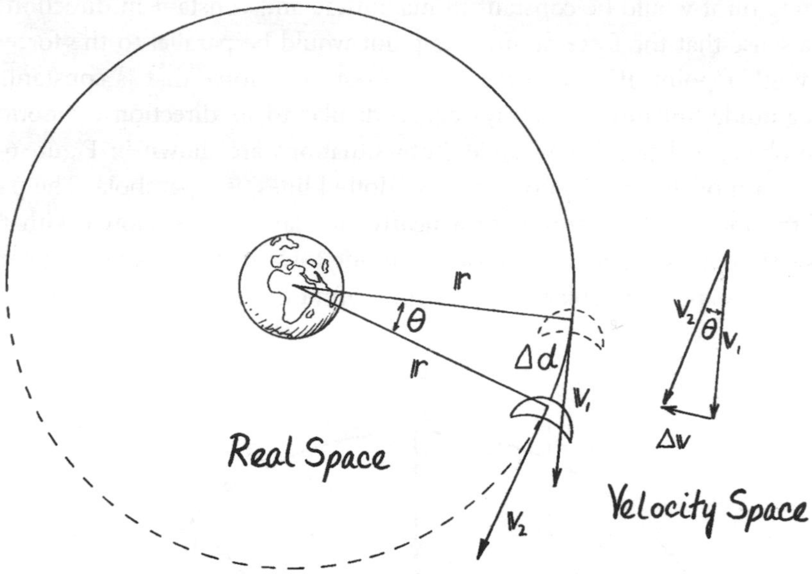

Figure 6-7

The velocity in circular motion changes in direction but not in magnitude. Two successive velocities and their change in velocity form an isosceles triangle similar to a triangle formed by two successive positions (measured from the center) and the change in position.

small enough time interval.) An isosceles triangle is also formed in velocity space by the two velocity vectors and the corresponding change in velocity. The velocity triangle cannot be congruent to the real space triangle—they aren't even in the same space—but it is similar. Each of the two velocities is tangent to the circle in real space and therefore perpendicular to the corresponding radius. Since two lines that are perpendicular to two other lines must meet at the same angle, the apex angle of both isosceles triangles must be the same. That makes them similar.

We can now use these two similar triangles to get the acceleration from the ratio of their sides. It is proper to drop the vector notation at this point, because we are interested in taking the ratio of their magnitudes. Using the fact that the ratio of the sides of two similar triangles are always equal, we

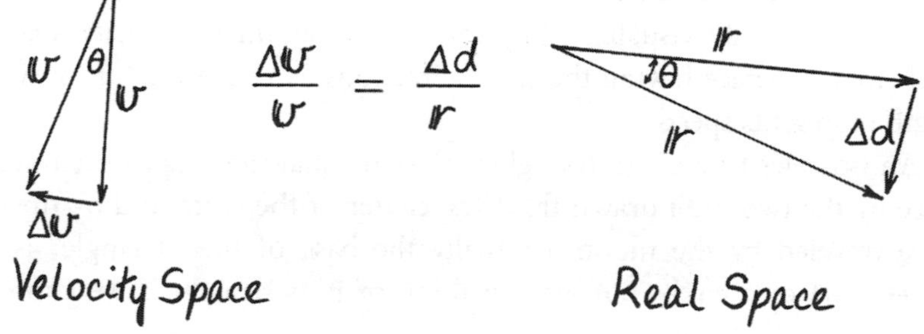

$$\frac{\Delta v}{v} = \frac{\Delta d}{r}$$

Figure 6-8

Similar triangles have proportional sides.

can say that the base of the velocity space triangle is to one side as the base of the real space triangle is to one side. This gives us the relationship:

$$\frac{\Delta v}{v} = \frac{\Delta d}{r}$$

We can eliminate the distance, Δd, by expressing it in terms of the velocity, v, of the moon and the time, Δt, during which the change of velocity takes place. From the definition of velocity, we have

$$\Delta d = v \cdot \Delta t$$

Combining expressions, we get

$$\frac{\Delta v}{v} = \frac{v \cdot \Delta t}{r}$$

Multiplying both sides by v and dividing both sides by Δt, we get

$$\frac{\Delta v}{\Delta t} = \frac{v^2}{r}$$

This is the moon's acceleration toward the center, since the definition of acceleration is the change in velocity over the change in time:

$$\boxed{a_c = \frac{v^2}{r}}$$

This is usually referred to as the **centripetal acceleration**. *Centripetal* comes from the Latin words meaning *to move toward the center.*

The force that causes this acceleration is, from Newton's second law,

$$F = ma_c$$

which is given as

$$\boxed{F_c = \frac{mv^2}{r}}$$

Since this force is toward the center, it is called **centripetal force.**

People often say mistakenly that an earth satellite is held up by something called **centrifugal** force. That cannot be, because *centrifugal* means *to flee the center*. It is true that there is a centrifugal force, but it acts on the earth, not on the satellite. If there were such a thing as centrifugal force acting on the moon in such a direction as to balance the force of gravity, the net force acting on the moon would be zero and the moon would travel in a straight line.[2] It is because the only force acting on the moon is *toward the center* of its path that the moon travels in a circle, being deflected away from straight motion. Therefore, nothing holds the moon up. It is falling.

CHECK QUESTION

An object is traveling at a velocity of 3.0 m/s around a circle having a radius of 1.0 m. (a) What is the centripetal acceleration of the object? (b) If the object has a mass of 2.0 kg, how much centripetal force is necessary to keep the object in its circular path?

Answer: (a) 9.0 m/s^2, (b) 18 N

---

[2]Centrifugal force can also refer to an imaginary force, called an inertial force, caused by our frame of reference accelerating. Something will seem to accelerate even though it doesn't if it is in a frame of reference where everything around it accelerates in the opposite direction. Such an object would require a force to keep it from accelerating in such a frame of reference. Examples are common. Passengers in a car feel a centrifugal force when the car goes around a curve. The car is really pushing them toward the center to deflect their bodies from a straight line. They might interpret this as a mysterious force pulling them away from the center, just as they might feel a mysterious attraction for the front of the car as it stops, but that is only because their frame of reference is accelerating.

PROBLEM SET 6

SUMMARY OF CONCEPTS

The problems in this chapter are related to motion along a curved path. Motion of this type is produced by a force having a component perpendicular to the velocity of the object.

When an external force acting on an object remains in the same direction, such as the force of gravity on the earth's surface pulling always downward on a projectile, the resulting path is a **parabola.** The horizontal motion of a projectile is independent of the vertical motion. A projectile moves at a constant horizontal velocity while experiencing a constant vertical acceleration under the influence of its weight. The horizontal component of its motion is described by

$$d_{\rightarrow} = v_{i\rightarrow}t$$

while the vertical component is described by

$$d_{\downarrow} = \frac{1}{2}at^2 + v_{i\downarrow}t$$

As long as we remain near the surface of the earth, we may regard g as a constant 9.8 m/s^2.

When an external force acting on an object remains always perpendicular to the direction in which the object is going, the resulting path is **circular.** The acceleration changes the direction, but not the magnitude, of the velocity. The acceleration is called **centripetal acceleration**, a_c, since it is always directed toward the center of the circle. Centripetal acceleration is related to the velocity v of the object and the radius r of the circle by

$$a_c = \frac{v^2}{r}$$

The force that produces this acceleration is called **centripetal force.** In accordance with Newton's second law, $F = ma$, the force acting toward the center in circular motion is given by

$$F_c = m\frac{v^2}{r}$$

Ch. 6 : Motion Along a Curved Path

EXAMPLE PROBLEMS

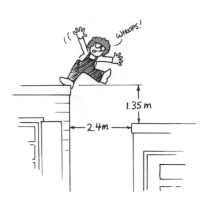

6-1 David Fall is rigging a hoist on the flat roof of his house and manages to knock himself over the edge. Fortunately, the flat roof of his garage some distance away is lower than the roof of his house. He makes a mighty jump in the horizontal direction.

Given: The roof of the garage is 1.35 m lower than that of the house. The garage is 2.4 m from the house.

Find: (a) How much time does David take to fall to the level of his garage?
(b) What horizontal velocity does David need to reach the roof of his garage in time?

BACKGROUND

This problem is an example of projectile motion of the simple kind, where the initial velocity is all in the horizontal direction. At the same time David is falling, he must be moving sideways at a certain rate if he is going to make it to the roof. The motions in the two directions may be thought of as independent. His downward motion is a constant acceleration with a zero initial velocity. His sideward motion is a constant velocity, which must be figured out from the data.

6-2 Daniel is admiring the view from a cliff and wishes to know how high it is above the ocean. Having taken physics, Dan casts a stone horizontally from the cliff and measures the time necessary for it to splash into the sea.

Given: The stone is in the air for 3.0 s and strikes the water 75 m horizontally from where it is thrown.

Find: (a) How high is the cliff above the ocean?
(b) How fast did Dan throw the stone horizontally?

BACKGROUND

This problem is another example of projectile motion of the kind where the initial velocity is all in the horizontal direction. In this case, however, both of the distances in the horizontal and vertical directions are unknown and must be found. You must know which is constant velocity and which is constant acceleration.

6-3 A handball player must meet the ball on the first bounce and return it with the same motion so that it strikes the front wall before hitting the floor again.

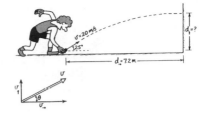

Given: A handball leaves the player's hand with a velocity of 20 m/s at 25° above the horizontal from a point just above the floor and 7.2 m from the front wall.

Find: (a) What are the horizontal and vertical components of the velocity?

(b) How much time will the ball take to reach the front wall, in seconds?

(c) How high up on the wall will it hit?

BACKGROUND

This problem is an example of trajectory motion in which the projectile has an initial velocity with both horizontal and vertical components.

The ball may still have a vertical component as it hits the front wall. A good handball shot is a bounce off the ceiling after the ball has hit the front wall.

6-4 A golf player uses a nine iron when he is close to the green. He wants to hit the ball with a high trajectory so that it will not roll much after landing.

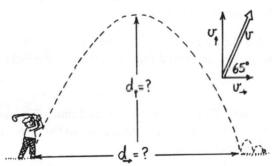

Given: A golf ball leaves the nine iron with a velocity of 35 m/s at an angle of 65° above the horizontal.

Find: (a) What are the horizontal and vertical components of the velocity?

(b) How far does the ball go horizontally before landing?

(c) What is the maximum height of the ball while it is in the air?

BACKGROUND

This problem is an example of a full trajectory, in which a projectile goes up and comes back to its original level while it is moving sidewards at a constant velocity. This special case assumes level ground.

6-5 A pilot can withstand ten times the acceleration due to gravity, 10 *g*, if she is wearing a special tight-fitting *g*-suit designed to keep the blood from rushing from her brain.

Given: An 85-kg pilot who is traveling east at 900 km/hr decides she wants to go west. She makes a circle with 10 *g* acceleration.

Find: (a) What is the radius of the smallest circle in which the plane can turn around?

(b) How much centripetal force acts on the pilot while she is in the turn?

BACKGROUND

This problem illustrates centripetal acceleration and force in circular motion. An object traveling in a circular path must experience a certain acceleration toward the center to deflect it from a straight line, the amount of the acceleration depending on the velocity of its motion and the radius of the circle. This force must result from an unbalanced force, in accordance with Newton's second law.

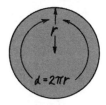

6-6 A little girl riding a merry-go-round thinks she feels a force pulling her away from the center. Actually, the merry-go-round horse is pushing her toward the center to keep her from going in a straight line.

Given: The merry-go-round takes her around in a circle of radius 3.0 m every 7.5 s. Her mass is 25 kg.

Find: (a) What is the magnitude of her velocity at any moment?

(b) How much centripetal force is acting on the little girl?

BACKGROUND

This is another problem illustrating centripetal acceleration in circular motion. The velocity of the object, however, is given in terms of the radius of the path and the time necessary to travel one circumference.

Solutions to EXAMPLE PROBLEMS

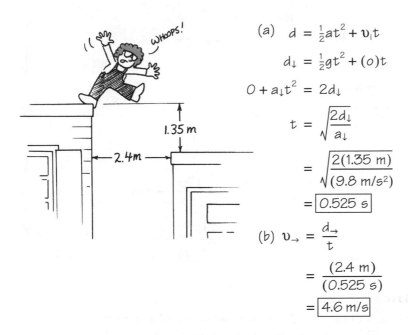

(a) $\quad d = \frac{1}{2}at^2 + v_i t$

$$d_\downarrow = \frac{1}{2}gt^2 + (0)t$$

$$0 + a_\downarrow t^2 = 2d_\downarrow$$

$$t = \sqrt{\frac{2d_\downarrow}{a_\downarrow}}$$

$$= \sqrt{\frac{2(1.35 \text{ m})}{(9.8 \text{ m/s}^2)}}$$

$$= \boxed{0.525 \text{ s}}$$

(b) $\quad v_\rightarrow = \dfrac{d_\rightarrow}{t}$

$$= \frac{(2.4 \text{ m})}{(0.525 \text{ s})}$$

$$= \boxed{4.6 \text{ m/s}}$$

Given:
The roof of the garage is 1.35 m lower than that of the house. The garage is 2.4 m from the house.

Find:
(a) How much time does David take to fall to the level of his garage?

(b) What horizontal velocity does David need to reach the roof of his garage in time?

DISCUSSION

The downward motion follows the rule for distance traveled under constant acceleration where the initial velocity is zero. The initial velocity is zero in the downward direction, since it is all sidewards and has no downward component. The formula for distance under constant acceleration is then solved for time. This time from part (a) can be used in part (b), since both vertical and horizontal motions, while independent, take place at the same time.

SAMPLE SOLUTION 6-2

Given:

The stone is in the air for 3.0 s and strikes the water 75 m horizontally from where it is thrown.

Find:

(a) How high is the cliff above the ocean?

(b) How fast did Dan throw the stone horizontally?

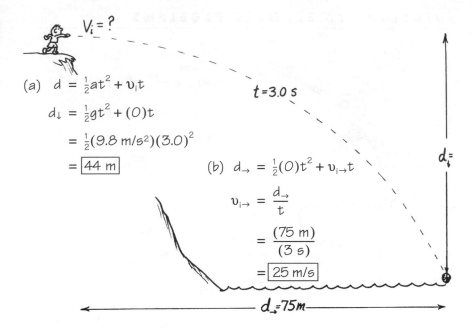

$$V_i = ?$$

(a) $d = \frac{1}{2}at^2 + v_i t$

$d_\downarrow = \frac{1}{2}gt^2 + (0)t$

$= \frac{1}{2}(9.8 \text{ m/s}^2)(3.0)^2$

$= \boxed{44 \text{ m}}$

$t = 3.0 \text{ s}$

(b) $d_\rightarrow = \frac{1}{2}(0)t^2 + v_{i\rightarrow}t$

$v_{i\rightarrow} = \dfrac{d_\rightarrow}{t}$

$= \dfrac{(75 \text{ m})}{(3 \text{ s})}$

$= \boxed{25 \text{ m/s}}$

$d_\downarrow =$

$d_\rightarrow = 75m$

DISCUSSION

The downward distance is found from the formula for motion under constant acceleration. The initial velocity is taken as zero in this part, because only the downward component counts, and the initial velocity is all in the horizontal direction as the problem is stated.

Part (b) may be thought of as being solved with the same formula, but the acceleration in the sideward direction is zero. From this point of view, this formula results in the definition of velocity in the case where velocity is constant.

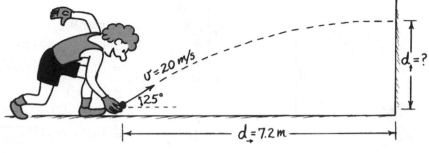

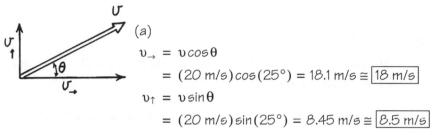

Given:
A handball leaves the player's hand with a velocity of 20 m/s at 25° above the horizontal from a point just above the floor and 7.2 m from the front wall.

Find:
(a) What are the horizontal and vertical components of the velocity?

(b) How much time will the ball take to reach the front wall, in seconds?

(c) How high up on the wall will it hit?

(a)

$$v_\rightarrow = v\cos\theta$$

$$= (20\text{ m/s})\cos(25°) = 18.1\text{ m/s} \cong \boxed{18\text{ m/s}}$$

$$v_\uparrow = v\sin\theta$$

$$= (20\text{ m/s})\sin(25°) = 8.45\text{ m/s} \cong \boxed{8.5\text{ m/s}}$$

(b) $v = \dfrac{d}{t}$

$$t = \dfrac{d_\rightarrow}{v_\rightarrow}$$

$$= \dfrac{(7.2\text{ m})}{(18\text{ m/s})} = \boxed{0.40\text{ s}}$$

(c) $d = \frac{1}{2}at^2 + v_i t$

$$d_\uparrow = \frac{1}{2}(-9.8\text{ m/s}^2)(0.40\text{ s})^2 + (8.45\text{ m/s})(0.40\text{ s})$$

$$= -0.8\text{ m} + 3.4\text{ m}$$

$$= \boxed{2.6\text{ m}}$$

DISCUSSION

The components of the initial velocity are found using the sine and cosine of the angle to the vertical. The time necessary to reach the front wall can be found from the definition of velocity, since the horizontal component of the motion is at a constant velocity.

The vertical motion follows the rule for distance under constant acceleration. The acceleration due to gravity is considered negative in this instance, because it is in a direction opposite to that of the initial velocity. The vertical component of the initial velocity must be used in this equation, since we cannot assume that the ball is at the top of its trajectory as it hits the front wall.

SAMPLE SOLUTION 6-4

Given:

A golf ball leaves the nine iron with a velocity of 35 m/s at an angle of 65° above the horizontal.

Find:

(a) What are the horizontal and vertical components of the velocity?

(b) How far does the ball go horizontally before landing?

(c) What is the maximum height of the ball while it is in the air?

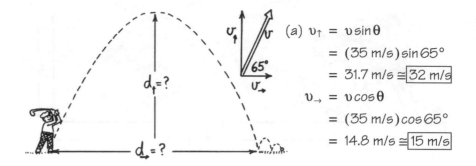

(a) $v_\uparrow = v\sin\theta$

$= (35\text{ m/s})\sin 65°$

$= 31.7\text{ m/s} \cong \boxed{32\text{ m/s}}$

$v_\rightarrow = v\cos\theta$

$= (35\text{ m/s})\cos 65°$

$= 14.8\text{ m/s} \cong \boxed{15\text{ m/s}}$

(b) $d_\rightarrow = v_\rightarrow t$

where we find t at the moment that $d_\uparrow$ returns to $= 0$.

From $d_\uparrow = \frac{1}{2}at^2 + v_i t$

$0 = \frac{1}{2}(-g)t^2 + v_\uparrow t$

$0 = (-\frac{1}{2}gt + v_\uparrow)t$

$t = 0$ at the start

$0 = -\frac{1}{2}gt + v_\uparrow$ at the finish

$t = \dfrac{2v_\uparrow}{g}$

$= \dfrac{2(31.7\text{ m/s})}{(9.8\text{ m/s}^2)} = 6.46\text{ s}$

back to $d_\rightarrow = v_\rightarrow t$

$= (14.8\text{ m/s})(6.46\text{ s})$

$= \boxed{96\text{ m}}$

(c) $v_f^2 = v_i^2 + 2ad$

$0 = v_\uparrow^2 + 2(-g)d_\uparrow$

$d_\uparrow = \dfrac{v_\uparrow^2}{2g}$

$= \dfrac{(31.7\text{ m/s})^2}{2(9.8\text{ m/s}^2)}$

$= \boxed{51\text{ m}}$

DISCUSSION

The time necessary for the ball to return to its original level is found by solving the formula for distance traveled under constant acceleration for the case where the distance is zero. This quadratic equation has two solutions. The trivial solution is when $t = 0$ at the start of the trajectory. The next time that d equals zero is the time that the ball has fallen back to its original level. This second equation should be solved for time.

An alternative approach to part (b) would be to use the general solution derived for this problem in the text of this chapter. That approach is not very good, however, as it involves memorizing an equation that holds true for only the special case of level ground. It would be better to derive that equation when you need it. In part (c), the maximum height of the ball is found from the equation relating the velocities, the distance and the acceleration that was derived in Chapter 4.

(a) $\quad a_c = \dfrac{v^2}{r}$

$\quad\quad r = \dfrac{v^2}{a_c}$

$\quad\quad = \dfrac{(900 \text{ km/hr})^2}{(10g)}\left(\dfrac{1.0g}{9.8 \text{ m/s}^2}\right)\left(\dfrac{10^3 \text{m}}{1 \text{ km}}\right)^2\left(\dfrac{1 \text{ hr}}{3600 \text{s}}\right)^2$

$\quad\quad = 638 \text{ m} \cong \boxed{0.64 \text{ km}}$

(b) $\quad F = ma$

$\quad\quad = (85 \text{ kg})(10g)\left(\dfrac{9.8 \text{ m/s}^2}{g}\right)\left(\dfrac{1 \text{ N}}{1 \text{ kg m/s}^2}\right)$

$\quad\quad = \boxed{8.3 \times 10^3 \text{N}}$

$a = 10g$

Given:

An 85-kg pilot who is traveling east at 900 km/hr decides she wants to go west. She makes a circle with 10 g acceleration.

Find:

(a) What is the radius of the smallest circle in which the plane can turn around?

(b) How much centripetal force acts on the pilot while she is in the turn?

DISCUSSION

The radius of the circle can be found by solving the centripetal acceleration formula. The acceleration is given in terms of the acceleration due to gravity and must be changed into meters per second squared using a conversion factor. The units of velocity must also be changed into meters per second to make things match.

The force necessary to produce this acceleration can be found from Newton's second law.

SAMPLE SOLUTION 6-6

Given:

The merry-go-round takes her around in a circle of radius 3.0 m every 7.5 s. Her mass is 25 kg.

Find:

(a) What is the magnitude of her velocity at any moment?

(b) How much centripetal force is acting on the little girl?

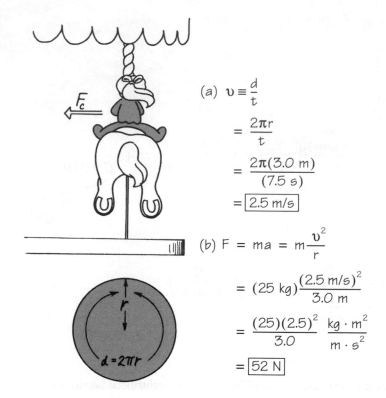

(a) $v \equiv \dfrac{d}{t}$

$\quad = \dfrac{2\pi r}{t}$

$\quad = \dfrac{2\pi(3.0\ \text{m})}{(7.5\ \text{s})}$

$\quad = \boxed{2.5\ \text{m/s}}$

(b) $F = ma = m\dfrac{v^2}{r}$

$\quad = (25\ \text{kg})\dfrac{(2.5\ \text{m/s})^2}{3.0\ \text{m}}$

$\quad = \dfrac{(25)(2.5)^2}{3.0}\ \dfrac{\text{kg} \cdot \text{m}^2}{\text{m} \cdot \text{s}^2}$

$\quad = \boxed{52\ \text{N}}$

DISCUSSION

Velocity is defined as distance over time. Since the time of one complete revolution is given, the proper distance to use is one circumference of the circle. The circumference of a circle is 2π times the radius of the circle.

Part (b) is found from Newton's second law, where the acceleration is given by the centripetal acceleration formula.

ESSENTIAL PROBLEMS

Be sure you can do the
Example Problems (without
peeking at the Sample
Solutions) before working
these Essential Problems.

6-7 The head of a hammer flies off the handle and sails horizontally from a tall building. The hammer head breaks a window 20 stories below on a neighboring building some distance away.

Given: The window is 50 m below and 35 m horizontally from the worker.

Find: (a) How much time elapses between the loss of the hammer head and the breaking of the window?
(b) What is the initial horizontal velocity of the hammer head?

6-8 A trapeze artist lets go at the bottom of his swing when his velocity is all in the horizontal direction. A trampoline has been placed on the floor at the correct horizontal distance to catch him.

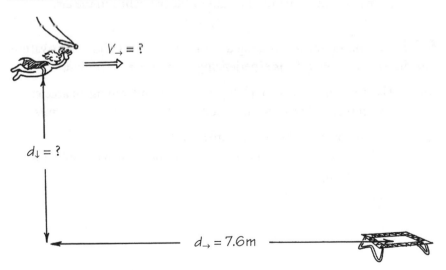

Given: He is in the air for 1.1 s. The trampoline is placed 7.6 m away horizontally.

Find: (a) What vertical distance does he fall in that amount of time?
(b) What was the trapeze artist's initial velocity in the horizontal direction?

6-9 A short basketball player shoots the ball upward with the correct horizontal and vertical velocity components to bounce it from the backboard into the net.

Given: The basketball player throws the ball at a velocity of 9.9 m/s at an angle of 32° from the horizontal. She is standing 3.4 meters from the hoop.

Find: (a) What are the horizontal and vertical components of the velocity?
(b) How much time will the ball take to reach the backboard?
(c) How high did the player throw the ball, or, what is the vertical distance between where the player let go of the ball and where the ball hit the backboard?

6-10 A football player starts the game by kicking the ball into the air from the ground. It is caught on the first bounce by a player on the other team, who was clever enough to judge the ball's initial velocity and forecast its landing place.

Given: The ball has an initial velocity of 25.0 m/s at an angle of 52.0° above the horizontal.

Find: (a) What are the initial horizontal and vertical components of velocity?
(b) How far does the ball go before landing?
(c) What is the maximum height of the ball while in the air?

6-11 An automobile is rounding a turn of constant radius of curvature. The driver notices that she is experiencing sidewards acceleration.

Given: The driver has a mass of 60 kg, and she is undergoing an acceleration of 0.25 *g*. The forward velocity of the automobile is 26 m/s.

Find: (a) What is the radius of curvature of the turn?
(b) How much centripetal force acts on the passenger while the car is turning?

6-12 A bug is sitting on a long-playing record going around at a constant speed.

Given: The record is turning at a constant rate of 33.3 revolutions per minute. The bug, whose mass is 1.4 *g*, is sitting 12 cm from the center of the record.

Find: (a) How long does it take the bug to go around once?
(b) How fast is the bug going? (Hint: Figure out the distance associated with part (a).)
(c) How much centripetal force acts on the bug?

MORE INTERESTING PROBLEMS

6-13 An automatic pistol fires a bullet horizontally at the same time that it kicks the brass shell casing out the side of the gun, also with a horizontal velocity. Both bullet and shell casing hit the ground at the same time.

Given: The pistol is 1.4 m above level ground. The bullet hits the ground 250 m downrange. The shell casing has an initial velocity of 0.8 m/s in the horizontal direction.

Find: (a) What is the time that both the bullet and shell casing are in the air?
(b) What is the muzzle velocity of the bullet?
(c) How far does the shell casing hit the ground from a point directly beneath the gun?

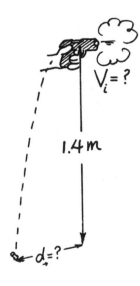

6-14 Another day at the bar. A flying shot glass hits a bottle of whisky behind the bar and topples it. The shot glass bounces off the bottle and lands some distance away, while the bottle falls off the shelf with a much smaller horizontal velocity.

Given: The shot glass lands 2.4 meters away from where it originally hit the whisky bottle. The bottle is propelled sidewards off the shelf at a velocity of 1.2 m/s. Both objects started from a height of 0.80 m.

Find: (a) How long, in seconds, are the shot glass and the whisky bottle in the air?
(b) What is the horizontal velocity of the shot glass after it leaves the whisky bottle?
(c) How far does the whisky bottle land horizontally from the shelf?

6-15 A juggler tosses a ball in the air while she is walking across the stage. The horizontal component of the ball's velocity is the same as that of the juggler, so that, to the juggler, the ball seems to simply go up and come back down.

Given: The initial upward component of the ball's velocity is 15 m/s. Both the ball and juggler have a horizontal velocity of 0.85 m/s.

Find: (a) How long does it take the ball to reach the top of its trajectory?
(b) How long does it take for the ball to go up and come back down?
(c) How far does the juggler walk between the time she tosses the ball and the time she catches it?

6-16 Rosa flips a coin while riding in an airplane. To her, it looks like the coin goes straight up and comes straight back down to her hand.

Given: The initial upward velocity of the coin is 2.6 m/s. The airplane is traveling at 179 m/s.

Find: (a) How long does it take for the coin to reach the top of its trajectory?
(b) How long does it take the coin to go up and come back down?
(c) How far do Rosa and the coin (and the airplane) travel horizontally in the time it takes her to throw the coin and catch it?

6-17 A cannon fires a cannon ball on level ground. Once the muzzle velocity has been determined by the gunpowder charge, the range may be adjusted by changing the elevation or angle above the horizontal at which the cannon is fired.

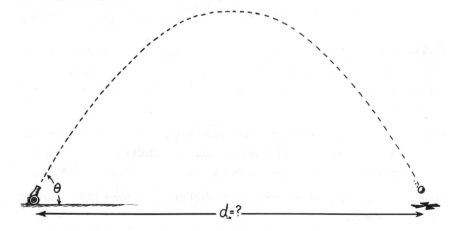

Given: The cannon ball has a muzzle velocity of 200 m/s at an angle of 27° above the horizontal.

Find: (a) How far does the cannon ball go horizontally before hitting the ground?
(b) What other elevation will give the same range?

6-18 Old Joe is watering his garden with a hose. He has found that, to water a particular azalea bush from where he is standing, he can hold the hose at either of two different angles.

Given: The hose has a "muzzle velocity" of 4.4 m/s. Joe is directing the stream from the hose at an angle of 69° above the horizontal.

Find: (a) How far is Joe from the azalea bush?
(b) At what other angle can Joe hold the hose to water the same bush?

6-19 Romeo decides to toss a note to Juliet while she is standing on her balcony. He has weighted it with a rock so that he can throw it properly. He stands back from her house and throws it at a slight angle so that she can catch it at the top.

Given: The note and attached rock have an initial horizontal velocity of 2.8 m/s and an initial upward velocity of 6.8 m/s. The note is in the air for 0.58 s.

Find: (a) How fast and in what direction did Romeo throw the rock?
(b) What is the vertical distance between Romeo and Juliet?
(c) What is the vertical velocity of the rock just before Juliet catches it?

6-20 Horace is vertically challenged. He is playing darts but is significantly shorter than the person who set up the dart board, so he has to throw the darts at an upward angle to hit the board.

Given: A typical one of Horace's darts has an initial horizontal velocity of 12.5 m/s and an initial upward velocity of 2.40 m/s. It is in the air for 0.28 s before it hits the board.

Find: (a) What is the magnitude and direction of the resultant initial velocity of the dart?
(b) How far upwards did the dart travel from Horace's hand to the dart board?
(c) What is the vertical velocity of the dart just before it hits the dart board?

6-21 An automobile is rounding a turn of constant radius of curvature. A passenger, even with his eyes closed, notices that the armrest is pushing him toward the center of the turn with a constant force.

Given: The passenger has a mass of 78 kg, and the force with which the armrest is pushing against him is 150 N. The forward velocity of the automobile is 21 m/s.

Find: (a) If the armrest is exerting the only horizontal force on the passenger, how much acceleration is produced?
(b) What is the radius of curvature of the turn?

6-22 There is an amusement park ride with a train that goes around a circular track very fast. On this ride, Esmeralda is pushed quite strongly against the side of the train by the centripetal force.

Given: The train is traveling at 16.8 m/s. Esmeralda is being pushed against the side of the train with a force of 2,519 N. Esmeralda's mass is 91.5 kg.

Find: (a) What is the centripetal acceleration acting on Esmeralda?
(b) What is the radius of the circular track?

6-23 A biologist uses a centrifuge to separate cell cultures. A centrifuge is a disc that spins very fast. Test tubes with cell cultures are placed on the edges of the disc. The cells separate because heavier cells are thrown to the outside of the disc with greater force than lighter cells.

Given: The centrifuge produces an acceleration of 980 m/s². The disc has a 15 cm radius. The sample in a particular test tube has a mass of 3.8 *g*.

Find: (a) What is the force on the sample?
(b) What is the velocity of the test tube containing the sample?
(c) How many revolutions per minute does the centerfuge produce?

6-24 Part of changing the piston rings of a car engine is to "hone" the cylinders. This is done with a tool that is essentially an electric drill with a bit attached to three abrasive stones which are spun against the inside surface of the cylinder.

Given: The drill produces an acceleration of 998 m/s². The stones are rotated at a radius of 5.0 cm. Each stone has a mass of 5.2 *g*.

Find: (a) What is the centripetal force on one stone?
(b) What is the velocity of the stones?
(c) How fast are the stones moving in revolutions per minute?

ANSWERS

Solutions to **ESSENTIAL PROBLEMS**

6-7　(a) 3.2 s; (b) 11 m/s

6-8　(a) 5.9 m; (b) 6.9 m/s

6-9　(a) 8.4 m/s, 5.2 m/s; (b) 0.40 s; (c) 1.32 m

6-10　(a) 15.4 m/s, 19.7 m/s; (b) 61.9 m; (c) 19.8 m

6-11　(a) 276 m $\cong 2.8 \times 10^2$ m; (b) 147 N $\cong 1.5 \times 10^2$ N

6-12　(a) 1.8 s; (b) 0.418 m/s $\cong$ 42 cm/s; (c) 2.0×10^{-3} N

Solutions to **MORE INTERESTING PROBLEMS,** *odd answers*

6-13　(a) 0.53 s; (b) 468 m/s $\cong 4.7 \times 10^2$ m/s; (c) 0.43 m

6-15　(a) 1.5 s; (b) 3.1 s; (c) 2.6 m

6-17　(a) 3.3 km; (b) 63°

6-19　(a) 7.4 m/s, 68°; (b) 2.3 m; (c) 1.1 m/s

6-21　(a) 1.9 m/s^2; (b) 229 m $\cong 2.3 \times 10^2$ m

6-23　(a) 3.7 N; (b) 12 m/s; (c) 772 rpm $\cong 7.7 \times 10^2$ rpm

CHAPTER 7 Momentum

FOR THE LAST SEVERAL CHAPTERS, we have been investigating the effects of something called force. In Chapter 3, we saw how force stretches things like springs and ropes, force being proportional to stretch according to the rule first stated by Robert Hooke. That elementary idea of this thing called force served while we looked at situations where forces combine, adding together as vectors. Then, in Chapter 5, we saw how force is related to change in motion according to a couple of rules named after Issac Newton. Those rules served in Chapter 6 while we looked at motion along curved paths, where the forces are constant in magnitude if not direction.

We now need to consider more sophisticated aspects of the force concept in order to deal with situations where force is anything but constant. Even with rapidly changing forces, there are quantities related to force, quantities called momentum and energy, that do remain constant if you look at the larger picture. These are different quantities, but they are so closely related that they are frequently confused with each other.

We will deal with momentum first. We might have taken the other approach. The advantage of talking about energy first is that, when you do get to momentum, there are some very nice applications that combine both ideas. The disadvantage is that discussing energy first tends to give the false impression that momentum is one of the many different forms that energy can take. To avoid that conceptual danger, we will deal with momentum before even introducing the idea of energy.

Looking at the larger picture, by the way, is not necessarily the best approach to understanding. Keeping the larger picture in mind, at least to be aware somewhere in the back of your mind that there is a larger picture, may be a good thing. But looking at the narrower picture is the way of focusing your attention on part of a problem at a time. That is what we have been doing by looking only at the object which has had a force applied to it. It is now time to enlarge our view to consider the entire interaction—the relationship between the thing that applies the force and the thing receiving it. It turns out that these two things affect each other equally. This is a lesson of Newton's third law of motion.

Newton's Third Law

Nothing can exert a force on an object without that object exerting an equal force back. We cannot touch something without that thing touching us. We cannot nudge someone without that person nudging us back. Forces always come in pairs. This comment on the very nature of force was first expressed by Sir Issac Newton and is called Newton's third law of motion:

For every force of action, there is an equal but opposite force of reaction on another object. It does not matter which is the **action** force and which is the **reaction,** as long as we realize that neither exists without the other. If the force exerted by a bat on a ball is the action, the force exerted by the ball on the bat is the reaction. Which is which depends on your point of view. You may wish to think of the bat as pushing the ball in the positive direction. Then the ball pushes the bat in the negative direction. That is actually the ball's view of the world. It got this big force that not only stopped it, but changed its direction. It got even, however, and pushed back on that bat in return. The bat, on the other hand, sees things in reverse. It was going along nicely when this ball came out of nowhere and gave it a big smack, slowing it down substantially. But it got even. It kicked that ball right out of the park.

If action is positive, then reaction is negative in direction, but equal in magnitude:

$$F_a = F_r$$

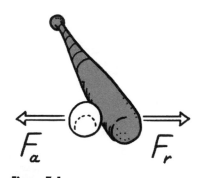

Figure 7-1

For every force of action there is an equal and opposite force of reaction.

It is important to realize that the action and reaction forces act on different objects. When we first considered forces, we looked at situations where forces are balanced and the acceleration is zero. We are now looking at unbalanced forces, and the acceleration is not zero.

For the ball and the bat, the acceleration is substantial. The two objects do not have the same acceleration, but the acceleration is definitely not zero for either object. Of course, the bigger object will experience the smaller change in acceleration. Applying Newton's second law to both the action and reaction forces, we see that the backward acceleration of the bat is proportional to the forward acceleration of the ball.

$$F_{\text{ball}} = -F_{\text{bat}}$$

$$m_{\text{ball}} a_{\text{ball}} = -m_{\text{bat}} a_{\text{bat}}$$

We can see that the accelerations produced by the collision are inversely proportional to the ratio of the masses. The more massive the bat, the less it will accelerate backwards relative to the forward acceleration of the ball.

$$a_{\text{bat}} = \left(\frac{m_{\text{ball}}}{m_{\text{bat}}}\right)a_{\text{ball}}$$

We can extend this argument to astronomical proportions. It leads to the most amazing conclusions when one object is astronomical and the other is not. When you stomp hard on the earth, you feel it solidly under your foot. The earth is in fact pushing back on your foot with the same magnitude of force that you pushed on it. Most people do not have too much trouble envisioning this. A more amazing phenomenon takes place when the earth pulls back on something that is not touching it. If a particular physics teacher were to leap off his desk, he would be accelerated downward by the force of gravity. The earth, in turn, will be accelerated upward by the same magnitude of force. Say our physics teacher has a mass of 100 kilograms. When he is in the air after leaping off his desk, he will be accelerated downward by his unbalanced weight of $(100 \, \text{kg})(9.8 \, \text{m/s}^2)$, or 980 N, and the unbalanced reaction force, then, is 980 N pulling the earth upward. The acceleration of the earth is not noticed because the earth is vastly more massive than the physics teacher. We can, however, calculate the upward acceleration of the earth in response to the force pulling the physics teacher to the ground:

$$F_{\text{teacher}} = -F_{\text{earth}}$$

$$m_{\text{teacher}} a_{\text{teacher}} = -m_{\text{earth}} a_{\text{earth}}$$

$$a_{\text{earth}} = -\left(\frac{m_{\text{teacher}}}{m_{\text{earth}}}\right)a_{\text{teacher}}$$

$$= -\left(\frac{10^2 \, \text{kg}}{10^{25} \, \text{kg}}\right)g$$

$$= -10^{-23} g$$

$$= -10^{-22} \text{m/s}^2$$

Since the mass of the earth is on the order of 10^{25} kg and the mass of the physics teacher is 100 kg, which is 10^2 kg, the acceleration of the physics teacher will be about 10^{23} times as great as that of the earth. The earth therefore accelerates upward at $10^{-23}\, g$, which is 10^{-22} m/s. Although this acceleration is far too weak to measure with even the most delicate instrument, it nonetheless exists.

Momentum

The mass of an object is of great importance in determining the effect of its interaction with other objects. Intuitively this is to be expected. When a massive truck interacts with a small car, we expect the larger vehicle to win the argument, so to speak, but we also expect the car to have at least some effect on the larger vehicle. Suppose they approach each other, each traveling with a velocity of 10 m/s; the smaller vehicle has a mass of 1,000 kg and the truck a mass of 5,000 kg. Further, suppose the wreckage fuses together after the collision and moves off with a single final velocity, v_f. Let us see how we can predict this velocity.

Figure 7-2

Two massive objects approaching each other each have a cer tain amount of momentum.

Newton's third law tells us that the force of action F_{car} on the small car is equal and opposite to the force of reaction F_{truck} on the truck:

$$F_{car} = F_{truck}$$

Applying Newton's second law, $F = ma$, to this expression, we obtain the following relationship between the accelerations:

$$m_{car}a_{car} = m_{truck}a_{truck}$$

We will call one velocity positive and the other negative. (This is just the same trick we have been using all along to deal with vector quantities that are in the same or opposite directions. Since vectors in the same direction add numerically, and those in the opposite direction subtract, we can treat them just like scalars as long as we choose a positive direction.) Let us say that right is positive and left is negative. That makes the initial velocity of the car v_{car} positive and the velocity of the truck v_{truck} negative. The acceleration of each vehicle can be expressed in terms of the change of velocity to its final value, v_f, from its initial value, v_i.

$$a = \frac{v_f - v_i}{t}$$

Thus, the average acceleration of the car and the truck can be stated as follows:

$$m_{car}\left(\frac{v_f - v_{car}}{t}\right) = -m_{truck}\left(\frac{v_f - v_{truck}}{t}\right)$$

The time of acceleration, t, is the same for the car as for the truck, since the force of action and the force of reaction both start and stop at the same instant. We can therefore eliminate the time term by multiplying both sides of the equation by t:

$$m_{car}(v_f - v_{car}) = -m_{truck}(v_f - v_{truck})$$

Thus, the mass times the change in velocity is the same for the car as it is for the truck. This happens to be a general result that holds for balls and bats or any other two things that interact. Always the mass times the change of velocity of one is equal to the mass times the change of velocity of the other. You might expect at this point that there is something special about the product of mass and velocity. If so, you are quite right—as we can see by continuing our example to get everything in terms of initial and final values.

Figure 7-3

The velocity at the time of collision can be found from the fact that the momentum after the collision is equal to the vector sum of the momentums before.

We can manipulate the preceding equation to express everything as the products of masses and velocities:

$$m_{car}v_f - m_{car}v_{car} = -m_{truck}v_f + m_{truck}v_{truck}$$

Moving those terms containing v_f to one side of the equation,

$$m_{car}v_f + m_{truck}v_f = m_{car}v_{car} + m_{truck}v_{truck}$$

If we factor out v_f,

$$(m_{car} + m_{truck})v_f = m_{car}v_{car} + m_{truck}v_{truck}$$

we see that the total mass of the car and truck, the final mass of the wreck, times its final velocity turns out to be the sum of the products of the initial masses and velocities of the car and the truck. This is an important expression. It is true in general. The final product of mass and velocity is always equal to the sum of the initial products of mass and velocity.

This expression allows us to predict, among other things, the final velocity of two things after they collide. We have been looking at the simple case in which they stick together after collision, but it can be extended to any case. Before doing so, however, let us finish our example. At this point we need to remember to use a negative value for the truck's velocity. (We assumed that it was negative to start with, but we only used the sign convention to work out the algebra. The algebraic result is therefore quite general, and we need to go back to the sign convention when we insert particular values.)

$$(1{,}000 \text{ kg} + 5{,}000 \text{ kg})v_f = (1{,}000 \text{ kg})(10 \text{ m/s}) + (5{,}000 \text{ kg})(-10 \text{ m/s})$$

$$(6{,}000 \text{ kg})v_f = (10{,}000 \text{ kg} \cdot \text{m/s}) - (50{,}000 \text{ kg} \cdot \text{m/s})$$

$$v_f = \frac{(10{,}000 - 50{,}000)\text{kg} \cdot \text{m/s}}{6{,}000 \text{ kg}}$$

$$v_f = -6.7 \text{ m/s}$$

The minus sign means that the final velocity of the wreck is in the original direction of the truck, since we had decided that the truck's initial velocity was negative. We could have made the truck's velocity positive and the car's negative, but the result would have been the same. The velocity of the wreck would still have been in the truck's direction.

The wreck does not have to go in the direction of the more massive object. The small car could have made up for the fact that it only has one-fifth the mass of the truck by going five times as fast. Then we would substitute this new value of velocity into the expression:

$$(1{,}000 \text{ kg} + 5{,}000 \text{ kg})v_f = (1{,}000 \text{ kg})(50 \text{ m/s}) + (5{,}000 \text{ kg})(-10 \text{ m/s})$$

$$(6{,}000 \text{ kg})v_f = (50{,}000 \text{ kg} \cdot \text{m/s}) - (50{,}000 \text{ kg} \cdot \text{m/s})$$

$$= 0$$

In this case, the velocity of the wreck would turn out to be zero:

$$v_f = 0$$

If the car were going faster than five times the velocity of the truck, the velocity of the wreck would be in the direction of the car. Perhaps this is the reason people drive small cars so fast.

We have seen that the ultimate result depends upon not only the mass of the colliding objects but upon the products of their masses and velocities. This product is called the **momentum** of the object.

$$\text{Momentum} = mv$$

One commonly used convention is to let the letter *p* stand for momentum (since *m* is already taken for mass). Rather than introduce a new symbol in this book, however, we will either spell it out or use "*mv*" to stand for momentum.

CHECK QUESTION

A woman whose mass is 60 kg is running with a speed of 2.0 m/s. What is her momentum?

   Answer: 120 kg · m/s

Impulse

Momentum is important, because it is absolutely conserved in all interactions that we know of. That means, if something like a ball gets momentum, something like a bat loses it, and always loses it in exactly the same amount. If a car and truck have momentums that are equal in magnitude but opposite in direction before a collision, when we take into account their directions, they have a total of zero momentum both before and after the collision. We know of no interactions where the total momentum in the system changes.

   In both of these examples, the ball interacting with the bat and the car interacting with the truck, the interaction is sudden and brief in time. The idea that momentum is conserved is particularly useful in situations where time is short. No matter how brief the period of time that two objects act on each other, no matter how complicated the relationship between force and time, we always find that the amount of momentum before the interaction is equal to the amount of momentum after.

   The fact that the total momentum in a system is always the same before and after an interaction may be seen as a direct consequence of Newton's laws of motion. In fact, Newton's second law was originally expressed in terms of momentum in such a way that it is a little easier to see why it must be conserved. Today, we are fond of expressing Newton's second law by saying that force equals mass times acceleration.

$$F = ma$$

Newton, however, didn't say it like that. Mass, after all, was not a concept his readers had firmly fixed in their minds. Momentum, however, was an idea that Newton's readers were discussing at the time. Using the language of his day, Newton stated his second law by saying that **force is proportional to the rate of change of momentum**. Using modern units, we can say "equal" instead of "proportional."

$$F = \frac{\Delta(mv)}{\Delta t}$$

We can see that this form of Newton's second law is equivalent to the conventional mass times acceleration form by expanding the change of momentum

$$F = \frac{mv_f - mv_i}{\Delta t}$$

and factoring out the mass, assuming that mass remains unchanged in the interaction.

$$F = \frac{m(v_f - v_i)}{\Delta t} = m\frac{\Delta v}{\Delta t} = ma$$

The result is that the force is mass times acceleration, the form of the equation that we are used to.

Newton's original form, force being the rate of change of momentum, has some distinct advantages over the more familiar form. For one thing, it is more general. Even when mass does change in an interaction, as is the case when we take Einstein's relativity into account, Newton's original form is still valid. Another advantage is that we can recast Newton's second law by putting the change of time on the same side of the equation as force.

$$F\Delta t = \Delta(mv)$$

This is such a nice way of looking at Newton's second law that people frequently give a name to the left-hand side, force times an interval of time. That name is **impulse**. Using this definition we can restate Newton's second law by saying that **impulse is equal to the change of momentum**.

$$\text{Impulse} = \Delta(\text{Momentum})$$

If we assume that Newton's third law, which states that the force of action is equal but opposite to the force of reaction

$$F_a = -F_r$$

is true for every moment in time, we can multiply both sides by the same increment of time to say that every impulse of action is equal to an equal but opposite impulse of reaction.

$$F_a\,\Delta t = -F_r\,\Delta t$$

It follows that every change in momentum of action is equal but opposite to a change in momentum of reaction somewhere else in the system.

$$\Delta(\text{Momentum})_a = \Delta(\text{Momentum})_r$$

Thus, the total amount of momentum in any system obeying Newton's laws of motion remains constant.

Impulse in Games of Hitting

The idea of impulse is particularly useful for understanding situations where a force acts for a very short period of time. To one who did not know better, such an interaction might seem instantaneous. It might seem that the time interval is so close to zero that its size does not matter. In fact, the opposite is true. The time interval is extremely important in these situations.

Consider, for example, a game in which you are supposed to hit a ball with a stick. There are a number of such games, but suppose there were one in which you are supposed to walk after the ball, hitting it again and again, using different sticks along the way, until you lose it. Then you put down another ball and hit it, making it follow, not too roughly, a prescribed path through an obstacle course.

$$F \cdot \Delta t = \Delta(mv) \qquad F \cdot \Delta t = \Delta(mv)$$

Figure 7-4

A golfer improves the momentum transference with a smooth follow-through.

As unlikely as it might seem to those with little experience in such games, many of us have carried whole bags of sticks for our elders to provide them with a wide variety of sticks from which to choose. Some of us even carried sticks so that we might afford to take lessons ourselves on how best to hit balls with sticks. It might seem an obvious thing, but people actually make a profession of teaching others this skill.

When the young and strong cannot hit the ball as far as the old and wise, there is clearly something to be learned from these professionals. Now imagine you are taking such a lesson and the professional criticizes your swing of the stick on the basis of something called "follow through." If one does not believe in psychokinesis (moving objects with the mind), one might tend to disregard what happens to the stick after the ball is gone, but a little experimentation will demonstrate that there is wisdom in the old professional's advice. The distance the ball goes does in fact seem to depend on what happens to the stick at the end of the swing. How can this be understood without resorting to theories based on mind control of distant matter?

The key to understanding is to appreciate the amount of time that colliding things are in contact with one another. Following through does not increase the amount of force, but it affects the length of time the force is applied. The longer the time, the greater the momentum imparted to the ball. The position of the stick long after the ball has left can reveal what was going on in the short period when they were in contact.

We can visualize the effect of follow-through with a graph of force versus time. The areas under the curves in Figure 7-5 represent the change in momentum $\Delta(mv)$ resulting from a force applied over a period of time. In

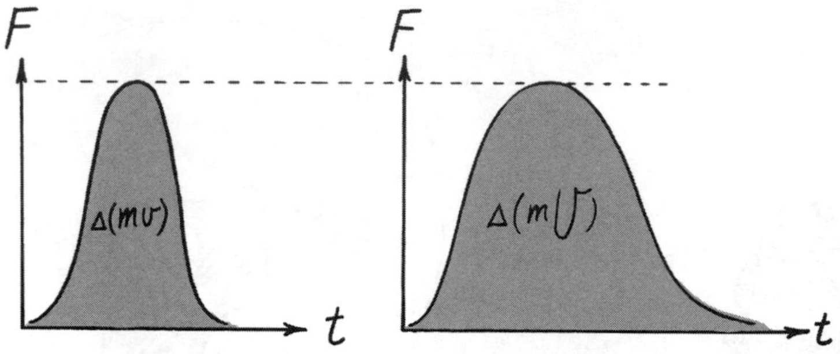

Figure 7-5

Although the maximum force is the same, the longer the club is held in contact with the ball, the greater the impulse.

the simplest approximation, we can think of the impulse area as being a rectangle bounded by time, Δt, and average force, F_{ave}. This approximation is adequate for calculating the average force in a situation. For example, suppose a full swing takes 1.3 seconds and results in a 0.1-kilogram ball traveling 200 meters in about 5 seconds. Let us see if we can use these reasonable figures to tell how much force had to be applied to the ball in the short time it was in contact with the head of the stick. We can figure out the momentum of the ball from the mass and velocity. Its mass is given, and its velocity is such that it travels 200 meters in 5 seconds. Thus, its velocity must be roughly 40 meters per second:

$$v \equiv \frac{d}{\Delta t} = \frac{200 \text{ m}}{5 \text{ s}} = 40 \text{ m/s}$$

Now that we know the momentum, at least in principle, we can estimate the time, Δt, that the stick head is in contact with the ball. If it is in contact while the stick swings through about 10°, and if the full swing goes through about $1\frac{1}{4}$ of a full circle—or about 450°, and if, as above, a full swing takes 1.3 seconds, then the time of contact is approximately 0.03 seconds:

$$\Delta t = \left(\frac{10°}{450°}\right)1.3 \text{ s} \cong 0.03 \text{ s}$$

We showed in the last section that one way to state Newton's second law is to say that impulse equals change in momentum:

$$F\Delta t = \Delta(mv)$$

We can solve this expression for force, F, since we have figured out everything else:

$$F = \frac{\Delta(mv)}{\Delta t}$$

$$= \frac{(0.1 \text{ kg})(40 \text{ m/s})}{1.03 \text{ s}} \cong 1.3 \times 10^2 \text{ N}$$

This force is the average force with which the stick pushes on the ball. The actual force would first increase to some peak value and then decrease to zero at the end of the time of contact. Although the actual force is a complicated function of time, we can approximate the momentum transfer produced by it with a rectangle having the same area in force-time space.

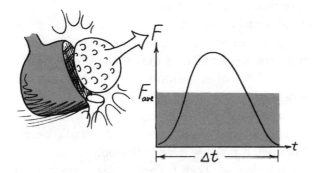

Figure 7-6

The impulse applied to a golf ball can be approximated by a rectangle on a force time graph.

(Since impulse equals force times time, area in force time space represents momentum transfer.) We should recognize that the actual peak value of the force might be something like twice the average value. Since the average force is about 130 N, which would be about 30 pounds, the peak force would be on the order of twice this much.

We therefore see that a stick hitter can achieve through skill a maximum momentum transfer by prolonging the time of contact as long as possible. This same skill becomes important in other ball-and-stick games such as baseball and tennis, where the player wants to increase the momentum, and hence the speed, of the ball while using a certain amount of force.

A related skill is involved in sports where a force is applied with the bare, or nearly bare, hand. A karate expert or boxer wants to increase the amount of force that is imparted through a blow. To increase the amount of force, the fighter will want to increase the amount of momentum available for transfer. A karate expert or boxer learns to put as much of the arm and body behind the blow as possible in order to increase the effective mass. Another way to increase the effective mass is to "think through" the point of impact—not, as for the stick hitter, to increase time of impact, but as a mental trick to keep the force of the blow throughout the entire karate chop or punch.

In sports where maximum force is the objective, there is an advantage in actually decreasing the time of impact. A karate expert, for example, can generate just a certain amount of momentum depending on the mass of his hand, arm, and body and his strength in setting those things in motion. Given a certain fixed amount of available momentum, the force of impact can be increased by delivering a sharp blow that transfers that amount of momentum in as short a time as possible. This can be seen by looking again at the equation for impulse. If the time decreases while the momentum stays the same, the force will increase. If the karate expert wants to break a board, he has the board solidly supported so that it will give as little as possible before breaking. This helps in reducing the time to transfer momentum. Delivering the blow with the hard part of the hand or foot also helps to shorten the time for the momentum transfer. Viewed in terms of a force-time graph, the idea is to squeeze the momentum area along the time axis to produce a high peak force.

The same idea can be used in reverse by the recipient of the blow. A boxer, for example, frequently wishes to reduce the average force of a blow. Not his own, of course, but that of the other fellow. A professional boxer learns to put a good deal of momentum of both his arm and body behind a punch he is delivering, just as does the karate expert. The average street fighter might be seriously injured, or even killed, by such a blow. A professional boxer's fist is therefore legally considered a deadly weapon outside the ring. Inside the ring, it is not so deadly. Not only do both contestants wear padded gloves but they are both trained to take punches with minimum injury. Both fighters know from considerable experience that they will be hurt more if they get hit flat-footed than if they roll with the punch.

Moving the head a meter per second or so one way or the other might seem to make little difference compared to the hundred or so feet per second that a fist is traveling. By moving his head away from the punch,

Figure 7-7

Force can be reduced for momentum transfer by prolonging the time of contact.

however, the boxer stretches his muscles and joints in the proper direction to give with the blow and spread the momentum transfer over a maximum amount of time.

If you have ever gone skating at a roller rink, you probably have had direct experience with the transfer of momentum. It does not seem to hurt as much to fall on an oak floor at a roller rink as it does to fall on a concrete sidewalk. The wooden floor may seem very hard to the touch, but it has considerably more give than concrete. If you happened to be resting in the middle of such a floor and another person tripped over you, you would feel the floor shake as he landed. The floor's small amount of give allows momentum to be spread out over more time, thus reducing the average force.

CHECK QUESTION

A roller skater falls and experiences a force of 600 N for 0.40 s on the first bounce. (a) How much momentum is transferred to the floor by this impulse? (b) If the skater has a mass of 60 kg, what is her change of velocity associated with this change of momentum?

Answer: (a) 240 kg · m/s, (b) 4.0 m/s

Impulse and Momentum Conservation

Falling is an art. The first thing you learn when you take judo lessons is how to fall. That is because you will be doing a lot of it, and you do not want to get hurt. You are taught to land flat on your side so that the force is spread out over as much area as possible. This reduces the pressure at any one point. This is not surprising to most people, but you are also taught to do something that does surprise many people. You are taught to slap the mat as hard as you can with your arm when you fall. A few people might think this is to make a loud sound to scare your opponent. Of course this isn't true. The arm slap has the effect of reducing the force on the rest of your body.

As you approach the mat, you have a certain amount of momentum. You cannot lessen this amount of momentum, but you can transfer as much as possible into your arm. The more momentum transferred to the mat by your arm, the less is transferred by the rest of your body.

Your arm has a less complicated internal structure than your rib cage and can better take the blow. A judo expert has toughened arms to take a mighty blow with little pain. A white-belt student, however, must live with a sore arm for many days. Many a neophyte has given up this martial art after the first time he slipped and fell on the sidewalk after having developed the proper reaction for falling on the mat. Concrete has much less give than those thin little mats used in judo schools, and this seemingly slight difference becomes painfully apparent as the student slaps hard with his arm as he hits the ground.

Figure 7-8

Either the top or bottom string can break, depending on how the bottom string is pulled.

Still another example of impulse and the conservation of momentum is a common physics demonstration used to demonstrate inertia. It consists of a massive ball attached to a support with string, with a second piece of string dangling from the bottom, as in Figure 7-8. A downward force is applied to the string attached to the bottom of the ball. The question is "Which string will break first?" This turns out to be a trick question. Either string may be made to break at the will of the person pulling the string.

The top string may be made to break by pulling with a slowly increasing force. It supports both the weight of the ball and the force in the bottom string. Having thus established that forces in the same direction add up to a larger force, the person pulling the string ties the ball back up and gives a hard jerk on the bottom string. The bottom string then breaks, leaving the top string intact. A short explanation for this phenomenon is that the inertia of the ball protects the top string. A fuller explanation involves the concepts of impulse and momentum.

When the bottom string is jerked, the force builds up suddenly to the point where the bottom string breaks. This impulse does in fact transmit some momentum to the ball, and the ball does move downward. If the impulse applied by the bottom string is sharp enough, however, the momentum transfer will be small and the ball will move downward only slowly. The force in the top string then increases as the ball stretches it, and this increased force tends to stop the ball by producing an impulse in the opposite direction. If the ball is moving slowly enough, however, the impulse applied by the top string will be spread out over enough time so that the increased force will never reach the point of breaking the string. The ball will return to the equilibrium position, bouncing up and down slightly due to the springiness of the string, and eventually come to rest.

Figure 7-9

The bottom string breaks if it is jerked suddenly; the same impulse is spread out over a longer time in the top string.

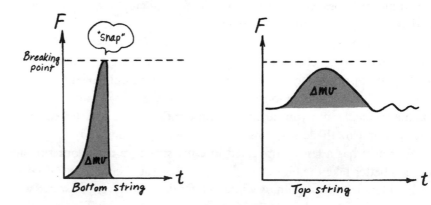

Momentum is conserved, although the forces are different. A sudden jerk of the hand transfers momentum to the ball through the bottom string before it breaks, and this same momentum is transferred over a longer period of time through the top string to the hook in the ceiling. The graphs of force versus time in Figure 7-9 represent this momentum as the area under the force curve and above the equilibrium position. Equilibrium for the lower string is zero when no force is applied. Equilibrium for the top string is the force of the weight of the ball. The areas are the same, because momentum is conserved. All the momentum gained by the ball is lost through the top string by the time the ball returns to its equilibrium position. Even a modest increase in time will keep the force in the top string below the breaking point, even though it has a head start—in terms of force—on the bottom string.

Velocity Before and After an Interaction

We have seen how the law of conservation of momentum can be derived from Newton's second and third laws of motion. Every bit of momentum that something receives has to be given up by something else, since every force of action must be accompanied by an equal and opposite force of reaction and since both forces act for exactly the same period of time.

Thus, the total momentum of a system remains unchanged. Different parts of the system can trade the momentum around by pushing on each other with equal and opposite forces of action and reaction, but the total momentum before an interaction is the same amount as after.

This assumes that we choose a large enough system to include everything that is in the act, so to speak. That includes everything pushing or pulling on anything else in the system with no unbalanced external forces that we would need to worry about. The little car and big truck were considered to be a single system, because we were concerned only with the forces of action and reaction between these two objects. We did not need to worry about the weight of these two vehicles, which would involve forces of action and reaction between the objects and the earth, because the force of the vehicles against the earth is entirely balanced by the ground pushing back on the tires. In any real case there would also be a force of friction between the wheels and the ground, but we ignored this little problem in our simple example, because its effect would be small during the short time that the wreck takes place compared to the vastly greater forces of collision. Thus, the car and the truck were taken together as an isolated system, because the only other forces acting on them either were balanced or could be ignored.

The law of conservation of momentum is particularly useful in the kind of situation where a system of objects is sufficiently isolated from external forces to allow us to worry only about internal forces acting between objects in the system. We looked at one such situation when we considered the wreck between the car and the truck—when two things come together and stick. That example developed the concept of momentum. We will now consider the opposite kind of situation to see how the conservation of momentum can be applied. We will look at two things that are together to begin with and consider their relative velocities after an internal force throws them apart. Let us consider the case where the two objects are on ice skates—one of the few real-life situations where the force of friction is negligible.

Suppose, for example, that you go to the Ice Follies and see the biggest clown pick up one of the smaller clowns and throw him across the ice. This supposedly humorous event comes to your attention because all the lights in the house are off except for one spotlight that follows the big clown as he is about to throw the little clown. As an artifice to build suspense, the spotlight follows the big clown and you are left to wonder what happens to the little clown.

Let us further suppose that you had already noticed from the program that the little clown has a mass of only 25 kg. You quickly estimate the mass of the big clown as about 75 kg and his velocity before and after he throws the clown. You observe him going along at about 3 m/s toward your right when he had the clown in his arms. After throwing the clown, you observe him going backward, toward your left, at about 1 meter per second.

The law of conservation of momentum tells you that the momentum before the big clown throws the little clown is equal to the momentum after. The momentum before, $(mv)_{before}$, is the total mass times the velocity before, v_{before}:

$$(mv)_{before} = (m_{big} + m_{little})v_{before}$$

The momentum after the big clown throws the little clown, $(mv)_{after}$, is the sum of the momentums of the big clown, $(mv)_{big}$ and that of the little clown, $(mv)_{small}$:

$$(mv)_{after} = (mv)_{big} + (mv)_{little}$$

The law of conservation of momentum states that the momentum before is equal to the momentum after:

$$(mv)_{before} = (mv)_{after}$$

Inserting the expressions for the momentums before and after,

$$(m_{big} + m_{little})v_{before} = (m_{big}v_{big}) + (m_{little}v_{little})$$

where m_{little} and m_{big} are the masses of the little clown and the big clown, v_{before} is their velocity before the big clown throws the little one, and v_{little} and v_{big} are the velocities of the little clown and big clown after the little clown is thrown. We can solve this expression for the velocity of the clown by rearranging terms.

$$m_{little}v_{little} = (m_{big} + m_{little})v_{before} - m_{big}v_{big}$$
$$= m_{big}v_{before} + m_{little}v_{before} - m_{big}v_{big}$$

U before = 3 m/s

Figure 7-10

An ice-skating clown carrying a smaller clown has a certain amount of momentum.

U after = 1 m/s U clown = ?

Figure 7-11

The total momentum after the small clown is thrown is equal to the total momentum before.

Dividing both sides of the equation by the mass of the little clown,

$$v_{\text{little}} = \frac{m_{\text{little}} v_{\text{before}} + m_{\text{big}} v_{\text{before}} - m_{\text{big}} v_{\text{big}}}{m_{\text{little}}}$$

This is the general solution for the final velocity of the little clown, but it can be simplified with a bit of factoring. Notice that the first term in the numerator contains the mass of the little clown, just as it appears in the denominator, while the other two terms contain the mass of the big clown. We can therefore separate the fraction and factor out the mass of the big clown from the last two terms.

$$v_{\text{little}} = \frac{m_{\text{little}} v_{\text{before}}}{m_{\text{little}}} + \frac{m_{\text{big}}(v_{\text{before}} - v_{\text{big}})}{m_{\text{little}}}$$

We can then divide the mass of the little clown into itself in the first fraction to show that the final velocity of the little clown is equal to the initial velocity, v_{before}, plus a term that is the ratio of the masses times the change in velocity of the big clown.

$$v_{\text{little}} = v_{\text{before}} + \frac{m_{\text{big}}}{m_{\text{little}}}(v_{\text{before}} - v_{\text{big}})$$

To see how this works in our numerical example, we put in values for mass and velocity. The masses of the big clown and little clown are, as you remember, 75 kg and 25 kg. If we say that the initial velocity of both big clown and little clown is +3 m/s, however, we must call the final velocity of the big clown −1 m/s. Assigning a positive value to the initial velocity assumes that the positive direction is from left to right. The final velocity of the big clown must then be negative, since it is in the opposite direction.

$$v_{\text{little}} = 3 \text{ m/s} + \frac{75 \text{ kg}}{25 \text{ kg}}[(3 \text{ m/s}) - (-1 \text{ m/s})]$$

The double minus in the last term gives a physically reasonable result. The big clown's change in velocity is truly 4 m/s. Not only did he stop going forward at 3 m/s but he kept on accelerating in the same direction to start going 1 m/s backwards.

$$v_{\text{little}} = 3 \text{ m/s} + (3)(4 \text{ m/s})$$

The big clown's change in velocity is multiplied by three before adding it to the little clown's initial velocity, because the big clown has three times as much mass.

$$v_{\text{little}} = 3 \text{ m/s} + 12 \text{ m/s}$$
$$= 15 \text{ m/s}$$

We should pause for a moment to see what is important about this solution. The numerical result might give us a nice solid, satisfied feeling that we know the answer, but it is not nearly as valuable as the general solution in terms of understanding what is going on. Just looking at the algebraic result, it is clear that the final velocity of the little clown would be increased by either increasing the mass of the big clown or decreasing the mass of the little clown. By just subtracting the initial velocity v_{before} from both sides of the general solution, we can see that the change in velocity of the little clown is directly proportional to the change in velocity of the big guy.

$$v_{\text{little}} - v_{\text{before}} = \frac{m_{\text{big}}}{m_{\text{little}}}(v_{\text{before}} - v_{\text{big}})$$

It is also directly proportional to the mass of the big clown, which is in the numerator on the right-hand side, and inversely proportional to the mass of the little clown, which is in the denominator.

The importance of this algebraic solution to the problem does not mean that nothing is to be learned from putting in the numbers. The trick of using a minus value for one of the velocities, for example, might well escape you unless you were to try it for the purpose of seeing whether or not the answer turns out to be reasonable. What is more, this trick is not a trivial thing, as its proper use depends on understanding the vector nature of velocity. The point is, however, that the algebraic solution has a great deal to say by itself and deserves your attention before you move on to putting in the numbers.

You should not, on the other hand, look upon the general solution of a problem like this as being so important that you try to memorize it as an equation you need to know. You should instead be able to derive it. There are many occasions when the law of conservation of momentum is useful. The preceding solution comes directly from that law as it is applied to a problem at hand. If we were to see another performance at the Ice Follies in which the spotlight followed the little clown, a slightly different application of the same law would yield the recoil velocity of the big clown. In yet another performance, if both the final velocities were observed, we could apply the law of conservation of momentum to calculate the mass of the big clown to see if the estimated value was correct.

If you were not lucky enough to attend more than one performance of the Ice Follies in one season, consider another example. Suppose you were sentenced to be shot for a political crime. In order to save yourself from this horrible fate, you might well wish to estimate the recoil of the gun with which you are to be shot. The momentum before the gun is fired is zero. The law of conservation of momentum tells us that the total momentum of the bullet and recoiling gun is also zero.

$$0 = m_{bullet} v_{bullet} + m_{gun} v_{gun}$$

where m_{bullet} is the mass of the bullet, m_{gun} is the mass of the gun, v_{bullet} is the velocity of the bullet, and v_{gun} is the unknown recoil velocity of the gun. Solving for v_{gun}, we get

$$m_{gun} v_{gun} = m_{bullet} v_{bullet}$$

$$v_{gun} = -\left(\frac{m_{bullet}}{m_{gun}}\right) v_{bullet}$$

Immediately upon realizing that the recoil velocity of the gun increases with the mass of the bullet, you would know the proper response to give when your antagonist offers you your last wish. You might claim that your crime demands more severe punishment than being struck by a tiny bullet. Insist that the mass of the bullet match the magnitude of your crime, that the bullet be much more massive than the gun, and that your antagonist should be the one to pull the trigger.

CHECK QUESTION

A cannon ball having a mass of 5×10^3 kg is fired from a cannon having a mass of 3×10^3 kg. The velocity of the cannon ball is 6 m/s. What is the velocity of the cannon?

Answer: −10 m/s

Figure 7-12

The velocity of recoil is proportional to the ratio of the masses.

SUMMARY OF CONCEPTS

The problems in this chapter have to do with impulse and momentum. The **momentum** of an object is defined as the product of its mass times its velocity.

$$\text{Momentum} = (mv)$$

In terms of momentum, Newton's second law of motion states that force is equal to the rate of change of momentum with respect to time.

$$F = ma = \frac{\Delta(mv)}{\Delta t}$$

This may be restated as **impulse** equals change of momentum, where impulse is defined as the average force times the change of time.

$$F\Delta t = \Delta(mv)$$
$$\text{Impulse} = \text{Change of momentum}$$

Momentum is a conserved quantity. The total momentum of a system can only change if there is an outside force applied to it. The law of conservation of momentum for systems isolated from outside forces states that the momentum before an interaction is equal to the momentum after.

$$(mv)_{\text{before}} = (mv)_{\text{after}}$$

The law of conservation of momentum agrees with Newton's second and third laws of motion. Newton's third law states that for every force of action there is an equal and opposite force of reaction somewhere else in the system. Assuming this is true for every instant of time, this third law is equivalent to saying that for every action impulse there is a reaction impulse. Thus, for every change of momentum there is an equal and opposite change of momentum of some other body in the system such that the total amount of momentum in the system is constant.

EXAMPLE PROBLEMS

Try to do the Examples yourself before looking at the Sample Solutions.

7-1 Two monkeys, named Ed and Marshal, are hanging from the opposite ends of a rope that goes up and over a pulley. The monkeys balance each other, because they both have the same weight. They are both stationary when Ed decides to scamper up the rope.

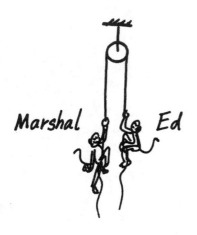

Given: Ed and Marshal each have a mass of 35 kg. When Ed decides to scamper up the rope, he gives himself an upward acceleration of 5.2 m/s².

Find: (a) How much does each monkey weigh?
(b) How great is the force of action necessary to make Ed accelerate up the rope?
(c) How great is the force of reaction and what does it act on?

BACKGROUND

This problem illustrates Newton's third law for two situations—one in which the forces all balance to produce a static situation and the other in which unbalanced forces cause acceleration.

Both situations involve equal but opposite forces of action and reaction. These equal and opposite forces of action and reaction called for in the third law are not the same as the forces that balance each other in equilibrium problems. The rules for equilibrium refer to forces that act on the same body; the third law refers to forces that act on different bodies.

Although the problem does not specifically call for it, you might address yourself to what happens to Marshal. Ignoring the mass of the rope and friction in the pulley, can Ed ever get away from Marshal?

7-2 A rifle exerts a constant force of action on a bullet while it is in the barrel.

Given: An 11.7-g bullet is accelerated at 3.8×10^5 m/s² through a 0.51-m rifle barrel.

Find: (a) How great is the force of action of the gun on the bullet?
(b) For how long does the bullet experience this force of action?
(c) For how long does the gun experience an equal but opposite force of reaction?

BACKGROUND

This problem points out that Newton's third law holds true at every instant of time. Not only do the action and reaction forces equal each other in magnitude, they are equal for the same period of time. The impulse produced by one is therefore equal and opposite to the impulse produced by the other.

7-3 An astronaut working in outer space tosses away a flashlight when he is finished using it. A few moments later, he tosses a screwdriver in the same direction.

Given: The astronaut exerts the same force of 20 N for the same period of time, 0.35 s, on both the 0.73-kg flashlight and the 0.28-kg screwdriver.

Find: (a) What is the momentum of the flashlight?
(b) What is the momentum of the screwdriver?
(c) How much faster is the screwdriver thrown than the flashlight?

BACKGROUND

This problem illustrates Newton's second law expressed in terms of impulse and momentum. Impulse is defined as the product of force and the time over which it is applied. The second law of motion can either be expressed as saying F = ma or as saying that impulse equals change of momentum.

7-4 A tennis ball, traveling north one fine day, encounters a tennis racket and reverses its journey and begins to go south.

Given: The racket exerts a force of 24 N on the 55-g tennis ball to change its velocity from 15 m/s north to 25 m/s south.

Find: (a) By how much does the momentum of the ball change?
(b) Over what period of time does this change of momentum take place?

BACKGROUND

This problem illustrates the vector nature of velocity and momentum. Note that the change of velocity, and hence momentum, is greater when the ball reverses its direction than if it had simply come to a stop.

7-5 A lumberjack, standing on the end of a stationary log floating in calm water, decides to take a little walk. The log reacts by floating in the opposite direction.

Given: The 100-kg lumberjack walks east at 1.5 m/s relative to the water. The log has a mass of 500 kg.

Find: (a) What is the momentum of the lumberjack?
(b) What is the momentum of the log?
(c) How fast does the log move in the opposite direction?

BACKGROUND

This problem illustrates the law of conservation of momentum. The log and the lumberjack form an isolated system in that the force applied by the log to the lumberjack is equal and opposite to the reaction force applied by the lumberjack to the log, assuming that outside forces can be ignored in the forward and backward directions.

7-6 During a snowball fight, a small girl is retreating across a nearly frictionless frozen pond toward her snow fort. She catches a snowball in the middle of her back. The snowball sticks.

Given: The child has a mass of 14.7 kg and an initial velocity of 3.0 m/s. The snowball has a mass of 0.30 kg and an initial velocity of 20 m/s.

Find: (a) What is the child's momentum before catching the snowball?
(b) What is the snowball's momentum before finding the child?
(c) What is the final velocity of the child and snowball combined?

BACKGROUND

This problem is an example of the law of conservation of momentum applied to an inelastic collision. The little girl and the snowball may be assumed to be an isolated system whose total momentum after collision is equal to the total momentum of the parts before collision.

Solutions to EXAMPLE PROBLEMS

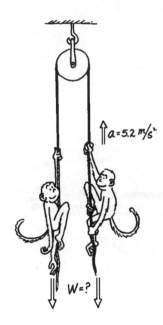

(a) $w = mg$

$$= (35 \text{ kg})(9.8 \text{ m/s}^2)\left(\frac{1 \text{ N}}{(\text{kg})(\text{m/s}^2)}\right)$$

$$= \boxed{343 \text{ N}}$$

(b) $\Sigma F = ma$

$$F_\uparrow - w = ma$$

$$F_\uparrow = ma + w$$

$$= (35 \text{ kg})(5.2 \text{ m/s}^2) + 343 \text{ N}$$

$$= 182 \text{ N} + 343 \text{ N}$$

$$= \boxed{525 \text{ N}}$$

(c) $F_{reaction} = -F_{action}$

$$F_{\downarrow(rope)} = F_{\uparrow(monkey)}$$

$$= \boxed{-525 \text{ N}}$$

SAMPLE SOLUTION 7-1

Given:

Ed and Marshal each have a mass of 35 kg. When Ed decides to scamper up the rope, he gives himself an upward acceleration of 5.2 m/s².

Find:

(a) How much does each monkey weigh?

(b) How great is the force of action necessary to make Ed accelerate up the rope?

(c) How great is the force of reaction and what does it act on?

DISCUSSION

Part (a) is solved by the now familiar relationship between mass and weight. If we think of the interaction between one monkey, say Ed, and the rope, we could call this the force of action that the rope must exert on Ed to balance his weight and hold him stationary. The force of reaction acting on the rope has the effect of balancing Marshal's weight. If the force of action acting on Ed is increased, as in part (b), to give him an upward acceleration, the equal but opposite force of reaction acting on the rope also increases, as shown in part (c). The pulley changes the direction of this downward force acting on the rope, represented by a negative sign in part (c), into an upward force acting on Marshal. This increased upward force acting on Marshal gives him the same upward acceleration as experienced by Ed.

SAMPLE SOLUTION 7-2

Given:

An 11.7-g bullet is accelerated at 3.8×10^5 m/s² through a 0.51-m rifle barrel.

Find:

(a) How great is the force of action of the gun on the bullet?

(b) For how long does the bullet experience this force of action?

(c) For how long does the gun experience an equal but opposite force of reaction?

(a) $F = ma$

$\qquad = (11.7 \times 10^{-3}\text{ kg})(3.8 \times 10^5\text{ m/s}^2)\left(\dfrac{1\text{ N}}{1\text{ kg} \cdot \text{m/s}^2}\right)$

$\qquad = \boxed{4.4 \times 10^3\text{ N}}$

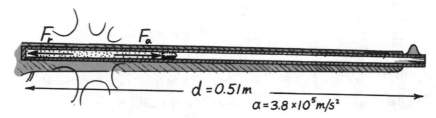

$d = 0.51\,m$

$a = 3.8 \times 10^5\,m/s^2$

(b) $\quad d = \frac{1}{2}at^2$

$\qquad t = \sqrt{\dfrac{2d}{a}} = \sqrt{\dfrac{2\,(0.51\text{ m})}{3.8 \times 10^5\text{ m/s}^2}} = \boxed{1.6 \times 10^{-3}\text{ s}}$

(c) $t_{gun} = t_{bullet} = \boxed{1.6 \times 10^{-3}\text{ s}}$

The gun is pushed by the bullet for exactly the same amount of time that the bullet is pushed by the gun.

DISCUSSION

In part (a) the unbalanced force applied to the bullet causes a certain acceleration, which can be calculated from Newton's second law. Assuming that this acceleration is constant and that the bullet starts from rest, the time it remains in the gun barrel can be calculated from the formula for distance traveled under constant acceleration. The gun exerts a force on the bullet only while it is in the gun barrel. The bullet, on the other hand, exerts a reaction force on the gun for exactly the same period of time. They cease to push against each other at the moment that the bullet departs from the muzzle.

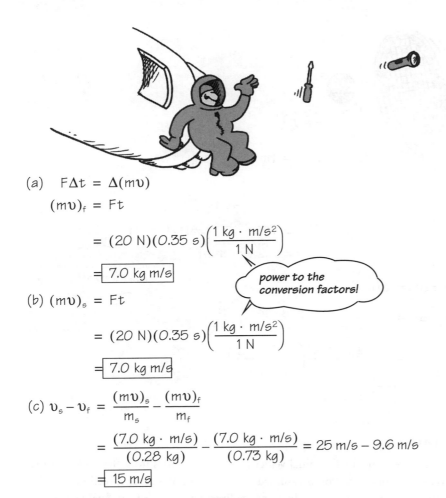

Given:

The astronaut exerts the same force of 20 N for the same period of time, 0.35 s, on both the 0.73-kg flashlight and the 0.28-kg screwdriver.

Find:

(a) What is the momentum of the flashlight?

(b) What is the momentum of the screwdriver?

(c) How much faster is the screwdriver thrown than the flashlight?

(a) $F\Delta t = \Delta(m\upsilon)$

$(m\upsilon)_f = Ft$

$$= (20 \text{ N})(0.35 \text{ s})\left(\frac{1 \text{ kg} \cdot \text{m/s}^2}{1 \text{ N}}\right)$$

$$= \boxed{7.0 \text{ kg m/s}}$$

power to the conversion factors!

(b) $(m\upsilon)_s = Ft$

$$= (20 \text{ N})(0.35 \text{ s})\left(\frac{1 \text{ kg} \cdot \text{m/s}^2}{1 \text{ N}}\right)$$

$$= \boxed{7.0 \text{ kg m/s}}$$

(c) $\upsilon_s - \upsilon_f = \dfrac{(m\upsilon)_s}{m_s} - \dfrac{(m\upsilon)_f}{m_f}$

$$= \frac{(7.0 \text{ kg} \cdot \text{m/s})}{(0.28 \text{ kg})} - \frac{(7.0 \text{ kg} \cdot \text{m/s})}{(0.73 \text{ kg})} = 25 \text{ m/s} - 9.6 \text{ m/s}$$

$$= \boxed{15 \text{ m/s}}$$

DISCUSSION

Parts (a) and (b) are found by solving Newton's second law for momentum. The delta notation can be ignored in this case, because the change of momentum is the momentum itself, since the screwdriver and flashlight are assumed to start from rest. The same is true of the time, since we can start measuring time at the instant that force is applied to each object. The result is the same for parts (a) and (b), because the same force is applied for the same period of time. Even though both objects have the same momentum, however, they have different velocities, since they have different masses.

SAMPLE SOLUTION 7-4

Given:

The racket exerts a force of 24 N on the 55-g tennis ball to change its velocity from 15 m/s north to 25 m/s south.

Find:

(a) By how much does the momentum of the ball change?

(b) Over what period of time does this change of momentum take place?

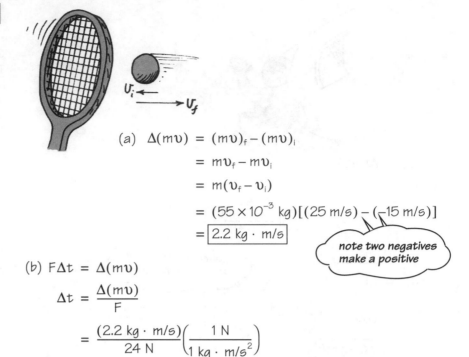

(a) $\quad \Delta(m\upsilon) = (m\upsilon)_f - (m\upsilon)_i$

$\qquad\qquad = m\upsilon_f - m\upsilon_i$

$\qquad\qquad = m(\upsilon_f - \upsilon_i)$

$\qquad\qquad = (55 \times 10^{-3}\text{ kg})[(25\text{ m/s}) - (-15\text{ m/s})]$

$\qquad\qquad = \boxed{2.2\text{ kg}\cdot\text{m/s}}$

note two negatives make a positive

(b) $F\Delta t = \Delta(m\upsilon)$

$\quad \Delta t = \dfrac{\Delta(m\upsilon)}{F}$

$\qquad = \dfrac{(2.2\text{ kg}\cdot\text{m/s})}{24\text{ N}}\left(\dfrac{1\text{ N}}{1\text{ kg}\cdot\text{m/s}^2}\right)$

$\qquad = \boxed{9.2 \times 10^{-2}\text{ s}}$

DISCUSSION

The negative value for the initial velocity (–15 m/s) comes from the vector nature of velocity. If one direction is assumed to be positive, the other direction must be negative. We can assume that the positive direction is that of the final velocity in this case, since that is the direction that the momentum is changing toward; we could as well assume the initial velocity to be positive, however, and the result would be the same but with a negative value because of the convention we would have adopted.

Part (b) is found by solving Newton's second law for the change in time associated with the impulse that must produce the change of momentum given by part (a).

Given:
The 100-kg lumberjack walks east at 1.5 m/s relative to the water. The log has a mass of 500 kg.

Find:
(a) What is the momentum of the lumberjack?

(b) What is the momentum of the log?

(c) How fast does the log move in the opposite direction?

(a) $(mv)_{Lj} = m_{Lj}v_{Lj}$

$\qquad = (100 \text{ kg})(1.5 \text{ m/s}) = \boxed{150 \text{ kg} \cdot \text{m/s}}$

(b) $\qquad F\Delta t = \Delta(mv)$

$\qquad\qquad \Sigma F$ is 0, since there is no outside force

$\qquad 0 = \Delta(mv)_{Lj} + \Delta(mv)_{log}$

$\qquad 0 = 150 \text{ kg} \cdot \text{m/s} + \Delta(mv)_{log}$

$\quad \Delta(mv)_{log} = \boxed{-150 \text{ kg} \cdot \text{m/s}}$

(c) $v_{log} = \dfrac{(mv)_{log}}{m_{log}}$

$\qquad = \dfrac{(-150 \text{ kg} \cdot \text{m/s})}{(500 \text{ kg})} = \boxed{-0.30 \text{ m/s}}$

DISCUSSION

Part (a) is solved from the definition of momentum. The momentum of the lumberjack is just the product of his mass and velocity.

Part (b) is solved by a special application of Newton's second law, where zero outside force is applied to the system as a whole. Any change in momentum of the lumberjack must therefore be accompanied by an equal and opposite change in the momentum of the log if the total change in the momentum of the system is to be zero.

The velocity of the log can be found from its momentum, as in part (c), by dividing momentum by mass, since momentum is defined as the product of mass and velocity.

SAMPLE SOLUTION 7-6

Given:

The child has a mass of 14.7 kg and an initial velocity of 3.0 m/s. The snowball has a mass of 0.30 kg and an initial velocity of 20 m/s.

Find:

(a) What is the child's momentum before catching the snowball?

(b) What is the snowball's momentum before finding the child?

(c) What is the final velocity of the child and snowball combined?

(a) $(m\upsilon)_c = m_c \upsilon_c$

$= (14.7 \text{ kg})(3.0 \text{ m/s}) = \boxed{44 \text{ kg} \cdot \text{m/s}}$

(b) $(m\upsilon)_{sb} = m_{sb} \upsilon_{sb}$

$= (0.30 \text{ kg})(20 \text{ m/s}) = \boxed{6.0 \text{ kg} \cdot \text{m/s}}$

(c) $\Sigma\upsilon = \dfrac{\Sigma(m\upsilon)}{\Sigma m}$

$= \dfrac{(m\upsilon)_c + (m\upsilon)_{sb}}{m_c + m_{sb}}$

$= \dfrac{(44.0 \text{ kg} \cdot \text{m/s}) + (6.0 \text{ kg} \cdot \text{m/s})}{(14.7 \text{ kg}) + (0.30 \text{ kg})} = \dfrac{50 \text{ kg} \cdot \text{m/s}}{15 \text{ kg}}$

$= \boxed{3.3 \text{ m/s}}$

DISCUSSION

The momentum of the child and the momentum of the snowball are found in parts (a) and (b) by the definition of momentum. Momentum is the product of mass and velocity.

The velocity in part (c) is found by dividing the total momentum after collision by the total mass after collision. The vector sum of the momentum is the same as the numerical sum of the momentums found in parts (a) and (b), since both momentums are in the same direction. The problem states that the little girl and the snowball were going in the same direction at the time of impact. The sum of the masses is always the numerical sum of the parts, because mass is a scalar quantity. The result indicates that the little girl is helped along by the snowball, her speed increasing by about 10%.

ESSENTIAL PROBLEMS

Be sure you can do the Example Problems (without peeking at the Sample Solutions) before working these Essential Problems.

7-7 Iowa Jones is hauling a golden idol out the top of a burial chamber. He has the idol attached to a rope which goes over a pulley and then back to Iowa, who is pulling on the rope. The idol is almost as heavy as Iowa. When the idol reaches eye level, a snake drops on it from above. When Iowa sees the snake eyeing him, he decides to crawl rapidly up the rope to get away. The combined weight of the snake and the idol matches Iowa's weight exactly.

Given: Iowa Jones has a mass of 101 kg. The combined mass of the idol and the snake is also 101 kg. Upon seeing the snake, Iowa crawls up the rope with an acceleration of 3.2 m/s^2.

Find: (a) How much does Iowa weigh?
(b) What force did he exert to crawl up the rope?
(c) How great is the reaction force? What does it act on?

7-8 An archer sets an arrow to his bow string, pulls it back to his ear, and then lets it fly toward the target.

Given: The 0.12-kg arrow has an average acceleration of 1,000 m/s^2 while it is being propelled by the bow. The bow acts on the arrow over a distance of 0.75 m.

Find: (a) What is the force applied to the arrow by the bow?
(b) For how long does the bow act on the arrow?
(c) For how long does the arrow act on the bow?

|← 0.75m →|

7-9 A fisherman has been casting a heavy spoon lure all morning without luck. He decides to switch to a light fly lure, casting it with the same force of action applied for the same amount of time.

Given: Both the spoon and fly lures experience a force of 5.3 N for 0.73 s. The spoon lure has a mass of 150 g, and the fly lure has a mass of only 25 g.

Find: (a) What is the momentum of the spoon lure?
(b) What is the momentum of the fly lure?
(c) How much greater is the velocity of the fly lure than the velocity of the spoon lure after being cast?

7-10 While being tossed around on the street, the ball hits the front of a moving car. The ball reverses direction and picks up additional momentum from the encounter.

Given: The baseball has a mass of 132 g and an initial velocity of 3.2 m/s. The moving car exerts a force of 32.5 N on the ball. After colliding with the car, the baseball has a velocity of 5.8 m/s.

Find: (a) What is the change in momentum of the baseball?
(b) Over what period of time does this change in momentum take place?

7-11 A woman in a bathing suit dives out of a rowboat into the water. The woman and the boat are initially stationary in calm water.

Given: The woman has a mass of 65 kg and the boat has a mass of 30 kg. The woman gives herself a horizontal velocity relative to the water of 0.85 m/s.

Find: (a) What is the woman's momentum?
(b) What is the momentum of the boat?
(c) How fast does the boat move in the opposite direction?

7-12 A pickup truck is rear-ended by a big American luxury car traveling in the same direction. After they hit, the two vehicles stick together and roll at a new velocity. Neither car has its brakes on at the time of impact, so there is a negligible amount of friction.

Given: The truck has a mass of 1,350 kg and is moving at a velocity of 1.2 m/s. The big car has a mass of 1,645 kg and an initial velocity of 3.5 m/s in the same direction as the truck.

Find: (a) What is the initial momentum of the truck?
(b) What is the initial momentum of the car?
(c) What is the final velocity of the linked cars?

MORE INTERESTING PROBLEMS

7-13 A horse is pulling a wagon equipped with nearly frictionless wheel bearings and excellent brakes. The horse exerts a constant force of action on the wagon.

Given: The force exerted by the horse is 2,200 N. The wagon has a mass of 500 kg.

Find: (a) How great is the reaction force with which the wagon pulls on the horse when the brakes are set and the wagon won't budge?
(b) When the brakes are released, how fast will the wagon accelerate?
(c) How great is the reaction force when the brakes are released?

7-14 John's car has run out of gas. His passenger, Jennifer, gets out to push the car to the side of the road while he steers, but he has left the brakes on.

Given: Jennifer pushes the car with a force of 46 N. The combined mass of the car and John is 403 kg.

Find: (a) How great is the reaction force of the car against Jennifer while the car has its brakes on?
(b) After the brakes are released, how fast will the car accelerate?
(c) How great is the reaction force of the car against Jennifer after the brakes are released and the car is rolling freely?

7-15 A karate expert is breaking a board. She concentrates momentum transfer into a short period of time to optimize the force of a blow. A beginning student can hit a board with the same mass as the expert but has a smaller effective force because she takes longer to hit the board.

Given: The karate expert can put an effective mass of 15 kg moving at 22 m/s behind the blow. The momentum transfer to the board takes place in 3.3×10^{-2} s. (Assume all the momentum is needed to break the board.) A novice, using the same effective mass, can make a blow with a velocity of 11 m/s, which takes 6.6×10^{-2} s.

Find: (a) How much momentum does the blow transfer?
(b) What is the average force of the blow?
(c) What is the average force the novice can produce?

7-16 A pianist is working on her piano technique. She is practicing different ways to play soft and loud. She is working on playing loud notes, not by hitting the note harder, but by hitting it faster so that she can play with less tension in her hands.

Given: To play a note softly she plays the note with an effective mass of 122 g moving at a velocity of 0.89 m/s. The momentum transfer takes place in 51 milliseconds (a millisecond is 10^{-3} s.) When she plays the same note loudly she uses the same effective mass, but strikes the note with a velocity of 1.19 m/s, which takes 35 milliseconds.

Find: (a) How much momentum does the blow transfer?
(b) What is the average force of the blow?
(c) What is the average force the second time?

7-17 In one of the nuclear reactions that produce solar energy, a neutron smashes into a deuteron and sticks to form a triton. A deuteron is just about twice as heavy as a neutron, so a triton is about three times as heavy.

Given: The deuteron is stationary and has a mass of 3.34×10^{-27} kg. The neutron is moving along at 2.5×10^6 m/s and has a mass of 1.67×10^{-27} kg.

Find: (a) What is the momentum of the neutron before the collision?
(b) What is the velocity of the triton after the collision?

7-18 Food fight in the faculty dining room! A flying blob of pudding, coming in horizontally, hits a dinner roll. They stick together and slide down the table, which is essentially frictionless because of the puddle of salad dressing that has already been applied to it.

Given: The blob of pudding has a mass of 95 g and comes in at a velocity of 2.3 m/s. The dinner roll is stationary before impact and has a mass of 138 g.

Find: (a) What is the momentum of the blob of pudding before it hits the dinner roll?

(b) Immediately after the collision, what is the velocity of the roll and pudding together?

7-19 Bernice Bowler is trying to popularize a new winter sport—ice bowling—where the bowler, on ice skates on a frozen lake, tries to hit bowling pins at the other end of the lake. When she tries it, however, she discovers a drawback to this new sport.

Given: Bernice has a mass of 59 kg. The bowling ball has a mass of 6.8 kg. She throws the ball so that it has a velocity of 4.2 m/s. The ball hits the bowling pins 2.2 s later.

Find: (a) What is Bernice's velocity after throwing the ball? In which direction is she moving?

(b) How far away is Bernice from where she started 2.2 s after she threw the ball?

(c) How far are Bernice and the ball away from each other when the ball hits the bowling pins?

7-20 At their year-end party, some Physics majors are playing with liquid nitrogen. They put some in an empty plastic soft drink bottle and wait for the nitrogen to warm and the bottle to explode.

Given: One piece of the bottle has a mass of 17.4 g and leaves the explosion horizontally at a velocity of 5.5 m/s. A smaller piece with a mass of 5.2 g flies off in the other direction. The students are curious to know how far the pieces will fly apart after 0.50 s.

Find: (a) Assuming the two pieces have the same momentum, what is the velocity of the smaller piece leaving the explosion?

(b) How far away is this smaller piece from the original site of the explosion 0.50 seconds later?

(c) How far apart are these two pieces of the bottle 0.50 s after the explosion?

7-21 A little girl is watering the weeds. As she points the hose at the weed bed, she notices that the force of action of the hose against the water is accompanied by a reaction force of the water against the nozzle, which she can feel in her hand.

Given: The water stream carries water at the rate of 0.30 kg/s. The nozzle velocity of the water is 40 m/s. She waters her dandelion for 0.5 s.

Find: (a) How much momentum is transferred to the water in 0.5 s?
(b) What is the reaction impulse necessary to transfer this momentum?
(c) What is the reaction force of the water against the nozzle?

7-22 It takes two fire fighters to hold a fire hose. The reaction force of the hose would be too great for one person alone.

Given: The hose carries water at a rate of 12.5 kg/s. The nozzle velocity is 40 m/s. The hose is directed at a particular window for 21 s.

Find: (a) How much momentum is transferred in 21 s?
(b) What is the reaction impulse necessary to transfer this momentum?
(c) What is the reaction force?

7-23 Carmen is skipping rope. When she jumps she pushes the earth away from her. The earth, in turn, pushes her away from it.

Given: Carmen jumps upward with a force of 676 N. She is in contact with the ground for 0.388 seconds, during which she lifts 15.5 cm before leaving the earth. Carmen has a mass of 57.0 kg. The earth has a mass of 5.98×10^{24} kg. Assume Carmen has a constant acceleration.

Find: (a) What is the unbalanced force acting on Carmen?
(b) What is the impulse acting on Carmen just as she leaves the earth?
(c) What is the reaction force against the earth due to Carmen's jump?
(d) What is the momentum of the earth due to Carmen's jump just as she leaves?
(e) What is the velocity of the earth as it moves away from Carmen as a result of her jump?

7-24 A kangaroo pushes off strongly from the earth to jump.

Given: The kangaroo has a mass of 28.5 kg. It pushes against the earth with a force of 9,878 N for 0.26 s to push itself 0.71 m before leaving the ground. The mass of the earth is 5.98×10^{24} kg.

Find: (a) What is the unbalanced force acting on the kangaroo?
(b) What is the impulse acting on the kangaroo just as it leaves the ground?
(c) What is the reaction force felt by the earth to the kangaroo's kickoff?
(d) What is the momentum of the earth in reaction to the push of the kangaroo?
(e) What is the velocity of the earth as it recoils from the jump of the kangaroo?

ANSWERS

Solutions to **ESSENTIAL PROBLEMS**

7-7 (a) 990 N; (b) 1,313 N $\cong$ 1.3×10^3 N; (c) −1,313 N $\cong$ −1.3 × 10^3 N against the idol

7-8 (a) 120 N $\cong$ 1.2×10^2 N; (b) 0.039 s; (c) 0.039 s

7-9 (a) 3.9 kg · m/s; (b) 3.9 kg · m/s; (c) 129 m/s $\cong$ 1.3×10^2 N

7-10 (a) 1.19 kg · m/s; (b) 0.037 s

7-11 (a) 55 kg · m/s; (b) −55 kg · m/s; (c) −1.8 m/s

7-12 (a) 1.6×10^3 kg · m/s; (b) 5.8×10^3 kg · m/s; (c) 2.5 m/s

Solutions to **MORE INTERESTING PROBLEMS,** *odd answers*

7-13 (a) 2,200 N; (b) 4.4 m/s^2; (c) −2,200 N

7-15 (a) 330 kg · m/s; (b) 10,000 N $\cong$ 1.0×10^4 N; (c) 2.5×10^3 N

7-17 (a) 4.2×10^{-21} kg · m/s; (b) 8.3×10^5 m/s

7-19 (a) 0.48 m/s backwards; (b) 1.1 m; (c) 10.3 m

7-21 (a) 6.0 kg · m/s; (b) −6.0 kg · m/s; (c) −12 N

7-23 (a) 117 N; (b) 45.5 kg · m/s; (c) −117 N; (d) 45.5 kg · m/s; (e) 7.6×10^{-24} m/s

CHAPTER 8 *Energy*

IN CHAPTER 7 WE DEALT WITH A QUANTITY which, while somewhat esoteric, is quite real in the sense that it cannot be destroyed or created. We saw that momentum is absolutely conserved in every interaction. One might still engage in a philosophical discussion on the true existence of this quantity. Certainly momentum is not as tangible as mass, nor is it as directly measurable as velocity. One might argue that momentum is merely the product of mass and velocity, that it cannot exist without either member of that product, and that it therefore has no independent existence of its own. Still, there does seem to be this thing that is passed from body to body, divided into parts, added together, and constantly manipulated, with the total amount remaining perfectly constant for all time. From a problem-solving point of view, there are certain advantages to thinking of such a conserved quantity as having an existence in its own right. Whether or not that existence is real or imagined may be contemplated when you have an idle moment with nothing more pressing to do. Until that luxury comes your way, you would do well to treat momentum as a *thing*. Momentum lives, as it were.

Another quantity just as abstract as momentum, and yet as real in the sense that it is absolutely conserved, is **energy**. As in the case of momentum, you might well argue that energy is more of a condition than a thing—a condition that remains absolutely constant in an isolated system, even as other things within the system change. Most people, however, prefer to think of this esoteric quantity as a nearly material substance. We buy the stuff from the electric company. We worry about whether or not there is going to be enough stored in fossil fuel to go around if we keep using it as fast as we do. Few people ever stop to wonder if the stuff really exists at all. The very fact that it is conserved, that you cannot create it out of nothing, that you have to get it from somewhere if you want some, makes it a commodity that is bought and sold like a material substance.

This idea of energy is new. It is so sophisticated that it has only been with us for about a century and a half. Neither Galileo nor Newton knew about this stuff. Of course, there were some general ideas kicking around concerning such things as vitality and work, but no one seems to have thought of tying these various types of related things together into the single quantity we call energy until about two centuries ago. Even then, it took half a century for the idea to catch on. As we will see, the key to this level of understanding was the fact that heat is just another form of this same kind of thing. It was not until the 1840s that James Joule was able to prove that the same amount of heat is always produced by the same amount of work, no matter how the work is done. This provided the missing link. Even the word *energy* was not in common use before that. The word was, in fact, coined to give a name to all the different kinds of things that describe the vitality of inert objects.

The word *energy* was derived from the Greek *energos*, meaning "active." This was a particularly appropriate name, since it is derived in turn from the older Greek term *ergon,* which stands for "work. "

Work

Energy is often defined as the ability to do work. That might seem like a fine definition until you discover that *work* is frequently defined as the transfer of energy. Such is the circular nature of definitions one runs across when dealing with such truly basic quantities as time, distance, and force. Like these other basic quantities, energy is so fundamental that it cannot be defined in terms of anything more basic. There just isn't anything more basic. Still, there are relationships between basic quantities that allow us to get a handle on one by using the others. We can get a handle on energy by looking at how it is transferred from one form to the other, or from one object to the other.

The only way that mechanical energy can be transferred is by a force acting through a distance. This **transfer of energy** is called **work**, and its relationship to force and distance agrees with most people's common understanding of the word.

If a horse is pulling on a cart, for example, most people would agree that the horse is doing work on the cart. Most people would also agree that the amount of work is proportional to the force with which the horse must pull:

$$\text{Work} \propto F$$

This seems reasonable, because one horse pulling at, say, 200 newtons, would do a certain amount of work, while two horses pulling together with twice the force, say, 400 newtons, would do twice as much work.

It also seems reasonable to most people that the amount of work done by the horse is directly proportional to the distance over which the force is applied.

$$\text{Work} \propto d$$

If the horse pulled the cart 3 meters, for example, most people would agree that he had done half as much work as if he had pulled it 6 meters. Here again, the reasoning proceeds from the consideration of the two horses which could divide the work evenly between them. We could let the horses run relays, having one horse pull for 3 meters, and then allowing the other horse to take over to pull the other 3 meters.

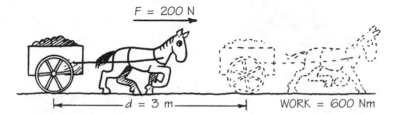

Figure 8-1

Work is proportional to both force and distance

Therefore, the work done (energy transferred) is directly proportional to both the force and the distance. Anything that is proportional to two separate things is also proportional to their product. We usually define work as the product of force and distance, with no proportionality constant.

$$\text{Work} \equiv F \cdot d$$

If the horse in our example pulled with a force of 200 newtons over a distance of 3 meters, it did 600 newton-meters of work. In the metric system this unit of work, the newton-meter, is called a **joule** (J):

$$\text{Work} \equiv F \cdot d$$
$$1\,\text{J} \equiv (1\,\text{N})(1\,\text{m})$$

CHECK QUESTION

Two horses wish to help each other pull a cart. They try teaming up in two different ways. In the first way, they take turns; in the second, they pull together. (a) How much work is done if one horse pulls the cart with a force of 200 N for a distance of 3 m and then the second horse pulls, also with a force of 200 N, for a second distance of 3 m? (b) How much work is done if they pull together with a combined force of 400 N for a distance of 3 m?
Answer: (a) 1,200 J, (b) 1,200 J

There are some peculiarities inherent in the definition of work. The first is that the force must act over some distance or else exactly zero work is done. It may not surprise you that you do no work on a wall when you merely lean against it, but it might be surprising to learn that the same thing is true even if you push forcefully against the wall. Sweat may stand out on your forehead, but unless the wall moves some, you are doing zero work on the wall. The wall will be in the same place where it was before you pushed. You will not have changed its condition at all. Any work you might have done, therefore, will have been only on yourself. Energy must be transferred when work is done, at least as the word *work* is used in physics. No force, however great, will transfer energy from one object to another unless that force acts through some distance.

Figure 8-2

Force times zero distance equals no work done.

Figure 8-3

Even a great force times zero distance produces no work. Here no work is done on the wall, as work is energy transfer and the wall remains in the same condition as it was before any force was applied.

Another peculiarity that must be included in our definition of *work* is the requirement that the force and the distance must be in the same direction. It would seem reasonable to assume that a horse would be concerned only in how hard he must pull in the forward direction to make a cart move. The same force might be needed to pull a heavily loaded cart having good wheel bearings or a lightly loaded cart having lousy wheel bearings. The amount of work the poor horse must do would be the same in either case although the weight, or force perpendicular to the ground, would be much greater in one case. The only part of the force that does any work is therefore the component that is **in the direction of the displacement**.

A woman pushing a lawnmower applies a force that pushes the mower forward and also against the ground. The part of the force that is forward does the work, while the part of the force that is perpendicular to the ground only helps to hold the wheels against the ground. Although this downward component serves a useful function in helping to keep the lawnmower's wheels from slipping, it still does no work. Although this

Figure 8-4

The resultant force applied by a woman pushing a lawnmover is made up of two components. Only the forward component, that component in the direction of the distance, does any work.

might seem strange at first, the same resultant force applied by the woman could be supplied as well by the combined efforts of a large stone to hold the lawnmower down and a small beast of burden to supply the forward force. Most people would agree that the small beast does all the work in this situation. The stone would not seem to do any work as long as the lawnmower is pulled on level ground. The same is true with the woman pushing downward at an angle. The downward component does no work; all the work is done by the forward component.

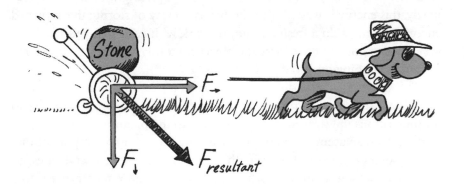

Figure 8-5

The forward component could be provided by a worker, while the downward component could be provided by the weight of an inert object, and the same resultant force would act on the lawnmower as was applied by the woman.

Figure 8-6

Little or no work is done in carrying something horizontally.

This applies also to work done on an object when you carry it across the room. You might think that you do a lot of work when you carry a very heavy object horizontally from one place to another. That is not true. You do work on an object when you lift it, because the force and the displacement are in the same direction. But you do no work in holding an object stationary, however tired you might become, and you do very little more work if you move it sidewards as you hold it up. The energy you use in holding an object at a stationary height is like the isometric exercises you might do in pushing against a stationary wall that refuses to budge. No energy is transferred to the object. No more energy is transferred if you happen to be moving sidewards, since the force and the displacement are perpendicular. We might understand this result by supposing that you are wearing roller skates while holding the heavy object. If someone were to give you a forward shove, you would do no more work on the object while coasting along than you would if you had just stood there holding your burden stationary.

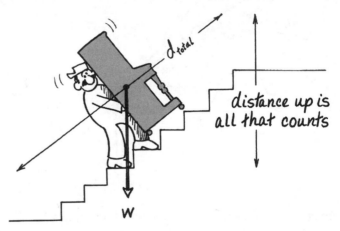

Figure 8-7

The only distance that counts for doing work in carrying something upstairs is the component that is in the upward direction.

Of course, a person who is walking with a heavy burden does a little more work than someone who is coasting on roller skates. A person taking a step bobs up and down a little bit. He is a little closer to the ground when his legs are at an angle as he steps out than when his legs are vertical as his feet come together. This work is not permanently transferred to the object, however, because the work you do on an object in bobbing upward is returned to you as you bob downward in completing the step. Unfortunately, this energy is wasted. Your body has no way of storing this returned energy; you must do a fresh amount of work with each step you take. The best you can hope to do is carry the object smoothly and reduce the energy wasted in bobbing up and down.

The situation is a bit different if we carry a piano upstairs. In this case there is a displacement in the direction of the force. The force necessary to lift the piano is upward, equal in magnitude to the piano's weight, and part of the total displacement is upward. It is only the upward component that counts for doing work, however; the sideward component still does no work. The same amount of work is done on the piano in carrying it to a certain height no matter how steep the stairs are or whether they are straight or spiraled. In any case, the work done is just the weight of the piano times the height, which is the upward component of the displacement. You would notice the additional energy given the piano in raising it up the stairs if you should chance to drop it from the greater height. If the piano were to make but one crashing bounce, the loudness of the sound would be equally great no matter what the steepness of the stairs.

In the example with the lawnmower, we split the force up into components. In the example with the piano, we split the displacement up into components. We used that component of the force in the direction of the

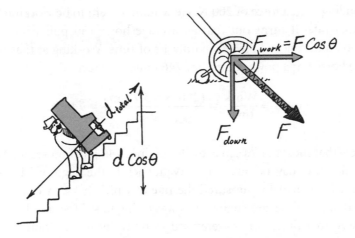

Figure 8-8

The force must be in the same direction as the distance or the distance in the same direction as the force. Either case leads to the same result.

distance in the one case and that component of the distance in the direction of the force in the other. We should note at this point that these two approaches are algebraically equivalent. They are just two different ways of looking at the same thing and give exactly the same result.

$$\text{Work} \equiv (F \cos\theta)d \quad \text{(Lawnmower case)}$$

$$\text{Work} \equiv F(d\cos\theta) \quad \text{(Piano case)}$$

$$\text{Work} \equiv F \cdot d \cdot \cos\theta \quad \text{(Either case)}$$

Power

Returning to the horse-and-cart example, notice that the amount of work done by the horse is independent of time. The horse does 600 joules of work when he pulls with 200 N over 3 meters, whether in 1 second or in 10 seconds. It might seem that the horse is working harder if he covers the distance in 1 second, but no more work gets done—at least not in the same sense that we have defined work.

We need to account for this idea of working harder when work is done faster. Because we live in a time-oriented society, we are frequently more interested in how fast work gets done than in the total amount that eventually is done. The rate of doing work is so important to us that we give it a separate name, **power**. Power is defined as the ratio of work done to the time necessary to do it.

$$\text{Power} \equiv \frac{\text{Work}}{\text{time}}$$

Power is analogous to velocity. Both concepts are rates. Whereas one is the rate of traveling distance, the other is the rate of doing work. The shorter the time, in both cases, the greater the rate. The horse would only be creeping along, for example, if he took 10 seconds to travel the 3 meters we were talking about. A velocity of only 0.3 m/s would be slow for a horse, even pulling with a force of 200 N. He wouldn't seem to be working hard at all at that rate. It turns out that an average horse can pull with a 200 N force over a distance of 3.73 m in only 1 s of time. Working at that rate, the average horse has a power output of 746 joules per second.

$$\text{Power} \equiv \frac{\text{Work}}{\text{time}} = \frac{(200 \text{ N})(3.73 \text{ m})}{(1.0 \text{ s})} = 746 \text{ J/s}$$

We know that most horses can actually produce 746 J/s of power for hours at a time, day after day, because James Watt, back in the reign of King George III of England, actually measured the rate at which horses work. Watt had invented improved steam engines and was trying to sell them to English coal miners. At that time, horses were used to pump water to drain the mines. The mine operators were interested in how many horses a particular steam engine would replace. Actual measurements on poor hard-working horses showed that an average horse could do 746 J of work each second. We therefore call this rate of doing work a *horsepower*, abbreviated hp.

$$1 \text{ hp} = 746 \text{ J/s}$$

James Watt built a one horsepower engine so that his steam engine would replace one horse.

James Watt's study of power as well as his technical contributions to the industrial revolution have led us to name the metric unit of power after him. As you recall, the metric unit of work is named after Sir James Joule. One joule of energy is the amount of work done when a force of one newton acts over a distance of one meter[1].

$$1 \text{ J} \equiv (1 \text{ N})(1 \text{ m})$$

A watt is then defined as one joule of energy per second.

$$1 \text{ W} \equiv 1 \text{ J/s} = 1 \text{ N} \cdot \text{m/s}$$

---

[1]A newton is, as you may recall, that amount of force required to accelerate a 1-kg mass at 1 m/s^2.

People rarely hear of a joule of energy except in a physics course, whereas the watt is a household word. A 100-watt light bulb uses 100 joules of energy per second. Our familiarity with Watt's name is further evidence that ours is a time-oriented society.

Figure 8-9

A 100-W lamp uses 100 J of energy each second.

CHECK QUESTION

A person pushes with a force of 300 N over a distance of 0.50 m in 1.0 s. (a) How much work is done? (b) If one horsepower is equal to 746 W, what is the power produced in horsepower?

Answer: (a) 150 J, (b) 0.20 hp

Forms of Energy

We started this chapter by considering the work done by a force acting through a distance. The definition of *work,* as the word is used in physics, was shown to be reasonable by an example in which a horse was pulling a cart at a constant rate. Other examples showed that the force must act through a distance greater than zero and that the force and the distance must be in the same direction in order to do any work. Also, the amount of work done is independent of time, so that we need to define power to handle situations in which time is a factor.

In all our examples, energy has been *transferred* from one object to another. We have thus defined work, but what is energy? Work changes the energy content of an object, but how can you tell? The fact is that there is no single measure of the energy content of an object, because energy can take on different forms. The energy added to the horse cart shows up as heat in the wheel bearings. Part of the energy added to the lawnmower goes into cutting the grass. Energy added to a piano can show up in its height as it is carried upstairs, or in its velocity as it falls back down. A moment later, the piano's energy might be liberated in sound if it should come to a crashing halt as it meets the ground.

Heat and sound are two forms of energy that will be considered in Chapters 10 and 11. For the remaining part of this chapter, we will discuss the forms of energy associated with an object's motion and position: kinetic energy and potential energy.

Kinetic Energy

Moving objects have energy by virtue of their motion. This energy of motion is called **kinetic energy** after the Greek word for "moving." The examples we have considered so far were selected to ignore the change in kinetic energy. The objects were moving at a constant velocity or were lifted from a position at rest to another position at rest. The forces were all balanced, so that all the work done would either go into frictional heating or gravitational potential energy. But what of the energy necessary to produce the motion in the first place? Kinetic energy is produced by an *unbalanced force* that accelerates the object from rest to its final velocity. The amount of kinetic energy an object has depends only on how much mass it has and on its velocity.

The relationship between kinetic energy, mass, and velocity can be found by considering the work done in accelerating the object from rest to its final velocity. Assume that the kinetic energy of an object comes entirely from a constant accelerating force acting over a distance.

$$E_k = \text{work done in accelerating the body}$$
$$= F_{\text{acceleration}} \cdot d$$

The amount of acceleration produced by the force is related to the mass of the object by Newton's second law:

$$F_{\text{acceleration}} = ma$$

And the amount of distance, *d,* over which this force is applied can be found from the average velocity of the object:

$$d = v_{\text{ave}}t$$

In terms of acceleration and average velocity, the kinetic energy of the object can be stated thus:

$$E_k = (ma)(v_{\text{ave}}t)$$

Assuming that the object started from rest and accelerated uniformly up to the final velocity, acceleration is simply final velocity over time, and average velocity is simply one-half the final velocity.

$$a = \frac{v_f}{t}$$

$$v_{\text{ave}} = \frac{v_f}{2}$$

Putting these relationships into the expression for kinetic energy,

$$E_k = m\left(\frac{v_f}{t}\right)\left(\frac{v_f}{2}\right)t$$

$$\boxed{E_k = \tfrac{1}{2}mv^2}$$

The unit for all forms of energy is the joule which, as we saw when we discussed work, is equal to a newton-meter.

Thus, the kinetic energy of an object is one-half the mass times the square of the velocity. This relationship is worth remembering. The one-half comes from using average velocity to find the distance over which the force acts, and the velocity is squared because it figures into both the distance and the acceleration produced.

The squared velocity in the kinetic energy expression has great importance in transportation energy uses. As the velocity doubles, the kinetic energy becomes four times as great. In an automobile, this energy is lost when it is changed into heat by the brakes. One hundred and five kilometers per hour (sixty-five miles per hour) is only 17% greater than 90 kilometers per hour (fifty-six miles per hour), but $(105)^2 = 11,025$ is 36% greater than $(90)^2 = 8,100$. Thus an 17% increase of speed is accompanied by a 36% increase in kinetic energy. That is one reason that slightly lower speed limits are more economical than you might think in terms of energy and, hence, fuel consumption.

Kinetic energy considerations also give rapid transit trains an advantage over private automobiles in terms of energy economy. Each time an automobile stops, its brakes convert the kinetic energy to heat. The kinetic energy of a rapid transit train, on the other hand, can be partly recovered when the train stops. The brakes on some types of commuter trains are actually electrical generators that feed power back into the system for use by other trains.

CHECK QUESTION

An elephant having a mass of 1,000 kg is traveling along at 3.0 m/s. What is the kinetic energy of the elephant?

Answer: 4,500 J

Potential Energy

An object can have energy by virtue of its position as well as by virtue of its velocity. A massive object elevated some distance above the ground clearly has some energy associated with its height. The amount of energy can be figured from the work that was necessary to get it up there. Work is, as always, force times distance.

$$\text{Work} \equiv F \cdot d$$

The force necessary to lift an object is its weight, *w*, and the distance is its height, *h*.

$$\text{Work} = w \cdot h$$

Figure 8-10

The kinetic energy of a moving body is $\frac{1}{2}mv^2$.

The energy imparted to an object by lifting it to a height *h* *is* called its **gravitational potential energy**. As long as the height is small compared to the radius of the earth, we can think of the weight of the object as constant, given by the product of mass and the acceleration due to gravity.

$$E_p = mgh$$

Figure 8-11

The gravitational potential energy of an object is its weight times its height.

CHECK QUESTION

An elephant having a mass of 1,000 kg is 3.0 m above the ground. What is the potential energy of the elephant?
  Answer: 29,400 J

This kind of energy is called *potential* because it can be converted to kinetic energy if the object should fall. Consider the example of a group of students transfixed on the edge of their seats by a physics lecture. If one of those students should happen to fall, there would be a little less energy stored in their collective positions.

Assume for a moment that you are the student. You fall asleep while listening to a physics lecture and we wish to know how hard you will hit the floor. We already know how to calculate the velocity of an object in free fall over a certain distance. You may recall the formula we derived by combining the definition of acceleration with the definition of average velocity:

$$v_f{}^2 = v_i{}^2 + 2ad$$

Solving this formula for final velocity, assuming that the initial velocity is zero, we can get the answer from our understanding of motion under constant acceleration:

$$v_f = \sqrt{2gh}$$

We can get the same result, however, from the idea of conservation of energy. We can assume that your kinetic energy as you near the ground will be equal to the potential energy you had when you were perched on your seat.

$$E_k = E_p$$

Kinetic energy can be expressed in terms of velocity, and potential energy in terms of height.

$$\tfrac{1}{2}mv^2 = mgh$$

Solving this expression for velocity, we get the same result as we did from the assumption of constant acceleration.

$$v = \sqrt{2gh}$$

The nice thing about the energy calculation is that it doesn't assume constant acceleration. Let us invent a situation in which this difference becomes important. We had you falling from your seat in your sleep before; let us construct a dream in which you slide limply out of your seat and on to the floor. Since you are dreaming, it would be easy for us to eliminate friction entirely and imagine what would happen if all of your potential energy were to change into kinetic energy as you slip from your seat. The conservation of energy would send you out the door and down the hall with the same velocity whether you slid limply to the floor with a non-constant acceleration or if you had fallen directly to the floor with the acceleration due to gravity as in the first example. Your kinetic energy at the floor level would be the same in either case.

The conservation of energy rule is always useful in situations where the acceleration is not constant. A common example in which frictional losses can be ignored for most purposes is the pendulum. The energy of the pendulum bob is traded back and forth from potential to kinetic and back to potential again as the pendulum swings back and forth. The acceleration is not constant, but the velocity of the pendulum bob can nonetheless be easily calculated from the conservation of energy.

The energy of a pendulum bob is all potential as the bob comes to a stop at the end of each swing, and it is all kinetic as the bob swings through the middle at the lowest point of its swing. The velocity of the bob as it swings through this lowest point can be calculated from the height of the bob at the ends of its path, measured vertically from its lowest level. Ignoring friction, we can assume that the kinetic energy at the bottom of the swing is equal to the potential energy at the top.

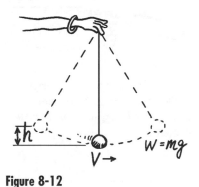

Figure 8-12

The energy of the pendulum changes back and forth between potential and kinetic.

$$E_k = E_p$$
$$\tfrac{1}{2}mv^2 = mgh$$

where *h is* the vertical height measured from the lowest point in the swing. This relationship can be solved for velocity by dividing both sides by $\tfrac{1}{2} m$.

$$v^2 = \frac{mgh}{\tfrac{1}{2}m} = 2gh$$

As in the previous example, an interesting thing happens. The mass divides out. Taking the square root of both sides of the equation:

$$v = \sqrt{2gh}$$

We see that the velocity at the bottom of the swing is proportional to the square root of the height of the swing and is *independent* of the mass of the pendulum. Most people find this to be an unexpected result. Your little brother, for example, probably wouldn't believe that light and heavy pendulums swing with the same velocity when raised to the same height unless you actually showed him. In fact, maybe you don't even believe it yourself. If not, try it and see. The mass in the kinetic energy expression $\frac{1}{2}mv^2$ really does balance out the effect of the mass in the potential energy expression mgh.

Elastic Potential Energy (Optional)[2]

Another form of potential energy is the kind that is stored in compressing and stretching a spring. This is a very common form of potential energy, because, as we pointed out in our discussion of Hooke's law in Chapter 3, nearly any solid object acts at least a little like a spring. The amount of force, F, that is needed to deform a solid object by stretching, compressing, bending, or twisting, is nearly proportional to the distance x through which the object is deformed, as long as the force and deformation remain within the so-called elastic limits of the object.

$$F = kx \quad \text{(Hooke's law)}$$

The letter k stands for the spring constant of the material and is unique to each material.

The energy stored in deforming a spring is called the **elastic potential energy** of the spring.

Let us consider the energy stored in the spring of a BB gun as an example of elastic potential energy. The force that must be applied to the cocking lever starts out small but increases in accordance with Hooke's law as the lever is pulled to its full extent. We can figure out the amount of work done, and hence the energy stored in the spring, if we look at a graph of force and distance, as shown in Figures 8-14 and 8-15. Such a graph may be looked upon as a diagram in a special mathematical space, where one dimension is represented by force while the other is represented by distance. Work, which is the product of force and distance, would be represented in this space by area. All the work situations considered until now have been ones where the force was constant. The energy transferred by a constant force acting over a distance would be represented by a rectangle in this force-distance space. The work done on a spring, however,

Figure 8-13

Part of the elastic potential energy stored in the spring of a BB gun is given to the BB in the form of kinetic energy.

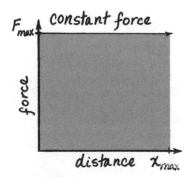

Figure 8-14

The shaded rectangular area is the work done by a constant force of F_{max} over a distance of x_{max}.

[2]There are no homework questions on this section. It is included because it is an important idea.

would be represented by a *triangular* area in this mathematical space. The work needed to cock the gun would be force times distance, as always, but the force actually needed to cock the gun is always smaller than the maximum value F_{max} that is reached when the gun is fully cocked. In accordance with Hooke's law, the actual force increases along a sloped line that forms the hypotenuse of a triangle in force-distance space. The area of this triangle, and hence the work actually done, is one-half the area of the rectangle formed by the maximum force acting over the same distance. It is therefore clear that the work done by the constantly increasing force is half as much as the work that would have been done by the maximum force if it had acted through the full distance.[3] From the definition of work,

$$\text{Work} \equiv F \cdot d$$

we find that the potential energy stored in a spring is

$$E_p = \left(\frac{F_{max}}{2}\right)x_{max}$$

$$= \tfrac{1}{2}(kx)x$$

$$\boxed{E_p = \tfrac{1}{2}kx^2}$$

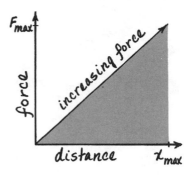

Figure 8-15

The triangular area represents the elastic potential energy stored by a spring, since the force increases as the spring compresses.

How to Calculate the Velocity of a BB (Optional)

We have already seen how to use the conservation of energy to figure out the maximum velocity of a pendulum bob from its potential energy at the end of each swing. We can now see how to modify our conservation of energy argument to calculate the velocity of a BB pellet from the elastic potential energy of the spring in the BB gun.

We must bear in mind that not all the energy stored in the spring goes into the kinetic energy of the pellet. A good deal of the energy goes into heat and sound as the spring is released. Some of the energy goes into the recoil of the BB gun. Even so, we can get some idea of the velocity of the BB as it leaves the gun if we estimate the gun's efficiency. Assume that only about 30% of the elastic potential energy stored in the spring finds its way into the BB pellet in the form of kinetic energy. This may be a crude estimate, but it will show us how to proceed.

$$E_k = 30\% E_p$$

[3]The situation is similar to the case of motion undergoing constant acceleration. In that case, the distance traveled was the average velocity times the time. The average velocity turned out to be half the maximum velocity when the initial velocity was zero. In the same way, we could use average force in the work definition, $\text{Work} = F_{ave} \cdot d$ where the average force is initial plus the final force divided by two. Since the initial force is zero, the average force is half the maximum force: $F_{ave} = \tfrac{1}{2}F_{max}$.

We now express the kinetic energy in terms of the mass, *m*, and velocity, *v*, of the BB pellet and the elastic potential energy in terms of the spring constant *k* and maximum stretch x_{max} of the spring:

$$\tfrac{1}{2}mv^2 = (0.30)(\tfrac{1}{2}kx_{max}^2)$$

Solving for the velocity of the BB, we get an algebraic result into which we could insert numerical data for a particular BB gun:

$$v^2 = \left(\frac{(0.30)(\tfrac{1}{2}kx_{max}^2)}{\tfrac{1}{2}m} \right)$$

$$v = \sqrt{\frac{(0.30)k}{m} \cdot x_{max}}$$

Of course, the estimate of 30% efficiency could be improved by making actual measurements on one particular model of BB gun, but the point is that the conservation of energy can still be used even when a known portion of the potential energy is lost because it changes into other forms of energy by such mechanisms as friction.

CHECK QUESTION

A BB gun with an efficiency of 30% has a spring with a spring constant of 533 N/m and shoots BB pellets with a mass of 1.0 g. When fully cocked, the spring is pulled back a distance of 20 cm. This gun can be fired from a half-cocked position, however, in which the spring is only pulled back a distance of 10 cm. (a) What is the velocity of the BB pellet as it leaves the gun after having been fired from the fully cocked position? (b) What is the velocity of the BB pellet when fired from the half-cocked position?
   Answer: (a) 80 m/s, (b) 40 m/s

Conservation of Energy and Momentum

People sometimes get the idea that energy and momentum are really different aspects of the same thing. Ask someone what happens to the momentum of a pendulum bob as the pendulum comes to the end of its swing. Our society is sufficiently sophisticated that most people seem to know that we should be able to account for the momentum that the pendulum bob has at the bottom of its swing but doesn't have when it comes to a stop at the end of its swing. The most frequent wrong answer is that the momentum changes into potential energy.

   The fact is that energy and momentum are separate things, which, though closely related, are separately conserved. Momentum turns out to be different from energy in the same sense that space is different from time. It is therefore not possible for the momentum of an ordinary pendulum bob to turn itself into some form of energy. While the kinetic energy of the pendulum changes into potential energy, something else must happen to the momentum. The proper answer to the question is that the momentum is transmitted through the string to the support and actually pulls the world along a little faster. This peculiar fact can be demonstrated by mounting a pendulum on a skate board. If started from rest, the skate board support will oscillate back and forth in the opposite direction from the pendulum bob, keeping the total momentum of the system constant.

One fine demonstration of the separate natures of energy and momentum is to allow one pendulum bob to bang into a whole row of stationary bobs. The apparatus for doing this classic physics demonstration was popularly available in toy stores and gift shops a while back. Perhaps you remember seeing a string of five metal balls, all separately hung in a straight line from a wooden frame. When one ball is lifted and allowed to swing down and bang into the others, its momentum is completely and dramatically absorbed. It just sits there, while one ball at the other end of the string bounces out with the same velocity, or at least nearly the same velocity, in the same direction.

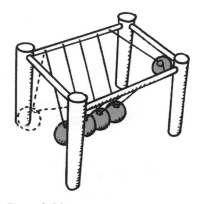

Figure 8-16

A series impact pendulum demonstrates the conservation of both energy and momentum.

This apparatus (which we might call a *series impact pendulum*) could be made to do other tricks as well. When two balls are lifted and allowed to collide with the others, two balls come out the other side. When three balls come in, three balls go out. Always the same number of balls come in as bounce out the other side. Furthermore, the number of balls that come out the other side does not seem to depend on the velocity of the incoming balls. It never seems to happen, for example, that two balls come in with a small velocity and one leaves with twice the velocity on the other side.

Some store clerks were quick to point out the educational value of this apparatus as a demonstration of the conservation of momentum, but there is more involved here. If you just look at the apparatus in terms of momentum, that does not entirely account for the way it works. One ball at a certain velocity has the same momentum as two balls traveling at half the velocity, but it never happens that when you drop two balls on the apparatus one ball comes shooting out the other end twice as fast. Somehow balls at one end seem to know how many balls came in at the other end. There must be some difference between the momentum contained in one fast-moving ball and the same amount of momentum contained in two slower-moving balls.

The difference turns out to be the energy. In the case where two balls come in with a velocity v, the same momentum is involved if one ball bounces out with twice the velocity, $2v$, as if two balls come out with the same velocity, v:

$$(2m)v = m(2v)$$

However, the kinetic energy of the one ball bouncing out would be twice as great as the energy of the two incoming balls (≠ means "is not equal to"):

$$\frac{1}{2}m(2v)^2 \neq \frac{1}{2}(2m)v^2$$

$$\frac{m \cdot 4v^2}{2} \neq \frac{2mv^2}{2}$$

$$2mv^2 \neq mv^2$$

Thus, two balls in and one ball out twice as fast would conserve momentum but not energy. The ball coming out would have twice the energy as the balls going in. There would be a net energy gain.

	momentum mv	E_k $\frac{1}{2}mv^2$
 (m m) $v \rightarrow$	$(2m)v =$ $2mv$	$\frac{1}{2}(2m)v^2 =$ mv^2
 (m) $2v \rightarrow$	$m(2v) =$ $2mv$	$\frac{1}{2}m(2v)^2 =$ $2mv^2$

Figure 8-17

One ball with twice the velocity of two balls would have the same momentum but twice the kinetic energy.

In a similar fashion, you would be surprised if you dropped one ball and two balls came out the other end, traveling half as fast. One ball coming in and two balls bouncing out with half the velocity would conserve momentum but not energy. The momentum of the two balls going out would be the same as the momentum of the incoming ball, but the energy would be only half as great. The balls are made of a hard material, such as steel, which is springy and doesn't absorb energy by permanently changing shape. Most of the kinetic energy of the ball as it comes in is therefore changed back into kinetic energy of the ball that comes out the other end, since little energy is converted into other forms. A little kinetic energy is lost to heat and sound, but nothing like half of it. Thus, you always observe the same number of balls coming out as going in.

Actually, you can sneak in a small change that will fix it so that two balls will come out when one ball goes in. If you put chewing gum between the last two balls, they will obviously stick together and they will both come out. You get the same result, however, even if you put the gum someplace down the line. The effect of the chewing gum is to cushion the impulse and absorb some of the energy. Momentum is conserved, but not the kinetic form of energy. Some of the kinetic energy is used in "chewing" the gum. While the total amount of energy is conserved, some of it is changed into the form of heat by the collision process.

The word *elastic* is used to describe how much of the energy retains its form as kinetic energy after a collision. A **perfectly elastic collision** would be a collision in which there would be just as much kinetic energy after the collision process as before. A perfectly elastic ball, for example, would be one that would bounce from a surface with the same kinetic energy after the bounce as before. Such a ball dropped on the floor would bounce up to the same height that you dropped it from. Obviously, because there is no such thing as a perfect exchange in real life, the steel balls in the toy-store-variety series impact pendulum are not perfectly elastic, but the proper operation of the device depends on the collision between balls being *nearly* elastic.

An **inelastic collision** is one in which some of the kinetic energy changes into heat, sound, or some other form of energy. Momentum is conserved in an inelastic collision, but not kinetic energy, since the energy of the system is transformed into other forms. One example of a perfectly inelastic collision would be our example in Chapter 7, where a little car collided with a truck and the wreckage fused together. As we saw in that example, not all the kinetic energy is lost in a completely inelastic collision. The wreckage may be moving after the collision. Much of the energy, however, does change form in an inelastic collision. The chewing gum between two balls of the momentum demonstration changes the largely elastic collision process into one that is largely inelastic.

PROBLEM SET 8

SUMMARY OF CONCEPTS

The problems in this chapter are related to work, power, and mechanical energy. **Work**, or energy transfer, is defined as the force times the distance over which the force acts:

$$\text{Work} = F \cdot d$$

The force and the distance must be in the same direction, because no work is done by any component of the force perpendicular to the distance. Thus, either the force or the distance must be resolved into components and only the parallel component be used to calculate work.

A special name is given to the unit of mechanical energy in the metric system: a newton-meter is called a **joule** and is abbreviated J.

$$1 \text{ J} = (1 \text{ N})(1 \text{ m})$$

A newton (N), remember, is a kilogram-meter per second squared:

$$1 \text{ N} \equiv 1 \text{ kg} \cdot \text{m/s}^2$$

The rate at which work is done is called **power**.

$$\text{Power} \equiv \frac{\text{Work}}{\text{time}}$$

The unit of power in the metric system is the watt, abbreviated W, and is defined as one joule per second. A commonly used measure of power is the horsepower, abbreviated hp.

$$1 \text{ W} \equiv 1 \text{ J/s}$$
$$1 \text{ hp} = 746 \text{ W}$$

The two common forms of mechanical energy are **kinetic** and **potential**. The kinetic energy of an object of mass m and velocity v is given by

$$E_k = \tfrac{1}{2}mv^2$$

The potential energy of an object is the energy stored in a restoring force, such as in a gravitational field or a spring. The potential energy of an object at a height h, near the surface of the earth (where its weight $w = mg$ is constant) is given by the definition of work, where the force is the weight and the distance is the height:

$$E_p = mgh$$

The potential energy stored in a spring, which obeys Hooke's law, $F = kx$, with a spring constant k, turns out to be

$$E_p = \tfrac{1}{2}kx^2$$

Energy, like momentum, is a conserved quantity. The relationship between velocity and the position of an object can be found from the conservation of energy in a whole class of problems where the loss of mechanical energy to other forms, such as heat, is either negligible or known.

$$E_{before} = E_{after}$$

In a completely elastic collision, for example, the frictional losses of energy are negligible; the kinetic energy before the collision equals the kinetic energy after. In a completely inelastic collision, on the other hand, much of the kinetic energy is changed into heat as the two objects stick together. In either case, however, momentum is conserved.

EXAMPLE PROBLEMS

Try to do the Examples yourself before looking at the Sample Solutions.

8-1 A small boy is helping his father change the tire on the family automobile. The car is not high enough for the fully inflated tire. The conservation of energy allows the boy's puny little force applied to the jack to lift the massive auto.

Given: The boy exerts a force of 80 N to move the jack handle a distance of 30 cm. Seventy per cent of the energy put into the jack comes out as useful work in lifting the automobile, which weighs 12,000 N, the rest going into heat.

Find: (a) How much work does the boy do in one stroke of the jack?
(b) How much work goes into lifting the automobile?
(c) How high does one stroke of the jack lift the car?

BACKGROUND

This problem illustrates the concept of work in a situation where a percentage of the energy is lost because of friction. From the percentage of the energy that does useful work, the distance over which the output force acts can be determined.

8-2 A carpenter's apprentice is dragging a wooden toolbox across a stone floor at constant velocity. (Most of the work he does goes into heating the floor and the toolbox, although some of it goes into sound, which is radiated in all directions.)

Given: The apprentice pulls with a force of 200 N on a rope that makes an angle of 30° with the horizontal floor. He pulls the toolbox a distance of 10 m along the floor.

Find: (a) How much of the force is in the direction the box actually moves?
(b) How much work does the carpenter's apprentice do?

BACKGROUND

This problem is an illustration of work done by a force that is not entirely in the direction of the distance over which the force acts. Part of the force does work, therefore, while the other does not.

8-3 A hoist is lifting concrete in the construction of a high-rise building. You wonder how much energy goes into lifting each bucket of concrete.

Given: The hoist lifts each bucket of concrete a vertical distance of 10 m before swinging it over a horizontal distance of 45 m. Each bucket of concrete has a mass of 2.0 metric tons (2,000 kg).

Find: (a) What is the weight of a bucket of concrete?
(b) How much work goes into lifting each bucket of concrete?
(c) How much work goes into swinging the concrete sidewards?

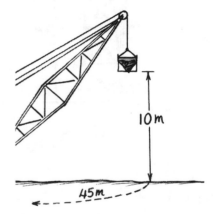

BACKGROUND

This is a work problem involving a transfer of energy into potential energy. The force involved is the weight of the bucket. Both vertical and horizontal displacements are separately considered.

8-4 A tractor is pulling a plow at a uniform constant velocity; all the energy developed by the tractor is used in frictional losses and in lifting the soil.

Given: The tractor is delivering power at the rate of 36 hp to pull the plow at a velocity of 2.6 m/s (1 hp = 746 W).

Find: (a) How much work is done each second, in joules?
(b) With how much force does the tractor pull?

BACKGROUND

This is a power problem in which work is being done at a constant rate. All the energy delivered by the tractor goes into frictional losses and in lifting the soil, so that no change in kinetic energy need be considered.

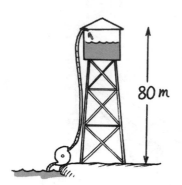

8-5 A large water tank is filled by pumping water into it from a reservoir below, using electric motor-driven pumps.

Given: 3.00×10^6 kg of water is lifted an average distance of 80 m in 12 minutes. The tank holds 1.5×10^7 kg of water.

Find: (a) How much work is done in the first 12 min?
(b) How much power is delivered by the pumps?
(c) How long will it take to fill the tank?

BACKGROUND

This is a power problem involving a continuous flow rate. A certain amount of energy is converted into the potential energy of the water in the tank in a certain amount of time. The amount of time it takes to fill the tank can be found from the power, or rate, at which work is being done.

8-6 A good part of the power developed by an automobile engine is lost in the transmission. A typical sports car might have an engine that produces 100 hp, for example, but the power available at the rear wheels to accelerate the automobile turns out to be only a small part of that.

Given: This sports car has a weight of 9,300 N and can accelerate to a velocity of 27 m/s in a measured time of 11 s.

Find: (a) What is the mass of the sports car?
(b) What is the kinetic energy of the car at the end of this acceleration?
(c) How much average power is delivered to the rear wheels during the acceleration (neglecting friction)?

BACKGROUND

This is a power problem in which work is being converted into kinetic energy at a constant rate. It is particularly important to distinguish between mass and weight when the kinetic energy of something is being considered.

8-7 A pile is a support driven into bedrock to support a structure. A pile driver consists of a large weight, called a hammer, that is repeatedly dropped from a considerable height onto the pile. Most of the kinetic energy of the hammer is converted upon impact into work used in forcing the pile into the ground.

Given: The pile hammer has a mass of 380 kg and is lifted 7.0 m above the top of the pile before being dropped. The impact drives the pile 20 cm into the ground.

Find: (a) What is the potential energy of the pile hammer before being dropped?
(b) What is the velocity of the hammer as it strikes the pile?
(c) With what average force does the hammer force the pile into the ground?

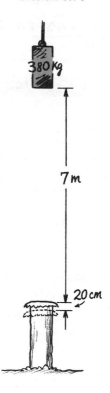

BACKGROUND

This problem illustrates the use of the work concept to find the force of impact from the energy delivered by the blow. The pile driver is simpler than some hammer problems in that this energy all comes from the potential energy of the hammer head before it is dropped. As in all hammer problems, however, the energy of the hammer head is all in the form of kinetic energy just before impact.

8-8 The brakes start to slip on a car parked on a hill without its wheels having been curbed. The automobile gains momentum as it travels down the hill in spite of the fact that the brakes absorb part of its energy.

Given: The automobile has a mass of 2,500 kg, and its brakes absorb 40% of its energy. The street is 160 m long and is inclined at an angle of 18° to the horizontal. (Thus only 60% of the car's potential energy changes into kinetic energy.)

Find: (a) What is the automobile's potential energy at the top of the street?
(b) What is the automobile's kinetic energy at the end of the street?
(c) How fast is the automobile going at the end of the street?

BACKGROUND

This problem illustrates conservation of energy in a situation where potential energy changes into kinetic energy with a known frictional loss. That portion of the potential energy that changes into kinetic energy determines the car's velocity.

Solutions to EXAMPLE PROBLEMS

Given:

The boy exerts a force of 80 N to move the jack handle a distance of 30 cm. Seventy per cent of the energy put into the jack comes out as useful work in lifting the automobile, which weighs 12,000 N, the rest going into heat.

Find:

(a) How much work does the boy do in one stroke of the jack?

(b) How much work goes into lifting the automobile?

(c) How high does one stroke of the jack lift the car?

(a) $\text{Work} = F \cdot d$

$$\text{Work}_{boy} = (80 \text{ N})(30 \text{ cm})\left(\frac{1 \text{ m}}{10^2 \text{ cm}}\right)\left(\frac{1 \text{ J}}{1 \text{ N} \cdot \text{m}}\right)$$

$$= \boxed{24 \text{ J}}$$

conversion factors

(b) $\text{Work}_{auto} = 70\% \text{ Work}_{boy}$

$$= (0.70)(24 \text{ J}) = \boxed{16.8 \text{ J}}$$

(c) $\text{Work}_{auto} = F_{auto} \cdot d_?$

$$d_? = \frac{\text{Work}_{auto}}{F_{auto}}$$

$$= \left(\frac{16.8}{12,000 \text{ N}}\right)\left(\frac{\text{N} \cdot \text{m}}{1 \text{ J}}\right)\left(\frac{10^3 \text{ mm}}{1 \text{ m}}\right)$$

$$= \boxed{1.4 \text{ mm}}$$

DISCUSSION

The work done by the boy can be found by taking the product of his force times the distance over which he acts. The units are changed into joules by the fact that a joule is defined as that amount of work done by a newton of force acting over a meter of distance.

The work coming out of the jack into the automobile is a simple percentage of the work done by the boy on the jack, as found in part (a).

The distance the automobile moves can be found by solving the definition of work, using the result of part (b) as the work and the weight of the automobile as the force.

Given:
The apprentice pulls with a force of 200 N on a rope that makes an angle of 30° with the horizontal floor. He pulls the toolbox a distance of 10 m along the floor.

Find:
(a) How much of the force is in the direction the box actually moves?

(b) How much work does the carpenter's apprentice do?

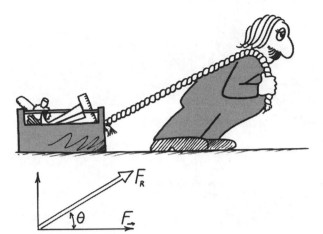

(a) $F_\rightarrow = F_r \cos\theta$

$\quad = (200\ N)(\cos 30°)$

$\quad = \boxed{173\ N}$

(b) $Work = F_\rightarrow \cdot d$

$\quad = (173\ N)(10\ m)\left(\dfrac{1\ J}{1\ N \cdot 1\ m}\right)$

$\quad = \boxed{1.7 \times 10^3\ J}$

DISCUSSION

The force acting on the toolbox is resolved into horizontal and vertical components, using the sine and cosine of the angle. Only that component of the force in the direction of the distance is of interest in this problem, however, since that is the only component that does work. The other component tends to reduce drag by helping to lift up on the toolbox, but it does not transfer energy, as there is no displacement in that direction.

SAMPLE SOLUTION 8-3

Given:

The hoist lifts each bucket of concrete a vertical distance of 10 m before swinging it over a horizontal distance of 45 m. Each bucket of concrete has a mass of 2.0 metric tons (2,000 kg).

Find:

(a) What is the weight of a bucket of concrete?

(b) How much work goes into lifting each bucket of concrete?

(c) How much work goes into swinging the concrete sidewards?

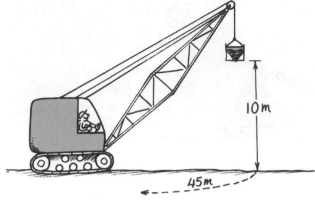

10 m

45 m

(a) $w = mg$

$$= (2,000 \text{ kg})(9.8 \text{ m/s}^2)\left(\frac{1\text{N}}{1 \text{ kg} \cdot 1 \text{ m/s}^2}\right)$$

$$= \boxed{19,600 \text{ N}}$$

(b) Work $= F_{\parallel} \cdot d_{\uparrow}$

$$= (19,600\text{N})(10\text{m})\left(\frac{1 \text{ J}}{1\text{N} \cdot 1\text{m}}\right)$$

$$= \boxed{2.0 \times 10^5 \text{ J}}$$

(c) Work $= F_{\parallel} \cdot d_{\rightarrow}$

$$= (0)(45 \text{ m})$$

$$= \boxed{0}$$

DISCUSSION

The weight in part (a) is the product of mass and the acceleration due to gravity.

The work done is the product of force times distance. A parallel symbol, ∥, is used as a subscript on the force to emphasize the fact that the force and distance must be in the same direction. When the vertical distance is considered, the vertical force, which is the weight found in part (a), must be used. When the horizontal distance is considered, as in part (c), there is no force to be used, since the weight is all in the vertical direction. The hoist transfers no energy to the concrete bucket when swinging it sidewards. The only work that is done is in lifting it.

Given:

The tractor is delivering power at the rate of 36 hp to pull the plow at a velocity of 2.6 m/s. (1 hp = 746 W.)

Find:

(a) How much work is done each second, in joules?

(b) With how much force does the tractor pull?

(a) $\text{Power} = \dfrac{\text{Work}}{\text{time}}$

$\text{Work} = \text{Power} \cdot \text{time}$

$= (36\ \text{hp})(1\ \text{s})\left(\dfrac{746\ \text{W}}{1\ \text{hp}}\right)$

$= \boxed{2.7 \times 10^4\ \text{J}}$

(b) $\text{Work} \equiv F \cdot d$

$F = \dfrac{\text{Work}}{d}$

$= \dfrac{2.7 \times 10^4\ \text{J}}{2.6\ \text{m}}$

$= \boxed{1.0 \times 10^4\ \text{N}}$

DISCUSSION

Power is defined as the rate at which work is done. The amount of work done, or change of work energy, in a unit of time can therefore be found as the product of power and the time increment involved. The units can be changed by using the relationship that 1 horsepower is equal to 746 watts.

The force exerted on the plow can be found by solving the definition of work, using the energy and distance for the same period of time. An alternative approach would be to express power in terms of force and velocity and solve for force.

$$\text{Power} = \dfrac{\text{Work}}{\text{time}} = \dfrac{F \cdot d}{t} = F\dfrac{d}{t} = Fv$$

The result is a little more than 1 ton of force either way.

SAMPLE SOLUTION 8-5

Given:
3.00×10^6 *kg of water is lifted an average distance of 80 m in 12 minutes. The tank holds* 1.5×10^7 *kg of water.*

Find:
(a) How much work is done in the first 12 min?

(b) How much power is delivered by the pumps?

(c) How long will it take to fill the tank?

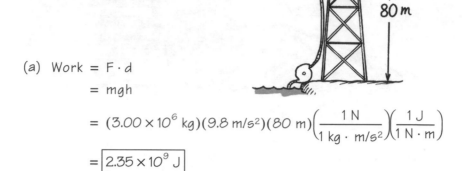

(a) Work $= F \cdot d$

$\qquad = mgh$

$\qquad = (3.00 \times 10^6 \text{ kg})(9.8 \text{ m/s}^2)(80 \text{ m})\left(\dfrac{1 \text{ N}}{1 \text{ kg} \cdot \text{m/s}^2}\right)\left(\dfrac{1 \text{ J}}{1 \text{ N} \cdot \text{m}}\right)$

$\qquad = \boxed{2.35 \times 10^9 \text{ J}}$

(b) Power $\equiv \dfrac{\text{Work}}{\text{time}}$

$\qquad = \dfrac{(2.35 \times 10^9 \text{ J})}{(12 \text{ min})}\left(\dfrac{1 \text{ min}}{60 \text{ s}}\right)\left(\dfrac{1 \text{ W}}{1 \text{ J/s}}\right)$

$\qquad = \boxed{3.26 \times 10^6 \text{ W}}$

(c) Power $\equiv \dfrac{\text{Work}}{t}$

$\qquad t = \dfrac{\text{Work}}{\text{Power}} = \dfrac{mgh}{\text{Power}}$

$\qquad = \dfrac{(1.5 \times 10^7 \text{ kg})(9.8 \text{ m/s}^2)(80 \text{ m})}{(3.26 \times 10^6 \text{ W})}\left(\dfrac{1 \text{ W}}{1 \text{ J/s}}\right)\left(\dfrac{1 \text{ J}}{1 \text{ N} \cdot \text{m}}\right)$

$\qquad = 3.6 \times 10^3 \text{ s} = \boxed{60 \text{ min}}$

DISCUSSION

The work done in a given amount of time is just the potential energy of that amount of water that is lifted in that amount of time. Potential energy is given by *mgh*, since the force involved in the definition of work is then weight, *mg*, and the distance is the height, *h*.

The power found in part (b) is the work done in part (a) divided by the given amount of time necessary to do that much work.

The time necessary to fill the tank of water can be found from this power by solving the definition of power, assuming that the work being done is the potential energy of a full tank of water.

Given:
This sports car has a weight of 9,300 N and can accelerate to a velocity of 27 m/s in a measured time of 11 s.

Find:
(a) What is the mass of the sports car?

(b) What is the kinetic energy of the car at the end of this acceleration?

(c) How much average power is delivered to the rear wheels during the acceleration (neglecting friction)?

(a)
$$w = mg$$

$$m = \frac{w}{g}$$

$$= \frac{9{,}300 \text{ N}}{9.8 \text{ m/s}^2} = \boxed{9.5 \times 10^2 \text{ kg}}$$

(b)
$$E_k = \tfrac{1}{2}mv^2$$

$$= \tfrac{1}{2}(9.5 \times 10^2 \text{ kg})(27 \text{ m/s})^2$$

$$= \boxed{3.5 \times 10^5 \text{ J}}$$

(c)
$$\text{Power} = \frac{\text{Work}}{\text{time}}$$

$$= \frac{(3.5 \times 10^5 \text{ J})}{11 \text{ s}}$$

$$= \boxed{3.2 \times 10^4 \text{ W}}$$

DISCUSSION

The mass of the sports car is found by solving the relationship between mass and weight.

The kinetic energy of the car at a given velocity can be found from this mass by using the formula for kinetic energy, which is derived in the text. All of the kinetic energy found in part (b) comes from the power delivered to the rear wheels. To find the answer in joules, keep in mind the conversion factors:

$$1 \text{ kg}\frac{\text{m}^2}{\text{s}^2} \cdot \left(\frac{1 \text{ N}}{\frac{\text{kg} \cdot \text{m}}{\text{s}^2}}\right) = \text{N} \cdot \text{m}$$

$$1 \text{ N} \cdot \text{m} = 1 \text{ J}$$

We can find the power since we know the work done, in this case the energy transfer, from part (b) and we are given the time. We are, of course, neglecting frictional losses that take place while the automobile is accelerating. The result of our power calculation can be converted to more familiar units by using the relationship between watts and horsepowers (1 hp = 746 W) as a conversion factor.

Since the answer to part (c) is only 42 hp, which is only 42% of the rated horsepower of the engine, over half the power is lost in the sports car transmission in this example.

SAMPLE SOLUTION 8-7

Given:

The pile hammer has a mass of 380 kg and is lifted 7.0 m above the top of the pile before being dropped. The impact drives the pile 20 cm into the ground.

Find:

(a) What is the potential energy of the pile hammer before being dropped?

(b) What is the velocity of the hammer as it strikes the pile?

(c) With what average force does the hammer force the pile into the ground?

(a) $E_p = mgh$

$\qquad = (380 \text{ kg})(9.8 \text{ kg} \cdot \text{m/s}^2)(7.0 \text{ m})\left(\dfrac{1 \text{ J}}{1 \text{N} \cdot \text{m}}\right)\left(\dfrac{1\text{N}}{1 \text{ kg} \cdot \text{m/s}^2}\right)$

$\qquad = \boxed{2.6 \times 10^4 \text{ J}}$

(b) $E_k = \frac{1}{2}mv^2$

$\qquad v = \sqrt{\dfrac{2E_k}{m}}$

$\qquad = \sqrt{\dfrac{2(2.6 \times 10^4 \text{ J})}{(380 \text{ kg})}} = \boxed{11.7 \text{ m/s}}$

(c) Work $= F \cdot d$

$\qquad F = \dfrac{\text{Work}}{d}$

$\qquad = \dfrac{(2.6 \times 10^4 \text{ J})}{(20 \text{ cm})}\left(\dfrac{10^2 \text{ cm}}{1 \text{ m}}\right)\left(\dfrac{1 \text{ N} \cdot \text{m}}{1 \text{ J}}\right)$

$\qquad = \boxed{1.3 \times 10^5 \text{N}}$

DISCUSSION

Potential energy is weight, *mg*, times height, *h*. This energy all shows up as kinetic energy at the time of impact, and the velocity of the hammer head can be found from it, as shown in part (b). An alternative approach would be to find the velocity from one of the equations for motion under constant acceleration. You may remember from Chapter 4 that

$$v_f{}^2 = v_i{}^2 + 2ad$$

It is instructive to try both approaches to show that they give the same result.

The units in part (b) work out to meters per second if you express joules in terms of kilograms, meters, and seconds:

$$\sqrt{\frac{\text{J}}{\text{kg}}} = \sqrt{\frac{\text{N} \cdot \text{m}}{\text{kg}}} = \sqrt{\frac{(\text{kg} \cdot \text{m/s}^2) \cdot \text{m}}{\text{kg}}} = \sqrt{\frac{\text{m}^2}{\text{s}^2}} = \frac{\text{m}}{\text{s}}$$

We find part (c) from the definition of work. In this case, we can assume that all the potential energy goes into the work of driving the pile.

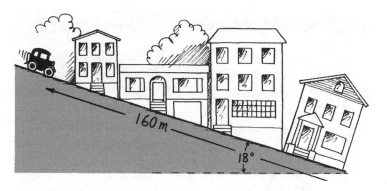

Given:
The automobile has a mass of 2,500 kg, and its brakes absorb 40% of its energy. The street is 160 m long and is inclined at an angle of 18° to the horizontal. (Thus only 60% of the car's potential energy changes into kinetic energy.)

Find:
(a) What is the automobile's potential energy at the top of the street?

(b) What is the automobile's kinetic energy at the end of the street?

(c) How fast is the automobile going at the end of the street?

(a) $E_p = mgh$

$= (2,500 \text{ kg})(9.8 \text{ m/s}^2)[(160 \text{ m})(\sin 18°)]$

$= \boxed{1.21 \times 10^6 \text{ J}}$

(b) $E_k = (100\% - 40\%)E_p = 60\% \, E_p$

$= (0.60)(1.21 \times 10^6 \text{ J})$

$= \boxed{7.27 \times 10^5 \text{ J}}$

(c) $E_k = \frac{1}{2} m \upsilon^2$

$\upsilon = \sqrt{\dfrac{2E_k}{m}}$

$= \sqrt{\dfrac{2(7.27 \times 10^5 \text{ J})}{(2,500 \text{ kg})} \left(\dfrac{1\text{N} \cdot 1 \text{ m}}{1 \text{ J}}\right)\left(\dfrac{1 \text{ kg} \cdot 1 \text{ m/s}^2}{1 \text{ N}}\right)}$

$= \boxed{24 \text{ m/s}}$

DISCUSSION

The potential energy of the automobile can be expressed in terms of the length of the roadway by taking the height as the hypotenuse of a right triangle times the sine of the angle opposite the height. The units of kilograms times meters per second squared times meters work out to joules, as follows:

$$(\text{kg})(\text{ m/s}^2)(\text{m}) = (\text{N})(\text{m}) = \text{J}$$

The kinetic energy is found in part (b) as a simple percentage of the potential energy found in part (a), as stated in the problem.

Velocity can be found by solving the kinetic energy expression, using the result of part (b).

ESSENTIAL PROBLEMS

Be sure you can do the Example Problems (without peeking at the Sample Solutions) before working these Essential Problems.

8-9 Lin-Ling is doing bicep curls. Her muscle exerts a force which pulls a ligament in her arm a small distance, which moves the arm a larger distance.

Given: Lin-Ling's muscle exerts a force of 791 N to move the ligament 2.5 cm. 78% of the energy of the muscle goes to lifting the 53.4-N barbell (the rest goes into lifting her arm and into heat.)

Find: (a) How much work does the muscle do on the ligament?
(b) How much of this work is used in lifting the barbell?
(c) How far does this ligament action move the barbell?

8-10 Paul Hew is cutting the grass again. The force he applies to the lawnmower handle performs two functions. It helps to hold the wheels down on the ground so they don't slip, and it pushes the lawnmower forward at a constant velocity.

Given: Paul is pushing with a force of 250 N at an angle of 38° to the (horizontal) ground. The lawnmower weighs 100 N. Paul pushes a distance of 8.0 m before stopping to wipe his brow.

Find: (a) What is the horizontal component of the force applied by Paul?
(b) How much work does Paul do before wiping his brow?

8-11 A circus acrobat climbs onto her brother's shoulders while he, in turn, is standing on his father's shoulders. They practice the feat while her father is standing on the ground, although the whole thing is to be done on the back of a galloping horse in an actual performance.

Given: The female acrobat has a mass of 50 kg and must climb a height of 3.0 m while the horse gallops a distance of 7.0 m around the ring.

Find: (a) How much does the female acrobat weigh?
    (b) How much work must she do in practice while her father is standing stationary on the ground?
    (c) How much additional work must she do in an actual performance while the horse is carrying her (and everyone else) sidewards?

8-12 A daddy is going for a run with his baby stroller and baby. The baby stroller has been designed to be low-friction.

Given: The daddy is using power at a rate of 2.6×10^{-2} hp to run at a velocity of 1.3 m/s. A horsepower is 746 watts.

Find: (a) How much work does he do each second, in joules?
    (b) With how much force is he pushing the stroller?

8-13 A gravedigger is shoveling dirt at a constant rate. He wishes to know how long it will take to dig a specifically sized hole.

Given: The gravedigger can lift one 7.6-kg shovelful of dirt at an average rate of 1.53 m in 7.8 seconds. He is trying to empty a hole of 2.91×10^4 kg of dirt.

Find: (a) How much work is done in the first 7.8 seconds?
    (b) What is the power delivered by the gravedigger in lifting one shovelful of dirt?
    (c) How long will it take to empty the hole at this rate?

8-14 Annette Racer is measuring her acceleration on her ten-speed bicycle.

Given: Annette can accelerate from rest to a velocity of 7.6 m/s in 20 s. Almost all the energy goes into the kinetic energy of her and her bicycle, which have a combined weight of 490 N. (1 hp = 746 W)

Find: (a) What is the mass of Annette and her bicycle?
(b) What is the kinetic energy at the end of this acceleration?
(c) How much average power did Annette produce in horsepower?

8-15 As a child, little Karen Terrible had a game she thought was a lot of fun. She would make little clay people and then drop rocks on them to smash them flat.

Given: One time, Karen dropped a 2.3-kg rock from a height of 1.44 m onto a clay person, which ended up 4.5 cm shorter than it had been originally.

Find: (a) What was the potential energy of the rock while Karen was holding it overhead, just before she dropped it?
(b) With what velocity did the rock hit the little clay person?
(c) If all the kinetic energy of the rock went into flattening the clay figure, with what average force did the rock push on the little clay person?

8-16 A little person is going down a slide at the playground. She discovers that she has some control over her velocity by pressing her feet outward against the side rails of the slide so as to change the rate at which her energy is converted into heat by friction.

Given: The little person has a mass of 20 kg and starts at the top of a 15-m slide, which is inclined at an angle of 50° to the ground. Only 25% of her potential energy is converted into kinetic energy.

Find: (a) What is the little person's potential energy when she is at the top of the slide?
(b) How much kinetic energy does she have at the bottom of the slide?
(c) What is her velocity at the bottom of the slide?

MORE INTERESTING PROBLEMS

8-17 A sports car is entered in a hill-climbing race. In a long straight section of the race the car gets up to a constant velocity, which it can just maintain in fourth gear.

Given: The car has a mass of 900 kg and is climbing a 12° incline (to the horizontal) at a velocity of 30 m/s. (One horsepower = 746 W.)

Find: (a) How much work is done in 1 s?
(b) What is the power delivered to the rear wheels, in watts?
(c) What is the power delivered to the rear wheels, in horsepower?

8-18 In a race, a professional runner is keeping a constant velocity uphill. Professional athletes, like bicycle racers or runners, train to put out power. Even so, the maximum power output of a highly trained human is only about one-tenth of a horsepower.

Given: The runner has a mass of 68 kg. He is moving at a constant velocity of 3.7 m/s on a hill with a 12.1° incline, measured to the horizontal. One horsepower is 746 W.

Find: (a) How much work does the runner do in 1 s?
(b) What is his power output in watts?
(c) What is his power output in horsepower?

8-19 A woodsman is using an ax to split a log. He swings the ax from above his head and maintains a constant downward force, which adds to the original potential energy in giving the ax head kinetic energy with which to split the log.

Given: The ax head has a mass of 4.5 kg, and the woodsman swings it from a height of 2.4 m through an arc having a length of 3.1 m with a relatively constant force applied to it of 80 N.

Find: (a) How much potential energy does the ax have at the top of its swing?
(b) How much work does the woodsman do on the ax head?
(c) With what velocity does the ax hit the log on the ground? (Hint: its kinetic energy comes from its potential energy added to the work done.)

8-20 A carpenter uses different-sized hammers for different jobs. She uses a 20-oz hammer for driving 16-penny nails into 2" × 4" lumber.

Given: A carpenter exerts a force of 35 N from a height of 69 cm to accelerate a 0.57-kg hammer toward a nail through an arc of 97 cm.

Find: (a) How much potential energy does the hammer have at the top of its swing?
     (b) How much work does the carpenter do on the head of the hammer?
     (c) How fast is the hammer going when it hits the nail?

8-21 An elevator is lifting two people at a constant velocity from the first to the twentieth floor of their apartment building.

Given: The elevator has a mass of 1000 kg, while each of the 2 people has a mass of 120 kg. The elevator lifts them 90 m in 30 s (each floor is 4.5 m tall).

Find: (a) How much energy goes into lifting the elevator this many floors when empty?
     (b) How much energy goes into lifting the two people?
     (c) What is the power required to lift the elevator with the two on board?

8-22 Some people are helping a friend move his refrigerator upstairs. Unknown to the people moving the refrigerator, the refrigerator has not yet been emptied of beverage bottles.

Given: The refrigerator has a mass of 169 kg. The refrigerator contains 78 bottles of beverage. Each bottle has a mass of 650 g. The refrigerator and beverage bottles are lifted 18 cm, or one stair, in 1.6 seconds.

Find: (a) How much energy goes into lifting the empty refrigerator up the stair?
     (b) How much energy goes into lifting the combined mass of the bottles of beverage?
     (c) What is the power required to lift the full refrigerator up the stair?

8-23 Water in a fountain is squirted up into the air by a motor-driven pump.

Given: The pump draws electrical power at 200 W and is 85% efficient at imparting this energy to the water. It pumps water at the rate of 400 kg/min.

Find: (a) How much energy is used by the pump in 1.0 min?
(b) How much kinetic energy is given to 400 kg of water?
(c) What is the maximum height to which the water will squirt?

8-24 A snow plow equipped with a snow blower picks up snow from the street and throws it upward and away from the street it is cleaning.

Given: The snow blower uses 7,000 W of power and is 58% efficient. The snow blower moves 6,000 kg of snow per min.

Find: (a) How much energy is used in one minute?
(b) How much kinetic energy is given to 6,100 kg of snow?
(c) What is the maximum height of the snow when it is thrown?

8-25 A rubber ball when dropped from a certain height will lose some of its energy on impact with a hard surface and will only bounce up to a certain percentage of its original height.

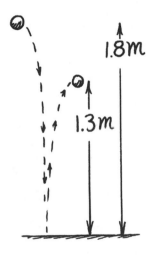

Given: A ball having a mass of 0.45 kg is dropped from a height of 1.8 m above a concrete sidewalk. It bounces back up to a height of 1.3 m.

Find: (a) What was the ball's initial potential energy?
(b) What percentage of this energy was changed into heat and sound by the impact?
(c) With what velocity did the ball bounce back up from the concrete?

8-26 An alarm clock goes flying out of an open window. It bounces once on the concrete below.

Given: The alarm clock has a mass of 545 g and is propelled from a window 7.73 m above the sidewalk. It bounces back up to a height of 24 cm.

Find: (a) What was the alarm clock's initial potential energy?
(b) What percentage of this energy was changed into heat and sound during the bounce?
(c) With what velocity did the clock bounce back up from the concrete?

8-27 Five minutes after the alarm clock in the previous example was thrown out the window, it was followed in its trajectory by the small backup alarm that the owner sets on days when waking up is particularly crucial. This backup alarm is thrown somewhat upward, but mostly outward.

Given: This alarm clock has a mass of 155 g. It is thrown with a speed of 9.10 m/s from an open window 7.73 m above the sidewalk.

Find: (a) What is the kinetic energy of the alarm clock when it is thrown?
(b) What is the potential energy of the alarm clock relative to the sidewalk when it is first thrown?
(c) What is the kinetic energy of the alarm clock just as it hits the sidewalk?
(d) What is the speed with which the alarm clock hits the sidewalk?

8-28 A stone is thrown from the edge of a cliff into the ocean below. It is thrown somewhat upward, but mostly outward.

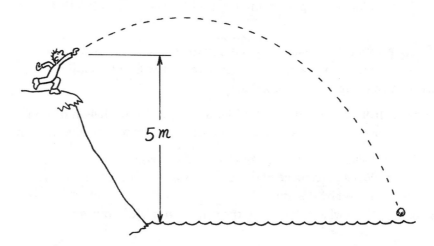

Given: The stone has a mass of 2.0 kg and is thrown with a speed of 10 m/s from an initial height of 5.0 m above the ocean.

Find: (a) What is the kinetic energy of the stone when it is thrown?
(b) What is the potential energy of the stone relative to the ocean when it is first thrown?
(c) What is the kinetic energy of the stone just as it hits the water?
(d) What is the speed with which the stone hits the water?

8-29 A small boy is flipping a piece of cake across the lunchroom at school. He uses a fork as a catapult by holding the handle on the table and pressing on the back of it with his thumb. Only part of the energy delivered to the fork goes into accelerating the cake, however, as the rest goes into moving the fork.

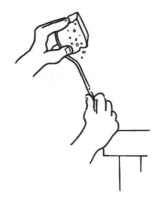

Given: The boy does 2 J of work on the fork with his thumb. The piece of cake has a mass of 34 g. The tines of the fork move through a distance of 15 cm before the cake becomes airborne. Only 51% of the work the boy does is transmitted to the cake.

Find: (a) How much energy is imparted to the cake?
(b) What force acts on the cake?
(c) With what velocity does the piece of cake leave the fork?

8-30 The first artillery pieces were catapults, which drew their energy from the deformation of twisted ropes or sinews that acted as a kind of spring. An engineering club is having a competition in building catapults to see who can launch a textbook the farthest.

Given: The spring does 1,130 J of work on the catapult arm. The book has a mass of 1.4 kg, and is accelerated through a distance of 5.15 m before it is launched. Only 42.0% of the energy of the spring is transmitted to the book, as the rest of the energy goes into moving the catapult.

Find: (a) How much energy is imparted to the book?
(b) What force acts on the book?
(c) What is the velocity of the book immediately after it is launched from the catapult?

8-31 Carmen is still skipping rope. When she jumps, she pushes the planet earth away from her. The earth, in turn, pushes her away from it. This problem is similar to problem 23 in the last chapter. When two objects of different sizes collide, the momentums are conserved, as in the last chapter, but the kinetic energy is not.

Given: Carmen has a mass of 57.0 kg. She is in contact with the ground for the first 15.5 cm of the jump. When she jumps upward she pushes against the ground with a force of 676 N over the entire 15.5 cm. The earth has a mass of 5.98×10^{24} kg.

Find: (a) What is Carmen's kinetic energy just as she leaves the ground?
(b) What is Carmen's upward velocity just as she leaves the ground?
(c) What is the earth's velocity in response to Carmen's jump?
(d) What small amount of kinetic energy is imparted to the earth through Carmen's jump?

8-32 An astronaut is repairing the surface of the intergalactic space station the "U.S.S. Deathstar," a spherical station about the size of a small moon. When a big and a small object collide, momentum is conserved but kinetic energy is not.

Given: The astronaut in her spacesuit has a mass of 152 kg. She makes a small leap by pushing against the station with a force of 73 N over a distance of 21 cm. The Deathstar has a mass of 7.24×10^{21} kg.

Find: (a) What is the astronaut's kinetic energy just as she leaves the surface of the space station?
(b) What is the astronaut's upward velocity just as she leaves the surface of the space station?
(c) What is the velocity of the Deathstar in response to the astronaut's jump?
(d) What small amount of kinetic energy is imparted to the Deathstar by the jump of the astronaut?

ANSWERS

Solutions to **ESSENTIAL PROBLEMS**

8-9   (a) 19.8 J; (b) 15.4 J; (c) 28.9 cm

8-10   (a) 197 N $\cong 2.0 \times 10^2$ N; (b) 1.6×10^3 J

8-11   (a) 4.9×10^2 N; (b) 1.5×10^3 J; (c) 0 J

8-12   (a) 19 J; (b) 15 N

8-13   (a) 114 J $\cong 1.1 \times 10^2$ J; (b) 15 W; (c) 8.3 hrs (8 hrs, 18 min.)

8-14   (a) 50 kg; (b) 1.44×10^3 J; (c) 9.7×10^{-2} hp

8-15   (a) 32.5 J; (b) 5.3 m/s; (c) 7.2×10^2 N

8-16   (a) 2.3×10^3 J; (b) 5.6×10^2 J; (c) 7.5 m/s

Solutions to **MORE INTERESTING PROBLEMS,** *odd answers*

8-17   (a) 5.5×10^4 J; (b) 5.5×10^4 W; (c) 74 hp

8-19   (a) 1.1×10^2 J; (b) 2.5×10^2 J; (c) 13 m/s

8-21   (a) 8.8×10^5 J; (b) 2.1×10^5 J; (c) 3.6×10^4 W

8-23   (a) 1.2×10^4 J; (b) 1.0×10^4 J; (c) 2.6 m

8-25   (a) 8.0 J; (b) 28%; (c) 5.0 m/s

8-27   (a) 6.42 J; (b) 11.7 J; (c) 18.16 J; (d) 15.3 m/s

8-29   (a) 1 J; (b) 7 N; (c) 8 m/s

8-31   (a) 18.2 J; (b) 0.80 m/s; (c) 7.6×10^{-24} m/s; (d) 1.73×10^{-22} J

CHAPTER 9

Fluids— Density and Pressure

Density

SUPPOSE YOU ARE INTERESTED in buying a water bed but are worried that it might be too heavy for your floor. One water bed manufacturer suggests that an easy way to test the strength of your floor is to have a party and arrange things so that most of your friends wind up standing in the part of the room where you plan to put the bed. If the floor holds, you would be safe to go out and buy a water bed. It is a simple matter to figure out how many friends you would need to invite to your party.

One cubic meter of water weighs in at 1,000 kg (2,205 pounds). Some people find it surprising that water is that heavy, but we can make it seem reasonable by thinking of the size and weight of a half gallon of milk. A carton of milk has a base of about 9 centimeters square and holds milk up to a level of about 20 centimeters. Each carton is about 2 kilograms. A cubic meter of milk would fill about 500 such cartons, since it occupies a volume a little less than ten times as great in both width and depth and about five times as high. That much milk would thus be about 500 times as heavy as one carton, or about 1,000 kg.

We can approximate the weight of water from the weight of milk because milk and water weigh about the same for the same volume. You can see the relative density of milk and water for yourself by putting a few drops of milk into a glass of water. If they sink slowly, milk is a little heavier than water. If they sink fast, milk is a lot heavier. If they float, it is lighter. Try it and see.

Sea water is a bit heavier than fresh water. That is because the salt dissolved in sea water is heavier than the water it displaces. One cubic meter of sea water weighs about 1.03 kilograms. We say that sea water is more **dense** than fresh water.

Density is the ratio of mass, *m*, to volume, *V*. The convention is to use the Greek letter *rho*, which is written ρ and pronounced "row." (It is not the letter "P," which is typically reserved for pressure.)

$$\rho \equiv \frac{m}{V}$$

$$\text{Density} \equiv \frac{\text{Mass}}{\text{Volume}}$$

The mass of a water bed can be derived from this definition by solving the equation for mass:

$$m = \rho \cdot V$$

Since the water bed is rectangular, its volume is length times width times height:

$$m = \rho \cdot (l \cdot w \cdot h)$$

Let us assume that the water bed is about 2 meters long ($l = 2$ m), 1.5 meters wide ($w = 1.5$ m), and 30 centimeters deep ($h = 0.3$ m). We have already decided that one cubic meter of water weighs 1,000 kg, so we know the density of water. The equation, then, will be:

$$m = (1000 \text{ kg/m}^3)(2 \text{ m})(1.5 \text{ m})(0.3 \text{ m})$$
$$= 900 \text{ kg}$$

We calculate that a water bed weighs about a ton, which is defined as 1,000 kilograms. To simulate this weight you would need to invite 18 friends who weigh 50 kilograms (about 110 pounds) each, or 9 friends who weigh 100 kilograms (220 pounds) each. It is true that the weightier friends tend to eat more, but you would need fewer of them.

If this experiment shows that the weight is a bit much for your floor, you might consider using something that is lighter than water. Ice, for example, is less dense than water, which is why ice floats. One cubic meter of ice only weighs 0.92 kilograms (1,980 pounds). Ice, however, tends to be lumpy. Alcohol would be better, since it is even less dense than ice. One cubic meter of alcohol weighs only 0.80 kilograms. Alcohol also has more social status than water, and it won't freeze in the winter if heating fuel becomes scarce.

Sometimes people mean by "density" the ratio of weight to volume

$$\rho_w \equiv \frac{w}{V}$$

as opposed to mass to volume. We will use the term **weight density** when we mean this kind of density and assume that ρ with no subscript stands for mass density.

These two kinds of density are related by the same proportionality constant that relates mass and weight, namely, the acceleration due to gravity, g

$$w = mg$$

If we divide both sides of this expression by volume, V,

$$\frac{w}{V} = \frac{mg}{V}$$

we have the proportionality between weight density and mass density

$$\rho_w = \rho g$$

We will find this idea of weight density useful in dealing with pressure in fluids.[1]

Densities of Common Substances

The mass density of pure water is exactly 1 kilogram per liter in SI units. We saw in Chapter 1 how this equivalence results from the way in which a kilogram was originally defined. A gram of water was originally the mass of the water that would fit in a volume 1 centimeter on a side. A kilogram was therefore the amount of water that would fill 1,000 cubic centimeters. When written by itself, liter is abbreviated L, since the lower case l is difficult to distinguish from the number 1.

They had to redefine the kilogram because of the tiny differences in the amounts of heavy water that pure water can contain. Before they redefined the kilogram, there was no difference between 1,000 cm³ and a liter. Now, 1,000 cm³ is defined in terms of the length standard based on the speed of light, whereas a liter is defined in terms of the mass standard based on the density of water. By definition, a liter is the volume occupied by one kilogram of pure water. That is why older glassware was marked with a cc, for cubic centimeter, and newer glassware is marked with an ml, for milliliter. There is almost no difference in the volume, but there is a difference in the standards.

Fixing things so that the density of water is 1, or unity, was a good idea because the density of other things is frequently compared to that of water. In British Imperial measure, there is something called *specific gravity*, which

---

[1]Many texts do not define a separate weight density, but just refer to it as ρg. We prefer the older conventions at this level because of the tendancy to confuse mass and weight.

is the ratio of the weight of something to the weight of the same volume of water. We don't need this idea when using SI units because density and specific gravity are the same thing except for the units. Density has units whereas specific gravity is a pure ratio.

Density is often used to help identify substances. Geologists use density to help distinguish between certain rocks that look very much alike. Jewelers use density to distinguish between different alloys of gold and silver.

The density of gold is one of the things that made it valuable. Used as money, gold was not only rare but easy to identify. In ancient times, it was the most dense of all substances by nearly a factor of two. You knew it was gold if it was dense. If a merchant suspected that the gold had been thinned out with a less valuable substance, he had only to measure both its weight and volume. Add almost anything to gold, and you will make it less dense.

Platinum and platinum-type metals are the only exception, but they were not discovered until 1735. When they were first discovered, they were promptly used to adulterate gold—until they became more valuable. By the 1920s platinum was about eight times as valuable as gold.

Substance	Mass Density (metric units) $\rho \equiv \dfrac{m}{v}$
Water, pure	1.00×10^3 kg/m^3
Water, salt	1.03×10^3
Ice	0.917×10^3
Alcohol	0.80×10^3
Wood, pine	$0.35 - 0.50 \times 10^3$
Rock salt	2.18×10^3
Aluminum	2.70×10^3
Iron	7.87×10^3
Silver	10.5×10^3
Lead	11.3×10^3
Mercury, 39° C	13.5×10^3
Gold	19.3×10^3
Platinum	21.5×10^3
Air, 0° C sea level, moist	1.29

Table 9-1

Don't use gold to occupy volume in your water bed. Oh, a little gold added to the water wouldn't hurt the feel, and it might help the social status under certain circumstances, but it's not going to help the weight. One of the best possible choices would be air. Not only is it cheap, but its density is only 1.3 kilograms per cubic meter. On the other hand, air lacks mass, and it is the inertia of the water that gives the undulating wave motion some people seem to enjoy. Besides, air lacks status. Nobody would invite somebody over to see their new air mattress.

CHECK QUESTIONS

1. You add a bit of alcohol to some fruit punch to make it lighter. To test the mixture, you hoist a 0.50 L mug of it in a toast. You judge the mass to be about 0.45 kg, your judgment at the start of the test being pretty good. What is the density of the alcohol-punch mixture?
    Answer: 0.9 kg/L

2. A king wishes to know the density of a metal crown. The crown has a mass of 7.5 kg and a volume of 0.50 × 10^{-3} m. What is the density of the metal?
    Answer: 15×10^3 kg/m^3.

Pressure

One of the advantages of a water bed is that it distributes the force of your weight evenly over your body. That is, it is soft. The larger the area over which you distribute the force, the less it is concentrated at any particular point. This idea of concentration of force is called pressure. Pressure, P, is defined as the ratio of force, F, to area, A:

$$P \equiv \frac{F}{A}$$

Just as you might expect, pressure is something that is directly proportional to force but inversely proportional to area. The greater the force, the greater the pressure; but the *smaller* the area, the greater the pressure. The discomfort you experience when your air mattress deflates is due to the same force being distributed over a smaller area.

The unit of pressure in SI units, the **Pascal**, is defined as a N/m^2. Pascal is abbreviated Pa.

$$1 \text{ Pa} \equiv \frac{1 \text{ N}}{\text{m}^2}$$

This unit is named after Blaise Pascal, a seventeenth century French mathematician and philosopher. Pascal did much to advance our understanding of pressure, and we owe him credit for the principle of operation of such modern devices as the hydraulic brakes in your car.

There are plenty of applications to the idea that pressure increases as area decreases. Legend has it that they passed a law back in the 1950's in Houston, Texas, prohibiting women from walking on the grass in shoes with the style of highly pointed heels fashionable in that era. The pressure generated by the tiny area of those heels when worn by an ordinary Texas-sized woman is huge.

We can calculate this pressure if we assume all this Texan's weight, say that of about 60 kg, is on one heel and that the heel comes down to a point about 1 centimeter across. Even without knowing the exact shape of the area presented to the ground, we can calculate the pressure approximately:

$$P \equiv \frac{F}{A}$$

$$= \frac{(60 \text{ kg})(9.8 \text{ N/kg})}{(1 \text{ cm})^2}$$

$$= 6 \times 10^2 \left(\frac{\text{N}}{\text{cm}^2}\right)\left(\frac{100 \text{ cm}}{\text{m}}\right)^2\left(\frac{\text{Pa}}{\text{N/m}^2}\right)$$

$$= 6 \times 10^6 \text{ Pa}$$

$$= 6 \text{ MPa}$$

Figure 9-1

Keep off the grass.

The story goes that the fire department was spending too much of its time capping the oil wells produced by this huge megapascal pressure. (Note in the solution that N/kg is just another way of writing m/s^2).

In this modern time, such a law might seem unfairly discriminatory. Men have worn high-heeled boots in Texas for a long time. We can examine this question in retrospect by calculating the pressure exerted by an ordinary 120-kg Texas-sized man wearing walking boots with 6 by 8 centimeter heels:

$$P \equiv \frac{F}{A}$$

$$= \frac{(120 \text{ kg})(9.8 \text{ N/kg})}{(6 \text{ cm})(8 \text{ cm})}$$

$$\cong \left(\frac{1200}{0.06 \times 0.08}\right)\left(\frac{\text{N}}{\text{m}^2}\right)\left(\frac{\text{Pa}}{\text{N/m}^2}\right)$$

$$= 2.5 \times 10^5 \text{ Pa}$$

$$= 0.25 \text{ MPa}$$

Certainly the 0.25 MPa pressure of the boot heel is substantially less than the 6 MPa pressure of the woman's heel—not because of the force, which is greater for the man—but because of the area over which the force is concentrated.

The relationship between pressure and area applies to many situations in our daily lives. The corners of desks are often rounded to increase the area of contact and decrease the pressure in case you bump against them. Automobile tires support the weight of a car with a relatively small amount of air pressure—two atmospheres, or twice as much pressure as outside the tire. Ice skates concentrate weight on an area so thin and small that the pressure is great enough to actually melt the ice under the skate so that you glide along on a thin film of water, which instantly re-freezes as soon as the pressure of the skate is removed.

One pressure we constantly experience is air pressure, which results from the weight of the atmosphere above us. Air pressure is the most common example of fluid pressure, but it is a bit difficult to visualize because air is invisible. The easiest example of fluid pressure to consider is the special case where the fluid happens to be a liquid such as water.

Figure 9-2

A Texan walking boot has a reasonably large heel.

CHECK QUESTION

Air pressure of 101 kPa is pushing in on one side of a rectangular box having an area of 0.05 m². What is the force produced on the side of the box?
Answer: 5.05×10^3 N

Liquid Pressure

Let us apply the idea of pressure to the example where we were worried about the floor beneath a water bed. The pressure on the floor is the weight of the water (ignoring the rest of the bed) divided by the area on which it is resting.

There is an old European folk tale in which a queen tested a princess by putting a pea under a whole stack of mattress just to see if she was sensitive enough to be disturbed by such a small discomfort in her bed. Great concern is expressed in this strange tale for the pressure exerted on the princess by the pea. How strange to worry about sensitive royalty. No one ever thinks about the pressure exerted on the poor pea.

If we had stacked two water beds, one on top of the other, the pea would feel twice as much pressure as with only one. Instead of 900 kg of water, we would have had 1,800 kg of water and the pressure would come to 6 kPa. With three and a third water beds, enough to bring the stack up to an even 1.0 m in height, we would have an even 3 m³ of water, which would have a mass of 3,000 kg.

We can now see how, from the pea's perspective, it no longer matters how big the water beds are—only how high they are stacked. With 3,000 kg of water up there, the pressure would be

$$P = \frac{F}{A}$$

$$= \frac{(3,000 \text{ kg})(9.8 \text{ m/s}^2)}{(2 \text{ m})(1.5 \text{ m})}$$

$$= (9,800 \text{ Pa})$$

$$= 9.8 \text{ kPa}$$

Remembering that the 3,000 kg came from the volume, which was a height of 1 m times an area of $(2 \text{ m})(1.5 \text{ m}) = 3 \text{ m}^2$, we can see that we would have gotten the same answer with a different area. Changing the height, on the other hand, changes the pressure. It seems reasonable that, as long as the fluid is the same, the pressure would be directly proportional to the number of beds, or what would be nearly the same thing, the combined height, h, of whatever fluid is in the water beds.

$$P \propto h$$

To express this proportionality as an equality, we need to find the proportionality constant.

From the definition of weight density

$$\rho_w \equiv \frac{w}{V}$$

we can see that weight is weight density times volume.

$$w = \rho_w V$$

Now we can take the definition of pressure and use the weight as the force in the equation, rewriting the volume as area times height:

$$P \equiv \frac{F}{A} = \frac{\rho_w A h}{A}$$

Then we can write the proportionality between pressure and height as an equality with weight density as the proportionality constant.

$$P = \rho_w h$$

Or, in words, the pressure in a fluid is equal to the weight density of the fluid times the height. If we had two water beds stacked on top of each other, we would have twice the height of water and thus twice the pressure—which is just what we expected.

CHECK QUESTIONS

1. A stack of water beds 3 m high is placed on top of a pea. The weight density of water is 9,800 N/m³. What is the pressure on the pea in kPa?
Answer: 29 kPa

2. The water in a toilet flush tank is 25 cm deep. The weight density of water is 9.8 N/L. What is the pressure of the water at the bottom of the flush tank?
Answer: 2.5 Pa

Figure 9-3
The force of the water pushing against two dams facing each other is independent of the distance between them.

We have shown that, as strange as it might seem, the pressure in a fluid does not depend on the *amount* of fluid, but only on its *height*. The force acting on a dam, for example, is independent of the amount of water backed up behind it; it depends only on the height of the water right next to the dam. The force acting on two dams facing each other would be the same if there were 50 kilometers of water in a reservoir between them or if there were only 1 centimeter of water between them. In either case, the height of the water is the same and the area over which the pressure of the water is exerted is the same. We may think of the role of a dam as being to hold in place only the layer of water right next to the dam. That layer holds the next layer in place, and so on.

On the other hand, the dependence of pressure only on height seems a little less strange if you think in terms of the pressure on your ears when you are swimming. Your ears will feel the same pressure when you swim 1 meter beneath the surface of a swimming pool as when you swim 1 meter under the waves of the ocean. The fact that there are thousands of miles of ocean out there has no effect on your ears.

Blaise Pascal, that fellow for whom we name the unit of pressure, liked to illustrate the relationship between pressure and height with some funny looking vases he designed for that purpose. The vases were different sizes and shapes but were all connected at the bottom. In spite of the differences, water would rise to the same level in each vase, demonstrating the fact that the pressure at the bottom of each vase was the same independent of the volume of water above. A pressure difference would cause water to flow from one vase to another and make the levels different. Thus, the principle that "water seeks its own level" is part of a much broader rule that **pressure equals weight density times height** independent of volume.

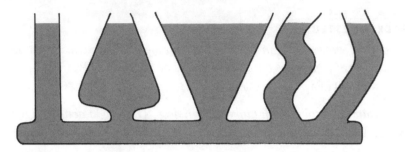

Figure 9-4

The Pascal vases demonstrate that pressure in a fluid is independent of volume.

This relationship between pressure, weight density, and the height of a fluid is an example of a deceptively simple rule with powerful ramifications. **Archimedes' Principle**, which itself has many unexpected results, can be entirely derived from the simple fluid pressure rule.

Archimedes' Principle

Once upon a time, King Hieron, king of the ancient Greek city of Syracuse, wanted a crown of absolutely pure gold. He suspected a traveling goldsmith of adding silver to the pure gold that had been refined by the court alchemist. When the king asked his alchemist to determine the purity of the gold, the alchemist proposed measuring its volume by melting down the newly fashioned crown. Reluctant to destroy his beautiful new crown, the king asked his wise man to calculate its volume from its measurements.

As it happened, Hieron's wise man was Archimedes, a master of geometry. The story goes that while Archimedes was swimming in public baths, he thought of a way of measuring the crown's volume with great accuracy, not relying on many measurements of size and the uncertainties that go with them. Instead, he conceived of using the buoyant force which would act on the crown if he were to place it under water. Legend has it that Archimedes was so enthusiastic about his new thought that, without bothering to dress, he ran through the streets naked, shouting, "*Eureka! Eureka!*" which means, in Greek, "I found it! I found it!"

Archimedes' principle can be stated, **the buoyant force is equal to the weight of the fluid displaced**. Buoyant force is the force with which the fluid pushes up on the object. If a rock is put under water, it seems to weigh less than it does in air. The water helps to lift up the rock and therefore decreases the rock's apparent weight.

Archimedes proved his principle without benefit of algebra, since algebra hadn't been invented yet. His thought process probably went like this: "If the rock were withdrawn and the resulting cavity allowed to fill with water, the latter would be in equilibrium under the action of its own weight

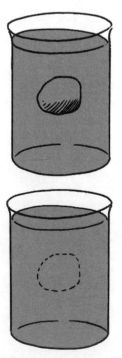

Figure 9-5

Water pushes up on a rock the same as it would on water of the same volume.

and the external forces formerly exerted by the surrounding water on the rock."[2] The water in the cavity may be thought of as the displaced water, since it has the same volume as the rock. The surrounding water pushes up with the same force on the rock as it would on the water in the cavity, because it is not in the nature of water to care what it is pushing on. The buoyant force acting on the rock is thus equal to the weight of the displaced water. The same consideration holds for any fluid, whether liquid or gas.

We will find it instructive to derive Archimedes' principle from the fluid pressure rule. The buoyant force is derived from the difference in the pressures acting on the bottom and the top of the object. It is convenient to select a rectangular shaped object to start with so that the pressure on the sides produces obviously symmetrical opposing horizontal forces that cancel each other out and do not enter into the buoyant force calculation. The pressure on the bottom of the object is greater than the pressure on the top because the bottom is at a greater depth in the fluid than the top.

The buoyant force is found by solving the definition of pressure

$$P \equiv \frac{F}{A}$$

for force to find the force on the bottom and force on the top of the object:

$$F_{\text{top}} = P_{\text{top}}A$$
$$\text{and}$$
$$F_{\text{bottom}} = P_{\text{bottom}}A$$

The buoyant force is the difference between the forces acting on the bottom and on the top.

$$F_{\text{bouyant}} = F_{\text{bottom}} - F_{\text{top}}$$
$$= P_{\text{bottom}}A - P_{\text{top}}A$$
$$= (\rho_w h_{\text{bottom}})A - (\rho_w h_{\text{top}})A$$

Pressure is equal to weight density times height (according to the fluid pressure rule). We factor out weight density and area:

$$F_{\text{buoyant}} = \rho_w A(h_{\text{bottom}} - h_{\text{top}})$$

This difference in the heights, $h_{\text{bottom}} - h_{\text{top}}$ is just the height of the body. The area, A, of the bottom times this height is the volume of the object:

$$F_{\text{buoyant}} = \rho_w V$$

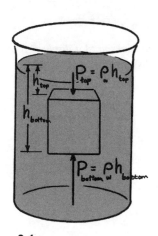

Figure 9-6

Archimedes' principle can be derived by thinking of the buoyant force in terms of the difference in pressures acting on the bottom and top of an object.

---

[2] *Van Nostrand's Scientific Encyclopedia*, 4th ed. (1968) p. 253.

But the weight density of the fluid times the volume of the object is just the weight of the fluid that would be there if the object wasn't. In words, we can say Archimedes' principle as

$$F_{\text{buoyant}} = \text{weight of the fluid displaced}$$

Our proof, in terms a rectangular box, can be generalized by approximating the shape of any object with a bunch of elongated boxes having rectangular horizontal and vertical surfaces. Archimedes' principle holds for each box, without regard to the sides of the boxes, no matter how many there are. By choosing an infinite number of boxes, we can generalize Archimedes' principle for an object of any shape.

If the buoyant force acting on an object is less than the object's weight, the object will seem to weigh less than it would in air but it will still sink. If the buoyant force equals the weight of the object, the object will float. A boat floats because it can displace more than its own weight of water. It sinks into the water until it displaces whatever water it needs to balance its weight. The more heavily a boat is loaded, the more water it must displace in order to float. One way to think of a submarine is as a boat that can adjust the amount of buoyant force acting on it so that it can just barely float or sink. The amount of water displaced by the submarine can be adjusted by allowing water to come into, or forcing water out of, ballast tanks on the sides of the submarine.

Figure 9-7

A boat sinks until it displaces its own weight of water.

Helium balloons float by displacing their own weight in air. We tend to disregard the changes in air pressure over short distances, yet the air pressure on the bottom of a helium balloon is sufficiently greater than the pressure on the top that the buoyant force more than compensates for the weight of the balloon. The helium in the balloon does not do anything other than hold out the sides of the balloon. Adding helium to the balloon without allowing it to expand would only make the balloon heavier. If it were some inherent tendency of helium to go up that made helium balloons float, balloon merchants would have a tough time holding down their tanks of compressed helium. (Think how nice it would be, though, to solve your parking problems by filling your tires with helium. Just put it in under enough pressure to make your car float, and you could just tie it to a flag pole when you went to class.)

A ship propeller, newly cast of aluminum, is dipped in water to cool it so that it can be worked on. If an aluminum casting has a volume of 1.0 m³, it will have a mass of 2,700 kg. (a) What is the mass of the water displaced? (b) What is the buoyant force acting on the casting? (c) What is the weight of the casting in air? (d) How much does the casting seem to weigh under water?

Answer: (a) 1,000 kg (b) 9,800 N (c) 26,500 N (d) 16,700 N

Pressure in a Gas

Our discussion so far has been mostly concerned with pressure in a liquid. We have also seen that Archimedes' principle applies to balloons floating in air, where the fluid is a gas instead of a liquid. Before leaving the topic of pressure, we should develop the general rule for pressure in a gas related to other variables such as temperature, volume, and the number of molecules of the gas that are present.

This relationship is called the "ideal gas law" because it only works for an "ideal" gas. Fortunately, the behavior of most gases approximates that of an ideal gas for ordinary temperatures and pressures. We will build up this general pressure rule for a gas by constructing proportional models.

PRESSURE AND AMOUNT OF GAS

Need pressure in your tire? No problem, you just add air. All else being equal, the more gas you have, the greater the pressure.

Gas can be visualized as a bunch of molecules (think of them as hard little balls like Ping Pong balls) flying around, bouncing off each other and anything they come in contact with. The pressure of a gas inside a tire is the measure of how hard these little molecules bounce against the inner surface of the tire. When you put more air molecules inside a tire, there are more molecules bouncing against the inside of the tire. The more molecules that are bouncing around inside the tire, the harder they push against the inner surface, and the greater the pressure in the tire.

Blow up a balloon and the balloon expands because more of those little molecules hit the inside surface of the balloon than the outside. The balloon will expand until the number of molecules hitting the inside of the balloon are equal to the number of molecules hitting the outside, or until the pressure inside the balloon equals the pressure outside. The pressure, *P*, on the inside surface of the balloon should be proportional to the number of molecules, *N*, in the balloon.

$$P \propto N$$

We should be careful to emphasize that everything else needs to remain constant. Our balloon would have to have a very tough surface, sort of like a beach ball, if its volume were not to change with increased pressure. We

Figure 9-8

Archimedes' principle also works when the fluid is air.

would also have to keep the temperature constant so that each molecule would strike the surface with the same average force of impact. We can consider the effect of allowing volume and temperature to change in a moment, but first we must look more carefully at the proportionality between pressure and the number of molecules in the balloon.

The pressure in our proportionality must be the kind of pressure that is equal to zero if the number of molecules is equal to zero. This is not the kind of pressure you would read on a tire gauge. When a tire gauge reads zero, when the tire is flat, there is still air in the tire. It is just that the air pressure inside the tire is no greater than that on the outside. When inflated, the pressure read on a tire gauge is understood to be the pressure over and above atmospheric pressure. The only difference between this gauge pressure and the absolute pressure used in our proportionality is the zero point. Whereas the gauge pressure is measured relative to the atmosphere, the absolute pressure is measured relative to a vacuum. To get from gauge pressure, P_{gauge}, to absolute pressure, P (with no subscript), just add the atmospheric pressure. At sea level, the atmospheric pressure is 101 kPa.

$$\text{Absolute pressure} = P = P_{\text{gauge}} + P_{\text{atmosphere}} = P_{\text{gauge}} + 101 \text{ kPa}$$

PRESSURE AND VOLUME

Let us now consider the relationship between pressure of a gas and the volume of the container. Instead of putting more air into our balloon or beach ball, we hold the number of molecules and their temperature constant and squeeze the volume down to one-half its original size. Here again, we expect a proportional relationship—but not a direct proportion. The relationship between pressure and volume should be an inverse proportion, since the pressure goes up when the volume goes down. Although the relationship is inverse, we still expect it to be basically proportional, because decreasing the volume increases the density of molecules in the same way that adding gas does. That is, if you try to crowd the same number of molecules into half the volume, the same number of collisions occur at the surface as if you had doubled the number of molecules in the original volume. This inverse proportionality can be expressed in terms of a reciprocal:

$$P \propto \frac{1}{V}$$

We pause here to note that, although these proportionalities seem obvious to the modern mind, that has not always been the case. The first person to suggest this inverse proportionality was Robert Boyle, an Irish scientist who is sometimes called the "father of chemistry," and he first published what we call Boyle's law in 1663.

PRESSURE AND TEMPERATURE

The remaining part of the equation, the relationship between pressure and temperature, was not discovered for over a century after Boyle's law. It came out of the work of Jaques Charles, the French experimentalist responsible for the first balloon assents using hydrogen rather than hot air. The results of his experiments with the expansion of gasses were first published by the French chemist and fellow balloonist Joseph Louis Gay-Lussac in 1802. Gay-Lussac's work on the volumes of reacting gasses, by the way, formed the foundation for Avagadro's hypothesis on the atomic theory of matter.

What Charles discovered was that the pressure of a gas increases in direct proportion to the absolute temperature of the gas. People had known for a long time that pressure and temperature were related to one another, however.

Let us think about temperature and pressure in terms of our mental model of gas being little molecules, like ping-pong balls, bouncing around. We discussed that pressure is the measure of how hard these molecules hit the inside of something, for example, an automobile tire. When you double the number of molecules in the tire, there are twice as many molecules hitting the inside of the tire, and there is twice as much pressure. The temperature, when looked at in this model, can be considered a measure of the energy of the molecules. When you heat molecules they gain energy, largely in the form of Kinetic Energy. Remember from the last chapter that kinetic energy is

$$E_k = \tfrac{1}{2}mv^2$$

So, if we increase the kinetic energy of a molecule its velocity will increase and it will bounce around faster. This means that when you heat the air in the tire, the molecules will bounce against the inner wall of the tire faster, and therefore harder, and the pressure inside the tire will increase.

Jaques Charles' contribution was that temperature and pressure, more than just being related, are directly proportional to one another. This relationship had not been discovered before because scientists, including Charles, were missing some key elements in the understanding of heat. In the first place, Charles did not know about molecules. At this time, heat was still understood in terms of caloric fluid theory, which, as we will see in the next chapter, was incorrect. The biggest problem in seeing the relationship between pressure and temperature, however, was that the concept of absolute temperature had not yet been invented.

Using an ordinary temperature scale, the relationship between pressure and temperature is linear but not proportional. Equal increases in one bring about equal increases in the other, but they don't go through zero at the same time. A graph of pressure against temperature produces a straight line. This relationship is like a proportion except that the line does not pass through the origin.

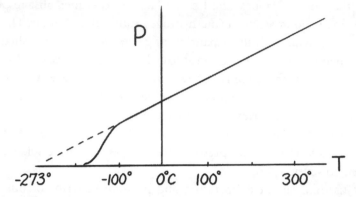

Figure 9-9

A graph of absolute pressure against temperature can be extrapolated to zero pressure at −273° C

The pressure is not equal to zero at 0 degree Celsius (formerly called centigrade). But if you decrease the temperature one degree, from 0 degrees Celsius to −1 degree Celsius (°C), the pressure will decrease by one part in 273 from the original pressure. If you decrease the temperature two degrees, say from 0°C to −2°C, the pressure in the gas decreases two parts in 273. This leads us to wonder what would happen if we were to decrease the temperature 273 degrees, from 0°C to −273°C.

Any real gas becomes a liquid before −273°C, but we could imagine an "ideal gas" that would follow the linear relationship all the way. For such a gas the pressure would equal zero at this temperature. This temperature is called **absolute zero**, because it forms a natural zero point just as a vacuum forms a natural zero point for pressure measurement. Pressure in a gas is directly proportional to temperature as long as temperature is measured on a scale that starts at absolute zero.

$$P \propto T$$

An absolute temperature scale that starts at absolute zero and has the same-size degrees as the Celsius scale is called the **Kelvin scale**. The freezing point of water is 273 K and the boiling point of water is 373 K on this scale. Temperatures are converted from degrees Celsius to degrees Kelvin by adding 273 to the Celsius reading.

The Ideal Gas Law

Now that we have discussed the way that all the elements affecting a gas are related to one another, these properties can be combined into a single equation. We have seen that pressure of a gas is directly proportional to the number of molecules of the gas, the temperature of the gas, and inversely proportional to the volume of the container holding the gas. These three relationships can be combined into a single statement:

$$P \propto \frac{NT}{V}$$

A proportionality can always be expressed in terms of an equality sign and a proportionality constant. With a proportionality constant k, called **Boltzmann's constant**, the pressure in an ideal gas can be expressed as

$$P = k\frac{NT}{V}$$

This proportionality constant turns out to have the same value for all gases:

$$k = 1.38 \times 10^{-23} \text{ newton} \cdot \text{meter/molecules} \cdot \text{K}$$

The proportional relationship between pressure, volume, temperature, and the amount of gas is called the **ideal gas law** and, using Boltzmann's constant, is written[3]

$$PV = kNT$$

All gases obey this relationship well at reasonably high temperatures and low pressures, where the gas is far from the liquid state. An ideal gas, a hypothetical gas that obeys the relationship exactly at all pressures and temperatures, is an approximation of most real gases under most conditions.

---

[3] In terms of R, the universal gas constant, the ideal gas law is usually expressed as

$$PV = nRT$$

where n is the number of moles of molecules. Avagadro's number, the number of molecules in a mole, is derived by combining these two versions of the ideal gas law. R can be determined in a well-equipped chemistry lab and Boltzmann's constant can be found because it is also the constant in the exponential decrease of pressure in the atmosphere as altitude increases.

SUMMARY OF CONCEPTS

The problems in this chapter are concerned with pressure in liquids and gases. The density, by which we mean the mass density, of a fluid is written "ρ" (rho) and is defined as the ratio of mass to volume:

$$\rho \equiv \frac{m}{V}$$

Weight density is written ρ_w (rho weight) and is defined as the ratio of weight to volume:

$$\rho_w \equiv \frac{w}{V}$$

Mass density and weight density are related by the proportionality

$$\rho_w = \rho g$$

where g is the acceleration due to gravity.

The density of water, almost exactly 1 gram per milliliter, follows from the original definition of a gram. A liter of water, which is 1,000 times as big as a cubic centimeter, therefore has a mass of 1 kilogram. A liter is abbreviated with a capitol L to avoid confusion with the lower case l and the number 1.

Pressure is defined as the ratio of the force to area:

$$P \equiv \frac{F}{A}$$

The unit of pressure in SI units is the Pascal, abbreviated "Pa" and defined as one newton per meter squared:

$$Pa \equiv \frac{1\,N}{m^2}$$

Pressure of a liquid is proportional to the height of the liquid, the proportionality constant being the weight density of the liquid:

$$P = \rho_w h$$

Archimedes' principle states that the buoyant force acting on an object is equal to the weight of the fluid displaced.

$$F_{buoyant} = \text{weight of fluid displaced}$$

$$= \rho_w V$$

$$= (\text{weight density})(\text{volume of object})$$

Pressure in a gas is directly proportional to the absolute temperature and the number of molecules of the gas, and inversely proportional to the volume. This is known as the **ideal gas law**:

$$PV = kNT$$

The value of the proportionality constant is the same for all gases:

$$k = 1.38 \times 10^{23} \text{ newton} \cdot \text{meter/molecules} \cdot \text{K}$$

K in this equation stands for "degrees Kelvin." What we refer to as "absolute temperature" is the temperature measured in Kelvins. To find the absolute temperature, add 273 to the celsius temperature:

$$K = {}^\circ C + 273$$

The pressure in the ideal gas law is the absolute pressure relative to a vacuum. Atmospheric pressure must be added to gauge pressure to get absolute pressure:

$$P_{absolute} = P_{gauge} + P_{atmosphere}$$

$$P_{atmosphere} = 1.01 \times 10^5 \text{ Pa}$$

Try to do the Examples
yourself before looking at
the Sample Solutions.

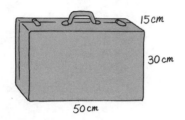

9-1 A suitcase filled with books is only about half as heavy as the same suitcase filled with water.

Given: A suitcase 50 cm long by 30 cm wide by 15 cm thick is filled with books having an average density of $0.50 \times 10^3 \, \text{kg/m}^3$. The density of water is $1.00 \times 10^3 \, \text{kg/m}^3$.

Find: (a) What is the volume of the suitcase in liters?
(b) What would be the mass of the suitcase filled with water?
(c) What is the mass of the suitcase filled with books?

BACKGROUND

This problem illustrates the relationship between mass and volume.

9-2 An automobile's weight is supported by the air pressure inside the tires spread out over the area that all four tires present to the ground.

Given: A 1946 Lincoln complete with driver has a mass of 2,300 kg. The inflation pressure recommended by one tire manufacturer is 200 kPa.

Find: (a) What is the weight of the car and driver?
(b) What is the area that each of the four tires presents to the ground?

BACKGROUND

This problem illustrates the relationship between pressure and force. It shows how a large force can be supported by a reasonably small pressure acting over a large area.

9-3 The height of water necessary to produce a given pressure is sometimes called the water head. If you live in a tall building or on a hill, your plumber needs to worry about this.

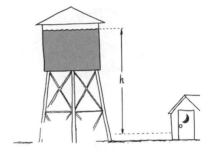

Given: A common water pressure available at a kitchen faucet is 4.01×10^5 Pa. Water has a density of 1.00×10^3 kg/m^3.

Find: (a) What is the weight density of water?
(b) How much water head is necessary to produce this pressure?

BACKGROUND

This problem illustrates one of the most common applications of the fluid pressure rule to our daily lives. Running water has become so much a part of our lives that we tend to take it for granted. Think of all that water up there someplace the next time you turn on a faucet.

9-4 Objects float when the buoyant force is equal to their weight. For example, a wooden plank floating in water sinks into the water up to the point where the buoyant force is equal to the weight of the plank.

Given: A plank floating in water is 2.0 m long and 15 cm wide. The density of the water is 1.00×10^3 kg/m^3. The mass of the plank is 11 kg.

Find: (a) When the plank is floating in water, what is the buoyant force?
(b) What is the volume of the water displaced?
(c) How far does the water come up on the edge of the plank?

BACKGROUND

Floating objects are applications of Archimedes' principle. If the buoyant force is equal to the weight of the water displaced by the object, then a floating object will sink until it displaces enough water to equal the buoyant force.

9-5 A balloon merchant is filling balloons to atmospheric pressure from a tank of compressed helium. As a first estimate of the number of balloons that can be filled, the merchant calculates the number of balloons that can be filled by the helium in the tank. As a second estimate, the merchant refines the calculation to correct for the fact that the tank will still have helium in it, at one atmosphere, when the last balloon is filled.

Given: Each balloon contains 1.40×10^{-2} m^3 at atmospheric pressure and temperature. The tank at the same temperature contains 1.4 m^3 of helium at 1.30×10^7 Pa. Atmospheric pressure is 1.01×10^5 Pa.

Find: (a) If the given pressure is the gauge pressure, what is the absolute pressure in the tank?

(b) What will be the volume of all the helium in the tank when it is at atmospheric pressure?

(c) As a first estimate, how many balloons can the merchant fill?

(d) As a second estimate, how many balloons can the merchant fill?

BACKGROUND

This is an application of the ideal gas law. Temperature is assumed to remain constant, since the gas in the balloons quickly adjusts to room temperature after the balloons have been filled

Solutions to EXAMPLE PROBLEMS

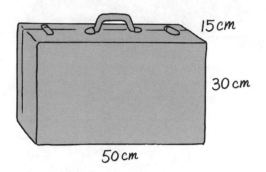

15 cm

30 cm

50 cm

(a) $V = l \cdot w \cdot h$

$$= (50 \ cm)(15 \ cm)(30 \ cm)\left(\frac{1 \ L}{10^3 cm^3}\right)$$

$$= \boxed{23 \ L}$$

(b) $\rho \equiv \dfrac{m}{V}$

$m = \rho_{water}V$

$$= (1.00 \times 10^3 \ kg/m^3)(23 \ L)\left(\frac{1 \ m^3}{1000 \ L}\right)$$

$$= \boxed{23 \ kg}$$

(c) $m = \rho_{books}V$

$$= \left(0.50 \times 10^3 \frac{kg}{m^3}\right)(23 \ L)\left(\frac{1 \ m^3}{1000 \ L}\right)$$

$$= \boxed{12 \ kg}$$

SAMPLE SOLUTION 9-1

Given:
A suitcase 50 cm long by 30 cm wide by 15 cm thick is filled with books having an average density of $0.50 \times 10^3 \ kg/m^3$. The density of water is $1.00 \times 10^3 \ kg/m^3$.

Find:
(a) What is the volume of the suitcase in liters?

(b) What would be the mass of the suitcase filled with water?

(c) What is the mass of the suitcase filled with books?

DISCUSSION

Twelve kilograms, about 24 lb California weight, is not heavy for a full suitcase. Get out a suitcase and try it. Use books.

SAMPLE SOLUTION 9-2

Given:

A 1946 Lincoln complete with driver has a mass of 2,300 kg. The inflation pressure recommended by one tire manufacturer is 200 kPa.

Find:

(a) What is the weight of the car and driver?

(b) What is the area that each of the four tires presents to the ground?

(a) $w = mg$

$\quad = (2{,}300 \text{ kg})\left(9.8 \dfrac{\text{N}}{\text{kg}}\right)$

$\quad = \boxed{2.25 \times 10^4 \text{ N}}$

(b) $\quad \rho \equiv \dfrac{F}{A}$

$A_{total} = 4 A_{per\ tire}$

$A_{per\ tire} = \dfrac{1}{4} A_{total}$

$\qquad = \dfrac{1}{4}\left(\dfrac{F}{\rho}\right)$

$\qquad = \dfrac{1}{4}\left(\dfrac{2.25 \times 10^4 \text{ N}}{200 \text{ kPa}}\right)\left(\dfrac{1 \text{ Pa}}{1 \text{ N/m}^2}\right)$

$\qquad = \boxed{2.8 \times 10^{-2} \text{m}^2}$

if each tire is about 17 cm wide, this corresponds to a length of about 17 cm touching the ground.

17 cm

DISCUSSION

The sample solution interprets the answer in terms of length of each tire on the ground, assuming a reasonable width. This is more work than is called for by the problem, and your solution might well not contain this information. Sometimes people get interested in a problem, however, and carry it a bit further than they really need to.

You might, for example, go out and measure the tires on your own automobile. With a ruler and a pressure gauge, you should be able to figure out the weight of your automobile.

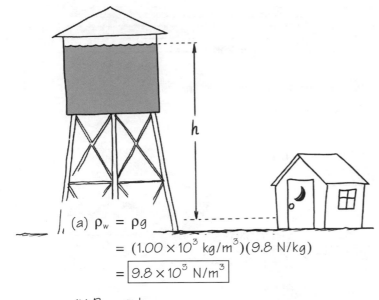

Given:
A common water pressure available at a kitchen faucet is 4.01×10^5 Pa. Water has a density of 1.00×10^3 kg/m³.

Find:
(a) What is the weight density of water?

(b) How much water head is necessary to produce this pressure?

(a) $\rho_w = \rho g$

$\qquad = (1.00 \times 10^3 \text{ kg/m}^3)(9.8 \text{ N/kg})$

$\qquad = \boxed{9.8 \times 10^3 \text{ N/m}^3}$

(b) $P = \rho_w h$

$h = \dfrac{P}{\rho_w} = \dfrac{4.01 \times 10^5 \text{ Pa}}{9.8 \times 10^3 \text{ N/m}^3}\left(\dfrac{\text{N/m}^2}{1 \text{ Pa}}\right) = \boxed{41 \text{ m}}$

DISCUSSION

Part (a) is found from the relationship between weight density and mass density. Part (b) is found from the fact that the pressure of a liquid is proportional to the height of the liquid, with the weight density of the liquid used as a proportionality constant.

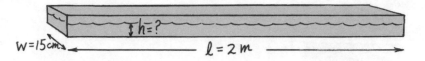

Given:

A plank floating in water is 2.0 m long and 15 cm wide. The density of the water is $1.00 \times 10^3 \, kg/m^3$. The mass of the plank is 11 kg.

Find:

(a) When the plank is floating in water, what is the buoyant force?

(b) What is the volume of the water displaced?

(c) How far does the water come up on the edge of the plank?

(a) Let F_b = buoyant force = weight of the fluid displaced

$$F_b = (\text{weight of plank})$$

$$= mg$$

$$= (11 \text{ kg})(9.8 \text{ N/kg})$$

$$= \boxed{108 \text{ N}}$$

(b) $$F_b = \rho_w V_{water} = \rho g V_{water}$$

$$V_{water} = \frac{F_b}{\rho g} = \frac{(108 \text{N})}{(1.00 \times 10^3 \text{ kg/m}^3)(9.8 \text{ N/kg})} = \boxed{1.1 \times 10^{-2} \text{ m}^3}$$

(c) $$V_{water} = l \cdot w \cdot h$$

$$h = \frac{V}{l \cdot w} = \frac{(1.1 \times 10^{-2} \text{ m}^3)}{(2.0 \text{ m})(15 \text{ cm})} = \boxed{3.7 \text{ cm}}$$

DISCUSSION

The buoyant force holding up the plank is equal to the weight of the plank. It must be. If the buoyant force were any less, the plank would sink.

Part (b) can be found from the definition of weight density, solved for the volume. Archimedes' principle states that the buoyant force on an object is equal to the weight of the fluid displaced. We can turn this around and say that the weight of the fluid, here water, must be equal to the buoyant force acting on the plank, which was found in part (a).

In part (b) we found the volume of the water displaced by the plank. This means that the plank sinks down to the level where it displaces that much water. The volume found in part (b), then, is equal to the volume of the plank under water. Since we know the length and width dimensions of the plank, the volume will give us the thickness h, which is actually how far the water comes up on the edge of the plank.

(a) $P_{absolute} = P_{gauge} + P_{atmospheric}$

$$= (1.30 \times 10^7 \text{ Pa}) + (1.01 \times 10^5 \text{ Pa})$$

$$= \boxed{1.31 \times 10^7 \text{ Pa}}$$

(b) $P_1V_1 = kN_1T_1$ for tank

$P_2V_2 = kN_2T_2$ for balloons

so $P_1V_1 = P_2V_2$ since $N_1 = N_2$ and $T_1 = T_2$

$$V_2 = \frac{P_1}{P_2}V_1 = \frac{(1.31 \times 10^7 \text{ Pa})}{(1.01 \times 10^5 \text{ Pa})}(1.4 \text{ m}^3) = \boxed{182 \text{ m}^3}$$

(c) $V_{total} = nV_{each}$

$$n = \frac{V_{total}}{V_{each}}$$

$$= \frac{182 \text{ m}^3}{1.40 \times 10^{-2} \text{ m}^3} = \boxed{1.30 \times 10^4}$$

(d) $V_{total} = nV_{each} + V_{tank}$

$$n = \frac{V_{total} - V_{tank}}{V_{each}}$$

$$= \frac{(182 \text{ m}^3) - (1.4 \text{ m}^3)}{(1.40 \times 10^{-2} \text{ m}^3)}$$

$$= \boxed{1.29 \times 10^4}$$

SAMPLE SOLUTION 9-5

Given:
Each balloon contains 1.40×10^{-2} m³ at atmospheric pressure and temperature. The tank at the same temperature contains 1.4 m³ of helium at 1.30×10^7 Pa. Atmospheric pressure is 1.01×10^5 Pa.

Find:
(a) If the given pressure is the gauge pressure, what is the absolute pressure in the tank?

(b) What will be the volume of all the helium in the tank when it is at atmospheric pressure?

(c) As a first estimate, how many balloons can the merchant fill?

(d) As a second estimate, how many balloons can the merchant fill?

DISCUSSION

When reading pressure from a gauge, you have to take into account the fact that you are not standing in a vacuum. The absolute pressure of something in a canister is the pressure you read on the gauge plus the pressure on it from the air surrounding, which is atmospheric pressure.

The volume can be found using the ideal gas law. We can divide the before and after situations of the ideal gas law by one another. For the gas in the tank we use subscript 1, and for the gas put in the balloons we use subscript 2. The number of molecules is the same before and after, hence $N_1 = N_2$. Also, the temperature is the same. Thus $T_1 = T_2$. Since the ideal gas law constant k is obviously the same for the gas when it is in the tank as when it is in the balloons, the right-hand sides of the two ideal gas law expressions are equal. The two left-hand sides are therefore equal, and this equality is solved for the volume V_2 of the gas after it is put into the balloons, which can then be expressed in terms of balloon volumes.

For a better number of balloons, we take into account that the tank isn't completely empty when it reaches atmospheric pressure, and therefore stops putting out compressed helium. We find the number of balloons the remaining air would fill, and subtract that from the number in part (b).

ESSENTIAL PROBLEMS

Be sure you can do the
Example Problems (without
peeking at the Sample
Solutions) before working
these Essential Problems.

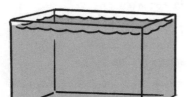

9-6 A rectangular fish aquarium filled with fresh water has more mass than the same aquarium filled with gin. Besides, most fish do poorly in gin.

Given: The inside dimensions of the aquarium are 70 cm by 35 cm by 40 cm. Fresh water has a weight density of 1.00×10^3 kg/m^3 while gin has a density of 0.91×10^3 kg/m^3. A liter is 1,000 cm^3.

Find: (a) What is the volume of the aquarium in liters?
(b) What is the mass of the aquarium filled with fresh water?
(c) What is the mass of the aquarium filled with gin?

9-7 Snowshoes allow you to walk through freshly fallen snow even carrying a backpack. By spreading your weight over a large area, you can walk on the surface of the snow without sinking in.

Given: Your mass with pack is 110 kg. Snow will support a pressure of 5.00 kPa.

Find: (a) What is the combined weight of you and your pack?
(b) What must be the area of each of your snowshoes?

9-8 When swimming under water one fine summer day, you notice that the pressure on your ears increases as you swim deeper. Close to the surface, you don't feel anything, but at a certain depth you begin to notice.

Given: You begin to notice the pressure when it gets up to 25 kPa. Water in a certain pool has a density of 1.01×10^3 kg/m^3.

Find: (a) What is the weight density of the water in this pool?
(b) How deep are you when you begin to notice the pressure?

9-9 A coal barge is floating in salt water. The water comes up to a place on the side of the barge called the water line.

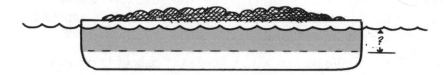

Given: The barge is rectangular, with a length of 10.0 m and width of 3.0 m. The density of salt water is $1.03 \times 10^3 \, kg/m^3$. The barge is loaded with coal and has a mass of $1.5 \times 10^4 \, kg$.

Find: (a) What is the buoyant force holding up the barge?
(b) What is the volume of the water displaced?
(c) How far does the water come up on the edge of the barge?

9-10 To breathe underwater, scuba divers bring their own air. The number of lungfulls of air in the tanks of compressed air they carry depends on the pressure at which they breath it. A first estimate of the number of breaths of air in a tank might ignore the air left in the tank when the last breath of air is taken. A closer estimate might correct for the fact that the tank will still have air in it when last lung-full has been breathed. No correction is needed, however, for change in temperature as the tank and the diver are at the same temperature. Atmospheric pressure is $1.01 \times 10^5 \, Pa$.

Given: Each breath for one particular diver has a volume of $0.82 \times 10^{-3} \, m^3$. Working at a particular depth, the air pressure in the diver's lungs is $1.51 \times 10^5 \, Pa$. The tank has a volume of $5.0 \times 10^{-3} \, m^3$ and holds air at a gauge pressure of $1.71 \times 10^7 \, Pa$.

Find: (a) What is the absolute pressure of the air in the tank?
(b) What will be the volume of the air in the tank when released at this depth?
(c) As a first estimate, how many lung-fulls of air are in the tank?
(d) As a closer estimate, how many lung-fulls of air are in the tank?

MORE INTERESTING PROBLEMS

9-11 A rectangular fish aquarium is filled with fresh water and one large fish. The fish does not swim; he hangs motionless in the water, neither sinking to the bottom nor floating to the surface.

Given: The aquarium is 22 cm long, is filled to a height of 15 cm, and is 10 cm thick. The fish has a mass of 1.7 kg.

Find: (a) What is the average density of the fish? How do you know?
(b) What is the volume of the fish?
(c) Find the combined mass of the water and the fish.

9-12 Submarines rise and fall in the water by compressing and decompressing their ballast tanks to change the average density of the submarine. A yellow submarine is submerged in a rectangular swimming pool. It is stationary—neither rising or falling.

Given: The swimming pool is 6.0 m long, is filled to a height of 3.7 m, and is 3.0 m long on the short end. The submarine has a mass of 16,000 kg.

Find: (a) What is the average density of the yellow submarine? How do you know?
(b) What is the volume of the yellow submarine?
(c) Find the combined mass of the submarine and the water in the swimming pool.

9-13 A metal crown seems to weigh less under water than in air because of the buoyant force of the water.

4.4 kg

Given: A crown weighed in air is found to have the same weight as a 5.0-kg mass, while the same crown under water seems to weigh only as much as a 4.4-kg mass.

Find: (a) How many kilograms does the buoyant force support?
(b) What is the volume of water displaced?
(c) Calculate the density of the metal with which the crown is made.
(d) Compare this to the densities in Table 9-1 on page 334. Is the crown gold?

9-14 Horace wants to see what kind of metal his earring is made of. He decides to find the density of his earring by using Archimedes' principle.

Given: When Horace weighed the earring in air on a counter-balance scale, he found that weighed the same as a 5.30-g mass. In water, he found that it weighed the same as a 5.05-g weight.

Find: (a) What was the buoyant force acting on the earring?
(b) What was the volume of water displaced?
(c) Calculate the density of the metal with which the earring is made.
(d) Compare this to the densities in Table 9-1 on page 334. Where does Horace's earring fit in?

9-15 The gravitational field of a neutron star is so great that ordinary matter is squeezed down to the density of nuclear matter. An ordinary student falls asleep in class and dreams of waking on a neutron star.

Given: The volume of a hydrogen nucleus is about 1.0×10^{-44} m^3 and has a mass of 1.67252×10^{-27} kg. The student has a mass of 65 kg.

Find: (a) What is the volume of the student back home on earth assuming her density is about 1.0×10^3 kg/m^3 (the same as that of water here on earth?)
(b) What is the density of nuclear matter?
(c) What would be the volume of the student if she really were on a neutron star?

9-16 When liquids freeze the density changes, which means that the volume changes as well. Water is one of the few liquids that expand when frozen. That is why it is a bad idea to put water in the freezer in a sealed container. Pete decides to test this idea. He sticks an unopened soft drink can in the freezer.

Given: In liquid form the soft drink has a density of 1.11×10^3 kg/m³, more dense than water because there is sugar dissolved in it. The soft drink in the can has a mass of 400.0 g. An ice cube made of the same kind of soft drink has a volume of 13.4 cm³ and a mass of 13.8 g.

Find: (a) What is the volume of the liquid soft drink?
(b) What is the density of the soft drink frozen into a cube?
(c) What is the volume of the soft drink after it has frozen?

9-17 Air pressure can be demonstrated by Magdeburg hemispheres, matching steel hemispheres that can be fitted together with air-tight edges and evacuated of air.

Given: Air pressure is 1.01×10^5 Pa, and the hemispheres have a diameter of 9.0 cm. (The area of a circle is $A = \pi r^2$).

Find: (a) What is the cross-sectional area of the vacuum chamber?
(b) What is the force necessary to pull the hemispheres apart when they are fully evacuated?

9-18 The water is being kept in a full bathtub by a round rubber plug covering the drain. The plug is kept in place by the downward pressure of the water in the tub.

Given: The water in the tub is 37 cm deep, so the water pressure on the top of the plug is about 3.6×10^3 Pa. The plug is round and has a diameter of 5.5 cm. (The area of a circle is $A = \pi r^2$).

Find: (a) What is the cross-sectional area of the bathtub drain?
(b) What is the force necessary to pull the plug?

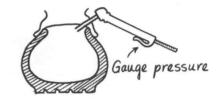

9-19 The air pressures recommended for automobile tires are listed by their manufacturers for the correct pressure when the tire is cold. When you have been driving for a while, the tires warm up and the air pressure inside increases. (You should *not* adjust the pressure downward when the tires are hot, as the recommended pressure is given in anticipation of the fact that it will be higher when the tire is hot.)

Given: The recommended pressure for a particular tire is 1.72×10^5 Pa (above atmospheric). You fill the tire when the temperature is $20°$ C.

Find: (a) What are the initial pressure and temperature on absolute scales in this case?

(b) What should the gauge pressure be after the tire has been driven for an hour on the freeway and has reached a temperature of $50°$ C?

9-20 House painters use compressed air to run their paint sprayers. The pressure of the tank will increase if it is left in the hot sun.

Given: A tank of compressed air filled at a temperature of $23°$ C has a gauge pressure of 656 kPa. Sitting in the afternoon sun, the tank reaches a temperature of $74°$ C.

Find: (a) What are the initial pressure and temperature on absolute scales in this case?

(b) What should the gauge pressure be in the afternoon sun?

9-21 A balloon will float in air if it weighs less than the air it displaces. It is common to use a gas lighter than air, such as helium or hydrogen, to hold the sides of the balloon out so that it will displace a great deal of air.

Given: The density of air is 1.29 kg/m³, while the density of helium is 0.18 kg/m³ and that of hydrogen is 0.09 kg/m³. A given balloon is inflated to a volume 6.1×10^{-3} m³. When empty, the balloon has a mass of 3.77 g.

Find: (a) What is the mass of the helium necessary to fill the balloon?

(b) What is the buoyant force acting on the balloon?

(c) How much force is necessary to hold the balloon down if it is filled with helium?

(d) How much force would be necessary to hold the balloon down if it were filled with hydrogen?

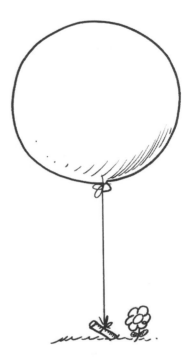

9-22 The Hindenburg was a German dirigible (a kind of blimp) filled with hydrogen, a highly flammable gas. It is remembered as a great disaster because it exploded. This would not have happened if it had been filled with helium.

Given: The density of air is 1.29 kg/m³, while the density of helium is 0.18 kg/m³ and that of hydrogen is 0.09 kg/m³. When inflated, the volume of gas held by the Hindenburg was 2.0×10^6 m³. When empty, the Hindenburg had a mass of 200,000 kg.

Find: (a) What was the mass of the hydrogen necessary to fill the dirigible?
(b) What was the buoyant force acting on the dirigible?
(c) How much force would have been necessary to hold it down if it were filled with hydrogen?
(d) How much force would have been necessary to hold the balloon down if it were filled with helium?

ANSWERS

Solutions to **ESSENTIAL PROBLEMS**

9-6 (a) 98 L; (b) 98 kg; (c) 89 kg

9-7 (a) 1.08×10^3 N; (b) 0.216 m^2

9-8 (a) 9.90×10^3 N/m^3; (b) 2.5 m

9-9 (a) 1.47×10^5 N; (b) 14.6 m^3; (c) 49 cm

9-10 (a) 1.72×10^7 Pa; (b) 0.57 m^3; (c) 695 lung-fulls; (d) 689 lung-fulls

Solutions to **MORE INTERESTING PROBLEMS,** *odd answers*

9-11 (a) 1.00×10^3 kg/m^3; (b) 1.7×10^{-3} m^3; (c) 3.3 kg

9-13 (a) 0.6 kg; (b) 6×10^{-4} m^3; (c) 8.3×10^3 kg/m^3; (d) No less than half that of gold

9-15 (a) 65×10^{-3} m^3; (b) 1.67×10^{17} kg/m^3; (c) 4×10^{-16} m^3

9-17 (a) 64 cm^2; (b) 6.4×10^2 N

9-19 (a) 2.73×10^5 Pa, 293 K; (b) 2.0×10^5 Pa

9-21 (a) 1.1 g; (b) 77 mN; (c) 29 mN; (d) 35 mN

10 *Heat*

Heat as a Form of Energy

WHEN WE TALKED ABOUT ENERGY IN CHAPTER 8, we introduced the idea that energy is always conserved. That means that the total energy before a collision or other interaction is exactly the same as the total energy afterward. This idea is relatively new. The idea of the conservation of energy did not arise until the 1850's, when James Joule did his experiments on heat. Before it was established that heat was a form of energy, it did not seem that there was the same amount of energy before and after a collision. If you only look at kinetic and potential energies of a system, it always looks like there is more energy before a collision than afterwards.

Before people understood heat, they had a basic understanding of energy, but they thought it had a finite lifetime. It seemed that you could make it live longer by reducing friction, but it always appeared to fade away eventually. They even understood that energy could change forms. They knew that the potential energy of a system could be converted into kinetic energy and, under the right circumstances, back again. It was obvious that energy could be stored in a coiled spring, as discussed in Chapter 3. The pendulum, as discussed in Chapter 8, is another classic example. As a pendulum swings back and forth, the potential energy of the pendulum bob changes into kinetic energy as it swings through the middle and then reappears again as potential energy as it reaches the end of its swing on the opposite side.

Even though people could see that energy could be transformed from one kind to another, it seemed that the mechanical energy, potential and kinetic, sort of dies away after a while. The pendulum always comes to a stop if you wait long enough. The piece of the puzzle that was missing was heat. The mechanical energy is not lost, but it is transformed into heat. If you carefully measure the heat given off by an interaction, you will find that the energy in the heat is equal to the amount of mechanical energy lost in the system. Without this relationship between heat and energy, you can not think of energy as a conserved quantity.

Early Theories of Heat

Until the 1840's most people thought of heat as a kind of fluid. They called it caloric. It permeated all objects and flowed from hot objects to cold ones in the same way that water seeks its own level. Under the old caloric fluid theory, heat was supposed to surround the atoms of both gases and solids to hold them apart by the mutual repulsion of their caloric atmospheres.

The first chink in the theory of caloric appeared in the late 1700's as the result of the careful investigations made by an American born military engineer working in Germany. Born in 1753 to a Tory family, young Benjamin Thompson was sent to England to get him out of harm's way as the American Revolution began to unfold. He eventually found his way to Bavaria where he was made Count Rumford by the Bavarian prince for whom he was making cannons.

In the making of a cannon, first you cast a large metal cylinder in a foundry. Then you turn it on a lathe and bore the barrel by advancing a stationary drill bit down the casting. Rumford's lathe was horse-driven, as was common in those days. What puzzled Rumford was the huge amount of heat given off in the process.

The quantity of heat produced limits the rate at which you can bore cannon without overheating the cutting tool. The old caloric fluid theory saw this heat as a material substance that was squeezed from the casting by the pressure of the cutting tool. Count Rumford carefully collected the metal cuttings, however, and discovered that there was no loss of weight due to the cutting process. He then tried using a dull drill bit. He found that the rate at which heat was produced was independent of how long the cannon had been drilled. If heat were a material substance, it would have been eventually drained out of the cannon casting. As long as the horses

Figure 10-1

Count Rumford's experiments with boring cannon led him to believe that the motion of the horses was the source of the large quantities of heat produced.

kept at their work, however, he found that more and more heat was produced. He therefore concluded that the motion of the horses was the source of the heat. The idea of energy as something that cannot be created or destroyed, however, still eluded him.

It was not until about 50 years later that anyone thought of the idea of conservation of energy. When this idea did come, it came to several people independently at about the same time. A number of people then set out to prove that energy is conserved. The most successful was James Joule, who did a whole series of experiments in the early 1850s. His experiments showed that the same amount of heat is always produced by a certain amount of frictional work, no matter how the work is done.

Historically, this equivalence between the mechanical work done and the change of heat, $\Delta Q = Q_2 - Q_1$, produced has been expressed as a proportionality.

$$\text{Work} \propto \Delta Q$$

What James Joule showed was that the proportionality constant is always the same for all kinds of work. In the past, heat and energy have been measured in different units and the proportionality constant between work and heat has been known as Joule's constant. These days, it is more common to simply express both mechanical and heat energy in terms of the same unit, a unit appropriately named a 'joule,' abbreviated J, where a joule is a Newton times a meter.

$$1\,J = 1\,N \cdot m$$

When heat and work are both expressed in joules, the proportionality constant can be taken as unity and we can replace the proportionality constant with an equal sign.

$$\text{Work} = \Delta Q$$

Joule's main contribution was to carefully measure the amount of heat produced by a given amount of work. In his central experiment, he used falling weights to turn a set of paddles in a well-insulated container of water. The idea was to measure the amount of heat produced by stirring the water and compare it with the known amount of potential energy stored in the weights at the start of the experiment.

Joule then went on to show that the same amount of heat was produced from all the other friction experiments he could think of. He tried stirring mercury instead of water. He tried rubbing iron rings together. He even tried rubbing iron rings together in a mercury bath. He went so far as to do experiments using electrical energy. He found that he always got the same result, as long as the work was frictional in the sense that it is "all used up" and not stored in some form of kinetic or potential energy.

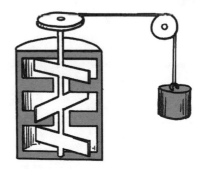

Figure 10-2

Joule measured the mechanical equivalent of heat by using falling weights to turn paddles in a container of water.

A few words of caution are needed when following the modern practice of expressing the equivalence between mechanical work and heat with an equal sign. We will see that there is a fundamental difference between work and heat. The difference is between order and disorder. The motion in work is all in the same direction whereas the motion in heat is in random directions.

Although there are no limits on the amount of work that can be converted into heat, there are fundamental limits on how much heat can be converted back into mechanical work. We will study those limits later in this chapter under the topic of thermodynamics. To help us keep in mind those fundamental differences in the meanwhile, we will use different symbols. We will use the word "Work" for ordered energy, and Q for heat, which is disordered energy.

Although Joule's experiments only looked at converting energy in one direction, they unlocked the secret of conservation of energy, a secret we take for granted these days. Count Rumford held the key to that secret, and you might wonder why he didn't put the key in the lock, so to speak. The answer must be that he didn't even recognize that the lock was there. The concept of heat as something that exists, almost as a material substance but not quite as material as the caloric fluid, is quite new to the human race.

Other Measures of Heat

Vestiges of the old fluid theory of heat remain in common usage, mostly in the way in which heat and energy are frequently measured. While the modern SI units are cleverly arranged so that separate units for heat are not needed, some people still use the calorie to measure heat energy and food energy. The British Thermal Unit (Btu) is still commonly found on heating appliances.

We should discuss these older units for measuring heat both because they are instructive and because they can be confusing. The calorie, as it was originally defined in the metric system, turns out to be a thousand times smaller than the calorie commonly found in dietary references. We will see how this came about by historic accident. Those who advocate use of SI units will find support for their preference in the fact that SI units avoid these points of confusion altogether.

The measurement of heat in both British and metric units was based on the change of temperature of a certain amount of water. In British units, water was measured in pounds. In metric units, they started out measuring water in grams. Both systems are based on 1° of temperature change, but the British use a degree on the Fahrenheit scale while the metric system uses a degree on the Celsius scale.

The British Thermal Unit, or the Btu as it is called, refers to the amount of heat necessary to raise the temperature of 1 lb of water 1° F. The formal definition adds some refinements as to the range of the temperature changes, because the heat capacity of water isn't exactly constant. Fortunately, these small differences can be ignored for all practical purposes.

CHECK QUESTION

Shower quick! Measured in British units, one of those little English hot water tanks holds only 100 lb of water. How much heat is required to change the cold 60° F water that comes into your house into the nice, steamy hot 180° F stuff you expect to see at the shower valve?

Answer: 12,000 Btu

The first metric unit of heat was the small **calorie**. It was defined as the amount of heat necessary to raise the temperature of 1 g of water 1° C. This original definition has been modified somewhat by popular usage. The small calorie represents such a small quantity of heat that people found the kilocalorie much more useful for most purposes. The **kilocalorie** may either be thought of as 1,000 small calories or as the amount of heat necessary to raise 1 kg of water 1° C. The kilocalorie became so common in usage that people just shortened its name to Calorie with a capital C to distinguish it from the small calorie. It is this big Calorie that is used to measure the energy content of various quantities of food. People used this Calorie so much for food energy values that they probably forgot why they capitalize the C and over the years just decided that Calorie should not be capitalized. Thus, caloric stands for both the amount of heat necessary to heat 1 g of water 1° C and for a thousand times that much heat, depending on who is using the word. The surprising thing about this state of affairs is that it seldom results in any permanent confusion. For clarity in this book, we will abbreviate the big C Calorie as kcal for "kilocalorie."

Figure 10-3

A kilocalorie is the amount of heat necessary to raise the temperature of 1 kg of water 1° C.

CHECK QUESTION

A peanut contains about 3 kcal of heat energy. If it is burned under a flask containing 0.5 liter of water and about half the heat goes into warming the water, how much will the temperature of the water increase?

Answer: 3° C

The Mechanical Equivalent of Heat

One reason why it took so long to understand that heat is a form of energy is that a little bit of heat is equivalent to a great big bunch of mechanical energy. You must do a huge amount of work before you begin to notice the little bit of heat you generate. Such huge quantities of work were first available in industrial processes like the cannon boring done by Count Rumford in about 1789.

It wasn't until 1839, however, that James Prescott Joule began a whole series of experiments to measure exactly the amount of mechanical energy in a certain amount of heat. His best known experiment, published in 1847, used paddle wheels driven by weights. The key to his investigation was to show that the same amount of heat is generated no matter how the work is done. Now we have found that the easiest way to measure heat is not with mechanical energy, like Joule used, but with electrical heat. These measurements put the best value at

$$1 \text{ kcal} = 4,186 \text{ J}$$

Let us use an example to illustrate the difference in size in the amount of energy represented by a normal quantity of heat as compared to a normal quantity of mechanical work. Suppose that you are served a luke-warm cup of coffee in a restaurant and that it occurs to you, after you fail to attract the waiter's attention by normal means, that you might be able to raise the temperature of the coffee by stirring it with a spoon. The question is whether you might not wind up by attracting the waiter's attention anyway if you were to try to stir the coffee violently enough to produce a significant temperature change, say, 4° C. That would amount to 1 kcal of heat in a quarter-liter cup of coffee.

Assume that you stir with a force of about 1 N. (That would be the amount of force you would need to hold a quarter-pound stick of butter on your spoon.) If you stir 5 cm in one direction and then 5 cm back again, you would stir a total distance of 10 cm—0.10 m—with each stroke. Ten such strokes would produce a joule of energy, but it would take almost 42,000 such strokes to produce a Calorie of heat energy. Therefore, the process of evaporation would cool your coffee more quickly than you could reasonably warm it by stirring.

We are fortunate in some ways that so much mechanical energy can turn into such a little bit of heat. In driving an automobile down the free-way, we have a tremendous amount of kinetic energy. All that energy must be dissipated if we should need to stop. Our brakes can handle all this energy only by converting it into the more concentrated form of energy known as heat.

Figure 10-4

A little bit of heat represents a great deal of work.

CHECK QUESTIONS

(a) What is the kinetic energy of a 837.2 kg automobile traveling at 10 m/s? (b) How many Calories of heat are produced by the brakes in bringing this automobile to a stop? (c) If this energy were used to warm 2 kg of water, what would be the temperature change?

Answer: (a) $\frac{1}{2} mv^2 = 41,860$ J, (b) 10 kcal, (c) 5° C

Perpetual Motion Machines of the First Kind

Philosophically, probably the most important thing about the mechanical equivalent of heat is that the proportionality works both ways. Not only can work be changed into a certain amount of heat but heat can also be changed into a certain amount of work in exactly the same proportion. In this case, however, we would be talking about a change of heat, ΔQ, since work is the change of, or flow of, energy.

$$\text{Work} = \Delta Q$$

The energy you can get out of any device in the form of work is just the mechanical equivalent of the amount of heat that went into the device as long as the internal energy of the device doesn't change. That is, the work done is the difference between the heat, Q_{in}, which flows into the device from a region of high temperature and the heat, Q_{out}, which flows out into a region of lower temperature.

$$\text{Work} = (Q_{in} - Q_{out})$$

Say you have a box that is set up in such a way that when you put heat into the box you get work out of the box.[1] This would be some sort of heat engine, as we will discuss later, but it does not really matter. No matter what is in the box, you must always put the same amount of heat energy into the box as the amount of energy you get out of the box. It is not possible to get more energy out of the box than went in. This statement seems reasonable to most people today, since it is really nothing more than the idea of the conservation of energy. You can't get more energy out of a box than you put in without the internal energy of the box changing. This is called the **first law of thermodynamics**.

Many very clever people have spent a good part of their productive life trying to get more energy out of a box than they put in. Some of these boxes contain such ingenious mechanisms that it is difficult to see why they wouldn't run forever without running down. Such a device would be very useful in that it would provide a limitless supply of work. Many inventors have applied for patents on perpetual motion machines of this kind, but the United States Patent Office has never issued one. The U.S. Patent Office has uniformly rejected claims to an invention of this kind as drawn to an inoperative device. An inventor can overcome a rejection on inoperativeness by submitting a working model, but not a single inventor has met this test. Perpetual motion machines of the first kind exist only in people's minds. They are a theoretical impossibility.

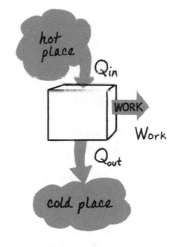

Figure 10-5

The work that comes out of a device on a continuous basis is just the mechanical equivalent of the difference between the heat energy in and the heat energy out.

---

[1] Physicists often use the idea of a "black box" to illustrate theoretical principles. The idea is that you can discuss a box having certain properties without having to specify exactly what is doing those things. The black box is an especially useful concept when discussing theoretical concepts without getting into the nitty-gritty of practical limitations.

Temperature

In the last chapter, we discussed the linear relationship, first proposed by Jacques Charles in the late 1700's, between temperature and pressure in an ideal gas. We introduced the idea of an absolute temperature scale so that we could view this linear relationship as proportional.

In this chapter, we have shown how temperature has been used to measure heat. We saw how both the French calorie and the British Thermal Unit were based on the temperature change of a convenient unit of water. While these older units have been replaced with the joule in SI units, the laboratory technique of measuring heat energy with a temperature change is still state-of-the-art.

In all of this discussion, we have assumed that everyone knows what temperature is. That is because everyone does. At least at the level of measurement, modern society teaches everyone to use a thermometer and to interpret the result. If the room temperature is 20° C, which is exactly the same as 68° F in British units, everyone understands that you are going to be much more comfortable than if the room temperature is a sweltering 35° C, or 95° F.

Translating between systems of units might seem to offer a challenge, but more of a language challenge than a conceptual challenge. Since some people may need to become bilingual in the two temperature scales, we will look at how to do this kind of calculation as an aside, in the Highlight on page 376. Mastering the technique of converting from one temperature scale to the other may help when traveling, and it serves as a good example for the mathematical understanding of linear functions. It adds little, however, to what our modern society teaches everyone at a young age about temperature.

While most people master the concept of temperature on one level, few appreciate it on a more sophisticated level. For example, most people know that temperature is a measure of heat energy in an object. Most people would believe, therefore, that the air in a room at a sweltering 35° C, or 95° F, contains more energy than the air in the same room at the same pressure and volume at a comfortable 20°C, or 65° F. The more the temperature, the more the energy, so to speak. That is not true!

Take air in a room, according to the ideal gas law discussed in the last chapter, for example.

$$PV = kNT$$

The air follows the rule that says that the pressure P times the volume V equals Boltzmann's constant k times the number of molecules N in the room times the absolute temperature T. According to this relationship, you cannot increase the temperature T in the room without either increasing the pressure P, the volume V, or decreasing the number of molecules N. If the room becomes uncomfortably hot, it is not because either the pressure

P or volume *V* has increased. The room will be hot even if those variables remain constant.

What does happen is that, as you heat the room, the air molecules that remain in the room gain energy and move around faster. As the air expands, it pushes exactly the number of molecules out of the room necessary to keep constant the total energy content of the room. Fewer fast moving molecules in the hot room have the same energy as more slower moving molecules in the nice comfortable room.

It is easy to calculate, by the way, the total energy of the air in the room. It is just the product of pressure *P* and volume *V* on the left hand side of the gas law. We can see that pressure times volume has units of energy by considering their definition. Pressure is force over area and volume is area times distance:

$$PV = \left(\frac{F}{A}\right)(Ad)$$
$$= Fd$$

therefore, pressure times volume has the dimensions of work (remember from the chapter on Energy, Chapter 8, that work is defined as force times distance). A change in either pressure or volume therefore does work in the sense that it transfers energy. If you maintain a constant pressure *P* while changing the volume by ΔV, as in operating your bicycle pump for example, the work you do will transfer ΔE energy.

$$P\Delta V = \Delta E$$

That is why the units for Boltzmann's constant are joules per molecule per Kelvin. If the left hand side of the gas law has units of energy, so must the right hand side. Temperature is, as everyone expects, a measure of energy, but it is proportional to energy in a gas only if the number of molecules remains constant.

A better definition of temperature will have to wait until we have discussed the **entropy** of a system. And if, as it commonly said, only chemists understand entropy, then only chemists have a prayer of understanding temperature. There is an essay at the end of this chapter, which you are encouraged to read, that develops the idea of entropy in full. For now we will understand it in the same way as did Sadi Nicholas Lenard Carnot, pronounced (CAR-no), (1796–1832), that young French soldier-scientist who invented entropy to develop the fundamental theory of heat engines.

Carnot took temperature to be defined by the operation of a gas thermometer. If the gas follows the ideal gas law exactly, with pressure times volume directly proportional to temperature, that is what is meant by temperature.

HIGHLIGHT
Temperature Scales

One of the obvious indications of temperature is the volume of mercury or alcohol in a glass thermometer that expands or contracts to fill the capillary tube to a certain point. By marking the tube at a point where the mercury or alcohol stops rising, we indicate a certain temperature. A temperature scale could be established by marking the tube at two reproducible temperatures, say, the freezing point and the boiling point of water under standard atmospheric pressure, and then putting a certain number of equally spaced marks in between. We might select that number to be 100, for example. Since the Latin prefix for one hundred is *cent,* we might well call our scale the centigrade scale. Actually, the first person to do this was a Swedish astronomer by the name of Anders Celsius (1701–1744). We therefore call this scale the Celsius scale.

The choices of 0° and 100° as representing the freezing and boiling points of water are arbitrary. You might argue that any other two numbers, such as 32 and 212, would do just as well. Subtracting 32 from 212 would give 180° between these two temperatures:

$$212 - 32 = 180$$

You could then think up a name for 180, say, call it the "fahren," and it would then seem natural to call such a temperature scale the fahrengrade scale. If the first person to think of doing such a thing had happened to have the name Gabriel Daniel Heit, you might well reason that it would be even better to call this scale the Fahrenheit temperature scale.

Actually, the first person to think of such a temperature scale was a German physicist and instrument maker named Gabriel Daniel Fahrenheit (1686–1736), who was also the first to use mercury in a thermometer. He did not really select arbitrary numbers for the freezing and boiling points of water. He just happened to select different temperatures for the 0 and 100 point marks on his scale. Perhaps Fahrenheit liked making hand-cranked ice cream or something, because the zero point he chose was the lowest temperature he could reach by mixing salt with ordinary ice. The salt forces the ice to melt at lower and lower temperatures, up to a point. That point, called the "brine point," is the temperature at which salt water freezes. The 100° mark, on the other hand, was selected to be near his own body temperature. Maybe he had a slight fever that day, but his temperature scale was well established and widely used before Fahrenheit heard of the fixed points used by Celsius. To his credit, Fahrenheit agreed that the temperatures of freezing and boiling of pure water were easier fixed points to use than the ones he had originally adopted, and he redefined his temperature scale in terms of them. The result is that average human body temperature is almost, but not quite, 100° on the Fahrenheit scale.

For people who want to convert from one temperature scale to the other, the conversion must take into account the differences in the degree size as well as the difference in the zero points on the two scales. The difference in size is easy. An ordinary conversion factor like we have been using all along will handle a difference in temperature very nicely. A temperature difference of 180° on the Fahrenheit scale, from 32° F for freezing to 212° F for boiling of water, is equal to a temperature difference of 100° on the Celsius scale. Any other temperature difference is in the same ratio. A temperature difference of 9° on the Fahrenheit scale is equal to a temperature difference of 5° degrees on the Celsius scale.

$$\Delta T = 9^\circ \ F\left(\frac{100^\circ \ C}{180^\circ \ F}\right) = 5^\circ \ C$$

There are nearly twice as many Fahrenheit degree marks between the freezing and boiling points of water as there are on the Celsius scale. For rough work, you can therefore just multiply or divide by 2 and add or subtract a degree or two to compensate for the fact that 180 divided by 100 is not exactly equal to 2.

When you convert from Fahrenheit degrees to Celsius degrees, you need to get down to a common starting point before you use a conversion factor. Since zero on the Fahrenheit scale is 32° warmer than zero on the Celsius scale, you can just subtract 32° from the Fahrenheit reading to get the reading relative to the freezing point of water. You could just as easily go the other way and add 32° to the Celsius reading to get the temperature relative to zero on the Fahrenheit scale. You can then compensate for the different sizes of the degrees using the above conversion factor. This can all be put into one equation as follows:

$$t_C = \frac{5}{9}(t_F - 32^\circ)$$

where t_C is the temperature in degrees Celsius, and t_F is the temperature in degrees Fahrenheit.

ice water

CHECK QUESTION

Photographic developing times are often given for 68° F. (a) How many Fahrenheit degrees is this above freezing? (b) What would this temperature read on a Celsius thermometer?
   Answer: (a) 36° F, (b) 20° C

To make the section on temperature scales complete, we should mention the absolute temperature scale, discussed in the last chapter. In SI units, we use the *absolute temperature scale*, which is measured in *kelvins*, abbreviated "K" with no degree symbol. Kelvins have as zero the theoretical lower limit of temperature, even though we have not been able to reach that temperature in any laboratory. Kelvins are the same size as Celsius degrees, so to convert to kelvins from degrees Celsius we only need to add the number of degrees that zero on the Celsius scale is above zero on the kelvin scale, which turns out to be exactly 273.15 degrees.

$$T = t_C + 273.15$$

where T is the absolute temperature, measured in kelvins and t_C is the temperature measured in Celsius degrees.

boiling water

It remained for Lord Kelvin in England, Rudolf Clausis in Germany, and Ludwig Boltzmann in Austria to show that the French soldier's conclusions were exactly correct even though his work was based on a primitive theory of heat. The success of these giants, building on Carnot's theory, was one of the crowning achievements of physics in the last century.

Thermal Expansion

Temperature is easily measured by the changes in various properties of materials. Different substances melt or boil at different temperatures. Other things undergo a change in color or electrical resistance as their temperatures change. However, one of the most familiar changes associated with an object's temperature is simply its size.

The expansion of mercury or alcohol in a glass thermometer is one of the best-known measures of temperature. It happens that the rate of expansion of neither one of these substances is exactly constant with increasing temperature, but they can be calibrated against a gas thermometer, which is much more accurate.

Ordinary solid objects also generally expand with increasing temperature. We do not ordinarily think of the length of various objects around us as depending much on temperature, but a careful look reveals many situations where it must be taken into account. Cracks in sidewalks are really thermal expansion joints put there so that forces generated by changes in temperature will not break the concrete. Freeway overpasses must be provided with much larger expansion joints. You can entertain yourself as you ride down the freeway by studying the overpasses to see if you can find the expansion joints. Sometimes one end of a bridge will be mounted on rollers, and sometimes the whole support structure will be on rockers. You can frequently spot the expansion joint from the top by the metal plate in the roadway used to cover the joint. Perhaps you have noticed the metal plates with a zig-zag tooth pattern running across the roadway and wondered what they were for. Everyone has noticed the cracks in a railroad track that make the wheels go clickity-clack on both trains and streetcars, but few people realize that these, too, are put there deliberately as expansion joints.

Let us consider an experiment to figure out how much of a crack you would need to leave in, say, a railroad track. Go down to your basement, at least in your mind's eye, and find an old piece of galvanized steel water pipe and a garden hose. Attach one end of the pipe to the drain on your hot water heater with the hose. Attach this end of the pipe to a block of wood and rest the other end on another block some distance away so that the end is free to slide. You might think that running hot water through the pipe would cause it to expand such a tiny amount that you would never be able to measure it, but there is a way to see the pipe expand before your very eyes. Just put a sewing needle under the free end of pipe to act as a roller, and stick a broom straw through the eye of the needle to act as a pointer.

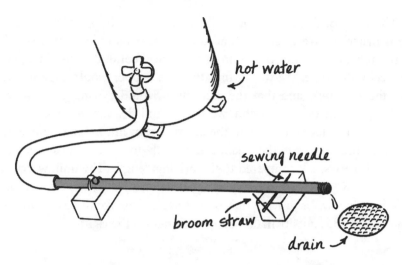

Figure 10-6

The thermal expansion of steel could be measured at home in your basement.

Surprisingly, a piece of pipe about a meter long will expand enough to make the broom straw turn through about 30 to 50 degrees, depending on the temperature of your hot water and the diameter of your sewing needle.

The distance that the pipe expands, Δl, (read "the change in length") is directly related to the angle through which the needle rolls. It would be reasonable to assume that this change in length of the pipe would be proportional to the temperature change, ΔT. The hotter the water flowing through the pipe, the more the pipe expands.

$$\Delta l \propto \Delta T$$

There is, however, something besides temperature that affects this change in length. You could verify that the change in length, Δl, is also proportional to the original length of the pipe, l. If you were, for example, to move the block with the sewing needle to the middle of the pipe, the angle through which the broom straw would twist would be only half as large as if it were at the end. The pipe would actually expand the same as before, but the rolling needle would only record the amount by which the first half of the pipe expands. Thus, we have a second proportionality, saying that the change in length is proportional to the length of the thing being measured.

$$\Delta l \propto l$$

We can put these two proportional relationships together in a single statement by just saying that Δl is proportional to the product of l and ΔT.

$$\Delta l \propto l\Delta T$$

The proportionality constant between Δl on one side of the equation and l times ΔT on the other is called the **coefficient of linear expansion**, and the symbol usually used for it is the Greek letter alpha, α.

$$\boxed{\Delta l = \alpha l\Delta T}$$

Substance	Coefficient of linear expansion, α
Aluminum	24×10^{-6} K^{-1}
Copper	17×10^{-6}
Steel	11×10^{-6}
Concrete	12×10^{-6}
Glass	9×10^{-6}
Ice	51×10^{-6}

Table 10-1

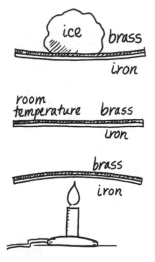

Figure 10-7

A bimetallic strip bends, because the two metals expand differently.

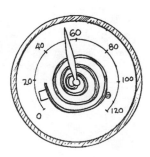

Figure 10-8

Some thermometers use a coiled bimetallic strip to turn a pointer.

The value of the coefficient of linear expansion, α, is different for different materials. We might look around in your basement for a piece of copper tubing or a piece of aluminum electrical conduit to use in place of the steel water pipe. We would find that the copper would expand more than the steel pipe and that the aluminum would expand over twice as much. If we did the calculation, we would find the values for α given in Table 10-1. The factor of 10^{-6} in the coefficient of linear expansion simply expresses the fact that the change in length for ordinary temperature changes is very small compared to the original length. The unit, which can be in either K^{-1} (kelvins in the denominator) or (°C)$^{-1}$, (degrees Celsius in the denominator,) is just the unit that α needs to make things on the left-hand side of the equality match up with things on the right.

CHECK QUESTION

By how much would a 1.0-m steel hot water pipe expand when its temperature is increased 50° C by running hot water through it?
    Answer: 0.55 mm

Some thermometers use the difference in α of different metals to measure temperature. A bimetallic strip is composed of two metals welded together. The two metals, such as brass and iron, expand differently. The difference in the expansion causes the strip to bend as it is heated. If you look inside a thermometer and see what seems to be a coil spring, the chances are that this spring is made of a bimetallic strip wound up into a coil. Heat the coil by breathing on it, and see if it doesn't unwind itself a little.

Bimetallic strips are frequently used in thermostats to open and close electrical contacts. The thermostat that controls the furnace in your house probably works this way.

Heat Capacity

The ideas of heat and temperature are so closely related that it is instructive to consider the difference. Let us suppose that, having abandoned our attempts to warm our cup of coffee by stirring, we proceed to warm it by holding it over the candle so thoughtfully placed on our table for just such emergencies. Right away we would notice that the amount of heat necessary to warm the coffee is directly proportional to the amount of coffee in the cup. If you were to drink half the coffee, you would need to warm only that which remains, and the temperature would go up much more quickly. That is, the change in heat, ΔQ, necessary to produce a certain change in temperature is directly proportional to the mass, m, of the coffee:

$$\Delta Q \propto m$$

Of course, the change in heat, ΔQ, is also proportional to the change in temperature, ΔT. Colder coffee would require more heat to warm to the same temperature:

$$\Delta Q \propto \Delta T$$

Since the heat, ΔQ, is proportional to both the mass of the coffee, m, and the change in temperature, ΔT, it is also proportional to their products:

$$\Delta Q \propto m\Delta T$$

The proportionality constant is called the **specific heat** of the material. People usually use c to stand for specific heat:

$$\Delta Q = cm\Delta T$$

The specific heat of coffee is very close to that of water, because that is what coffee mostly is. If you think about it, we already know all the elements we need to know the specific heat of water. The Celsius degree, in fact, is defined in terms of the specific heat of water. As we have discussed before, one kilocalorie is defined as the amount of heat that raises the temperature of 1 kg of water 1° C:

$$1 \text{ kcal} \equiv c_{\text{water}}(1 \text{ kg})(1° \text{ C})$$

In writing it this time, we have written in the specific heat of water, c_{water}, as the conversion factor. Since the left side is equal to the right, the specific heat of water must be 1, with units to make the sides equal.

$$c_{\text{water}} = 1\frac{\text{kcal}}{\text{kg}° \text{ C}}$$

In SI units specific heat is measured in joules, kilograms and kelvins. A kilocalorie is defined as 4,184 joules, and a kelvin is the same size as a Celsius degree, so we can use these to write the specific heat of water in SI units.

$$\boxed{c_{\text{water}} = 4,184\frac{\text{J}}{\text{kg} \cdot \text{K}}}$$

Actually, it would be safe to say that the specific heat of coffee is a little less than that of water, even without knowing the exact organic chemicals that make it into coffee. Irish coffee, for example, would have a specific heat smaller than that of regular coffee. Water happens to have the highest specific heats of all common substances. The specific heat of alcohol is less than six-tenths that of water. Just for interest, the specific heats of several common substances are listed in Table 10-2.

Figure 10-9

The amount of heat necessary to warm a cup of coffee depends on the amount of coffee in the cup and the temperature of the coffee to begin with, as well as on the kind of coffee you are drinking.

Substance	Specific Heat
Alcohol (ethyl)	2400 J/kg · K
Aluminum	900
Brass	380
Copper	387
Glass	837
Ice (−5° C)	2090
Iron	448
Lead	128
Marble	860
Mercury	140
Water	4184

Table 10-2

Calorimetry Problems

Many situations in daily life involve combining things and allowing them to come to the same final temperature. Suppose that the waiter in our example finally notices you and offers to warm your coffee. He adds piping hot coffee to your half-filled cup of lukewarm coffee, and the mixture comes to a temperature somewhere in the middle depending on how much piping hot coffee was added. It would seem obvious that the final temperature would be exactly halfway between the temperatures of the lukewarm and the piping hot parts if the parts of the mixture are equal to begin with. Also, the final temperature would be closer to that of the larger part if they were not equal. Let us assume that the parts are not equal, and we will figure out how to predict the final temperature of the mixture using the law of conservation of energy.

All the energy to be conserved in this case is in the form of heat. If we assume that none of the heat gets away from us by ignoring any cooling of the overall mixture during the short time that we do our experiment, the principle of conservation of energy comes down to the heat gained by the lukewarm coffee being equal to the heat lost by the piping hot coffee that was added.

$$\boxed{\text{Heat gained } = \text{ Heat lost}}$$

This particular form of the energy conservation principle turns out to be the key to solving a whole class of physics problems involving thermal equilibrium. Careful measurements are frequently done in a well-insulated container called a **calorimeter**, and the whole science of measuring heat in this way has come to be known as **calorimetry**. (If you reach into your pocket and pull out a thermometer with which to stir your coffee, the waiter will recognize you as a calorimetrist.)

We can express the heat gained by the cold coffee, ΔQ_{gained}, in terms of its mass, m_{cold}, and change in temperature in going from the cold temperature, T_{cold}, to the final temperature, T_{final}:

$$\Delta Q_{\text{gained}} = cm_{\text{cold}}(T_{\text{final}} - T_{\text{cold}})$$

Similarly, we can express the heat lost, ΔQ_{lost}, by the hot coffee in terms of its mass, m_{hot}, and the change in temperature going from hot, T_{hot}, to the final temperature, T_{final}:

$$\Delta Q_{\text{lost}} = cm_{\text{hot}}(T_{\text{hot}} - T_{\text{final}})$$

Setting the heat gained equal to the heat lost, we can solve for the final temperature:

$$cm_{\text{cold}}(T_{\text{final}} - T_{\text{cold}}) = cm_{\text{hot}}(T_{\text{hot}} - T_{\text{final}})$$

First we need to multiply the temperatures and then move all terms containing the final temperature, T_{final}, to one side of the equation:

$$m_{cold} T_{final} + m_{hot} T_{final} = m_{hot} T_{hot} + m_{cold} T_{cold}$$

Factoring out the final temperature and solving, we see that the final temperature turns out to be a sort of average of the initial and final temperatures:

$$T_{final} = \frac{m_{hot} T_{hot} + m_{cold} T_{cold}}{m_{hot} + m_{cold}}$$

In fact, it is easy to see that the final temperature will be halfway between the initial and final temperatures if $m_{hot} = m_{cold}$. If the mass of the hot coffee, m_{hot} equals the mass of the cold coffee, m_{cold}, there is no difference between the masses and we can ignore their subscripts.

$$T_{final} = \frac{m_{hot} T_{hot} + m_{cold} T_{cold}}{m_{hot} + m_{cold}} = \frac{m(T_{hot} + T_{cold})}{2m} = \frac{T_{hot} + T_{cold}}{2}$$

This is exactly what you would expect with equal amounts of coffee. The temperature change of the two half cups of coffee are equal, and they come to an equilibrium temperature halfway in the middle.

If the two masses m_1 and m_2 are not equal, a so-called weighted average results. If two-thirds of a cup of hot coffee is added to one-third of a cup of cold coffee, the contribution of the hot coffee to the temperature change is twice as important as that of the cold. The final temperature would still be between the hot and the cold, but it would be two-thirds of the way toward the hot end of the scale.

$$T_{final} = \frac{(\frac{2}{3}m) T_{hot} + (\frac{1}{3}m) T_{cold}}{\frac{2}{3}m + \frac{1}{3}m} = \frac{2 T_{hot} + T_{cold}}{3}$$

CHECK QUESTION

A good husband adds 3.0 kg of boiling hot water from a tea kettle to the 12.0 kg of 30° C warm water in his wife's bath. What is the final temperature of the resulting 15.0 kg of water?

Answer: 44° C

The problems so far discussed are of the type where both the hot and cold substances have the same specific heat. If they didn't have the same specific heat, say the waiter was feeling remorseful for having neglected you and decided to top your cup with steaming hot Irish coffee, you would need to repeat the above reasoning with subscripts on the specific heats.

The specific heat c would not drop out of the equation in this case, since the c_{hot} on one side of the equation would have a different value from the c_{cold} on the other. The general solution would therefore be similar to that shown except that it would contain the specific heats.

You could also fix up this reasoning to account for the change in temperature of the cup itself. All you need to do is add in another heat term, ΔQ_{cup}, on the heat gained side of the equation to take care of the heat used in warming the cup.

There are many other variations on this same theme. Sometimes you are trying to find the final temperature, and sometimes you are trying to use the final temperature to find the specific heat of an unknown substance or the heat generated in a given process. The same approach can be extended to account for the heat of fusion and vaporization as water or other substances freeze or evaporate. All these problems, however, are solved in the same general way, starting from the premise that the heat gained equals the heat lost.

Heat Flow

There is another class of problems in which the focus of attention is on the *rate* at which heat is gained or lost. The rate at which heat flows from a region of high temperature to a region of low temperature is found to follow a rule known as Newton's law of cooling. Sir Isaac Newton noticed that the rate at which something cools is proportional to how hot it is relative to the ambient room temperature. We can use algebraic notation to express the rate of heat flow, $\Delta Q/\Delta t$, as the ratio of the change of heat, ΔQ, to the change in time, Δt. According to Newton's law of cooling, this rate of heat flow is proportional to the difference in temperature, ΔT, between the object and its surroundings:

$$\frac{\Delta Q}{\Delta t} \propto \Delta T$$

The proportionality between heat flow and temperature difference is the reason that conservationists recommend lowering the thermostat in your house. The rate at which heat escapes through your walls and roof is proportional to the difference between the inside and outside temperatures. Another way to save heat is to add insulation, which is a layer of material having a low thermal conductivity. The meaning of this term will become apparent if we consider the other factors, besides temperature, that affect heat flow.

Consider a slab of insulating material such as you might buy at a lumber yard for use in insulating your house. It is easy to see that the rate of heat flow is directly proportional to the area of this slab of material. If you had two slabs of the same size side by side and maintained the same temperature difference across both of them, you would expect the same heat to flow through each of them. By doubling the area through which the heat can flow, you double the rate of heat flow. If you triple the area, you would expect triple the heat flow rate. Thus, we consider the rate of heat flow, $\Delta Q/\Delta t$, to be directly proportional to the area, A, through which the heat can flow:

$$\frac{\Delta Q}{\Delta t} \propto A$$

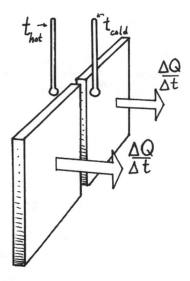

Figure 10-10

Heat flow is directly proportional to area as well as to temperature difference.

Instead of laying the two slabs of insulating material side by side, we could stack them one on top of the other. Instead of doubling the area, we would then be doubling the thickness. The result would be to cut the heat flow rate in half. This result seems reasonable if we stop to consider that laying the insulation on twice as thick has the effect of reducing the temperature drop across each slab of insulation. We would expect a thermometer stuck between the two blankets of insulation to read a value halfway between the hot temperature, T_{hot}, on one side of the stack and the cold temperature, T_{cold}, on the other. That is true because of the way in which we have these two slabs of insulation stacked. All the heat that passes through one slab must pass through the other. It therefore requires the same temperature difference across each slab to cause the same amount of heat to flow. If we stacked three slabs, one on top of another, we would expect the temperature across each one to be one-third of the difference between that on the hot side of the stack and that on the cold. Thus, the heat flow would only be one-third as great. We are therefore led to the conclusion that heat flow is inversely proportional to the thickness of the insulation. Expressed algebraically, an inverse proportion between the heat flow rate, $\Delta Q/\Delta t$, and the thickness, l, of the material is the same as a direct proportion between $\Delta Q /\Delta t$ and the reciprocal of l.

$$\frac{\Delta Q}{\Delta t} \propto \frac{1}{l}$$

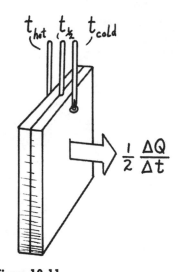

Figure 10-11

Heat flow is inversely proportional to thickness.

Combining this proportion with the ones for area, A, and temperature difference, ΔT, we get the proportionality

$$\frac{\Delta Q}{\Delta t} \propto \frac{A\Delta T}{l}$$

This can be expressed in equation form using a proportionality constant k, called the **thermal conductivity** of the material.[2]

$$\boxed{\frac{\Delta Q}{\Delta t} = k\frac{A\Delta T}{l}}$$

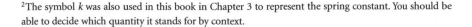

[2] The symbol k was also used in this book in Chapter 3 to represent the spring constant. You should be able to decide which quantity it stands for by context.

This is the general expression by which the rate of heat flow through a solid may be calculated. The thermal conductivity of various materials may be found in reference handbooks. Just to give you an idea of the range of values, the thermal conductivities of a few common substances are listed in Table 10-3. In SI units, thermal conductivity is expressed in watts, W, per meter-kelvin. A watt, remember, is a joule per second (J/s). You will often see thermal conductivity expressed in W/m · °C, but since Celsius degrees are the same size as kelvins the numbers in Table 10-3 will be the same.

Material	Thermal Conductivity Constant k W/m ·K (or W/m · °C)
Copper	401
Steel	46
Glass	0.8
Wood, oak	0.15
Wood, pine	0.11
Concrete	0.19 – 1.3
Water	0.6
Air	0.0234

Table 10-3

The Thermal Conductivities of Various Substances, in Metric Units at ordinary temperature.

CHECK QUESTION

A certain glass window has an area of 1.0 m² and a thickness of 0.5 cm. Glass has a thermal conductivity k = 0.8 W/m ·°C. How much heat goes through this glass in 1 minute (60 s) if the temperature inside is 20° C and the temperature outside is 10° C?

Answer: 96,000 J

If you buy insulation for your house, a much more useful figure to consider is the **thermal resistance per unit area**, known as the R factor, of the insulating material. This depends not only on the thermal conductivity of the material but also upon its thickness. As you might expect, the R factor increases as the thickness of the material increases and as the thermal conductivity decreases. In fact, the R factor as it is used in the building trade is simply defined as the ratio of the thickness, l, of the material to its thermal conductivity as defined in the preceding equation.

$$R \equiv \frac{l}{k}$$

In terms of the R factor, the expression for heat conduction may be stated as follows:

$$\frac{\Delta Q}{\Delta t} = \frac{A\,\Delta T}{R}$$

It is easy to see why people talk about this R factor when they are buying insulation to improve energy conservation in their homes. The greater the R factor, the smaller will be the heat flow through the area protected by this insulation.

Entropy—The Ghost of the Caloric Fluid Theory

In his brief life, the French soldier Sadi Carnot was able lay the foundation for a theoretical understanding of heat engines. His *Reflections On Machines That Do Work* (1824) introduced the idea of an absolutely perfect heat engine. Carnot's engine not only formed a model upon which real engines were designed, but it anticipated the invention of such modern devices as the refrigerator and the heat pump. He showed why such machines have a fundamental limit on their efficiency.

All this he did without understanding heat as a form of energy. To Carnot, heat was the caloric fluid that we mentioned earlier. He saw temperature as the pressure that pushes this caloric fluid from a hot place to a cold place. We saw earlier that pressure P times the change in volume ΔV does work ΔE

$$P\,\Delta V = \Delta E$$

We discussed this in terms of the pressure and change in volume in a bicycle pump, but Carnot probably thought of it in terms of driving a mill with a water wheel. With his training as an engineer, he would have thought of P in terms of the height of the water and ΔV in terms of water flow. We know that he made the analogy that implied that he could get work out of heat by allowing it to flow from a hot place to a cold place. In this analogy temperature T times the flow of caloric fluid ΔS does work ΔQ.

$$T\,\Delta S = \Delta Q$$

We know that Carnot thought of the flow of his caloric fluid ΔS as the change of stuff in a physical sense. He called it **entropy**, and we call it that today. We no longer believe in caloric fluid, but we still use ΔS to stand for the change of entropy as if it were the flow of actual physical stuff. Now we think of entropy as a measure of disorder in a system.

Today, it is common to define entropy in the same way mathematically except that we believe that what is really flowing is heat energy ΔQ at a certain temperature *T.* Then we define the change of entropy of a system as the ratio of the two.

$$\Delta S \equiv \frac{\Delta Q}{T}$$

It is remarkable that, although we no longer agree with Carnot's model, his conclusions were exactly correct. They formed a foundation upon which not only heat engines are understood, they became a model for the development of electrical theory of the day. When we get to the subject of electricity, we will see that voltage times the flow of charge is also equal to the change of energy in a system. Carnot's followers were many. It is no accident that voltage is like fluid pressure and like the temperature of a hot place in that it pushes stuff around and does work in the process. That too is the ghost of Carnot's deceased theory.

Entropy is the more thorough way to understand heat. It is a harder concept to conceptually grasp than the caloric fluid theory, but is an extremely important idea. It addresses the issue of chaos—the idea that nature prefers disorder—and is the basis for many new ideas in modern physics. Since you do not need entropy to understand heat at this level or solve the problems in this chapter (after all, the founders of thermodynamics did not understand entropy either), we have left the discussion out of the main text and included it as a separate essay at the end of this chapter. You are encouraged to read it.

Perpetual Motion Machines of the Second Kind

When speaking of the principle of conservation of energy earlier in this chapter, we pointed out that no one has invented a machine that would put out more work energy than it takes in. Such a machine is known as a perpetual motion machine of the first kind, and it would violate the principle of conservation of energy, which when expressed in terms of heat flow, is called the **first law of thermodynamics.**

There is nothing in this law, however, that would prevent *all* the energy taken in by a machine from being converted into work. Energy would be conserved; it would just be converted from heat into mechanical energy.

A machine of this second kind would be even more useful in some ways than a perpetual motion machine of the first kind. Not only would it provide a nearly limitless supply of work, but in absorbing this energy in the form of heat from its environment, it would make a nice refrigerator. As an air conditioner, it would absorb the heat from a building and use the energy to operate the fans that circulate the cooled air—and no outside power would be needed. As an ice cream machine, it would be great. Not only would it freeze the ice cream, but it would turn the crank as well.

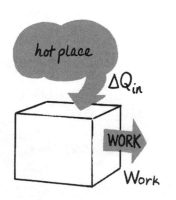

Figure 10-12

A perpetual motion machine of the second kind would convert all the heat coming into it into work.

Needless to say, people have tried to invent one of these perpetual motion machines of the second kind, and they have met with the same lack of success as they did with the first. Their failure comes from the fact that heat is disordered energy, whereas work is ordered energy. To get work out of heat, you need to make order out of disorder. Thus, their failure is not due so much to practical considerations as to theoretical ones. If you believe that disorder prevails over order simply because the number of disordered states in a system is vastly greater than the number of ordered states, you have to believe that it is quite impossible to build a perpetual motion machine of the second kind.

Just as the first law of thermodynamics prohibits the construction of a perpetual motion machine of the first kind, the impossibility of building a perpetual motion machine of the second kind is called the **second law of thermodynamics**. Another statement of this law is to say that the **entropy of a closed system always increases**.

You can stall. You can do things that will slow down the increase of entropy. Under ideal situations, it is theoretically possible to keep the entropy of a system constant. Nothing you do, however, will cause the total entropy of a closed system to decrease. You can get the entropy of part of a system to go down, but only by letting the entropy of another part of the system increase by as much or more to compensate. All heat engines do in fact work in this way. Changing heat into work represents entropy decrease in part of the system, but this is only possible by allowing an equal or greater entropy increase when some of the heat goes out, still in the form of heat, at a lower temperature. This lower temperature allows the entropy increase to more than compensate for the other decrease.

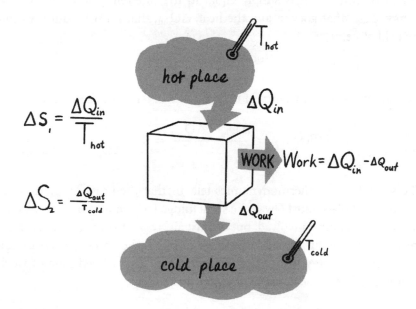

Figure 10-13

All heat engines that really work have to exhaust some heat to a cold place so that the total entropy of the system can increase.

The entropy decrease, ΔS_{in}, associated with the heat flow into the machine from the hot place, ΔQ_{in}, is, as you recall, inversely proportional to the temperature of the hot place, T_{hot}:

$$\Delta S_{in} = \frac{\Delta Q_{in}}{T_{hot}}$$

The entropy increase associated with the heat dumped into the cold place, ΔS_{out}, can be as large or larger than the preceding entropy increase, even though the energy, ΔQ_{cold}, is smaller, because the temperature, T_{cold}, of the cold place is smaller in as much or greater proportion.

$$\Delta S_{out} = \frac{\Delta Q_{out}}{T_{cold}}$$

The heat that must be dumped into the cold place (called a **heat sink**) to satisfy the second law of thermodynamics represents an inherent inefficiency of the heat engine. By the efficiency, we mean the ratio of the work that we get out of the machine to the heat energy that we put in:

$$\text{Efficiency} = \frac{\text{Work}}{\Delta Q_{in}}$$

The first law of thermodynamics tells us that the efficiency can never be greater than 1. You can never get more work out of a machine than the heat energy you put in. In fact, the conservation of energy tells us that the work that comes out is simply equal to the difference between the heat energy, ΔQ_{in}, that goes in and the heat, ΔQ_{out}, that must be dumped into the cold place.

$$\text{Work} = \Delta Q_{in} - \Delta Q_{out}$$

We can express the efficiency of the machine in terms of these heats:

$$\text{Efficiency} = \frac{\Delta Q_{in} - \Delta Q_{out}}{\Delta Q_{in}} = 1 - \frac{\Delta Q_{out}}{\Delta Q_{in}}$$

The second law of thermodynamics tells us that the heat, ΔQ_{out}, dumped into the cold place must always be something. The amount of heat depends on the temperatures involved, but it must always be big enough to keep the entropy increase due to the heat dump as big as the entropy decrease due to converting energy from disordered heat into the ordered state of work. That is, ΔS_{out} must be equal to or larger than ΔS_{in}:

$$\Delta S_{out} \geq \Delta S_{in}$$

Even if you had a perfectly ideal engine such that the entropy of the whole system remained constant, the best you could do would be to make these two changes in entropy equal. Let us assume the equality case for the moment and express these entropies in terms of heat and temperature:

$$\frac{\Delta Q_{\text{out}}}{T_{\text{cold}}} = \frac{\Delta Q_{\text{in}}}{T_{\text{hot}}}$$

From this we can see that the ratio of the heats in the ideal case is equal to the ratio of the absolute temperatures:

$$\frac{\Delta Q_{\text{out}}}{\Delta Q_{\text{in}}} = \frac{T_{\text{cold}}}{T_{\text{hot}}}$$

We can use this equivalence to get the maximum ideal efficiency of a heat engine in terms of the temperatures between which it is working:

$$\boxed{\text{Efficiency} = 1 - \frac{T_{\text{cold}}}{T_{\text{hot}}}}$$

The maximum efficiency of even a perfect heat engine is therefore limited by the temperature of the cold place, T_{cold}, as well as the temperature of the heat source, T_{hot}. At any temperature of the cold place above absolute zero, the efficiency must necessarily be less than one.

The first law of thermodynamics may be interpreted as saying, "You just can't win." You cannot get any more energy out of something than you put in. The second law of thermodynamics tells us, "You can't even break even." You can't even get out as much as you put in.

The inherent need to waste part of their heat energy as well as the need for a cold place to dump it becomes a serious problem for large generating plants. Water-cooled steam electric generating systems can cause severe thermal pollution of local water resources. By 1980 the electric power industry will be using between 15% and 20% of the nation's flowing fresh water for cooling its power plants.

The earth itself is a heat engine operating on energy from the sun. Most people realize that the sun is our source of energy, but few people stop to think about the heat sink, which is just as essential in thermodynamics as the source. That heat sink is the cold night sky. It is essential to the thermodynamics of the situation that the dark side of the earth radiate its waste energy into the universe. We are fortunate that the universe does not radiate back. Some say that the reason it does not, why the average temperature of the night sky is but a few degrees Kelvin, is that the universe is expanding. From that point of view, the expansion of the universe is of great importance to us. Let's all hope it keeps going.

ENTROPY *Boltzmann's Statistical Model* (optional)

The fall from grace of the caloric fluid as physical reality left a great challenge to those for whom entropy was an essential tool in thermodynamics, the theory of heat engines. "If it's not a physical fluid," pondered those attempting to base all of physics on mechanical principles, "what can it be? It works too well not to be really *something*."

One of those searching for a physical basis for entropy was Ludwig Boltzmann (1844–1906), whose students at the University of Vienna believed to be "a 'pure soul,' full of goodness of heart, idealism, and reverence for the wonder of the natural order of things."[3] Boltzmann's powerful command of mathematics allowed him to connect the idea of entropy with the probability of finding a certain degree of order or disorder in the universe.

Some people like to think that there is a certain order in the universe and that this is the natural state of things. It is certainly true that there is order in the universe, but it seems that nature actually prefers disorder. Natural processes almost always seem to make disorder out of order. A new deck of cards, for example, comes out of its box in ordered suits. Shuffle it once and you have disorder. Shuffle it again and you have even more disorder.

Think of the probability of shuffling a deck of cards enough to spontaneously get some degree of order out of disorder. Although it might seem unlikely, there must be at least some tiny probability that some random arrangement will produce upon shuffling some arrangement we could call ordered suits. The number of possible arrangements of the cards is $52! = (52)(51)(50) \ldots (3)(2)(1)$. This number turns out to be something very big, like 8×10^{67}. Even with $4! = 24$ possible arrangements of ordered suits, the chances of your getting any one of these ordered arrangements in any one shuffle would be about 1 out of 3×10^{66}. With 52 cards the probability of spontaneously getting even some degree of order out of disorder is so slight as to be hard to imagine.

With a deck of cards, we are dealing with numbers much smaller than the number of particles in an ordinary object, say, a breath of air. Yet we can extend our arguments concerning the likelihood of order over disorder to a breath of air if we think of the molecules as cards. The value of these "cards" might represent the energy of the molecules. Blowing a breath of air into a small balloon could then be thought of as shuffling the cards, in the sense that you would be randomly distributing molecules having different energies into different parts of the balloon.

There might be an arrangement such that we could get the energies of the molecules to do some work for us. Work, after all, is the transfer of energy. It is just that mechanical energy is ordered and heat is disordered. All we have to do is find an orderly arrangement of molecules in the balloon.

---

[3] Ruth Sime, *Lise Meitner, a Life in Physics,* Univ. of California Press, 1996.

Of course, the number of possible ways these molecules could be distributed is much greater than for the pack of cards. Instead of 52!, we are talking about 10^{22}!.

Don't try to figure this out on your calculator. Most calculators can handle the statistics for a single deck of cards. For a double deck, however, say for a two deck shoe at a blackjack table, the number is too big for most calculators. They give up and come back with an error message. You didn't do anything wrong, it's the calculator that runs out of space. If the "deck" had ten thousand million million million "cards," as in the case of molecules in a balloon, just figuring out the number of possibilities could take several life times for even a very fast computer. A big number indeed!

You might think that it would be possible to accidentally get more "hot" molecules, molecules having a high value of energy, on one side of the balloon and thereby leave the air on the other side of the balloon relatively cold. While it is true that such a possibility does exist, the number of random energy distributions is so vastly great in comparison that even a slight spontaneous departure from a state of disorder can be entirely disregarded.

It is therefore virtually certain that the energy of the molecules will start off being randomly distributed throughout the balloon, all parts of the balloon thus being at the same temperature. You might stop to consider what the chances would be that heat would spontaneously flow from one part of the balloon to the other. The molecules are, after all, in constant motion. They are constantly bouncing off each other, trading their kinetic energy around. Isn't it possible that the molecules on one side of the balloon would wind up at least temporarily with more than their average share of the heat? Here again, a spontaneous departure from disorder is so vastly improbable as to have vanishingly small probability.

As an example of what is meant by vanishingly small probability, we can use Boltzmann's thinking to consider the chances that an airmail stamp would jump up off the table and take flight because of thermal agitation. The airmail stamp is, after all, composed of molecules in constant motion. It is only the law of averages that says that half these molecules are moving downward at any one moment in time when the other half are moving upward. Certainly it would seem that there would be a finite probability that more than half the molecules would jump upward at the same moment and that the stamp would then leap up into the air. The fact is that such a probability does exist and can be figured out using a law of statistical mechanics known as Boltzmann's principle.[4] This same theory works very well for gas molecules and gives the right result for the density of air in the atmosphere of the earth. When applied to objects as massive as postage stamps, the theory also gives us results in keeping with

---

[4]Kerson Huang, *Statistical Mechanics* (New York: Wiley, 1963), p. 91

our observations, although it pushes those results beyond what we might ever expect actually to see. This theory has it that the probability of finding a postage stamp at some height above the table decreases exponentially with the height, but the exponential curve is so sharp that the probability is virtually zero for any measurable height. Even if we choose the smallest imaginable height that would be consistent with the idea of flight, one atomic diameter (about 10^{-10} m), the probability turns out to be smaller than the time it takes a molecule to complete one jump compared to the life of the universe. Nature so favors disorder that even the tiny degree of spontaneous order that would be necessary for all the molecules in an airmail stamp to get it together long enough to jump one atom high is not likely to have ever occurred.

The same argument that applies to motion of the molecules in the postage stamp can be extended to apply to the flow of energy of the molecules in a balloon filled with air, or to the energy of molecules, for that matter, in any object. One of the best ways to look at heat flow in a solid is to think of the heat energy as diffusing through the solid—like gas molecules moving about in a container. Heat energy becomes evenly distributed in the same way that gas in a balloon becomes evenly distributed. The even distribution is the state of maximum disorder, and maximum disorder is by far the most favored by probability. It is true that there are many ways for energy to be concentrated in only part of a system, just as there are many ways for all the red cards to be arranged in half a deck, but there are vastly more ways for that same energy to be evenly distributed. Just as you expect the cards to become mixed when you shuffle them, you expect heat to flow from a hot to cold place in a system for the same reasons of probability.

Temperature difference is associated with a concentration of energy in a system. If one side of the balloon filled with gas is hotter than the other, the molecules on that side have a greater average energy than on the others. This is a state of order that is ripe to disintegrate into disorder if the system is left to itself. Heat spontaneously flows from a hot place to a cool place until it becomes evenly distributed. Then there is no hot place or cool place, and that is the state of maximum disorder of the system.

The entropy of a system, according to Boltzmann's entropy hypothesis, is defined as something that is proportional to the natural logarithm of the statistical probability of finding that system in that particular state. Boltzmann's constant k, which we have seen in the ideal gas law, is just the proportionality constant between entropy and the natural log of probability. (In algebra, just so you can say you saw it here first, we can say $S = k \ln W$, where W is the probability of that particular arrangement of the system.) Thus, as a system tends to evolve to its most probable state, the entropy increases. Then, if we define the most probable state as being maximum disorder, less probable states have relative order.

Being a measure of the disorder of the system, entropy is a function of the energy of the system and how it is distributed. Its philosophical importance comes from the fact that it always increases in any closed system. If you want to lower the entropy of a system, you have to exert some sort of outside influence on the system.

You have to sort the cards if you want to start out with some low degree of disorder. You have to get all the hot molecules on one side of the balloon, wind the spring of the clock, or put fuel in your car. You have to somehow concentrate the energy to start with, because all systems run on energy transfer.

Once the system is isolated from outside influences, the energy can only flow from the places of concentration to become more evenly distributed in such a way as to increase the entropy of the system. The only way that the entropy of a part of a system can decrease is for the entropy of the rest of the system to increase by as much or more. In fact, exerting outside influence to initially lower the entropy of the system may be looked upon as making the system a part of a larger overall system in which the total entropy increases.

Our real concern with entropy is how it changes. For now, you need only worry about how the change in entropy, ΔS, is related to the flow of heat energy, ΔQ, and the temperature. As you might well expect, the change in entropy is directly proportional to the flow of heat from one part of the system to another:

$$\Delta S \propto \Delta Q$$

This means that, all else being equal, twice the entropy change is associated with twice the heat flow, and three times the entropy change will occur with three times the heat flow, and so on.

As it turns out, the "all else" that must be equal is the temperature. The more the heat is concentrated in a system, the greater is the degree of order in the system. But the entropy of the system is a measure of the *disorder* of the system, so that, as you might expect, a higher temperature is associated with a *lower* change in entropy. The change in entropy is inversely proportional to temperature:

$$\Delta S \propto \frac{\Delta Q}{T}$$

The temperature, T, involved in this proportionality must be measured on an absolute temperature scale, which starts off at absolute zero ($-273°\,C$). The use of temperature scales that start off at the freezing point of water or brine would otherwise produce an artificial discontinuity at the zero point, since division by zero is undefined. At true absolute zero, on the other hand, the change of entropy, ΔS, should actually be undefined, according to the theory of thermodynamics.

By choosing units for entropy to be joules per kelvin, we can make the proportionality constant be unity (1), so that we can just replace the proportionality sign with an equality sign:

$$\Delta S = \frac{\Delta Q}{T}$$

Thus we come to the same mathematical formulation that Carnot derived with a less sophisticated theory of heat.

CHECK QUESTION

A heat engine accepts 900 kcal of heat from a heat source having a temperature of 450 K. Some of this heat energy is converted to work and some is released as waste heat dumped into a cold place at a temperature of 300 K. (a) How much entropy decrease is associated with accepting heat from the heat source? (b) How much heat must be dumped into the cold place if this entropy decrease is to be balanced by an equal increase of entropy due to heat flow at the lower temperature?

Answer: (a) 2.00 kcal/K, (b) 600 kcal

This relationship between entropy, heat, and temperature turns out to be more than an equality in the usual development of the theory of thermodynamics. In those more advanced courses, temperature is usually defined in terms of the ratio of heat flow, ΔQ, to entropy change, ΔS, in the limit where the latter is very small.

$$T \equiv \frac{\Delta Q}{\Delta S}$$

For our purposes, however, we can think of temperature as that which would be measured on an ideal gas thermometer. A good enough definition of the change of entropy of a system is therefore the ratio of the flow of heat, ΔQ, to the temperature, T. If the flow of heat is from a hot place, the entropy change represents a decrease, which must be compensated someplace else in the system by at least as great an entropy increase, represented by heat flow into a cold place.

SUMMARY OF CONCEPTS

The problems in this chapter are related to heat as a form of energy. The relationship between heat and energy was only established once it was shown that mechanical energy, or work, is exactly proportional to heat, Q. If we write work and heat in joules, they can be said to be exactly equal in energy

$$\text{Work} = \Delta Q$$

where the main difference is in the degree of order and disorder. Work is ordered energy, and Q is heat, which is disordered energy.

A calorie (cal), still in common use although not an SI unit, is defined as the amount of heat necessary to raise the temperature of 1 g of water $1°\,C$.

The commonly used temperature scales are Celsius, Fahrenheit, and Kelvin. All are defined in terms of the freezing point ($0°\,C = 32°\,F = 273.15\,K$) and boiling point ($100°\,C = 212°\,F = 373.15\,K$) of water. Absolute temperature is measured in kelvins, K. The temperature in kelvins is 273.15 more than the temperature in Celsius degrees, but the size of the degrees are the same.

Most solid objects increase in size as their temperatures increase. The linear expansion, Δl, of an object is proportional to its original length, l, as well as to the change in temperature, ΔT:

$$\Delta l = \alpha l\, \Delta T$$

The proportionality constant α is called the **coefficient of linear expansion** and depends on the composition of the material involved.

A change in the heat content, ΔQ, is also generally associated with a change in temperature, ΔT, but the proportionality also includes the mass, m, of the stuff being heated.

$$\Delta Q = cm\, \Delta T$$

The proportionality constant c is called the **specific heat** of the substance. Almost everything else has a smaller specific heat than water. The specific heat of water is

$$c_{\text{water}} = 1\ \frac{\text{kcal}}{\text{kg}^\circ\text{C}} = 4{,}184\frac{\text{J}}{\text{kg} \cdot \text{K}}$$

Heat flows from a hot object to a cold one until they both reach the same temperature. The final temperature can be predicted by saying that the heat gained equals the heat lost:

$$\Delta Q_{\text{gained}} = \Delta Q_{\text{lost}}$$

and expressing these heats in terms of the specific heats, masses, and changes in temperatures of the various objects involved. The rate at which heat flows by conduction from one object to another is proportional to the temperature difference, ΔT, and to the area, A, through which the heat can flow, but it is inversely proportional to the thickness, l, of the material conducting the heat:

$$\frac{\Delta Q}{\Delta t} = k\,\frac{A\,\Delta T}{l}$$

The proportionality constant k is called the **thermal conductivity** of the material. Its reciprocal times the thickness is sometimes known as the thermal resistance, or R factor, of a certain thickness of insulation.

A heat engine is a device that takes some of the heat flowing from a hot place and turns it into work, a more ordered form of energy. To balance order with at least an equal increase in disorder, the heat engine must dump some of its heat into a cold place. The maximum theoretical efficiency possible under the constraint that a closed system never becomes more orderly is determined by the temperature, T_{cold}, of the cold place as well as by the temperature, T_{hot}, of the hot place:

$$\text{Efficiency} = 1 - \frac{T_{\text{cold}}}{T_{\text{hot}}}$$

These temperatures must be expressed on an absolute temperature scale, such as the kelvin scale.

Try to do the Examples yourself before looking at the Sample Solutions.

10-1 A Girl Scout is making fire by rubbing two sticks together. She gets a Brownie Scout to stand on one of the sticks so that she can concentrate her efforts on the frictional force. An unusual effort is needed because the leaves, wood, and other fire starting kindling is damp.

Given: The Girl Scout exerts a frictional force of 120 N in pushing the stick a distance of 40 cm in one direction and then the same distance back again. With damp kindling, it requires 50 kcal to start a fire. A kcal of heat is 4,184 J of energy.

Find: (a) How much work does the Girl Scout do on each complete stroke going back and forth?
  (b) How many kcal of heat energy are generated on each stroke?
  (c) How many strokes would be necessary to start a fire?

BACKGROUND

This problem illustrates that a little bit of heat represents a great deal of work. Starting a fire in this way is always something of a feat.

10-2 Many swimming pools are heated to make them more pleasant, but the cost of heating a pool may get unpleasantly high. Of course, the cost varies depending on the cost of energy. In relative terms, the present cost of heating a pool can be expressed in terms of the gasoline needed by a gasoline fired burner to do the job.

Given: A gasoline-fired water heater is used to warm a swimming pool to raise its temperature 4.4° C. The pool contains 140,000 kg of water. A gallon of gasoline produces about 8.4×10^7 J of heat energy, and the specific heat of water is 4,184 J/kg · °C.

Find: (a) How much heat is necessary to warm the pool if none is lost?
(b) How many gallons of gasoline does this represent?

BACKGROUND

This problem illustrates the concept of heat capacity in a system containing only water.

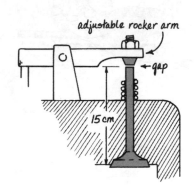

adjustable rocker arm

gap

15 cm

10-3 The valves in a sports car engine can be adjusted to compensate for wear. The gap that should be left between the valve stem and the rocker arm (which opens the valve by pushing on the valve stem) is given by the manufacturer in anticipation of the thermal expansion of the valve.

Given: The proper setting of this gap is 0.025 cm when the engine is cold, assuming the temperature will increase by 22° C. The valves are 15 cm long and have a coefficient of linear expansion of 1.1×10^{-5}/°C.

Find: (a) What is the change in length of the valve as the engine warms up?
(b) What would be the new gap after thermal expansion if the engine did not also expand as it got hot?

BACKGROUND

This problem is an example of thermal expansion in an old fashioned sports car with solid rod valve lifters. Since this is only one part of an engine that expands as it warms up, the conventional wisdom has it that a driver should let the car engine warm up before driving. Click and Clack, the Tappet brothers, on the popular Car Talk radio show, say this is nonsense. Decide for yourself.

10-4 You get to wondering how much it costs to heat an auditorium in preparation for a performance. You figure the main things to heat are the air, plaster in the walls, and steel in the seats. Wood in the structure is not hard to heat, nor is the fabric in the seats, so you leave them out of your first estimate.

Given: The auditorium is heated from 12° C to 20° C. It contains 9.7×10^4 kg of air, 1.2×10^4 kg plaster, and the seats contain 3.0×10^4 kg of steel. The specific heats of air, plaster, and steel are 1.0×10^3 J/kg·°C, 8.4×10^2 J/kg·°C, and 4.6×10^2 J/kg·°C, respectively. A joule of heat costs about 2 microcents.

Find: (a) How much heat is necessary to heat the air in the auditorium?
(b) How much heat does it take to warm the walls and the seats?
(c) How much does it cost to heat the auditorium?

BACKGROUND

This problem demonstrates the idea of heat capacity in systems having several different kinds of material. People who rent auditoriums to other people frequently list the cost of heating among other things that seem to make it necessary to charge as much as they do. This problem gives you an opportunity to estimate this cost for yourself.

10-5 A good treatment for sinus congestion is to inhale the steam from a pan of hot water containing a few drops of eucalyptus oil. Don't use boiling water direct from a tea kettle, however, as it is too hot and will cause more problems than it cures. Start with cool water in the pan and add hot water a little at a time.

Given: The tea kettle holds 3.75 kg of boiling hot water at 100° C. You pour part of it into a 1.00 kg aluminum pan containing 0.23 kg of cool water and oil mixture at 19° C with a combined specific heat of 4,050 J/kg·°C. You pour until the ideal combined temperature is 88° C is reached. Aluminum has a specific heat of 900 J/kg·°C while water has a specific heat of 4,184 J/kg·°C

Find: (a) How much of the boiling water should be put into the pan to get the ideal temperature?
(b) What would be the temperature if you just poured in all the hot water?

BACKGROUND

This is a calorimetry problem in a situation where several things change temperature at the same time. Both the cool water and its container are warmed by the addition of hot water to the system. Algebra becomes a very useful tool in problems of this sort, because enough things are going on so that, while the reasoning is straightforward, a pencil and paper are needed to keep track of everything.

10-6 A good part of the heat lost from a house goes right out the window, so to speak. The glass in the window may stop the wind from blowing through, but the difference in temperature across the glass is surprisingly small. Double glazed windows, which cut the heat loss through the windows by trapping a layer of dry air between two layers of glass, saves money in the long term.

Given: The average heat loss through your single-glazed windows is 1.4×10^8 J per day, assuming your windows have a combined area of 18 m^2 and thickness of 6.1 mm. The thermal conductivity of window glass is 1.04 W/m° C. You can save 75% of this heat by retrofitting with double glazed windows. A joule of heat costs about 2 microcents.

Find: (a) What is the temperature difference between the inside and outside of your single-glazed windows?

(b) How much does it cost you per year not to replace your windows?

BACKGROUND

The rate of heat flow through a window pane is proportional to the area and the temperature difference between the faces and inversely proportional to the thickness.

10-7 An inventor designs a heat engine that runs off of the waste heat from the cooling system of an automobile, turning part of it into useful work.

Given: The cooling system of an automobile operates at 85° C whereas the temperature of the ambient atmosphere is 20° C. The inventor claims that, for every 1000 J of waste heat, only 750 J must truly be wasted and that his invention is capable of converting the rest into useful work.

Find: (a) What is the maximum theoretical efficiency possible?

(b) What efficiency does the inventor claim?

BACKGROUND

One comes across schemes for using or saving energy from time to time. Some are practical and some are not. The rules of thermodynamics provide a useful estimate of what is possible and what is impossible under perfectly ideal circumstances.

Solutions to EXAMPLE PROBLEMS

(a) Work = F · d

$= (120 \text{ N}) \cdot 2(40 \text{ cm})$

$= \boxed{96 \text{ J}}$

(b) $(\text{Work})_1 = 96 \text{J}\left(\dfrac{1 \text{ kcal}}{4184 \text{ J}}\right)$

$= \boxed{2.3 \times 10^{-2} \text{ kcal/stroke}}$

(c) $(\text{Work})_{total} = n(\text{Work})_1$

$n = \dfrac{(\text{Work})_{total}}{(\text{Work})_1}$

$= \dfrac{50 \text{ kcal}}{2.3 \times 10^{-2} \text{ kcal/stroke}}$

$= 2174 \text{ strokes}$

$\cong \boxed{2.2 \times 10^{3} \text{ strokes}}$

SAMPLE SOLUTION 10-1

Given:
The Girl Scout exerts a frictional force of 120 N in pushing the stick a distance of 40 cm in one direction and then the same distance back again. With damp kindling, it requires 50 kcal to start a fire. A kcal of heat is 4,184 J of energy.

Find:
(a) How much work does the Girl Scout do on each complete stroke going back and forth?

(b) How many kcal of heat energy are generated on each stroke?

(c) How many strokes would be necessary to start a fire?

DISCUSSION

Mechanical energy is being converted into heat in given increments, in this case strokes. The number of increments, or strokes, in a given amount of heat is to be found. The energy in a single stroke is found from the definition of work, where the frictional force is the effort exerted by the Girl Scout and the distance is the total displacement, forward and back, in a single stroke. This distance is two times 40 cm, or 2×0.40 m, since a complete stroke includes motion in both directions.

To find the heat generated for part (b), the work from part (a) is converted into calories using the known conversion factor.

Part (c) starts from the assumption that the number of strokes is the ratio of the total amount of energy necessary to start the fire divided by the amount of energy in one stroke.

SAMPLE SOLUTION 10-2

Given:

A gasoline-fired water heater is used to warm a swimming pool to raise its temperature 4.4° C. The pool contains 140,000 kg of water. A gallon of gasoline produces about 8.4×10^7 J of heat energy, and the specific heat of water is 4,184 J/kg · °C.

Find:

(a) How much heat is necessary to warm the pool if none is lost?

(b) How many gallons of gasoline does this represent?

(a) $\Delta Q = cm\,\Delta T$

$= \left(4{,}184\dfrac{J}{kg°C}\right)(140{,}000\ kg)(4.4°C)$

$= \boxed{2.6 \times 10^9\ J}$

(b) $\Delta Q = 2.6 \times 10^9\ J\left(\dfrac{1\ gal\ of\ gas}{8.4 \times 10^7\ J}\right)$

$= \boxed{31\ gallons\ of\ gas}$

DISCUSSION

The heat necessary to warm the pool is found from the proportionality between heat on the one hand and mass and the change of temperature on the other.

One approach to the solution of part (b) is to treat it as a conversion factor problem. Since one gallon of gas represents 8.4×10^7 joules of heat when it is burned, it can be said to be, in a restricted sense, "equal" to that much heat. An alternative approach would be to construct another proportional model using a proportionality constant called the "heat of combustion" of gas. This constant is in common use and may be found listed in reference tables for various substances.

The result for this problem shows that gas is a poor way to heat a swimming pool. This is one of those cases where using solar heat is not only the correct thing to do, but financially prudent as well.

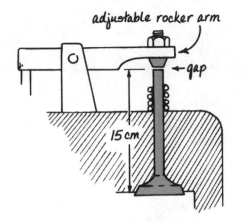

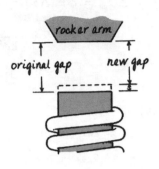

Given:
The proper setting of this gap is 0.025 cm when the engine is cold, assuming the temperature will increase by 22° C. The valves are 15 cm long and have a coefficient of linear expansion of $1.1 \times 10^{-5}/°C.$

Find:
(a) What is the change in length of the valve as the engine warms up?

(b) What would be the new gap after thermal expansion if the engine did not also expand as it got hot?

(a) $\Delta \ell = \alpha \ell \, \Delta T$

$= (1.1 \times 10^{-5}/°C)(15 \text{ cm})(22°C)$

$= \boxed{3.6 \times 10^{-3} \text{ cm}}$

(b) $Gap_{hot} = Gap_{cold} - \Delta \ell$

$= [(0.025 \text{ cm}) - (3.6 \times 10^{-3} \text{ cm})]$

$= \boxed{0.021 \text{ cm}}$

DISCUSSION

The change in length due to thermal expansion is given by the proportionality between it and the original length and the change in temperature.

The new gap will be smaller than the original gap by the amount that the valve expands, assuming that is the only thing in the engine to expand as it warms up. This calculation therefore represents an outside limit to the possible change in relative dimensions. In an actual case, the change in gap would be much smaller, since almost everything in the engine expands at nearly the same rate.

SAMPLE SOLUTION 10-4

Given:

The auditorium is heated from 12° C to 20° C. It contains 9.7×10^4 kg of air, 1.2×10^4 kg plaster, and the seats contain 3.0×10^4 kg of steel. The specific heats of air, plaster, and steel are 1.0×10^3 J/kg · °C, 8.4×10^2 J/kg · °C, and 4.6×10^2 J/kg · °C, respectively. A joule of heat costs about 2 microcents.

Find:

(a) How much heat is necessary to heat the air in the auditorium?

(b) How much heat does it take to warm the walls and the seats?

(c) How much does it cost to heat the auditorium?

(a) $\Delta Q_{air} = c \, m \, \Delta T$

$$= \left(1.0 \times 10^3 \frac{J}{kg\,°C}\right)(9.7 \times 10^4 \text{ kg})(20°C - 12°C)$$

$$= \boxed{7.8 \times 10^8 \text{ J}}$$

(b)

$\Delta Q_{W7} = \Delta Q_{walls} + \Delta Q_{seats}$

$\quad = c_{walls} m_{walls} \Delta T + c_{seats} m_{seats} \Delta T$

$\quad = (c_w m_w + c_s m_s) \Delta T$

$\quad = \left[\left(8.4 \times 10^2 \frac{J}{kg}\right)(1.2 \times 10^4 \text{ kg}) + \left(4.6 \times 10^2 \frac{J}{kg}\right)(3.0 \times 10^4 \text{ kg})\right]$

$\qquad \times [(20°C) - (12°C)]$

$\quad = \boxed{1.9 \times 10^8 \text{ J}}$

(c) $\Delta Q_{total} = \Delta Q_{air} + \Delta Q_{ws}$

$$= (7.8 \times 10^8 \text{ J}) + (1.9 \times 10^8 \text{ J})\left(\frac{2 \times 10^{-6} \frac{\$}{100}}{1 \text{ J}}\right)$$

$$= \boxed{\$19}$$

DISCUSSION

Part (a) is a simple application of the idea of heat capacity as the proportionality constant between the heat necessary to warm some substance on the one hand and the mass of that substance and the change in temperature on the other. Part (b) is just an application of the same proportionality to a system containing two different materials.

You might be interested to note that this estimate indicates that it costs nearly five times as much to warm the air in the auditorium as it does to heat the walls and seats. The conversion factor given in the problem, the amount a joule of heat costs, is only given to one significant figure because of the fluctuating price of heat.

(a)
$$\Delta Q_{lost} = \Delta Q_{gain}$$

$$c_{hot}m_{hot}(T_{hot} - T_{final}) = c_{A\ell}m_{A\ell}(T_{final} - T_{cold}) + c_{cold}m_{cold}(T_{final} - T_{cold})$$

$$m_{hot} = \frac{(c_{A\ell}m_{A\ell} + c_c m_c)(T_f - T_c)}{c_h(T_h - T_f)}$$

$$= \frac{\left(\left(900\frac{J}{kg\,°C}\right)(1.00)\ kg + \left(4,050\frac{J}{kg\,°C}\right)(0.23\ kg)\right)\left((88°C) - (19°C)\right)}{\left(4,184\frac{J}{kg\,°C}\right)((100°C) - (88°C))}$$

$$= \boxed{2.51\ kg}$$

(b)
$$c_h m_h(T_h - T_f) = c_{A\ell}m_{A\ell}(T_f - T_c) + c_c m_c(T_f - T_c)$$

$$c_h m_h T_h - c_h m_h T_f = c_{A\ell}m_{A\ell}T_f - c_{A\ell}m_{A\ell}T_c + c_c m_c T_f - c_c m_c T_c$$

$$c_{A\ell}m_{A\ell}T_f + c_c m_c T_f + c_h M_h T_f = c_h m_h T_h + c_{A\ell}m_{A\ell}T_c + c_c m_c T_c$$

$$T_f = \frac{c_h m_h T_h + (c_{A\ell}m_{A\ell} + c_c m_c)T_c}{c_{A\ell}m_{A\ell} + c_c m_c + c_h m_h}$$

$$= \frac{\left(4,184\frac{J}{kg\,°C}\right)(3.75\ kg)(100°C) + \left(\left(900\frac{J}{kg\,°C}\right)(1.00)\ kg + \left(4,050\frac{J}{kg\,°C}\right)(0.23\ kg)\right)(19°C)}{\left(900\frac{J}{kg\,°C}\right)(1.00\ kg) + \left(4,050\frac{J}{kg\,°C}\right)(0.23\ kg) + \left(4,184\frac{J}{kg\,°C}\right)(3.75\ kg)}$$

$$= \boxed{92°C}$$

SAMPLE SOLUTION 10-5

Given:
The tea kettle holds 3.75 kg of boiling hot water at 100° C. You pour part of it into a 1.00 kg aluminum pan containing 0.23 kg of cool water and oil mixture at 19° C with a combined specific heat of 4,050 J/kg · °C. You pour until the ideal combined temperature of 88° C is reached. Aluminum has a specific heat of 900 J/kg · °C while water has a specific heat of 4,184 J/kg · °C

Find:
(a) How much of the boiling water should be put into the pan to get the ideal temperature?

(b) What would be the temperature if you just poured in all the hot water?

DISCUSSION

The first part of the problem is easier than the second in spite of the fact that algebraically they both start with the same two first steps. Both start from the assumption that heat gained equals heat lost, and this general rule of calorimetry is applied to both situations in about the same way. These heats are expressed in terms of the specific heats, masses, and initial temperatures, using the subscripts for the hot water, aluminum and cold water-oil mixture. In part (a), however, the only unknown is the mass of the hot water. The temperature of boiling water, assuming sea level, is 100° C. The algebra is simpler, because the unknown is only on one side of the initial equation. In part (b), the unknown quantity is the final temperature, T_f, which occurs on both sides of the equation in every term. More algebra is necessary to solve for this unknown.

SAMPLE SOLUTION 10-6

Given:

The average heat loss through your single-glazed windows is 1.4×10^8 J per day, assuming your windows have a combined area of 18 m^2 and thickness of 6.1 mm. The thermal conductivity of window glass is 1.04 W/m °C. You can save 75% of this heat by retrofitting with double glazed windows. A joule of heat costs about 2 microcents.

Find:

(a) What is the temperature difference between the inside and outside of your single-glazed windows?

(b) How much does it cost you per year not to replace your windows?

(a) $\dfrac{\Delta Q}{\Delta t} = k \dfrac{A \, \Delta T}{L}$

$\Delta T = \dfrac{L \, \Delta Q}{kA \, \Delta t}$

$= \dfrac{(6.1 \text{ mm})(1.4 \times 10^8 \text{ J})}{(1.04 \text{ W/m °C})(18 \text{ m}^2)\left(\dfrac{1 \text{ J/s}}{1 \text{ W}}\right)(1 \text{ day})} \left(\dfrac{1 \text{ day}}{24 \times 3600 \text{ s}}\right)$

$= \boxed{0.53°C}$

(b) $\dfrac{\Delta Q_2}{\Delta t} = 75\% \left(\dfrac{\Delta Q_1}{\Delta t}\right)$

$= 75\% \left(1.4 \times 10^8 \dfrac{\text{J}}{\text{day}}\right)\left(\dfrac{2 \times 10^{-6}¢}{\text{J}}\right)\left(\dfrac{\$1}{100¢}\right)\left(\dfrac{365 \text{ day}}{1 \text{ year}}\right)$

$= \boxed{\$767/\text{year}}$

DISCUSSION

The temperature difference between the inside and outside of the window pane for a reasonable rate of heat flow is so small that it can not be responsible for keeping your house warm on a cold day. Instead, it is the layers of air on both sides of the window that keep you warm. A second layer of glass not only doubles the thickness, but adds layers of air.

(a) Efficiency $= 1 - \dfrac{T_{cold}}{T_{hot}}$

$= 1 - \dfrac{20°C + 273°C}{85°C + 273°C}$

$= \boxed{18\%}$

SAMPLE SOLUTION 10-7

Given:
The cooling system of an automobile operates at 85° C whereas the temperature of the ambient atmosphere is 20° C. The inventor claims that, for every 1000 J of waste heat, only 750 J must truly be wasted and that his invention is capable of converting the rest into useful work.

(b) Efficiency $= \dfrac{W_{out}}{Q_{in}}$

$= \dfrac{Q_{in} - Q_{lost}}{Q_{in}}$

$= 1 - \dfrac{Q_{lost}}{Q_{in}}$

$= 1 - \dfrac{750 \ J}{1000 \ J}$

$= \boxed{25\%}$

Find:
(a) What is the maximum theoretical efficiency possible?

(b) What efficiency does the inventor claim?

DISCUSSION

The ideal thermodynamic efficiency is less than unity (1) by the ratio of the temperatures of the cold and hot places available for heat flow. Under ideal circumstances, the best one can do is 18% at these temperatures. And that is before the practical losses due to friction and the need for heat to actually flow. The inventor is therefore claiming greater effficiency than permitted by the second law of thermodynamics. Don't put money into this one.

ESSENTIAL PROBLEMS

Be sure you can do the Example Problems (without peeking at the Sample Solutions) before working these Essential Problems.

10-8 In their continued adventures, Ratman and his sidekick Sparrow find themselves traped in a freezer. Sparrow is underdressed for the occasion and finds that his bare legs require frictional heating to keep them warm. Ratman to the rescue!

Given: Ratman exerts a frictional force of 40 N in rubbing Sparrow's legs 35 cm in one direction and then the same distance back again. It requires 1.5 kcal to warm Sparrow's legs. A kcal of heat is 4,184 J of energy.

Find: (a) How much work does Ratman do on each complete stroke going back and forth?
(b) How many kcal of heat energy are generated on each stroke?
(c) How many strokes would be necessary to warm Sparrow's legs?

10-9 Jerry Meek has a large hot water heater installed in his house so he can take a long shower in the morning. His wife expresses concern, however, about the cost of the energy bill from the utility company. Rather than discuss it, he converts his water heater to burn gasoline.

Given: The water heater heats the water 42° C and holds 180 kg of water. A gallon of gasoline produces about 8.4×10^7 J of heat energy energy, and the specific heat of water is 4,184 J/kg · °C.

Find: (a) How much heat is necessary to bring one tank of water up to shower temperature?
(b) How many gallons of gasoline does this represent?

10-10 A bridge made of concrete has an expansion joint to allow for thermal expansion with daily fluctuations in temperature. One end of the bridge rests on rollers while the other is solidly attached.

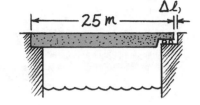

Given: In the morning, the expansion joint had a gap of 3.2 cm. The bridge is 12° C warmer in the afternoon, and therefore a bit longer than its original morning length of 25 m. The bridge is made of concrete having a thermal expansion of $2.1 \times 10^{-5}/°C$.

Find: (a) How much longer is the bridge that afternoon?
(b) What is the gap of the expansion joint that afternoon?

10-11 You might not be able to see the air in your house, but it has half again the mass of the plaster in your walls. The air is the hardest to heat; the second hardest to heat is the plaster. Even considering all those steel nails holding the house together, most of your heat goes into the air.

Given: Your furnace warms your house from 15° C to 20° C. The air in your house has a mass of 6,000 kg, and the plaster has a mass of 4,000 kg. The nails holding your house together have a mass of 200 kg. The specific heats of air, plaster, and steel are 1.0×10^3 J/kg·°C, 8.4×10^2 J/kg·°C, and 4.6×10^2 J/kg·°C, respectively. A joule of heat costs about 2 microcents.

Find: (a) How much heat is necessary to heat the air in your house?
(b) How much heat does it take to warm the walls, both plaster and nails?
(c) How much does it cost to warm your house?

10-12 Rick, finding his coffee a bit too hot to drink, adds some Irish whiskey to cool it off. It works so well, he is tempted to pour in the rest of the bottle.

Given: The bottle holds 0.55 kg of Irish whiskey at 18° C. Rick pours part of it into a 0.37 kg mug containing 0.22 kg of hot coffee at 95° C. Coffee, the way he makes it, has a specific heat of 4,100 J/kg·°C. He pours until the ideal combined temperature is 77 °C is reached. Rick's mug has a specific heat of 800 J/kg·°C, while his brand of whiskey has a specific heat of 3,200 J/kg·°C.

Find: (a) How much of the whiskey should be added to the coffee to get the ideal temperature?
(b) What would be the temperature if Rick just poured in all the whiskey?

10-13 A good part of the heat lost from a house goes through the roof. Not most, but a good part. Given today's insulation, that heat loss is easy to reduce.

Given: An average of about 1.8×10^8 J of heat per day goes through your roof. Your roof has an area of 65 m², most of which is wood having a thickness of 25.4 mm. The thermal conductivity of wood is 0.104 W/m °C. A layer of insulation saves 85% of the heat loss through the roof. A joule of heat costs about 2 microcents.

Find: (a) What is the temperature difference between the inside and outside of your roof?
(b) How much does it cost you per day not to insulate your roof?

10-14 Experimental engines have been developed, not so much for practical purposes, but just to see how close we can get to converting all of the heat from a hot place into mechanical energy. The trick is to minimize the heat that must be dumped into the cold place, usually called a heat sink.

Given: One experimental engine operated between a heat source at 1,727° C and a heat sink at 427° C. Even at these temperatures, for every 1000 J of heat from the heat source, 600 J of heat goes into the heat sink.

Find: (a) What is the maximum theoretical efficiency of this engine?
(b) What efficiency did the engine actually produce?

MORE INTERESTING PROBLEMS

10-15 The limiting factor in the speed of metal working tools is the rate of heat production. If the cutting tool becomes too hot, it will lose its temper—and we all know what happens when you lose your temper.

Given: A steel cutting tool has a mass of 0.35 kg and a heat capacity of 460 J/kg · °C. It can be heated up to 218° C before it loses its temper. It starts at room temperature, 20° C, and is acted upon by a frictional force of 50 N.

Find: (a) How many joules of heat can the tool absorb before getting too hot?
(b) Over what distance would the frictional force need to work to convert this much energy into heat?

10-16 The brakes on your car convert all that kinetic energy into a little bit of heat. Of course, if you make repeated stops from high speed, or you are going down hill in the mountains, your brakes may get hot. Then they will fade, or start to lose their ability to stop your car.

Given: The steel brake drums on your car have a mass of 7.5 kg and a heat capacity of 460 J/kg·°C. It can be heated up to 148°C before your brakes start to fade. Starting at room temperature, 20°C, they are acted upon by a frictional force of 4.4 kN.

Find: (a) How many joules of heat can the brakes absorb before getting too hot?

(b) Over what distance would the frictional force need to work to convert this much energy into heat?

10-17 The steel cables on the suspension section of the San Francisco Bay Bridge expand and contract with changes in temperature. One cold summer day, you get to wondering how much the cables would expand on a hot summer day.

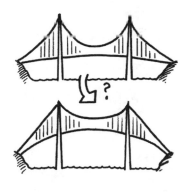

Given: It is 12°C that cool foggy day, and you are hoping it would get up to 31°C. The coefficient of linear expansion of steel is 1.1×10^{-5}/°C. Assume the suspension cables are 3.234567 km long.

Find: (a) What change in length would the cables undergo?

(b) What would be the new length of the cables?

10-18 San Francisco street cars run on steel tracks put together with joints that make a clickety-clack as the street cars run over them. A bright young engineer gets the idea of welding the tracks together to improve the ride. Unfortuantely, she forgets about the expansion of the rails. This could be a problem even with the mild temperature changes in the city by the bay.

Given: The tracks are laid on a warm day when they are 33°C. Then the fog rolls in and the temperature drops to 10°C. The coefficient of linear expansion of steel is 1.1×10^{-5}/°C. The longest straight surface run in San Francisco could be no longer than 12.34567 km.

Find: (a) What change in length would the rails undergo?

(b) What would be the new length of the rails?

10-19 Lead shot for old fashioned shotgun shells is made by pouring molten lead through a sieve above a pan of water. The trick is to get the lead droplets to harden in the air before they hit the water so that they will be round instead of tear-drop shape. For this reason, the sieve is usually placed some distance above the water at the top of a shot tower.

Given: 0.45 kg of lead shot is dropped above 5.0 kg of water at 20.00° C. The lead is just barely solid at 325.00° C and has a specific heat of 125 J/kg· °C as it enters the water. The lead has 280 J of kinetic energy as it hits the water.

Find: (a) To what temperature would cooling the lead alone warm the water?

(b) To what temperature is the water and lead brought by the kinetic energy of the shot?

10-20 The Ahwahnee people, native to California, heated water by adding hot stones. The only containers they had for holding water were water-tight baskets closely woven by hand. It was important to stir because a scorched basket was a disaster. Each basket took nearly a year to weave.

Given: An Ahwahnee cook puts 12 kg of stones into a basket containing 5.5 kg of water at 20° C. The stones start with a temperature of 280° C and a specific heat of 837 J/kg· °C Then the Ahwahnee cook stirs vigorously, adding 30 kJ of work to the mixture.

Find: (a) To what temperature would the stones alone warm the water?

(b) To what temperature is the mixture warmed by stirring?

10-21 A good tea kettle for stove top use has a copper bottom. Copper is outstanding both in absorbing heat from the flame and in conducting the heat through to the water inside.

Given: A tea kettle has a round copper bottom 20 cm in diameter and 1.2 mm thick. The thermal conductivity of copper is 385 W/m· °C. The flame keeps the outside surface of the bottom at 106 °C, while the inside surface is at 105 °C when the water is boiling. It requires 2.26 MJ to boil away 1 kg of water. (K = 9.2×10^{-2} kCal/s· m· °C) After 2 minutes, the flame is turned off.

Find: (a) What is the rate of heat flow through the bottom of the kettle?

(b) How much water boils away before the flame is turned off?

10-22 Will a bucket of hot water put out on the back porch in cold weather freeze before a bucket of cold water? You betcha! Put two buckets out after dinner and check them before going to bed. Melting snow on the back porch quickly brings the hot water down to freezing. Within a few minutes, you have one bucket of cold water set in ice and another set in snow. The one in snow had a bit of a head start, but ice conducts heat much faster than snow.

Given: A bucket has a round galvanized steel bottom 25.4 cm in diameter and 2.8 mm thick. The thermal conductivity of steel is 40 W/m · °C. Ice keeps the outside surface of the bottom at 3.75° C, while the inside surface is at 4.00° C as water inside is freezing. It requires 335 kJ to freeze 1.0 kg of water. After 45 min, measured from the time the hot water got cold, the buckets are checked.

Find: (a) What is the rate of heat flow through the bottom of the bucket?
(b) How much water freezes before the buckets are checked?

10-23 The basal metabolism of a person is defined as the rate of heat production of that individual at rest some 12 to 15 hours after having eaten a meal, although it is frequently measured indirectly in terms of the rate at which oxygen is absorbed by the lungs and carbon dioxide given off.

Given: A classroom containing 45 people and 500 m³ of air becomes stuffy after 10 minutes if the ventilation system is on the blink. The basal metabolism of an average person is 75 W. Air has a density of 1.2 kg/m³ and a specific heat of 1.0×10^3 J/kg · °C.

Find: (a) How much heat is given off by all these people during these ten minutes of the lecture, assuming they remain at rest?
(b) By how many degrees Celsius does the temperature of the room change?

10-24 Small mammals generally have a high metabolic rate compared to larger mammals largely because they have a higher ratio of surface to volume. Rats, for example, use energy about four times as fast, for their weight, as dogs and eight times as fast as cows. If you keep rats, don't do it in a closed box.

Given: A box holding 4 rats and 0.8 m³ of air becomes stuffy after a 20-minute trip to a former friend's house. The basal metabolism of an average rat is 2.6 W. Air has a density of 1.2 kg/m³ and a specific heat of 1.0×10^3 J/kg · °C.

Find: (a) How much heat is given off by our ratty friends during the trip?
(b) By how many degrees Celsius does the temperature of the air in the box increase?

ANSWERS

Solutions to **ESSENTIAL PROBLEMS**

10-8   (a) 28 J; (b) 6.7×10^{-3} kcal; (c) 224 strokes

10-9   (a) 32 MJ; (b) 0.38 gal of gasoline

10-10 (a) 0.63 cm; (b) 2.6 cm

10-11 (a) 3.0×10^7 J; (b) 1.7×10^7 J; (c) $0.94

10-12 (a) 0.114 kg; (b) $49°$ C

10-13 (a) $7.8°$ C; (b) $3 per day

10-14 (a) 65%; (b) 40%

Solutions to **MORE INTERESTING PROBLEMS,** *odd answers*

10-15 (a) 32 kJ; (b) 640 m

10-17 (a) 68 cm; (b) 3.235243 km

10-19 (a) $20.82°$ C; (b) $20.83°$ C

10-21 (a) 10 kW; (b) 0.54 kg

10-23 (a) 2.03 MJ; (b) $3.4°$ C

Sound and Other Oscillations

CHAPTER **11**

DELIGHTS OF THE EAR RANK with delights of the eye in the collection of enjoyments generally referred to as culture. Just as artists have been supplemented and aided in modern times by the technical skill of the photographer, so have musicians seen their art modified and magnified by the electronics technician. Some of the very shortcomings of electronic distortion have even grown into positive contributions at the hands of the artist in that advanced and esoteric form of music called rock.

Figure 11-1

Electronic distortion has influenced the art of rock music.

The modern electronics enthusiast will spend thousands of dollars and weeks of time working on a high fidelity system whose chief function is to wiggle the eardrums back and forth in just the right way. We will start our study of this elegant and complex wiggling of the eardrums by considering the simpler motion associated with a pure tone. We will then study the tone itself in terms of the propagation of a wave through a medium. Finally, we will be in a position to see how waves can be added together to make any motion imaginable—or even no motion at all.

Simple Harmonic Motion

The kind of motion associated with a pure tone turns out to be very common. Usually called **simple harmonic motion**, it is also sometimes referred to as **sinusoidal motion**, because a graph of position against time for this kind of motion is a sine curve. If you were to look at the position of your eardrums, for example, when you were listening to a single note produced by flute or a tuning fork, you would find that your eardrums move back and forth in very much the same way that the sine of an angle increases and decreases as the angle increases. An easy way to produce a graph of this kind of motion would be to attach a marking pen to a weight that is bobbing up and down on the end of a spring and to use this pen to mark the position of the weight on a card as the card is pulled past at a constant velocity.

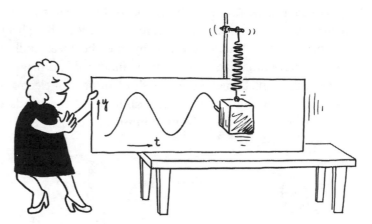

Figure 11-2

Sinusoidal motion results when a mass oscillates on a spring.

Simple harmonic motion is produced whenever you have a mass on a spring. That is important because in nature just about anything may be regarded as a mass on a spring. A leaf on a tree swings back and forth because of the springiness of the leaf stem. For small enough displacements all supports, from building beams to spider webs, obey Hooke's law—which is to say that force in a spring is proportional to stretch but in the opposite direction. Any object restored to its equilibrium position by a force proportional to displacement will undergo simple harmonic motion. If the frictional force is small, just about anything will bounce around undergoing simple harmonic motion—sometimes several simple harmonic motions at the same time.

A pendulum bob swinging back and forth on a string also undergoes simple harmonic motion. The pendulum bob is restored to its lowest position by the horizontal component of the force acting in the string. As long as the bob swings only a small distance compared to the length of the string, this horizontal component of force, as well as the resulting acceleration, is nearly proportional to the displacement.

Velocity and Acceleration

Think about the way a pendulum swings. Think about how the velocity and acceleration of a pendulum bob do different things at different times. The bob comes to a stop as it reaches the end of its swing. That is to say, it decelerates until the velocity is zero.

Some people, who may have not given much thought to the difference between the ideas of velocity and acceleration, might get confused at this point. They might be inclined to think that both the velocity and acceleration are zero at the end of the swing. Here lies a common pitfall.

Although velocity goes through zero at the end of a pendulum's swing, acceleration does not. Quite the contrary. Whether called acceleration or deceleration, the rate of change of the pendulum's velocity reaches a maximum at the top of the swing and only begins to decrease as the pendulum returns to the bottom. When the pendulum bob changes direction and starts back down, the velocity changes direction while the acceleration does not. The direction of the acceleration always points back to the bottom of the swing.

Certainly one would not expect a sudden change in the acceleration of a pendulum at the end of the swing from Newton's laws of motion. The part of the force that accelerates the pendulum bob, the horizontal component of the force in the string, reaches a maximum at that point. It continues to accelerate the bob back toward the equilibrium position. The acceleration only decreases as the bob gets closer to this midpoint, picking up velocity along the way.

When the bob finally reaches the lowest point and has its maximum velocity, the acceleration does momentarily go through zero. The bob coasts on through the equilibrium position as the acceleration goes negative. As the pendulum swings back and forth, the acceleration and the velocity go through zero at different times. The acceleration goes through zero at the middle of the swing, while the velocity goes through zero at the ends of the swing.

A mass bounces up and down on a spring with the same kind of motion just described for a pendulum. Instead of the acceleration being produced by one component of the force acting in the pendulum string, the acceleration is produced by the changing Hooke's law force produced as the spring stretches. (Hooke's law, you remember, states that the force in a spring is proportional to the amount of stretch or other deformation of the spring.) In both cases, the acceleration is proportional to the displacement and opposite in direction. That relationship between displacement

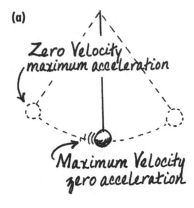

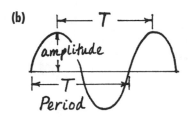

Figure 11-3

(a) A pendulum has maximum acceleration when it has minimum velocity, and vice versa.
(b) The amplitude describes the size of the motion in space and period describes its size in time.

and acceleration turns out, in fact, to be a necessary and sufficient condition for simple harmonic motion.

If you want to describe the motion of a pendulum or a mass on a spring, you would need to specify how far it swings or bounces and how much time it takes. The usual way of describing distance in simple harmonic motion is to measure from the equilibrium position to one of the peaks. This is called the **amplitude** of the motion. The letter A is usually used to express this distance. Since the equilibrium position is halfway between the peaks in opposite directions, the amplitude is half the total peak-to-peak distance traveled by the pendulum or the mass.

The time necessary for one complete cycle of the simple harmonic motion is called the **period** of the motion. The letter T is usually used to signify this length of time. The only point of caution is to measure a complete cycle with the motion going in both directions. If the period is measured between peaks, make certain that the peaks are on the same side of the equilibrium position. If the period is measured between times that the motion passes through the equilibrium position, make certain that the direction of motion is the same at the end as it is at the start. Simple harmonic motion passes through the equilibrium position twice in one period, but it is going the other direction as it passes through at the halfway point.

An alternative to describing the amount of time for one cycle is to describe the number of cycles for a given period of time. The **frequency**, f, of a simple harmonic motion is the reciprocal of its period.

$$f \equiv \frac{1}{T}$$

If the period of a pendulum is $\frac{1}{3}$ s, for example, its frequency is three times per second.

CHECK QUESTIONS

1. A grandfather clock is a fancy old full length floor standing clock with a pendulum about as long as grandfather's arm. The clock ticks off 1.0 s each time the pendulum starts back, first one side and then the other, from a maximum point on its swing. The two maximums are 4.5 cm apart. What are the period, T, and the amplitude, A, of the simple harmonic motion involved?

 Answer: $T = 2.0$ s $A = 9.0$ cm

> (The clock ticks twice each time the pendulum makes a complete trip, traveling through twice the amplitude from one peak to the peak on the opposite side.)

2. An AM rock radio station broadcasts at a frequency of 750 kHz (750,000 cycles per second). What is the period of the simple harmonic motion associated with this radio frequency?

 Answer: $\frac{1}{750,000}$ s

Years ago, the widely accepted unit for frequency was the cycle per second (cps). It seemed to make a lot of sense to say that the frequency of a radio station was 750,000 cycles per second if, in fact, the carrier wave went through 750,000 complete cycles every second. In the 1960s, however, some electrical engineers noticed that we had named a lot of units after great scientists who had pioneered different fields, but there was one great man who seemed to have been left out. They thought that Heinrich Hertz was a very important man because he had discovered radio waves. So they decided to name a cycle per second after him. Henceforth, the unit Hz, pronounced hertz, will stand for a cycle per second.

$$1 \text{ Hz} \equiv \left(\frac{1 \text{ cycle}}{1 \text{ second}} \right)$$

That leaves the unit for period unaccounted for. We don't have a good name for a second per cycle. Henceforth the unit Wl, pronounced Wall, will stand for the second per cycle.

Sinusoidal and Circular Motion

You may have wondered why there seem to be connections between trigonometry, the behavior of a pendulum and the behavior of a mass bouncing up and down on a spring. Were you to do experiments with things that bounce and swing, you would find even more similarities. You would find, for example, that the period of an eraser hanging from a string is directly proportional to the square root of the length, *l*, of the string and inversely proportional to the square root of the acceleration due to gravity, *g*. The proportionality constant turns out to be 2π.

$$T_{\text{pendulum}} = 2\pi \sqrt{\frac{l}{g}}$$

By doing experiments with the same eraser bouncing up and down on a rubber band, you would find that the period of oscillation is directly proportional to the square root of the mass, *m*, and inversely proportional to the square root of the spring constant *k* of the spring. The proportionality constant turns out to be the same 2π.

$$T_{\text{mass on spring}} = 2\pi \sqrt{\frac{m}{k}}$$

We shall see that the similarity of these two relationships, as well as the fact that both motions trace out a sine curve, is more than a coincidence. The two equations result naturally from the relationship between simple

Figure 11-4

A pendulum swinging in a circle is undergoing two simple harmonic motions in perpendicular directions at the same time.

harmonic and circular motion. The fact that there is a relationship can be easily seen by imagining a simple experiment with a pendulum.

Suppose you had a pendulum swinging back and forth from an overhead support, and you were to stop it and start it swinging instead from side to side. As you would expect, not having changed either the length of the pendulum or the acceleration due to gravity, the period of oscillation would be exactly the same in both directions. But then suppose you decide to verify this result by setting the pendulum swinging in both directions at once. Is that possible? Such an arrangement is called a **conical pendulum.** The pendulum bob will sweep out a circle as it undergoes two simple harmonic motions of the same amplitude in two perpendicular directions.

We can, on the other hand, look at simple harmonic motion as one component of circular motion. Imagine the handle of a wheel turning at a constant velocity. If this handle were illuminated from the side with parallel light, the shadow would move up and down with simple harmonic motion—just as the conical pendulum might seem to be swinging only from side to side if you viewed its circular path from the edge in such a way that you could not tell that it was also swinging toward and away from you.

Figure 11-5

As a wheel turns at constant velocity, the shadow of the handle undergoes simple harmonic motion.

The fact that simple harmonic motion is one component of circular motion makes simple harmonic motion very common in the mechanical gadgets of our modern society. The horses on a merry-go-round, for example, go up and down in simple harmonic motion because they are raised and lowered by a crankshaft that turns at constant velocity. The pistons in an automobile engine also move in a near approximation to simple harmonic motion, since they are attached to a crankshaft that travels in circular motion.

We can use its relationship to circular motion to see why simple harmonic motion traces out a sine curve. Let us return to our example of the handle on the wheel and plot the height of the handle as a function of the angle through which the wheel turns.

The shadow of the handle starts off at zero distance from the equilibrium, or midpoint of its travel, when the angle θ is zero. This distance increases with the angle, at first rapidly and then more slowly, until the angle θ reaches 90° (which is the same thing as π/2 radians). The shadow then comes back down, at first slowly and then more rapidly, to cross the equilibrium position when θ reaches 180°, or π radians. It then does the same thing again in the opposite direction, returning to the equilibrium position to start the cycle over again at 2π radians.

The amplitude of the simple harmonic motion, or maximum distance that the shadow travels, is the radius of the circle involved. The distance from the equilibrium position only reaches this peak at π/2 and 3π/2 radians. At all other angles, the distance is something smaller than the radius of the circle. In fact, you could find this distance by dropping a vertical line to the horizontal axis that passes through the equilibrium position. By doing this, you would form a right triangle whose hypotenuse is the radius of the circle. The distance from the equilibrium position, the length of the vertical line, is the side of the right triangle opposite the angle θ. The length of this line is, of course, related to the radius of the circle by the sine of the angle θ:

$$y = r \sin \theta$$

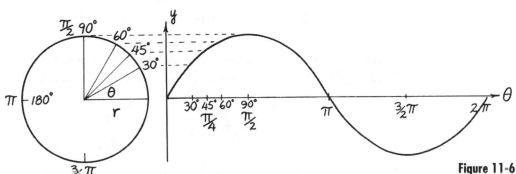

Figure 11-6

Sinusoidal motion is a one-dimensional projection of circular motion.

As the angle θ increases at a constant rate, the simple harmonic motion of the shadow naturally traces out the path of a sine curve.

The close relationship between simple harmonic and circular motion can also be used to derive a relationship between the acceleration and amplitude of simple harmonic motion and its period. We will find this relationship useful, because the equations we referred to earlier as giving the periods of a pendulum and a mass bouncing on a spring turn out to be special cases of this more general relationship.

Acceleration in simple harmonic motion may also be thought of as a one-dimensional projection of acceleration in circular motion. As you recall, the acceleration in circular motion, centripetal acceleration, is given by the square of velocity divided by the radius of the path.

$$a_c = \frac{v^2}{r}$$

This acceleration is a vector quantity that is always pointed toward the center of the circular path. If we are only looking at one component of

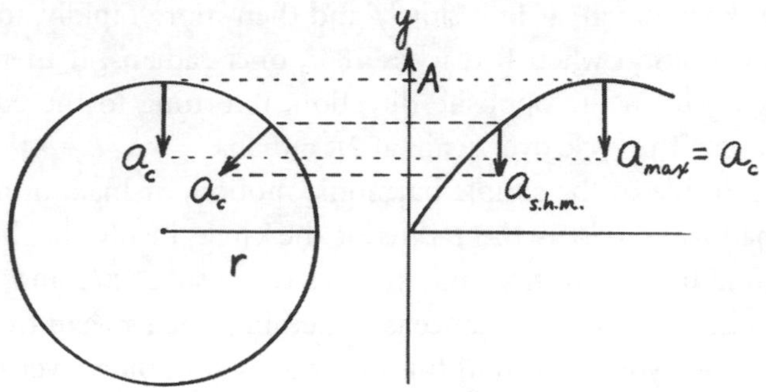

Figure 11-7

Acceleration in simple harmonic motion is a projection of centripetal acceleration.

displacement, say the vertical component, then only that component of acceleration is of interest. This vertical component of acceleration is zero when the angle θ is zero. (The centripetal acceleration is then all in the horizontal direction.) As the angle increases, the vertical component of the acceleration grows to reach a maximum value as the displacement reaches the full amplitude, A, which is the same as the radius for circular motion.

$$a_{max} = a_c \qquad \text{when } y = r = A$$

Velocity in the centripetal force expression can be expressed in terms of the period and amplitude of simple harmonic motion, using the definition of velocity:

$$v \equiv \frac{d}{t} = \frac{2\pi r}{T}$$

The distance, d, is equal to one circumference, $2\pi r$, when the time, t, is equal to one period, T. The centripetal acceleration then becomes

$$a_c = \frac{\left(\frac{2\pi r}{T}\right)^2}{r} = \frac{4\pi^2 r^2}{T^2 r} = \frac{4\pi^2 r}{T^2}$$

Replacing the radius r of circular motion with the amplitude, A, of simple harmonic motion, we get an expression for the maximum acceleration at the point where the simple harmonic motion goes through maximum displacement:

$$a_{max} = \frac{4\pi^2 A}{T^2}$$

This relationship is usually expressed in the form where it is solved for the period, T, in terms of the amplitude, A, and the maximum value of the acceleration, a_{max}, that occurs in simple harmonic motion.

$$T = 2\pi \sqrt{\frac{A}{a_{max}}}$$

This expression states that, in general, the period of oscillation of any object undergoing simple harmonic motion is directly proportional to the square root of the maximum displacement, or amplitude, of the motion and is inversely proportional to the square root of the maximum value of the acceleration that the object undergoes as it passes through the maximum displacement. The proportionality constant turns out to be 2π.

As you may remember, the form of the preceding expression is very similar to the equations mentioned earlier as giving the period of a mass bouncing up and down on a spring and the period of a pendulum. That is because the latter two expressions are just special applications of the general result. In both cases, the maximum acceleration is found in terms of the maximum force applied to the object in agreement with Newton's second law:

$$F_{max} = ma_{max}$$

In the case of a mass bouncing up and down on a spring, the force applied to the object is given by Hooke's law:

$$F = kx$$

The maximum value of the force is reached when the stretch of the spring reaches the maximum value, which is the amplitude, A, of the simple harmonic motion:

$$F_{max} = kA$$

Combining this expression with Newton's second law, we can see that the ratio of amplitude to maximum displacement is equal to the mass, m, over spring constant k in the case of a mass on a spring:

$$ma_{max} = kA$$

$$\frac{m}{k} = \frac{A}{a_{max}}$$

The period of oscillation is therefore 2π times the square root of the ratio of mass to spring constant:

$$T = 2\pi \sqrt{\frac{m}{k}}$$

A similar argument is used for the period of a pendulum, but the argument must be restricted to small angles. In this case, the force restoring the pendulum to its equilibrium position is one component of the weight, *w*:

$$F = w \sin\theta$$

The maximum acceleration is produced, in accordance with Newton's second law, when the pendulum has swung to its maximum displacement:

$$(ma_{max}) = (mg)\sin\theta_{max}$$

The sine of the angle can be expressed in terms of this maximum displacement, or amplitude *A*, using the small angle approximation. This approximation states that the sine of an angle and the tangent of the angle are nearly equal to the angle itself, expressed in radian measure, if the angle is sufficiently small. This approximation holds to three significant figures if the angle is less than about 0.1 radian or about 6°.

$$\sin\theta \cong \tan\theta \cong \theta$$

The angle measured in radians is equal to the arc length over the radius. In terms of the pendulum, this is the amplitude of the displacement over the length of the pendulum:

$$\theta = \frac{A}{l}$$

We can therefore use this small angle approximation to replace the sine of the angle in the earlier expression:

$$(ma_{max}) = (mg)\left(\frac{A}{l}\right)$$

Thus, we see that the ratio of the length of the pendulum to the acceleration due to gravity is equal to the ratio of amplitude to maximum acceleration in the case of the pendulum:

$$\frac{l}{g} = \frac{A}{a_{max}}$$

Using this ratio in the general expression for period in simple harmonic motion gives the period of the pendulum equal to 2π times the square root of length, *l*, of the pendulum divided by the acceleration due to gravity, *g*.

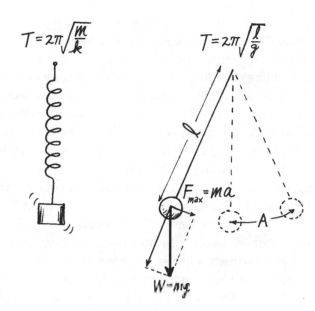

$$T = 2\pi\sqrt{\frac{m}{k}} \qquad T = 2\pi\sqrt{\frac{l}{g}}$$

$$F_{max} = ma$$

$$W = mg$$

Figure 11-8

The period of a mass on a spring is given by an expression similar to that which gives the period of a pendulum.

Complex Periodic Motion

The simple harmonic motion we have considered so far is the motion your eardrums experience when you hear a pure tone. It is possible for an object to undergo many simple harmonic motions at the same time, just as it is possible for you to hear several tones at the same time. Most sounds are, in fact, many tones mixed together. A harmony that is pleasing to the ear is a mixture in the right proportions of various tones that have frequencies related in a simple manner. A guitar note, for example, is made up mostly of the basic, or *fundamental,* tone plus the first *overtone,* a tone having exactly twice the frequency of the fundamental. It is a particular mix of these tones that gives a guitar its distinctive mellow sound. There are, moreover, tiny amounts of other overtones mixed in as well. The proportion of these vibrations is the factor that distinguishes an expensive guitar from a cheap one.

This same difference is more exaggerated when different instruments are involved. A piano sounds very different from a flute, even though they are both playing exactly the same note. One difference is that the piano note starts loud, as the hammer hits the string, and then dies out, while the flute note continues at an even level. That is called the *attack* of the note. But even if you could control the level of a flute note, you could never make it sound like a piano. The sounds are clearly distinguishable by the overtone mix you hear. A flute note is a much purer sound with almost all the vibration being in the fundamental tone. A piano note is made with two or three strings being struck simultaneously at a point which is one-seventh to one-ninth of their length from their end. These strings are thus made to vibrate in several ways at the same time. What is more, other

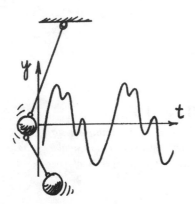

Figure 11-9

A double pendulum illustrates the idea of adding several simple harmonic motions together.

strings on the piano pick up these vibrations to reinforce their sound in the mixture you hear. That is why a piano that is badly out of tune sounds all wrong from the first note that is played on it.

The idea of adding vibrations on top of other vibrations can be visually understood by tying one pendulum on to another to make a double pendulum. The second pendulum will then add its influence to the first. They will both wobble back and forth under this mutual influence even as they swing back and forth together. This compound motion may be looked upon as the sum of several simple harmonic motions added together. Different lengths of string connecting the two pendulum bobs will change the frequency of the wobbling motion that is superimposed upon the swinging motion and produce different kinds of compound motion. The kinds of resulting motion that are associated with harmonies, in musical terms, are motions that repeat themselves in a noticeable pattern. Such a motion can be generated by the double pendulum by adjusting the lengths of the strings so that the wobbling motion is at a frequency that is some simple multiple of the general swinging motion. Then the two motions get back together and repeat their motion after a certain number of swings.

As you might well imagine, you can get even more complicated motion by adding more pendulum bobs. The more pendulums you string together, the more simple harmonic motions you add into the mix. In fact, you can get just about any motion you like, no matter how complex, if you are willing to add in enough simple harmonic motions with the right frequencies, amplitudes, and starting times.

There are many physical situations in which a number of independent influences are superimposed upon the same system. If these influences are linear in the sense that their effects add, a rule called the **principle of superposition** may be used to understand the result as simply the sum of the individual influences. One of the great values of this rule is that it can also be used in reverse. Not only can you add several simple harmonic motions together to get a more complex motion but you can look upon any complex motion as being made up of simple harmonic components.

As an example of the principle of superposition, let us consider the problems encountered in high fidelity sound systems. A funny-shaped electrical signal is generated by, say, a microphone or phonograph cartridge. Depending on the quality of the pickup device, this signal may be a more or less accurate representation of the sound you want to get out. The amplifier must be able to take this little funny-shaped signal and turn it into a big signal having nearly the same funny shape. To do this, it must be able to amplify a vast range of frequencies equally so that all of the frequency components in the original signal will remain in the finished product. The speaker must then take this big electrical signal and produce corresponding air pressure waves.

The demands of handling such a vast range of frequencies and giving them all equal treatment is just too great for any one speaker. Instead, high frequencies are split off and sent into a little speaker, called a tweeter, which has a low enough mass to respond to the rapidly changing signals associated with high frequency sound. The low frequencies go to a woofer, which is large enough to handle the long wavelengths associated with low frequency sound. Frequently, there is even a mid-range speaker to handle the middle frequencies. The point is to get an even response to all frequencies. But a single sound, if it is very complex, can have parts that come through on all three speakers.

One way to test a high fidelity sound system is to see what it does to a complex signal with lots of different frequency components. Such a signal need not necessarily look complex, however. One simple looking signal, which is easy to generate but which has a huge range of frequency components, is called a square wave. This signal just switches back and forth between a positive and a negative voltage at even intervals of time. To be a good square wave, the signal need only go along at a constant voltage until it suddenly changes to the negative voltage and remains there, going along until it suddenly changes back to repeat the cycle. This signal looks simple, as shown in Figure 11-10(a). It has, however, a large number of frequency components.

Electronically, it is easy to generate a square wave signal, but just to see how the principle of superposition applies, let us show how you would go about making a square wave out of simple harmonic components. First you would need a sine wave having the same fundamental frequency as the

Figure 11-10

A square wave (a) is made up of odd harmonics having the correct amplitudes and starting times. If only the fundamental signal and one having a frequency three times as great are added together, the result looks like (b). Add in a signal with five times the frequency, and (c) the result starts to look a little more like a square wave. The more frequency components you add (d), the more the result looks like a square wave.

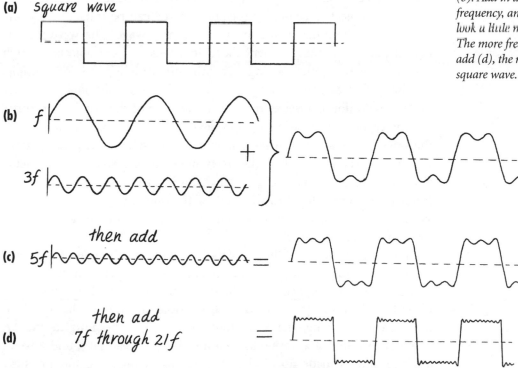

square wave. Then you would need to add another sine wave having three times the frequency of the fundamental but one-third the amplitude. If you adjusted the starting time of this so-called third harmonic so that it goes positive and negative at the same times as the fundamental, the sum of these two would go positive extra fast and then wobble before going negative extra fast to wobble again. If you then add still another sine wave having five times the frequency, one-fifth the amplitude, and the same starting time, you can get the signal to go up still faster and then "wibble-wobble" at the top before going down to wibble-wobble again at the bottom. The more odd-numbered harmonics of the right amplitude you add in, the flatter you can get the top and bottom and the faster you can make the signal change from positive to negative. A square wave therefore contains only odd harmonics, but it contains an infinite number of them—at least theoretically. The more of them that are in an actual signal, the sharper will be the corners and the flatter will be the tops and bottoms of the square wave shapes.

If you put a square wave signal into a decent high fidelity amplifier, you should find that you get a pretty good square wave out. If you put a square wave into a speaker system, however, even a good one, you could easily see the distortion in the output. Speakers, at their present state of development, are the weakest link in most home sound systems.

Waves

We have been discussing simple harmonic motion of a single object or of several objects coupled together, as in the double pendulum. If a larger number of objects are coupled together, energy can be transmitted along from one to the other in the form known as a **wave**. The wave may be thought of as something independent, in a sense, of the objects through which it moves. Ripples moving across the surface of a pond, for example, are thought of as going from one place to the other, although the water molecules do not themselves travel from one side of the pond to the other but merely bob more or less up and down as the wave passes through. As the wave moves along at what is known as the **velocity of propagation**, which is the velocity at which the crests and troughs travel, the water actually travels in an elliptical path. Water that is in a trough of the wave at one moment moves up to form the crest of the oncoming wave at the next. While this is happening, nearby water that starts at a crest falls down to form the advancing trough.

This elliptic motion of the water molecules may be thought of as a composite of two simpler motions—one parallel to the direction of wave travel and the other perpendicular. To the extent that the actual motion of the water is back and forth in the same or opposite direction to that in which the wave is traveling, the wave motion is said to be **longitudinal**. A common example of a wave that is entirely longitudinal is a sound wave in air. The wave is made up of alternate regions of high and low pressure, which move through the air as the air molecules move back and forth in the same or opposite direction to form these regions of high and low density.

To the extent that a water wave involves motion perpendicular to the direction in which the wave travels, the wave motion is said to be **transverse**. One example of purely transverse wave motion is the propagation of a wave on a string. The string moves from side to side as the wave moves along its direction of extension. Next time you are cleaning house, give the vacuum cleaner cord a sudden sideward jerk and watch the wave move out across the room while the cord itself remains in your hand. Think "transverse wave" as you watch this fine demonstration.

An earthquake is usually made up of both kinds of wave motion propagating at different velocities. If you are any significant distance from the epicenter, or origin, of the earthquake, you feel two separate shock waves.

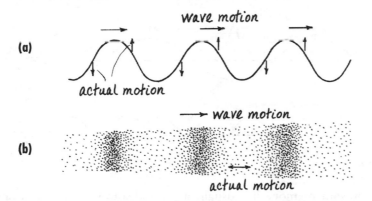

Figure 11-11

(a) A transverse wave is one in which the actual motion is perpendicular to the direction of wave propagation. (b) The particles propagating a longitudinal wave move in the same or opposite direction as the wave itself.

The great rumbling you first feel is the longitudinal pressure wave, which seismologists call the P wave. This is followed by a swaying motion from side to side as the transverse shear wave passes through. This is called the S wave. Since they know the velocities of propagation for these two kinds of waves, seismologists can tell the distance to the epicenter of the earthquake by the separation in time between the P and S waves.

Wavelength

The distance between adjacent crests, or between adjacent troughs, of a wave is aptly called the **wavelength**. Wavelength is to a wave as a period is to simple harmonic motion. Both indicate a separation, but period indicates a separation in time, while wavelength indicates an actual separation in space. The two are related by the velocity of propagation.

The relationship between wavelength, period, and the velocity of propagation results directly from the definition of velocity:

$$v \equiv \frac{d}{t}$$

For a wave, the appropriate distance and time are wavelength, λ, and period, *T*.

$$v = \frac{\lambda}{T}$$

That is all there really is to the relationship. It is usual, however, to express it in terms of frequency instead of period. Since frequency, *f*, is simply the reciprocal of period, the relationship becomes a product instead of a ratio:

$$v = f\,\lambda$$

Many people prefer not to memorize this relationship. They feel that they can be more reliable about getting it straight if they simply start with the definition of velocity and work it out each time they use it. If you prefer

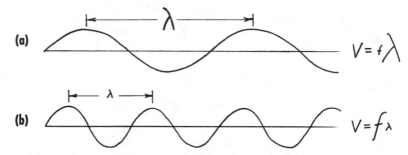

Figure 11-12

The higher the frequency, the shorter the wavelength.

to rely on your memory, it is usually a good idea to write the relationship down and then stop for a moment to be sure that it seems reasonable. A high frequency should mean a short wavelength if the velocity of propagation remains unchanged.

Reflection

Wave propagation requires a continuous medium of some sort. The medium can actually be lumpy, but it must be lumpy in a continuous manner. Air, for example, forms a continuous medium even though it is made up of hard little molecules. The molecules aren't even all the same size, but the mix of light molecules to heavy molecules is evenly distributed. It therefore behaves like a continuous medium in that wave energy is passed along from molecule to molecule without disruption.

For a medium to be continuous, each part of the medium must have the same amount of inertial give and take, at least on the average, as the rest of the medium. We saw an example of this when we considered a toy store demonstration of energy and momentum conservation. At the end of Chapter 8, you may recall, we considered the behavior of a string of metal balls all hung in a line from a wooden frame, called the series impact pendulum. If one ball is lifted and allowed to collide with the others, one ball will bounce out the other side. Each ball in the string has the same mass as the last and is therefore able to absorb all the energy and momentum of the ball colliding with it and pass it on to the next ball. The exception is the ball on the end. It has nothing to do with its energy and momentum but to swing out at the end. This end ball, however, does swing back and send its energy and momentum back down the line.

Whenever there is some interruption in a medium, at least part of the energy in a wave will be reflected at the boundary. If you give a sideward jerk to a vacuum cleaner cord, for example, and the other end of the cord is plugged into the electrical outlet, the wave energy will not disappear when it reaches the plug. Assuming the plug remains in the outlet, you will see the wave reflect at the boundary and come back toward you along the cord.

Even more fun than cleaning house, however, would be to build your own wave demonstration machine from a box of soda straws, paper clips, and a roll of cellophane tape.[1] Stretch a meter or so of the tape out on the floor with the sticky side up. Lay straws perpendicular to the tape with even spacing between them and with their centers stuck to the center line of the tape. An easy way to get the spacing even is to use spacers made by cutting a couple of straws into short sections. These spacers can be put between the straws and left there to add strength to the final assembly.

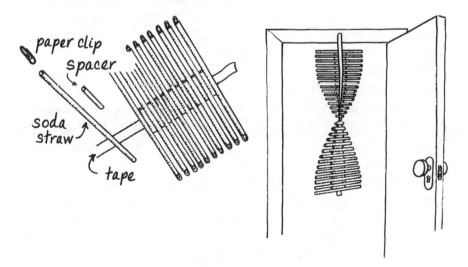

Figure 11-13

A wave machine can be made from soda straws and tape. Spacers can be made by cutting a straw into short sections. Paper clips can be inserted in the ends to add mass. A door frame makes a convenient support.

Place another strip of tape, sticky side down, over the straws and spacers to hold the whole thing together. Insert a paper clip in both ends of each straw to give them a significant amount of inertia so that the kinetic energy

---

[1] A form of this type of machine using steel rods attached to a horizontally supported torsion wire was developed by John N. Shive of Bell Laboratories to demonstrate microwave transmission theory. A model suitable for classroom demonstrations is available through Ealing Corporation.

stored in the wave is large compared to frictional losses. Hang the whole thing up by one end of the tape from a door frame and proceed to do experiments on waves.

The first experiment is to twitch the top straw and watch the pulse travel along the machine until it gets to the bottom. Instead of getting lost, the pulse seems to turn around and return in the direction from which it came. One way to look at this is to say that the end straw (commonly called the last straw) must do something with its kinetic energy and momentum, these being conserved quantities. It therefore gives them back to the straw from which it got them, this being the only straw at hand. From a force point of view, the last straw is not restrained by another straw on down the line so that its motion continues under its own inertia until it is restrained by the very straw that gave it a shove in the first place. From a wave point of view, the motion of the last straw is greater than the motions of the other straws by virtue of the discontinuity, and this larger wave motion may be looked upon as an additional wave superimposed upon the original. The reflected wave in this abstract way of thinking is looked upon as the wave needed to make the last straw wiggle more than the others; a wave that seems to come out of nowhere to produce the observed result.

This business of waves that seem to come out of a boundary where there are no more straws to wiggle might seem obtuse. Indeed it is, but you might find it a bit easier to use this way of thinking in dealing with the opposite kind of discontinuity. Instead of allowing the last straw be free to wiggle as much as it will, hold it so it cannot wiggle at all.

If you clamp the last straw in position, either with your fingers or by anchoring it to the door frame with a couple pieces of tape, you might well expect that no wave will come back. Surprisingly, a wave will be reflected by a fixed boundary but the reflected wave will be upside down relative to the incoming wave. From a force point of view, the fixed boundary applies whatever force is necessary to hold the last straw stationary. This force is greater than the force between straws, since the straws up the line have some give to them. The reflected wave is then seen as a result of this reaction force. From a wave point of view, the fixed boundary supplies a wave that exactly cancels the wave coming in. The reflected wave is then seen as this canceling wave supplied by the fixed boundary.

This picture of the reflected wave as a new wave that comes out of nowhere to satisfy the boundary conditions agrees with what happens when two waves traveling in opposite directions run into each other. You can do this experiment by sending a second wave into the wave machine at the moment that the first is reflected from a fixed end. As we just mentioned, the reflected wave will be upside down after having been reflected from a firmly attached fixed boundary. If you twitch the straw the same way the second time, the new wave you send in will be right side up relative to the upside down pulse traveling in the opposite direction.

You might expect these two waves to cancel each other out somehow, since one is sort of the negative of the other. A point midway between them is going to see a pulse coming from one direction pulling it up at the same time that the pulse coming from the other direction pulls it down. The two waves should therefore cancel each other's effects and the midpoint should remain stationary.

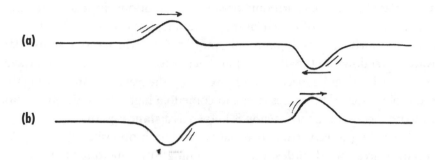

(a)

(b)

Figure 11-14

(a) Two pulses approach each other from opposite directions.
(b) The two pulses just seem to cross over each other and depart from the midpoint, although their joint effect on that midpoint may have been simply to hold it stationary.

Instead, what you actually observe is that these two waves seem to cross right over each other and continue in the same directions as they have been going. The poor straw at the midpoint may indeed get opposite orders from opposite directions and remain stationary as a result, but the energy in the two conflicting wave pulses cannot be so easily destroyed. For one brief moment in time, the straws on both sides of the midpoint may all line up in a straight line so that you might think both pulses have disappeared. That would be to forget, however, the kinetic energy of the straws. The straws on one side of the midpoint are traveling downward as they pass through their equilibrium position, while those on the other side are traveling upward. Both sets of straws coast on through their equilibrium positions to form pulses in the other direction. It is exactly as though the pulse coming in from the right has crossed over the one coming in from the left.

As far as the straws on one side of the midpoint are concerned, the result is the same if you hold the midpoint stationary by attaching it to some solid object. Those straws don't know what is holding that midpoint straw stationary, and they don't particularly care. It's all the same to them whether there is some upside down pulse coming in from the other side of the boundary or whether the same thing is being accomplished by a clamp. The main thing is that the fixed boundary condition is satisfied.

This argument can be repeated for a wave reflected from a free end of the wave machine. The last straw has nothing to restrict its motion and therefore moves farther than its neighbors. The situation is the same as if a positive pulse, one pulling in the same direction, were coming in from the opposite direction and superimposed on the first pulse. The straws down the line respond to the extra large motion of the last straw by passing this reflected pulse back in the other direction with no regard as to whether it was caused by a boundary or by an actual pulse coming from beyond the boundary. The only difference is that a pulse reflected by a free-end boundary is right side up relative to the incoming pulse.

Standing Waves and Interference

A rather pleasant thing happens when the pulses fed into the wave machine are replaced by simple harmonic or sine wave motions. Instead of jerking the straw at one end of the chain from one side to the other, try pushing it back and forth in simple harmonic motion. You will see this sinusoidal wave travel down to the other end and get reflected. As the reflected wave comes back along the machine and mixes with the wave you are sending in, the progression of the waves seems to come to a halt. The waves just seem to stand there. This phenomenon is known as a **standing wave**.

A standing wave is really two waves moving in opposite directions. At certain places, called **nodes**, the wave coming from one direction cancels out the wave coming from the other direction. At one moment in time, one wave tells the straw to go up while the other tells it to go down. Later, when the first wave gets around to telling the straw to go down, the second wave is telling it to go up. The result is called **destructive interference** of the waves. The straw remains stationary as the two waves cross over. At other points the two waves **constructively interfere,** or combine to produce a motion greater than that which would be produced by either wave independently. These places are distinctly the opposite of nodes. They are called **antinodes.**

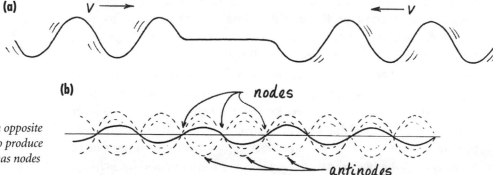

Figure 11-15

Two waves approaching from opposite directions (a) will combine to produce a standing wave, (b) which has nodes and antinodes.

The wavelength of the two waves that make up a standing wave is, of course, closely related to the distance between nodes. In fact, you might think at first that this distance is the wavelength itself. The standing wave does seem to repeat itself after going from a node to an antinode and back to a node again. Actually, this distance is only half a wavelength. When one antinode is moving in a positive displacement in one direction, the next is moving in a negative displacement in the other direction. Just as a full cycle of simple harmonic motion involves both positive and negative peaks, a full wavelength includes two antinodes. A wavelength can therefore be measured between *alternate* nodes.

Not just any frequency of oscillation will produce standing waves when applied to the top straw of the wave machine. Generally, you can get short wavelength standing waves with a high frequency wiggle and long wavelength standing waves with a low frequency wiggle, but these wavelengths have to fit on the wave machine if they are to produce stable standing waves. If the end straw is left free, the only waves that will produce standing waves on the machine are ones that produce an antinode at that point. If you wiggle the top straw with just the right amount of force, you can force an antinode at that point also. Depending on the frequency of the wiggle, you could produce standing waves having one, two, three, or any other integral number of nodes in between, but the ends must be antinodes.

If you attach the end straw to something solid so that it can't move, you will find that a different set of frequencies produce standing waves. By holding the end straw stationary, you force that end to be a node. You thereby change the boundary conditions that waves must satisfy if they are to produce standing waves.

Resonance

Nearly everything around you has its own sound. If you tap your fist on the chair you are sitting in, it makes a different sound than if you reach over and bang on the wall. If someone in the other room starts banging on the wall in response, you would probably recognize right away what is happening. The sound of someone pounding on the wall is much different from, say, the sound of that person banging on the radiator pipes. The difference comes from the ability of different objects to support different kinds of standing waves.

Your chair, for instance, is made up of various lengths of different materials attached together at certain points and free at others. Any standing wave induced in your chair will be forced to have nodes at certain points and antinodes at others. Depending on the velocities of sound in the various materials, only certain frequencies of vibration are capable of producing standing waves that fit these boundary conditions. They are called the **resonant frequencies** of the object. Any vibration close enough to one of the resonate frequencies of an object will excite the corresponding standing wave vibration in the object.

A shock wave, such as you might make by rapping your chair with your fist, is really a complex sound having a vast number of simple harmonic components. Only those components that are near a natural resonate frequency will be picked up by the chair and reinforced to produce the chair's distinctive sound. If you reach over and pound on the wall, the shock wave will excite a whole different set of resonate frequencies, depending on the boundary conditions and the velocity of wave propagation in the wall.

There is also a difference in how fast friction dampens out a particular vibration in different material. The wall is likely to go "thud" while the radiator pipe is likely to "ring" for a longer time. The characteristic sound of an object is not only recognized by its frequency components but by how long they last.

Musical Sounds

Musical instruments produce their characteristic sounds by means of various types of standing waves. Stringed instruments such as pianos and guitars use strings of various lengths and under various tensions to support standing waves having the proper frequency. Wind instruments such as flutes, on the other hand, use a column of air. In both cases, different notes are played by forcing nodes and antinodes to select the length of the proper standing wave. The wavelength, together with the velocity of propagation of the wave, determines the frequency in accordance with the relationship we derived when we first started talking about wavelength:

$$v = f \, \lambda$$

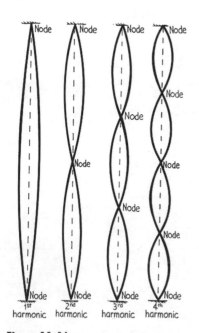

Figure 11-16

A string, with a node at each end, can vibrate in different ways.

which, as you remember, is really nothing more than the definition of velocity, v, applied to the case of a wave where distance is wavelength, λ, and time is the period, which is the reciprocal of frequency, f.

For wind instruments, the velocity is fixed. The velocity of sound in air depends on the average speed of the molecules. That, in turn, depends on the temperature and the mass of the molecules, neither of which changes much on the absolute scale of things. As we shall see, the changes in frequency of a wind instrument due to such things as temperature and carbon dioxide concentrations are small enough to be ignored or compensated for by the musician.

For stringed instruments, the velocity is far from fixed. As we shall see the instrument is tuned by setting the velocity of the waves on the strings. Once tuned, however, the instrument is played by selecting the lengths of strings.

Music Theory

The notes that are played on the instrument are, of course, a matter of culture, but culture is influenced by the way that the standing waves fit together on the instrument. The music of Western tradition is heavily influenced by Pythagoras who explored the principles of harmony based on string instruments. The music of Eastern tradition is heavily influenced, on the other hand, by Ling Luan, Minister of Music in the court of the

Yellow Emperor, about 2000 B.C. Ling Luan explored the principles of harmony based on wind instruments. Most musicologists believe that both traditions are based on an earlier African foundation.

Pythagoras found that certain notes were pleasing to the ear when played together if the ratio of the string lengths that produced the notes were the ratios of whole numbers. For example, one note will sound pleasing to the ear when played together with a note having half the wavelength.

The lowest note that can vibrate a string is the note that produces a node at each fixed end of the string. That note is called the fundamental or first harmonic. The next higher note that will vibrate the string produces a node in the middle of the string in addition to the nodes on the ends. That note is called the second harmonic. (For musically minded folks, the second harmonic sounds an octave higher than the first harmonic.) Add another node or two and you have the third or fourth harmonic, and so on. Musically, these modes of vibrations sounded together sound like a plucked string. In fact, that is how a plucked string vibrates. The amplitude of each mode of vibration depends on how the string is plucked, the resonance of the instrument, and many other factors, but all of the harmonics are present.

What Pythagoras found experimentally was that these harmonics sounded good to his ear when played together on separate strings. We now understand his rule about the ratio of whole numbers in terms of the number of nodes you can have in a standing wave.

We can also understand this rule in terms of the shape of the waveform produced when sine waves are added together. Sine waves of harmonic frequencies, when added together, make a stable and relatively simple waveform. Your ear drum dances in a repeated pattern and your brain recognizes order. The particular pattern depends on the instrument and the relative strengths of the various harmonics. Your brain recognizes the harmonic content as a piano or clarinet.

The first and second harmonics when played separately sound rather far apart. Pythagoras and his followers called this interval an octave and divided it into seven steps—the familiar *do, re, mi, fa, sol, la, ti, do,* where the first *do* and the last *do* are an octave apart. Octave actually means eight, but the terminology predates the invention of zero by Arabian mathematicians.

These seven notes, which sound evenly spaced to the ear, were in reality based, for many hundreds of years, on the ratios of whole numbers 16 and less. Two of the intervals, between the third and fourth note and between the seventh and eighth note are the closest together. They increase by a factor of 16/15ths. Three of the others were 9/8ths apart and two were 10/9ths apart. This is the diatonic scale and seems very natural for a violinist. Even today, a violinist playing unaccompanied will tend to use the diatonic scale.

This unequal spacing can still be seen on a piano keyboard by the groupings of the black notes. There are five black notes per octave, grouped in twos and threes. In the modern equal tempered scale, there are seven

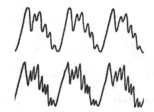

Figure 11-17

Different instruments playing the same note will sound different.

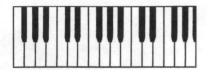

Figure 11-18

The unequal spacing of the diatonic scale can be seen on a piano keyboard.

white notes and five black notes for twelve notes altogether. The spacing is therefore the twelfth root of two, or $2^{1/12}$ so that an "octave" of twelve notes will double the frequency.

The principles of harmony in the tradition of Eastern music, based on wind instruments, developed in a similar fashion to that of the ancient Greeks but preserved the five tones of African traditions. If you play about on a modern keyboard, playing only on the black notes, you will find that you can make music that sounds either Chinese or African depending mostly on the notes you start on and finish on.

The tunes played in the African pentatonic or five tone scale together with syncopated rhythms played over a European march beat in the left hand, led to the first wildly successful cross cultural musical phenomenon in the early twentieth century known as ragtime. The classical blues that led to modern popular music got its sophistication from this technically demanding and highly structured musical form.

Musical Instruments

STRINGED INSTRUMENTS

String instruments are first tuned by adjusting the velocity, *v*, of wave propagation along the strings. Two factors determine the velocity of a transverse wave on a string. One is the mass of the string and the other is its tension.

Imagine a wave pulse traveling along a string. As the pulse passes a certain point on the string, it jerks the string up and then down again. The force that causes this accelerated motion is really nothing more than the tension in the string, since the motion of the pulse along the string results from the way that the string pulls on itself. It therefore seems reasonable that the wave would move along more quickly if the tension in the string were increased. That is exactly what happens. Piano and guitar strings are both tuned by adjusting their tension. Increasing the tension raises the pitch, or frequency, *f*, of the standing waves by increasing the velocity, *v*, with which waves propagate along the string.

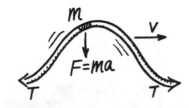

Figure 11-19

The velocity of a wave pulse depends on the tension in the string as well as on the mass of the string.

Some of the tuning is built into the string itself, however. Bass strings on both guitars and pianos are big and massive compared to the strings for higher notes. They are even wrapped with wire to increase their mass. The purpose of this is to slow down the waves. The greater the mass of a piece of string, the less will be its acceleration under a given force. Increased mass of the string therefore has the opposite effect of increased tension. The bass strings vibrate at a low frequency, *f*, because they have a slow velocity, *v*, of wave propagation.

Once a guitar string is tuned by adjusting its tension, it can still be used to play different notes by changing its effective length. In playing different chords, the guitarist holds the strings down on different frets to change the boundary conditions, which, in turn, changes the allowable wavelengths that can produce standing waves. Clamping the string between the finger and the fret forces that point to be a node. Another node is forced on the string by the bridge. The longest distance between nodes is therefore the distance from the bridge of the guitar to the guitarist's finger. Since a wavelength is the distance between alternate nodes, the distance from the bridge to the finger is half a wavelength of the fundamental frequency being played.

The guitar string can also have one or more nodes between the two that are forced by the boundary conditions. If there is one node in the middle of the string, the frequency is called the first overtone. If there are two nodes between the ones on the end, it is called the second overtone, and so forth. These different ways of supporting standing waves are called **modes** of vibration. The same string can actually vibrate in several modes at the same time. In a guitar the first overtone is strongly excited by strumming the string at a point one-fourth of the way along the string's length. That induces an antinode at that point, a node in the middle, and an antinode one-fourth of the way from the other end of the string. The string also vibrates in its fundamental mode with an antinode in the middle. The actual motion is a superposition of these two main modes of vibration with much smaller amounts of other vibrations thrown in.

One technique of guitar playing, called playing the harmonics, is to dampen out the fundamental mode by touching the string lightly in the middle and letting it vibrate in the higher frequency modes, mainly the first overtone. This is actually known as the "second" harmonic because, as it happens, the "first" harmonic is a name assigned to the fundamental.

WIND INSTRUMENTS

Boundary conditions also determine which note is played by a wind instrument. Instead of forcing a node at a particular point as in a string instrument, however, the boundary condition is frequently such as to force an antinode. A flute, for example, is a hollow pipe that has a mouthpiece at one end and is open at the other. Both ends force an antinode in the standing wave that is generated when the lowest note is played. Blowing across

Figure 11-20

The distance from the bridge of a guitar to the finger holding the string on a fret is one-half the wavelength of the fundamental mode of vibration and equal to the wavelength of the first overtone.

open

Figure 11-21

A flute is played by directing a stream of air across the opening in the mouthpiece to reinforce standing waves in the tube.

the mouthpiece produces a swirling flow of vortices that make a soft whispering sound. The air molecules are displaced in a complex waving motion that disturbs the molecules next to them, causing them to be displaced, and so forth as the wave moves on down the tube. Much the same as a pulse is reflected by a free end on the wave machine, the displacement wave in the air molecules is reflected when it gets to the free end of the flute. Just as the inertia of the last straw on the wave machine leads to an extra large motion, so the inertia of the air in the end of the tube leads to an extra large displacement as the displacement wave comes to the sudden freedom of the open end of the flute. This extra large motion leads to a reflected wave that travels back toward the mouthpiece.

When the reflected wave gets back to the mouthpiece, it not only is reflected again but is amplified. An outward displacement wave deflects the air stream away from the opening to form a decreased pressure that is even greater than that caused by the inertia of the air at a simple open end. An inward displacement wave deflects the air stream into the opening to form an increase in pressure that is even greater than that caused by the inertia of the air alone. Thus the waves move back and forth along the tube, reflected at the open end and both reflected and driven at the mouthpiece end. The resulting standing wave has an antinode at each end since the displacement wave motion is forced to be extra large at these points.

The note that the flute is playing can be controlled by opening various holes on the side of the tube to force antinodes at the proper points. The flute can easily be shifted into the upper register by simply blowing harder to excite the second harmonic. Interestingly, the overtones that are present in the lower register almost completely disappear in the higher octave and the notes turn out to be almost pure sine waves.

We have examined resonance in musical instruments in which boundary conditions force both ends of the standing wave to be either a node or an antinode. In both cases, the distance between the boundaries is half a wavelength for the fundamental mode of vibration. Depending on the wave velocity in the instrument, that puts a lower limit on the frequency that the instrument can play. The lowest note is one with a wavelength equal to twice the distance between the boundaries. There is a way, however, to set boundary conditions so that the fundamental has a wavelength equal to four times the length of the instrument. This trick is sometimes used to make very low notes in organ pipes. The pipe is closed at one end and open at the other. The fundamental mode of vibration then has a node at one end and an antinode at the other. The length of the pipe is therefore equal to one-fourth of a wavelength.

A good amount of musical instrument design has been done by people who knew very little about the physics of sound. People just tried different things and used the things that seemed to work. In fact, much of what we know about the physics of sound has come from people who were interested in studying musical instruments. Once our understanding of theory caught up to our practical knowledge, however, substantial improvement in instrument design was possible. The transverse flute, the instrument we know today, was designed by Theobold Boehm in the 1860s, after he first studied physics for two years to better understand musical sounds.

PROBLEM SET 11

SUMMARY OF CONCEPTS

The problems in this chapter are related to vibration and sound. **Simple harmonic motion**, the kind of vibration associated with a pure tone of a single frequency, is the sinusoidal motion that results from a mass oscillating on a spring or a pendulum bob swinging back and forth on a string. The time required for the motion to repeat itself is called the **period**, T, of the motion, but people often speak of the **frequency**, f, of the motion to convey the same information, the frequency simply being the reciprocal of the period:

$$f \equiv \frac{1}{T}$$

The unit for frequency is the **hertz**, abbreviated Hz, which is one cycle per second:

$$1 \text{ Hz} = \frac{1 \text{ cycle}}{1 \text{ second}}$$

Amplitude, A, of the simple harmonic motion is defined as the distance from the equilibrium position to the maximum displacement. It is

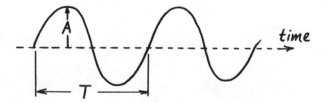

equal to one-half the distance from one peak to the other along the displacement axis. The period, T, is related to the amplitude, A, of the simple harmonic motion by

$$T = 2\pi \sqrt{\frac{A}{a_{\text{max}}}}$$

where a_{max} is the maximum acceleration associated with the maximum displacement of the motion. The ratio of A to a_{max} turns out to be equal to the ratio of mass, m, to the spring constant k for a spring and the ratio of the length, l, to the acceleration due to gravity, g, for a pendulum.

$$T_{spring} = 2\pi \sqrt{\frac{m}{k}}$$

$$T_{pendulum} = 2\pi \sqrt{\frac{l}{g}}$$

The **principle of superposition** holds that any motion can be made up of the sum of simple harmonic motions of different amplitudes, frequencies, and starting times.

Sound and other wave motion results from a vibration moving through a medium. The **wavelength**, λ (the greek letter *lambda*) associated with such motion is just the length corresponding to one period, T, of the motion as it moves alone at the velocity, v, of propagation through the medium:

$$v \equiv \frac{d}{t} = \frac{\lambda}{T}$$

This relationship is usually stated in terms of frequency: velocity equals frequency times wavelength, since frequency, f, is the reciprocal of period, T.

$$v = f\lambda$$

This relationship has frequent application to resonant systems, where a half wavelength refers to the distance between nodes or antinodes of a standing wave oscillating in the system. The boundary conditions of the system determine the positions of various nodes or antinodes and thereby determine the possible wavelengths that will fit. The only frequencies at which the system will resonate are those that produce the proper wavelengths to fit these boundary conditions.

EXAMPLE PROBLEMS

Try to do the Examples yourself before looking at the Sample Solutions.

11-1 Tarzan is swinging from tree to tree at the end of a long vine attached to a tall tree located midway between his starting point and his destination. He steps off one tree and swings to the same height on the second tree, where he stops.

Given: The two trees are 12 m away from one another. It takes him 4.0 s to swing from one tree to the other.

Find: (a) What is the period of Tarzan's motion?
(b) What is the amplitude of this swing?
(c) What is Tarzan's acceleration at the bottom of the swing, when he is half-way between the two trees?

BACKGROUND

In working problems on period and amplitude, it is often helpful to draw a sine wave to decide how much of the wave is being used in any given situation.

11-2 A mother is feeding her baby in a rocking chair. The gentle acceleration in alternate directions soothes both mother and child and makes them sleepy.

Given: The baby's head moves back and forth over a distance of 7.3 cm, making the trip 45 times in 1 min. The motion is simple harmonic.

Find: (a) What is the amplitude of the simple harmonic motion?
(b) What is the period of the motion?
(c) What is the maximum acceleration of the baby's head?

BACKGROUND

This problem illustrates the relationship between the amplitude and period of simple harmonic motion and the maximum acceleration. If you find yourself wanting to take a little nap, think instead of the acceleration of students' heads as they shake themselves awake.

11-3 A radio speaker has a small coil of wire mounted in a magnetic field and attached to a flexible paper cone. A changing electric current through the coil shoves it back and forth against the cone to set the air in motion and make sound. The loudness of the sound is greater, however, if the frequency of the electric signal is near the natural frequency of the speaker, determined by the mass of the coil and cone and the springiness of the cone.

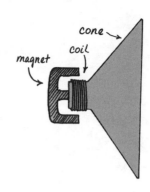

Given: A speaker has a peak response at 25 Hz. The mass of the coil and cone is 45 g.

Find: (a) What is the natural period of oscillation of the speaker?
(b) What is the spring constant of the flexible paper cone?

BACKGROUND

This problem is an application of the relationship between the natural period and the mass and spring constant of a spring oscillator. This natural resonance of a speaker cone causes the speaker to sound "boomy" because of its exaggerated response to bass notes at near the resonance frequency. Acoustic suspension speaker systems attack this problem by weakening the spring constant of the speaker cone and allowing the air pressure in the speaker enclosure to hold the speaker cone generally in place.

11-4 A boy takes his sister to the park to use the swings. Although they usually use the little swing set, they decide to use the big swings today.

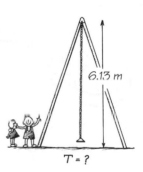

Given: They are interested in what the period will be on the big 6.13 m swings. The period of the little swings was 2.8 s when the boy was swinging. The boy weighs 321 N, and his younger sister weighs 289 N.

Find: (a) What will be the period of the boy on the big swings?
(b) What will be the sister's frequency on the big swings?
(c) What was the length of the little swings?

BACKGROUND

This problem illustrates the relationship between the period of a pendulum and its length and the acceleration due to gravity. Recall that the condition for simple harmonic motion is that the acceleration is proportional to the displacement but in the opposite direction. This condition is met for a pendulum only for small angles of oscillation, much smaller angles than are common for small children on a swing set. Still, simple harmonic motion gives a reasonable approximation to the period and frequency in this situation.

11-5 The velocity of an ocean wave depends on its wavelength compared to the depth of the ocean floor. A wave generated by a seismic disturbance, called a tsunami (soo-nah-mee), generally moves with a very high velocity, because its extremely long wavelength makes it a shallow water wave even in ocean depths of four kilometers.

Given: A typical wave generated by wind action has a period of 8.0 s and a wavelength of 85 m. A tsunami, on the other hand, might have a period of 15 min and move with a velocity of up to 200 m/s (400 mph) because of its wavelength.

Find: (a) What is the velocity of the wave generated by wind action?
(b) What is the wavelength of the tsunami?

BACKGROUND

This problem illustrates the relationship between velocity, frequency, and wavelength of wave propagation. The difference in velocity for waves of different wavelength is of interest to the surfer, who is waiting for just the right wave to ride on a surfboard. The change in the ocean depth relative to the wavelength of the wave as it comes into the beach also changes the wave's velocity and causes it to break.

11-6 A guitar is tuned by adjusting the tension to change the velocity of waves along the strings so that the ends are nodes for the proper fundamental frequency. Different notes are then played on the same string by effectively shortening it by pressing down on a fret.

Given: The E string of a guitar is 61.7 cm long and vibrates at 330 Hz when properly tuned. The A note, at a frequency of 440 Hz, can be played on this string by holding it on the proper fret.

Find: (a) What is the proper velocity of the wave along the E string such that the distance between nodes corresponds to the length of the string?
(b) What string length at this same velocity will produce an A note?

BACKGROUND

This problem illustrates resonance on a string where both ends are clamped and thereby forced to be nodes. A guitar note is made up mostly of the fundamental and first over-tone. The fundamental has a wavelength equal to twice the length of the string, and the first overtone has a wavelength equal to one times the length of the string.

11-7 Some organ pipes are open at one end and closed at the other. The length of such a pipe is the distance from a node to the first antinode of a wave having the proper frequency and traveling at the velocity of sound in air.

Given: A certain organ pipe is 27.9 centimeters long. The velocity of sound in air is 343 m/s at room temperature.

Find: (a) What is the wavelength in meters of the note produced by this organ pipe?

(b) What frequency does this note correspond to?

BACKGROUND

This problem illustrates resonance in a pipe closed at one end and open at the other. This boundary condition forces a node at one end and an antinode at the other.

Solutions to **EXAMPLE PROBLEMS**

Given:
The two trees are 12 m away from one another. It takes him 4.0 s to swing from one tree to the other.

Find:

(a) What is the period of Tarzan's motion?

(b) What is the amplitude of this swing?

(c) What is Tarzan's acceleration at the bottom of the swing, when he is half-way between the two trees?

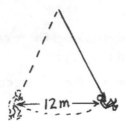

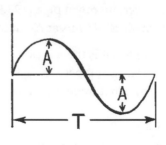

(a) $\quad T = 2t$

$\qquad = 2(4.0s)$

$\qquad = \boxed{8.0s}$

(b) $\quad d = 2A$ $\qquad$ (c) $\quad F = ma$

$\qquad A = \frac{1}{2}d$ $\qquad\qquad\qquad a = \dfrac{F}{m}$

$\qquad\quad = \frac{1}{2}(12m)$

$\qquad\quad = \boxed{6.0m}$ $\qquad\qquad\qquad = \dfrac{(0)}{m} = \boxed{0}$

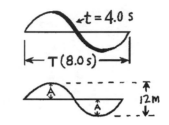

DISCUSSION

Tarzan swings from one peak to another peak. On the graph of a sine wave, his journey is from the top of one peak to the bottom of the next one. This is two amplitudes on the vertical displacement axis and half of the period on the time axis.

The period is the time that it would take to swing all the way there and back, so the period is twice the time that it takes Tarzan to complete half a cycle.

The amplitude is the distance that it takes to go from equilibrium to one peak of a sine wave. Since Tarzan's swing goes from one peak through equilibrium and out to a peak on the other side, his swing covers two amplitudes. The amplitude of his swing, then, is half the distance that he traveled.

The acceleration at the bottom of the a pendulum swing is zero. At the bottom of the swing the force on the vine is vertical and there is no horizontal force acting on Tarzan, so that Newton's second law

$$F = ma$$

tells us that there can be no acceleration. Tarzan may be moving, but he is in equlibrium.

(a) $2A = d_{total}$

$$A = \frac{d_{total}}{2}$$

$$= \frac{7.3 \text{ cm}}{2} = 3.65 \text{ cm} \cong \boxed{3.7 \text{ cm}}$$

(b) $f = \dfrac{1}{T}$

$$T = \frac{1}{f}$$

$$= \frac{1}{(45 \text{ cycles/min})}\left(\frac{60 \text{ s}}{1 \text{ min}}\right)$$

$$= \boxed{1.3 \text{ s}}$$

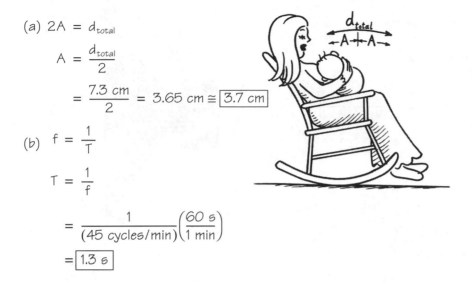

(c) $T = 2\pi\sqrt{\dfrac{A}{a_{max}}}$

$$a_{max} = A\frac{4\pi^2}{T^2}$$

$$= (3.65 \text{ cm})\frac{4\pi^2}{(1.3 \text{ s})^2}\left(\frac{1 \text{ m}}{10^2 \text{ cm}}\right)$$

$$= \boxed{0.85 \text{ m/s}^2}$$

SAMPLE SOLUTION 11-2

Given:
The baby's head moves back and forth over a distance of 7.3 cm, making the trip 45 times in 1 min. The motion is simple harmonic.

Find:
(a) What is the amplitude of the simple harmonic motion?

(b) What is the period of the motion?

(c) What is the maximum acceleration of the baby's head?

DISCUSSION

The amplitude of simple harmonic motion is measured from the equilibrium position to one peak. It is therefore equal to one-half the peak-to-peak displacement. Period is found, as in part (b), as the reciprocal of the given frequency.

The maximum acceleration can be found by solving the relationship between period, amplitude, and maximum acceleration. The general form of this relationship is useful to remember, because it works as well for the period of a pendulum and a spring oscillator if you just change the variables under the square root sign to be reasonable in light of the physics of the situation. You might expect the maximum acceleration, a_{max} to be in the denominator, for example, since greater acceleration would reasonably produce a shorter period of oscillation. The resulting acceleration is a bit less than a tenth of that due to gravity.

SAMPLE SOLUTION 11-3

Given:

A speaker has a peak response at 25 Hz. The mass of the coil and cone is 45 g.

Find:

(a) What is the natural period of oscillation of the speaker?

(b) What is the spring constant of the flexible paper cone?

(a) $f = \dfrac{1}{T}$

$T = \dfrac{1}{f}$

$= \dfrac{1}{(25 \text{ Hz})}\left(\dfrac{1 \text{ Hz}}{\text{cycle/s}}\right)$

$= \boxed{4 \times 10^{-2} \text{s}}$

(b) $T = 2\pi\sqrt{\dfrac{m}{k}}$

$\dfrac{T^2}{4\pi^2} = \dfrac{m}{k}$

$k = \dfrac{4\pi^2 m}{T^2}$

$= \dfrac{4\pi^2 (45 \text{ g})}{(4 \times 10^{-2}\text{s})^2}\left(\dfrac{1 \text{ kg}}{10^3 \text{ g}}\right)\dfrac{N}{\text{kg} \cdot \text{m/s}^2}$

$= \boxed{1.1 \times 10^3 \text{N/m}}$

cone

coil

magnet

DISCUSSION

The period is found in part (a) as simply the reciprocal of the frequency. The standard unit of frequency is the hertz, defined as a cycle per second. The resulting period is in seconds per cycle, but the units of cycles in the denominator may be left out as they are implicit in the meaning of period. Units such as cycles or revolutions or other angular measure can often be inserted or left out as the occasion demands. They are sort of unitless units, whose presence or absence is implied by the thing being measured.

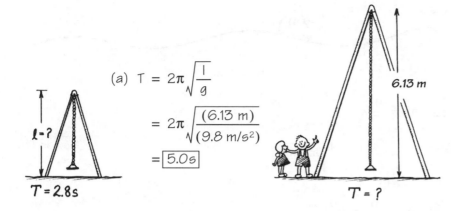

Given:
They are interested in what the period will be on the big 6.13 m swings. The period of the little swings was 2.8 s when the boy was swinging. The boy weighs 321 N, and his younger sister weighs 289 N.

Find:
(a) What will be the period of the boy on the big swings?

(b) What will be the sister's frequency on the big swings?

(c) What was the length of the little swings?

(a) $T = 2\pi\sqrt{\dfrac{l}{g}}$

$= 2\pi\sqrt{\dfrac{(6.13\ m)}{(9.8\ m/s^2)}}$

$= \boxed{5.0s}$

$T = 2.8s$

(b) $f = \dfrac{1}{T}$

$= \dfrac{1}{(5.0s)}\dfrac{Hz}{s^{-1}}$

$= \boxed{0.20\ Hz}$

(c) $T = 2\pi\sqrt{\dfrac{l}{g}}$

$l = g\left(\dfrac{T}{2\pi}\right)^2$

$= (9.8\ m/s^2)\left(\dfrac{2.8\ s}{2\pi}\right)^2$

$= \boxed{1.95\ m}$

DISCUSSION

The period for the little girl in part (b) is the same as that of the little boy in part (a), in spite of the difference in their masses, because mass doesn't enter into the relationship for the period of a pendulum. If the lengths of two pendulums are the same, they will have the same period in the same gravitational field, since the weight effects of mass balance the inertial effects for the same reason that light and heavy objects in free fall have the same acceleration.

SAMPLE SOLUTION 11-5

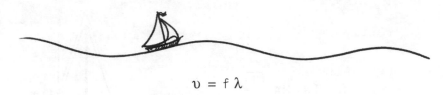

Given:

A typical wave generated by wind action has a period of 8.0 s and a wavelength of 85 m. A tsunami, on the other hand, might have a period of 15 min and move with a velocity of up to 200 m/s (400 mph) because of its wavelength.

Find:

(a) What is the velocity of the wave generated by wind action?

(b) What is the wavelength of the tsunami?

$$\upsilon = f\,\lambda$$

(a) $\upsilon = \dfrac{1}{T}\lambda$

$ = \dfrac{85\text{ m}}{8.0\text{ s}} = \boxed{10.6\text{ m/s}}$

(b) $\lambda = \dfrac{\upsilon}{f} = \dfrac{\upsilon}{\frac{1}{T}}$

$ = \upsilon T$

$ = (200\text{ m/s})(15\text{ min})\left(\dfrac{60\text{ s}}{1\text{ min}}\right)\left(\dfrac{1\text{ km}}{10^{3}\text{ min}}\right)$

$ = \boxed{180\text{ km}}$

DISCUSSION

The basic relationship, velocity is equal to frequency times wavelength, is easy to derive if you aren't quite certain that you have it right. The definition of velocity is distance over time:

$$v \equiv \frac{d}{t}$$

The appropriate distance in this case is a wavelength, and the appropriate time is a period. The relationship in this form is useful in part (a) just the way it is. It may also be solved for wavelength to be used in part (b).

The wavelength of a tsunami as found in part (b) turns out to be a bit over 100 miles.

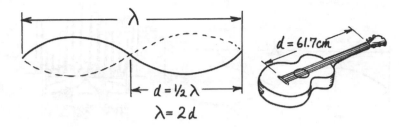

$\lambda = 2d$

(a) $\upsilon = f\lambda = f(2d)$

$$= (330 \text{ Hz})2(61.7 \text{ cm})\left(\frac{\text{cycle/s}}{\text{Hz}}\right)\frac{1 \text{ m}}{10^2 \text{ cm}}$$

$$= \boxed{407 \text{ m/s}}$$

(b) $\upsilon = f(2d)$

$$d = \frac{\upsilon}{2f}$$

$$= \frac{(407 \text{ m/s})}{2(440 \text{ Hz})}\left(\frac{\text{Hz}}{\text{cycle/s}}\right)\left(\frac{10^2 \text{ cm}}{1 \text{ m}}\right)$$

$$= \boxed{46.3 \text{ cm}}$$

DISCUSSION

The distance between nodes of a standing wave is equal to half of a wavelength, since a full wave contains part of the string above the equilibrium position as well as part below at any one time. Part (a) gives the fundamental frequency times the wavelength as the velocity of the wave along the string, the wavelength of the fundamental being expressed as twice the length of the string.

SAMPLE SOLUTION 11-6

Given:
The E string of a guitar is 61.7 cm long and vibrates at 330 Hz when properly tuned. The A note, at a frequency of 440 Hz, can be played on this string by holding it on the proper fret.

Find:
(a) What is the proper velocity of the wave along the E string such that the distance between nodes corresponds to the length of the string?

(b) What string length at this same velocity will produce an A note?

SAMPLE SOLUTION 11-7

Given:

A certain organ pipe is 27.9 centimeters long. The velocity of sound in air is 343 m/s at room temperature.

Find:

(a) What is the wavelength in meters of the note produced by this organ pipe?

(b) What frequency does this note correspond to?

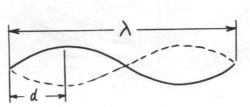

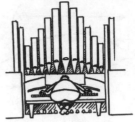

(a) $d = \dfrac{1}{4}\lambda$

$\lambda = 4d$

$\quad = 4(27.9\ cm)$

$\quad = \boxed{1.12\ m}$

(b) $v = f\lambda$

$f = \dfrac{v}{\lambda}$

$\quad = \dfrac{(343\ m/s)}{(1.12\ m)}\left(\dfrac{Hz}{s^{-1}}\right)$

$\quad = \boxed{306\ Hz}$

DISCUSSION

The distance between a node and the first antinode is one-fourth of a wavelength, since it is half the distance between nodes, which in turn is half a wavelength. A wavelength of the sound generated by this organ pipe is therefore four times the length of the pipe.

   The frequency of this sound can then be calculated from the velocity of sound in air, using the relationship between velocity, frequency, and wavelength of wave propagation.

ESSENTIAL PROBLEMS

Be sure you can do the Example Problems (without peeking at the Sample Solutions) before working these Essential Problems.

11-8 A grandfather clock has a pendulum that swings from one side to another to mark off seconds.

Given: It is 4.5 cm from the peak of one swing to the peak of the next, measured horizontally. Naturally, the swing from one side to another takes 1.0 s.

Find: (a) What is the period of the clock pendulum?
(b) What is the amplitude of the pendulum?
(c) What is the acceleration of the pendulum at the bottom of it's swing?

11-9 A gambler is reciting an incantation about his baby needing a new pair of shoes as he shakes a dice cup with simple harmonic motion.

Given: The gambler shakes the dice cup back and forth over a distance of 20 cm, from one end to the other, 7 times in 4.0 s.

Find: (a) What is the amplitude of the motion?
(b) What is the period of the simple harmonic motion?
(c) What is the maximum acceleration of the dice cup?

11-10 A little girl playing with her kitten ties a ball to a rubber band and allows it to dangle near the floor. The kitten bats it with her paw and discovers that it bounces up and down at a particular frequency.

Given: The ball at the end of the rubber band bounces with a frequency of 1.6 Hz. The ball has a mass of 125 g.

Find: (a) What is the period of oscillation of the ball on the rubber band?
(b) What is the spring constant of the rubber band?

11-11 The heroine of a Japanese monster movie is dangling at the end of a long rope that is held in one hand by the monster. A close-up shot showing the heroine screaming, with the monster's face in the background, is followed by a long shot showing the whole monster holding a doll on the end of a string. A later shot shows the hero screaming at the end of the same rope, with the monster's face in the background.

Given: The length of the rope in the two close-up shots was 5.0 m long. The heroine looked as if she had a mass of about 65 kg, while the hero seemed to have a mass of 85 kg. The doll on the end of the string in the long shots swings back and forth with a period of 1.0 sec, because the producer did not bother to change the speed of the motion to correspond to the change in scale. It all turned out happily in the end.

Find: (a) What was the period of the heroine's swing at the end of the rope?

   (b) What was the frequency of the hero's swing in the same situation?

   (c) What length of string did the movie producers use in filming the long shot?

11-12 Earthquakes travel outward from an epicenter in three waves. The fastest, called P-waves, are pressure waves that travel through the ground like sound through air. You feel those first. Then come the S-waves, or shear waves. They shake the ground from side to side. Last, but certainly not least, are the so-called Rayleigh waves. They are ripples in the surface of the earth that travel like ocean waves. That's what makes the telephone poles dance. Like counting the seconds between thunder and lightning to judge the distance to a storm, you can judge the distance between you and the epicenter of an earthquake by noticing the time between the shaking and the wiggles.

Given: An S-wave might well have a period of 0.50 s and a wavelength of 7.0 km. A Rayleigh wave, on the other hand, might have a period of 0.217 s and move with a velocity of up to 800 m/s.

Find: (a) What is the velocity of the S-wave?

   (b) What is the wavelength of the Rayleigh wave?

11-13 A xylophone is a percussion instrument consisting of a series of wooden or metal bars that are graduated in length to sound the musical scale. These bars are generally supported on felt pads so that their ends are free to vibrate. This boundary condition forces an antinode at both ends and thereby fixes the fundamental frequency at which a bar will ring when struck with a small wooden hammer.

Given: A xylophone has a middle-C bar that is 28.0 cm long and has a fundamental frequency of 261.6 Hz. The G bar of the same octave has a fundamental frequency of 392 Hz (American Standard pitch).

Find: (a) With what velocity does the wave in the C bar travel?

(b) Assuming the same velocity for the wave in the G bar, what is the length of the G bar?

11-14 A physics instructor is using a Shive wave machine to demonstrate standing waves. He has one end clamped and is driving the other end up and down in simple harmonic motion to produce the longest possible wave on the machine. The clamped end is a node and the other end is the first antinode.

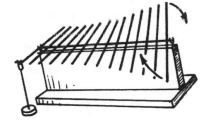

Given: The wave machine is 1.0 m long. The velocity of waves along the machine is 2.5 m/s.

Find: (a) What is the wavelength of the fundamental mode of oscillation where one end is clamped and the other is free to move?

(b) With what frequency should the physics instructor push the other end up and down to excite this fundamental mode?

11-15 An automobile is a mass on springs and has a natural frequency of oscillation. The function of the shock absorbers is to dampen this motion so that the car does not bounce up and down as it goes down the road. You may have noticed a car with worn-out shock absorbers doing just that.

Given: An automobile having a mass of 1,700 kg must have worn-out shock absorbers, because you notice that it is bouncing up and down slowly with a period of about 2.5 s as you follow it down the freeway.

Find: (a) What is the effective spring constant of this automobile's suspension system?
(b) How many centimeters would this car settle if a person having a weight of 800 N got in?

11-16 Anna has a fancy desk chair. It has a slightly springy action. She finds that when she sits on it by dropping herself on it from a distance it bounces up a down a couple of times in simple harmonic motion.

Given: Anna and the seat of the chair have a combined mass of 76 kg. The chair bounces with a period of 0.44 s. Anna's office mate, David, weighs 88 N more than Anna.

Find: (a) What is the spring constant of the office chair?
(b) How much lower is the office chair when David sits on it than when Anna sits on it?

11-17 A swimmer is poised on the end of a diving board about to start bouncing up and down to gain altitude for a back flip. You estimate her weight and notice how much she makes the diving board sag when she is just standing there.

Given: Her weight seems to be about 530 N, and this weight makes the diving board bend a distance of about 37 cm.

Find: (a) What is the spring constant of the diving board?
(b) What is the diver's mass?
(c) With what period would you expect the diver to start bouncing when she first starts and is bouncing in simple harmonic motion?

11-18 Iowa Jones is hanging from a tree branch. As he tries to pull himself up, the branch starts to bounce in simple harmonic motion.

Given: Iowa weighs 990 N. Before he began bouncing, his weight displaced the branch 1.41 m.

Find: (a) What is the spring constant of the branch?

(b) What is Iowa Jones' mass?

(c) With what period would you expect Iowa to bounce up and down once he begins moving in simple harmonic motion?

11-19 The limiting factor in determining the "red line," or maximum safe rotational velocity, of an automobile engine is the forces that develop between parts that go up and down, such as the pistons, and other parts that go around, such as the crankshaft. A common failure resulting from over-revving an engine is to break the connecting rod that connects the piston to the crankshaft.

Given: The piston in an experimental engine has a mass of 1.00 kg and is operated almost in simple harmonic motion with an amplitude of 4.00 cm. The engine is red-lined at 6,000 revolutions per minute (rpm).

Find: (a) What is the period of the simple harmonic motion at the red-line rotational velocity?

(b) What is the maximum acceleration of the piston at redline?

(c) With what maximum force does the connecting rod push on the piston when the engine is operated at its red-lined rpm?

(d) What would be the force acting on the piston if the engine were over-revved to 8,000 rpm?

11-20 Clever Coyote is in a railroad car where he pushes on a bar in simple harmonic motion to roll the car along the tracks.

Given: The bar being pushed up and down has an effective mass of 30 kg and is operated in simple harmonic motion with an amplitude of 0.87 m. At first, he is pushing himself along the tracks by making 14 complete cycles per minute. Later, in a hurry, he pushes the bar at 24 cycles per minute.

Find: (a) What is the period of the simple harmonic motion when he is first pushing himself along?

(b) What is the maximum acceleration of the bar when Clever Coyote is moving at this speed?

(c) What is the maximum force that Clever Coyote uses to push the bar at the original speed?

(d) When Clever Coyote decides he is in a hurry, what is the maximum force he uses to push on the bar?

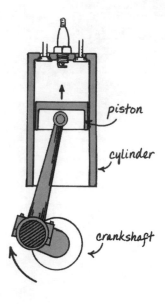

11-21 The pistons in an automobile engine go up and down each time the crankshaft revolves one time, but they only produce power on alternate down strokes. The other down strokes between power strokes are used to draw air-gas mixture into the cylinder in preparation for the next power stroke.

Given: In a four-cylinder automobile engine, the crankshaft is turning at a frequency of 2,000 revolutions per minute when the car is going 20 meters per second in top gear.

Find: (a) What is the period of revolution of the crankshaft in seconds?

(b) What is the frequency of the power strokes for one cylinder, in Hz?

(c) What is the frequency of the power strokes for all four cylinders in Hz?

11-22 A *555 timer* is a common electronic device sold in local neighborhood stores to hobbyists. For less than a dollar in parts and a few minutes in time, anyone can hook up a circuit that will put out pulses to control whatever. If the pulses come too fast, just add a *7490 decade counter,* which also costs less than a buck and which puts out one pulse for every ten that come in. The decade counter is a two-stage device. The first stage divides by two and the second stage takes the output of the first stage and divides by five.

Given: A 555 timer, wired to produce 80 pulses per minute, is connected to a 7490 decade counter.

Find: (a) What is the period, in seconds, of the pulses from the timer?

(b) What is the frequency of the pulses from the divide by two output of the counter?

(c) What is the frequency of the pulses from the second stage of the counter?

11-23 An FM radio receiver must be able to receive radio signals having wavelengths anywhere within the FM broadcast band.

Given: The FM band extends from 88 to 108 Mc on your dial. Mc stands for megacycles per second, or 10^6 Hz. The velocity of radio waves is the same as that of light, 3.00×10^8 m/s.

Find: (a) What is the length of the longest radio wave your FM receiver must be able to handle?

(b) What is the length of the shortest radio wave your FM receiver must be able to handle?

11-24 A healthy human ear can pick up a very wide range of frequencies.

Given: A normal ear of a young person can hear sounds with a frequency as low as 15 Hz and as high as 20,000 Hz. The velocity of sound in air is about 335 m/s.

Find: (a) What is the length of the longest wave that can be heard by the human ear?

(b) What is the length of the shortest wave that can be heard by the human ear?

11-25 Bats hunt and navigate by listening for the echoes of high-pitched chirps they emit from several times to hundreds of times a second, depending on how hard they are "looking." The sounds must be of a very high frequency to have a wavelength of about the same size as the insects the bats feed upon.

Given: The speed of sound in air is about 335 m/s. A bat directs a beam of high-pitched sound toward a mosquito that is 0.9 cm long when flying in the air. One chirp lasts 1/400 of a second.

Find: (a) At what frequency will the wavelength of the sound emitted by the bat correspond to the length of the mosquito in flight?

(b) What is the period of this sound?

(c) How many of these high-pitched sound waves are produced during the duration of one chirp?

11-26 Television signals can be delivered by fiber-optic cables. For this to be done, the television signal may be used to control the brightness of a laser light, which can then be sent through a fiber of optical-quality glass.

Given: Laser light has a wavelength of 680 nm (a nanometer is 10^{-9} m.) The speed of light in a particular optical fiber is 2.1×10^8 m/s. One period of a television signal is about 5.0×10^{-6} s.

Find: (a) What is the frequency of this laser light in the fiber-optic cable?

(b) What is the period of this laser light?

(c) How many light waves are delivered by the laser during the transmission of one television wave?

11-27 The speed of sound in air increases with temperature. This is why temperature fluctuations can affect the tune of a musical instrument.

Given: A particular organ pipe is 2.3 m long and is open at both ends. The speed of sound in air at 0° C is 331.4 m/s, but it increases by 0.61 m/s for each degree Celsius above zero.

Find: (a) What is the fundamental resonant frequency of the pipe at 0° C?
(b) What is the velocity of sound at 25° C?
(c) What is the resonant frequency of the pipe at 25° C?

11-28 The exhibit in San Francisco's Exploratorium called "The Pipes of Pan" consists of a set of glass pipes, open at both ends. Natural standing waves are produced by ambiant noise. There is no need to blow into them to make sounds—you can just put your ear up to them to hear a different pitch for each pipe.

Given: A particular pipe is 1.2 m long. The speed of sound in air at 0° C is 331.4 m/s, but it increases by 0.61 m/s for each degree Celsius above zero.

Find: (a) What is the fundamental resonant frequency of the pipe at 0° C?
(b) What is the velocity of sound at 19° C?
(c) What is the resonant frequency of the pipe at 19° C?

11-29 The modern transverse flute is a woodwind instrument developed by Theobold Boehm in the 1860s, after he had studied physics for two years to better understand musical sound. It is a hollow tube closed at one end and open at the other, but the placement of the mouthpiece makes it act as if it were open at both ends.

Given: The velocity of sound in air is 343.6 m/s at 20° C. The lowest note that can be played on older flutes is C, at 261.6 Hz, and the highest note is C three octaves higher, at 2,093 Hz.

Find: (a) What is the wavelength of the lowest C that can be played on a flute?
(b) What is the distance from a node to an antinode for the lowest C on this instrument?
(c) What is the wavelength of the highest C that can be played on the flute?
(d) What is the distance from antinode to antinode for the highest C on the flute?

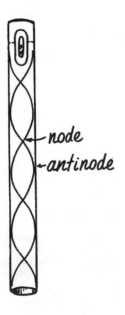

node
antinode

11-30 Vince Musical is playing his electric guitar. He plays one note then takes his finger off the string to play a lower note.

Given: The velocity of sound in air is 343.6 m/s at 20° C. Vince plays a low B, with a frequency of 123.5 Hz, then takes his hand off the string to play a low E, the lowest note on an electric guitar, which has a frequency of 82 Hz.

Find: (a) What is the wavelength of the low B on a guitar?
(b) How long is the distance from node to antinode for the B?
(c) What is the wavelength of the low E?
(d) How long is the distance from node to node for the E?

ANSWERS

Solutions to **ESSENTIAL PROBLEMS**

11-8   (a) 2.0 s; (b) 2.25 cm; (c) zero

11-9   (a) 10 cm; (b) 0.57 s; (c) 12 m/s^2

11-10 (a) 0.63 s; (b) 12.6 N/m

11-11 (a) 4.5 s; (b) 0.22 Hz; (c) 25 cm

11-12 (a) 14 km/s; (b) 174 m

11-13 (a) 146 m/s; (b) 18.7 cm

11-14 (a) 4.0 m; (b) 0.625 Hz $\cong$ 0.63 Hz.

Solutions to **MORE INTERESTING PROBLEMS,** *odd answers*

11-15 (a) 1.1×10^4 N/m; (b) 7.5 cm

11-17 (a) 1.4×10^3 N/m; (b) 54 kg; (c) 1.2 s

11-19 (a) 0.01 s; (b) 1.58×10^4 m/s^2; (c) 1.58×10^4 N; (d) 2.81×10^4 N

11-21 (a) 3.0×10^{-2} s; (b) 17 Hz; (c) 68 Hz

11-23 (a) 3.4 m; (b) 2.8 m

11-25 (a) 3.7×10^4 Hz; (b) 2.7×10^{-5} s; (c) 93

11-27 (a) 72 Hz; (b) 346.7 m/s; (c) 75 Hz

11-29 (a) 1.313 m; (b) 32.84 cm; (c) 16.42 cm; (d) 8.21 cm

12 *Electricity*

Statics

ONLY TWO KINDS OF FORCES ACCOUNT for the majority of everyday phenomena—forces of a gravitational nature and forces of an electrical nature. We have already discussed the gravitational force between two objects. It is directly proportional to the product of the masses and inversely proportional to the square of the distance between the two objects. As you recall from our discussion of Newton's laws of motion, gravitational force is given by this formula.

$$F = G\frac{m_1 m_2}{d^2}$$

The electrical force between any two stationary objects obeys a similar inverse square relationship with distance. The thing about an object that produces an electrical force, however, is something called **charge.** Charge plays a role in electrical phenomena much like the role that mass plays in gravitational phenomena. The electrical force between any two objects is directly proportional to the product of their charges q_1 and q_2 and inversely proportional to the square of the distance between them.

$$F = k\frac{q_1 q_2}{d^2}$$

The letter k in this equation stands for the proportionality constant which in the electrical force law is similar to the universal gravitational constant G in Newton's law of gravitation. Instead of being a very small number, as in the gravitational case, however, the electrical proportionality constant k turns out to be a very large number if the charges are of ordinary size in human terms. This proportionality constant is called the **Coulomb constant** and has the value

$$k = 9 \times 10^9 \text{ N} \cdot \text{m}^2/\text{C}^2$$

where C stands for the usual unit for measuring electrical charge, the **coulomb**. We may think of it as the amount of charge that is on a certain number of electrons—or on the same number of protons, for that matter. That number turns out to be 6.25×10^{18}. This might seem like a great number of electrons, but it only represents the amount of charge that passes through a 100-watt light bulb in a little over a second.

If two charges of one coulomb each were one meter apart, the previously stated value of k would mean that the force of repulsion between the two charges would be 9 billion newtons. That would be about twelve times the weight of the *Queen Elizabeth* ocean liner. Obviously, such amounts of pure charge do not exist in room-sized objects around us.

In spite of the vast difference in size between electrical and gravitational forces, they still look a lot alike. They are both inverse square law forces in that they both decrease inversely with the square of distance. The striking similarities between these two forces have made some people think that they might really be different aspects of the same thing. Albert Einstein was one of those people. He spent the latter part of his life working with little success on this problem. Although there are similarities between these two kinds of force, there are also some very basic differences between them—and the vast difference in size is only one of these. Another difference is that electrical forces can be balanced out. They can be either attractive or repulsive, whereas gravitational forces can only be attractive.

The old saying that opposites attract, usually referring to people, was first popularized by public lecturers who traveled about the countryside on horseback to entertain people by demonstrating the scientific marvels of electricity. An essential part of these demonstrations was the charging and discharging of pith balls. Pith is a light spongy plant tissue resembling Styrofoam, and balls made of it were coated with silver paint so that their outside surfaces would conduct electricity. Such a ball, hung from a silken thread, would exhibit a strong attraction for a rubber rod that had been rubbed on cat's fur, but this force of attraction would suddenly change to a force of repulsion as soon as the ball touched the rod. Another ball charged in the same manner would exhibit the same behavior. The two balls charged the same way would be found to repel each other. Pith balls would also be first attracted and then repelled by a glass rod, but a ball charged with a glass rod would be attracted to a ball charged with a rubber rod. The

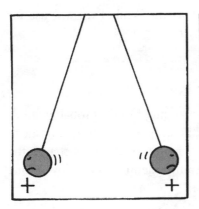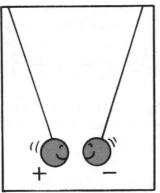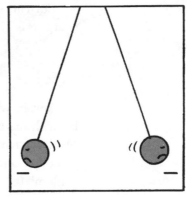

Figure 12-1

Opposites attract.

lecturer could make much of the fact that nature provides two kinds of charge just as it provides two sexes.

Benjamin Franklin named these two kinds of charge positive and negative in accordance with his feeling that an electrical charge represented either an abundance or lack of some kind of electrical fluid. For some reason, he decided to call the rubber negative and the glass positive. Benjamin Franklin's fluid theory was remarkably like our present understanding of electrical charge in terms of the excess or lack of electrons. He missed, however, when it came to guessing which was which. It turns out that the rubber rod gains electrons in the rubbing process, while the glass rod loses them. We compensate for Franklin's wrong guess by saying that electrons have a negative charge and protons have a positive charge. Thus, a positive charge does wind up representing an excess of *something* even if that thing is not the electrons that normally flow through a wire.

Current

The fluid theory of electricity was based on the observation that some things, such as the silver paint on the pith balls, allow electrical charge to flow from one place to another in much the same way as a pipe allows water to flow from one place to another. These things, usually metallic objects such as watch chains or wires, were called electrical conductors. Other things, which did not allow electrical charge to flow, such as the silken thread attached to the pith balls, were called insulators.

The idea of an electrical current arose naturally from the fluid theory of electricity as an analogy to water flow in a pipe. Just as the rate at which water flows through a garden hose might be measured in gallons per minute, the electrical flow rate, or current, is the ratio of the amount of charge that passes a particular point in a wire to the amount of time that it takes to pass.

$$\text{Current} \equiv \frac{\text{charge}}{\text{time}}$$

Current is usually represented by *I*, while the symbol we will use for charge is *q*. Time is, as usual, represented by *t*.

$$I \equiv \frac{q}{t}$$

Anyone who has had experience with changing fuses is probably familiar with the unit for current. It is called the ampere, or "amp" for short. Much less familiar to the average person is the unit for electrical charge, the coulomb. One ampere (A) of electrical current is defined as the flow of one coulomb (C) of charge per second.

$$1 \text{ A} \equiv \frac{1 \text{ C}}{1 \text{ s}}$$

You might wonder at this point where the coulomb unit of charge came from. We mentioned earlier that this is the charge of about 64 billion billion electrons. That's a lot of electrons, and the electrostatic forces between charges of this size are, as we also mentioned, huge. Still, this quantity of charge was originally defined in terms of about the smallest amount that could, in a sense, be weighed. When an electric current is passed through an electrolytic cell containing silver nitrate, pure silver is plated out on one electrode. The amount of silver is directly proportional to the charge that passes through the cell, since individual electrons deposit individual atoms in the plating process. One coulomb of charge was defined as the amount that would deposit about 1 mg of silver—a small amount, but one that could be measured on an analytic balance.

While it is true that this unit of charge is too large for use in most electrostatic force problems, it turns out to be just right for problems involving the flow of current through household appliances. Just a little less than a coulomb of charge flows, for example, through a 100-Watt light bulb each second. An electric toaster typically passes 11 coulombs per second, or 11 amperes, when it is in operation. These units of charge and current have therefore been widely adopted, because they are so convenient for common use.

The direction of current has also been established by common usage. The plus and minus sign convention was, as we have mentioned, arbitrarily

Figure 12-2

Current is from positive to negative, although electrons flow the other way.

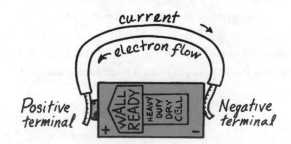

assigned by Benjamin Franklin on the basis of what he thought was going on in charging rubber and glass rods by friction. This convention was well established and widely used long before the electron was discovered. People just naturally assumed that current comes out of the positive (+) terminal of a battery and into the negative (−) terminal. It seems so sensible that current should flow from a high to a low potential that this convention has stuck. In fact, positively charged ions do move in this direction when they are free to move. In a wire, however, the positive charges are tightly bound in the nuclei of the metal atoms and therefore don't move. The charges that do move, the electrons, actually travel in the other direction. Sometimes the term *electron current* is used to indicate the direction of electron flow. It is from negative to positive. The word *current* alone, however, is simply defined as something that is imagined to flow from positive to negative.

Batteries and Bulbs

Some people find the subject of electricity somewhat abstract because they have had little direct experience with making electrical things work. Mechanical things seem more concrete because, as children, most people have had experience playing with blocks and mechanical toys. If you find yourself among the many who have had far less direct experience with the inner workings of electrical devices, as compared to mechanical gadgets, you are encouraged to do the following experiment with your own hands.

An ordinary flashlight has a simple electrical circuit, or hookup, but many people never stop to figure out how it operates. It is instructive to take one apart and make the bulb light up in your fingers. The wires you will need can be made either from paper clips or strips of aluminum foil

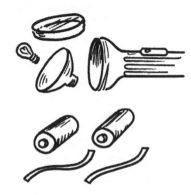

Figure 12-3

An experiment using a flashlight bulb and batteries and strips of aluminum foil.

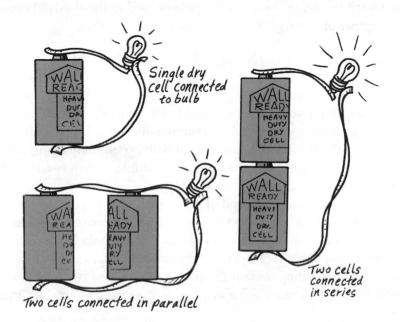

Single dry cell connected to bulb

Two cells connected in parallel

Two cells connected in series

Figure 12-4

Various configurations of one and two batteries lighting a flashlight bulb. With the batteries in series, the bulb is significantly brighter.

cut from that roll you probably keep in the kitchen for wrapping food. Remove the bulb from the flashlight and try to make it light up first with one battery and then with two.

One pleasant diversion is to take the bulb, batteries, and wires in hand and seek out children of various ages to see how long it takes them to light the bulb.

After some smart kid makes the bulb light with one battery in less time than it took you, remind yourself that children come equipped with less intellectual baggage than adults and that it is therefore easier for them to experiment with random combinations. In order to save face, produce the other battery from your pocket and suggest that there are two ways to light the same bulb with two batteries and that one of these ways will make the bulb light brighter. If the wise kid beats you at those combinations, start pulling out additional bulbs. Eventually your superior experience with mechanical things will prevail as you start producing, from the other pocket, rubber bands with which to hold things together.

Voltage

The whole point of fooling around with flashlight bulbs and batteries is to gain a practical feeling for what pushes electricity around a circuit, or, to say it another way, to gain a practical understanding of something called voltage. We have been using the term *current* to refer to the flow of electrical charge. Voltage is the thing that causes current; it is the thing that pushes electrical charge around and makes it flow.

You might like to think of voltage as something a little like force in mechanical situations or like pressure in fluid flow situations. These are good mental models for the behavior of voltage, but actually voltage is more related to energy. Voltage is formally defined as the electrical potential energy per unit charge. The unit for voltage is the volt, V.

$$1V \equiv \frac{1\ J}{1\ C}$$

Where V stands for volt, J for joule, and C for coulomb. Still, the analogy with fluid pressure is very useful in understanding what is going on in an electrical circuit. This mental model, even with certain limitations, allows you to gain a qualitative understanding, for example, of series and parallel circuits.

Thinking of a battery as a kind of pump that pushes current through a light bulb by developing some sort of electrical pressure across it, we can see how two batteries connected in series would light the bulb brighter than a single battery acting alone. Two water pumps would, after all, develop more pressure than one acting alone if the output of one were connected to the input of the other. Picture a water fountain with a stream of water squirting up in the air because of the pressure developed by a pump under the fountain, as in Figure 12-5. If you connected two pumps in series, the water would certainly squirt higher in the air than with one pump, all else being the same, because the second pump would take the pressurized water from the first pump and boost its pressure still more. The volume of water would also increase, but that would depend on the size of the opening in the fountain. All the water that passes through one pump in this arrangement must also pass through the other, and so the maximum volume of water is restricted by the capacity of a single pump. If you restrict the opening, you wouldn't necessarily get a lot more water, but the water you would get would squirt higher.

Figure 12-5

Two water pumps connected in series to one figure in a water fountain and another two pumps connected in parallel to another. The series connection can develop more pressure while the parallel connection can develop more volume.

Much the same thing happens when two batteries are connected in series to a flashlight bulb. The voltage of one battery is boosted still higher by the second battery; there is more voltage across the bulb, so it lights brighter. The maximum current would be limited by the capacity of a single battery cell, since all the current that passes through one must pass through the other. Although the series connection of batteries doesn't necessarily increase the current, it puts more energy into that current.

If you want more volume in a water flow system, or more current in an electrical system, the better connection would be parallel. Suppose, for example, that you had a double figure in a water fountain and you needed a combined volume of water greater than could be carried by a single pump. You could obtain this volume of water by connecting two pumps in parallel so that they both draw from the same source and deliver to the same figure. You wouldn't get much increase in pressure with this arrangement, but you would get a lot more water.

Much the same thing happens when you connect two batteries in parallel to the flashlight bulb. Both negative terminals of the batteries are connected to one side of the bulb, and both positive terminals are connected to the other. This parallel connection does not seem to light the bulb much brighter, as long as the batteries are fresh, because voltage is not much increased by this arrangement, but the batteries will stay fresh longer. Only part of the current in the bulb comes from each battery, so that the current drain on each battery would be less. The parallel connection therefore increases the capacity of the batteries for delivering current.

The analogy between electrical voltage and fluid pressure can also be used to understand the relative nature of voltage. Since voltage is an energy concept, it has meaning only in terms of a reference level. As you know from our discussion of energy, the potential energy of this book you are holding has meaning only if you specify whether you are talking about its height relative to the floor, to the ground outside, or to the center of the earth. It is even negative if you are talking about the ceiling over your head as the reference level. That is why you need to connect both terminals of the battery to the bulb if you are to get it to light. The voltage of the battery has meaning only in reference to the other terminal. For this reason, voltage is also known as **potential difference**. Anybody who tries the experiment with the flashlight bulb and batteries will soon discover that the key to lighting the bulb is to provide a difference in electrical potential across the bulb by making a closed circuit from the battery, through the bulb, and back to the other terminal of the battery. This is very much like needing to provide a difference in pressure to get fluid to flow. No amount of pressure from a pump will make water squirt out of a fountain if that same pressure exists at the opening of the fountain. A stopped-up water fountain doesn't squirt water. Unstop the fountain and water flows because of the difference in pressure between the output of the pump and the atmosphere.

Have you ever wondered how a bird can perch comfortably on a high tension wire without getting shocked? Power lines for trolley cars, for example, are typically at 400 V relative to ground. This is nearly four times the voltage considered safe for house use. A bird can land on this wire, however, without even noticing the high voltage. That is because this high voltage is relative to the potential of the ground, and the bird does not touch the ground. Both legs of the bird are on the wire and are therefore at the same potential. So long as there is no potential difference, or voltage, across the bird, no current will flow and the bird will not feel anything.

Perhaps you might wonder what would happen if the bird should happen to land on the wire with one foot on either side of an insulator. The overhead trolley power wires are supported by other wires, which are attached to lamp posts. Since the lamp posts are at ground potential, these

Figure 12-6

Voltage is relative. A bird sitting comfortably on a high tension wire will not get shocked because both legs are at the same potential.

support wires are interrupted by porcelain or glass insulators to keep them from conducting current to ground. In fact, a close examination will show that each wire is double insulated. It is interrupted with an insulator in two different places just in case something should happen to one of the insulators. If the bird should happen to get across one of the insulators, nothing would happen. The bird would still be protected by the other insulator. It would take two birds to get across both insulators simultaneously to produce fried bird.

Ground is a very common reference potential for use in power supplies like the large 400-V generators used to power the trolleys, particularly streetcars that run on metal rails set in the street. These rails are used as a return conductor to complete the circuit from the streetcar back to the power generator. They are said to be *grounded* since they are electrically at the same potential as the earth as a whole. That is very convenient so that you don't get shocked if you should step on one of these tracks or board a streetcar. It is a good idea to have electrical appliances you use grounded. Otherwise, you might become a little bit fried yourself.

Ohm's Law

If we think of voltage as something that pushes charge through a conductor like pressure difference pushes water through a pipe, it seems reasonable that a greater voltage would push a greater current through the conductor. That is exactly what happens. The voltage even turns out to be directly proportional to the current for many conductors, including most metals, as long as the temperature and everything else is kept constant.

$$V \propto I$$

This proportionality was first noticed experimentally by George Simpson Ohm in 1825. It is therefore called Ohm's law. The proportionality constant between the voltage, V, and the current, I, is called the resistance, *R*, of the conductor.

$$V = IR$$

The resistance of an electrical conductor depends upon its length and thickness as well as what it is made of. Silver is the best electrical conductor at ordinary temperatures, in that the resistance of a silver wire of certain length and thickness will be less than the same size wire made of another material. You might reason that the electrical wiring in your house would be better if the wires were made out of silver. Copper, however, comes in as a close second. A copper wire of a certain length and thickness has only about 6% more resistance than the same size silver wire. Gold and aluminum would be next, in that order. Gold, of course, would be too expensive for house use, although it does seem a shame for it to be lying around Fort

Knox not doing anything. Actually, most of the gold in industrial use is being used for conducting electricity. Not only does it conduct electricity with low resistance but it is highly resistant to corrosion. That combination makes gold an excellent choice for electrical contacts. It is only used in very small quantities for this purpose, however, because of its expense.

Aluminum, on the other hand, is cheaper than copper. For that reason, it is sometimes used in house wiring, even though it has over one and a half times as much resistance for a wire of the same size. From Ohm's law, a higher resistance means that a higher voltage is needed to induce a certain amount of current in the wire.[1]

$$V = RI$$

In house wiring, this effect is countered by using thicker wire. A thick wire has less electrical resistance than a thin wire of the same length and composition. The usual copper wire used in a 15-amp house circuit is No. 14 gauge. Aluminum wire for the same circuit would be No. 12 gauge, which has about 1.6 times as much cross-sectional area. The greater thickness brings the resistance back down to about the same value as the thinner copper wire has.

To be specific about the numbers involved, we must define units for this thing we call resistance. The common unit is the **ohm**, commonly represented by the Greek letter omega, Ω. One ohm of resistance is that amount of resistance that requires one volt of potential difference to cause a current of one ampere:

$$1 \text{ V} = (1\Omega)(1 \text{ A})$$
$$1\Omega = \frac{1\text{V}}{1\text{A}}$$

A piece of No. 14 gauge copper wire would be about 120 meters long to have a resistance of 1 Ω. An aluminum wire of the same diameter would only be about 75 meterslong to have the same resistance.

CHECK QUESTION

An electric toaster draws 11.25 A at 120 V. What is its resistance?
Answer: 10.7 Ω

Our discussion of Ohm's law assumed that the resistance of a wire, or of anything else, is constant. It is actually a variable that can depend on

---

[1]Notice that here we have chosen to place the *R* next to the equality sign rather than in the reversed order shown in the proceeding equation, because we wish to emphasize that it is a proportionality constant. Ohm's law is usually written *V = IR*. It makes no difference algebraically which way you write it.

several factors. The resistance of your skin, for example, can change over a wide range depending on moisture. Should you tell a little lie, the tiny amount of sweat your body produces as an emotional response can be easily measured by the change in the resistance of your skin. Skin resistance is, as a matter of fact, one of the three parameters commonly measured in polygraph lie detector tests.

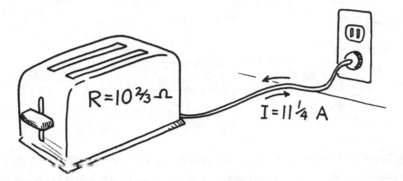

Figure 12-7

The resistance of an electric toaster is the ratio of voltage, or potential difference, across it to the current induced by the voltage in the toaster.

The electrical resistance of something can even depend, directly or indirectly, on the current. The resistance of a wire, for example, depends on its temperature. The hotter the wire gets, the more the thermal agitation of the atoms disrupts the flow of electrons. The flow of charge, in turn, tends to heat the wire and thus raise its resistance. This effect is quite pronounced in the case of a light bulb, where the resistance of the tungsten wire filament is several times greater when the lamp is lit than when the wire is cold. A graph of voltage against current would therefore produce more of a curve than the straight-line characteristic of a direct proportion.

There are many cases, however, in which the departure from a strict proportionality is of little consequence in a given situation. Ohm's law may then be used as a close approximation to reality. As with other mental models, what is close enough depends on the accuracy you need. If the situation demands it, you can easily obtain an electrical device, called a precision **resistor**, whose resistance is constant to five or six significant figures over a wide range of current. The resistance of the common radio resistor is constant to within two to four significant figures, depending upon its quality. Such a device may be nothing more than a little chunk of carbon with electrical leads attached to opposite ends, or it may be a length of special alloy wire wrapped into a coil in a special way so as to eliminate magnetic effects. A resistor may be defined as any electrical device whose resistance may be considered constant. Even a light bulb may be thought of as a resistor for some purposes.

The idea of a resistor is a useful mental tool for understanding electrical circuits, although, if you stop to worry about such things, it is a simplification of real life situations. The electric toaster, for example, could be thought of as a resistor connected to the power lines by resistanceless wires. This

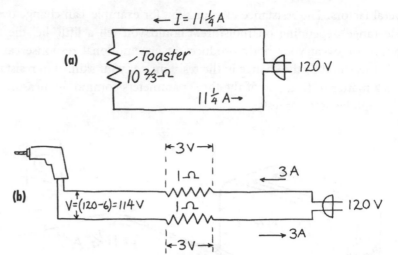

Figure 12-8

(a) A diagram showing a toaster as a pure resistor ignoring the resistance of the power cord. (b) A diagram showing the resistances in long extension cords for an electric drill.

simplified picture would ignore any voltage across the toaster cord due to the resistance that must be present in even a good heavy toaster cord. It also ignores any voltage difference in the power lines themselves. The voltage difference is there, and you might notice it by the slight dimming of a light bulb connected to the same circuit when you start the toaster; but these are only secondary effects and may be safely ignored for many purposes.

It is easy to imagine a situation, however, in which the voltage across the the power cord assumes major importance. Let us see how the idea of a resistor can be used to upgrade our mental model of what is going on in such a situation. Suppose you want to drill a hole in the dashboard of your automobile parked in the alley behind your house. You string out extension cords from your house so you can use your electric drill. You happen to own three 15-meter light-duty extension cords. Actually, you don't need that much length, but you decide to go ahead and use all three just in case. (If you've got them, you might as well use them.) Besides, you find a little label on the cords that says they are rated to carry 5 A of current, and the label on the drill says that it draws only 3.0 A. It seems that you should be okay, but when you try it you find that the electric drill just doesn't have as much power as when you plug it directly into the outlet in the house. Obviously, there must be a significant loss of some kind in the extension cords.

One way to picture the electrical circuit is to think of the two wires in the extension cord as two resistors connected in series with the drill. Just to make the numbers easy to work with, let us assume that each wire has a resistance of 1 Ω That would correspond to 45 meters of light-duty extension cord of No. 18 gauge.

If the drill draws 3.0 A, this amount of current must pass through each of the two wires. It requires 3.0 V to push 3.0 A through 1.0 Ω of resistance, in accordance with Ohm's law.

$$V = IR$$
$$= (3.0 \text{ A})(1\Omega)$$
$$= 3 \text{ V}$$

The same voltage, or potential difference, is needed to push the current through the other wire. The voltage available to the electric drill is thereby reduced by a total of 6.0 V below that available at the wall outlet. Even though the extension cords will handle the current, you shouldn't use longer extension cords than you need to reach the project .

Schematic Diagrams

Electrical circuits are frequently described in terms of simplified diagrams, using a few common symbols to represent ideal wires, ideal resistors, and other idealized electrical components to approximate what really happens in a practical situation. The preceding diagrams for a toaster and an electric drill are examples of this kind of diagram, called a **schematic diagram**. Resistance is shown as a zig-zag line, and ideal resistanceless wires are shown with straight solid lines.

Although real wires have resistance, and real resistors may not follow Ohm's law precisely, you can more easily understand the current in a circuit if you ignore these little difficulties. You can always go back and add in the complications if they turn out to matter. Let us return to the flashlight example to see how a schematic diagram may be used to describe a simple circuit. The simplest way to light the bulb, as you remember, was to connect one side of the bulb to one terminal of the battery at the same time that the other side is connected to the other terminal. This would be represented by showing ruler-drawn straight lines connecting the bulb with a single-cell battery, represented by long and short parallel lines. The convention is to represent the positive terminal of the battery with a long line and the negative terminal with a short line.

Let us now see how this circuit is similar to what is actually used in a two-cell flashlight. The two cells would be represented by two pairs of long and short parallel lines. One side of the bulb is connected, as before, to one terminal of the battery. In most flashlights one terminal of the bulb actually rests upon one terminal of one battery; but this fact is not important in understanding the current in the system. This connection is still shown as a ruler-straight line drawn from one side of the bulb to one terminal of the pair of batteries. The other side of the bulb is connected to the other terminal of the batteries through a switch.

The purpose of the switch is, of course, to interrupt the circuit to prevent current from flowing when the flashlight is turned off. For our purposes, we may think of this as an ideal switch, which either has zero or

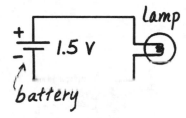

Figure 12-9

A schematic diagram of the circuit in which a flashlight bulb is lit with a single battery.

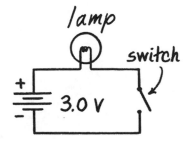

Figure 12-10

A schematic diagram of a flashlight.

infinite resistance. A switch is said to be closed when the contacts touch each other and open when the contacts do not touch. In this respect, the commonly accepted terminology is exactly the opposite as that used for a water valve. When you close a switch, you allow current to flow in much the same way as you allow water to flow by opening a faucet. Opening a switch should give it infinite resistance to electrical flow in much the same way that closing a faucet should shut off the flow of water.

CHECK QUESTIONS

In Figure 12–10, (a) what is the voltage across the bulb when the switch is closed? (b) What is it when the switch is open? (c) What is the voltage across the switch when the switch is closed? (d) What is it when it is open?

Answer: (a) 3 V, (b) 0, (c) 0, (d) 3 V

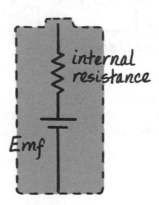

Figure 12-11

The internal resistance of a battery may be pictured as a resistor in series with the emf of the battery.

The fact that the resistance of the light bulb changes when the bulb is lit has little effect on our understanding the operation of the circuit. For the purposes of comparing the voltage across the bulb with that across the switch, we would do as well to draw the circuit using a resistor to represent the bulb. This kind of simplification allows you to ignore unimportant factors.

One complication frequently ignored in a simplified analysis of a circuit is the fact that the voltage of a battery decreases as the current delivered by the battery increases. This effect is called the **internal resistance** of the battery. If it turns out to be important in a given situation, it may be taken into account by representing the battery as a constant voltage source, called the **electromotive force**, or emf, of the battery, in series with a resistor. The emf of a battery is defined as the maximum voltage the battery can deliver when there is no load connected to it. A battery that has run down can usually still provide a voltage in the absence of a load, but the internal resistance has become so high that the voltage available at the terminals drops to a low value when you attempt to draw current from it. You might look upon this high internal resistance of a dead battery as having resulted from most of the available current paths inside the battery having been exhausted. This is just another example of creating a mental model to account for additional factors as they become important.

The value of a schematic diagram is that it allows us to focus our attention on essential matters. The actual structure of the flashlight, for example, may well be more complicated than our diagram would indicate. Parts of the flashlight may serve both to hold things together and to act as wires to conduct current. Just as in other mental models of reality, we need only be aware of simplifying assumptions so that we are capable of taking them into account should our judgment indicate that a more sophisticated model is warranted.

Resistors in Series and Parallel

A battery can light two flashlight bulbs in two ways. The bulbs are said to be connected in **series** if they are attached to the battery in such a way that all of the current from the battery passes through each of the bulbs. The bulbs are said to be connected in **parallel** if they are attached in a branching circuit such that part of the current passes through one bulb and the rest passes through the other.

If you get a spare bulb for your flashlight and try this experiment, you will find that the bulbs connected in series will not be as bright as the two bulbs connected in parallel. The two bulbs connected in series have more resistance than either taken independently and will not draw as much current as would either one if it were connected across the battery by itself. Only enough current can flow, in fact, for the sum of the voltages across these two bulbs to add up to the voltage of the battery. This kind of series arrangement of resistors is sometimes called a **voltage divider**, because it can divide a voltage into two parts. It is useful when you need a smaller voltage than that of the source.

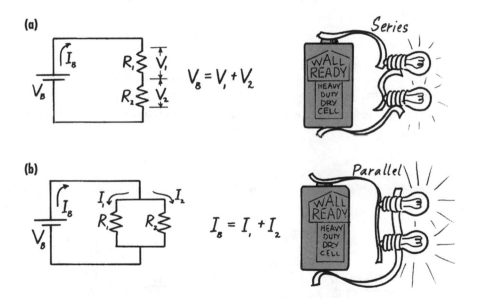

Figure 12-12

Two ways to light two bulbs with the same battery. The bulbs connected in series (a) will not light as brightly as the bulbs connected in parallel (b), because they will not draw as much current.

The two bulbs connected in parallel, on the other hand, are equivalent to one, having less resistance than either taken independently. The entire voltage of the battery is across each bulb so that there is more current than if only one bulb were connected. The current produced by the battery is, in fact, the sum of the currents in each bulb.

We can generalize this argument to find the **equivalent resistance** of any two resistors connected in series or in parallel. The rule for adding resistance in series is that the equivalent resistance is simply the sum of the

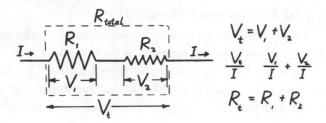

Figure 12-13

The resistance of a series circuit is the sum of the resistances.

resistance of each of the separate resistors. This rule follows from Ohm's law by solving for resistance:

$$V = IR$$

$$R \equiv \frac{V}{I}$$

The total voltage V_t, across both resistors is the sum of the voltages V_1 and V_2 across each of the resistors. Since each resister carries the same current, we can divide the total voltage, V_t, on one side of the equation and the sum of the voltages V_1 and V_2 on the other with this current and we have proved the rule.

$$R_{total} = R_1 + R_2$$

The rule for adding resistance in parallel, on the other hand, to take the *reciprocal* of the sum of the *reciprocals* of each of the resistances. This rule comes from the fact that the reciprocal of resistance, again from Ohm's law, becomes

$$V = IR$$

$$\frac{1}{R} = \frac{I}{V}$$

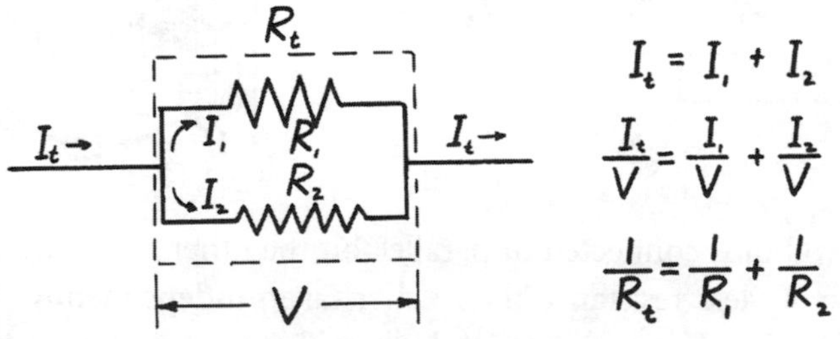

Figure 12-14

The reciprocal of the total resistance of a parallel circuit is equal to the sum of the reciprocals of the resistances in the separate branches.

In the parallel circuit, the current I_t in the total circuit is the sum of the currents I_1 and I_2 in the separate branches. Since the voltage across the total circuit is the same as across each of the resistors, we can divide I_t, I_1, and I_2 by this voltage and we will have proved the rule.

$$\frac{1}{R_{total}} = \frac{1}{R_1} + \frac{1}{R_2}$$

CHECK QUESTION

Find the equivalent resistance for each circuit.

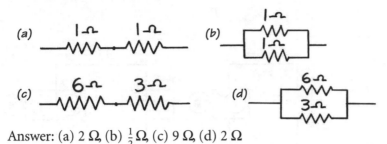

Answer: (a) 2 Ω, (b) $\frac{1}{2}$ Ω, (c) 9 Ω, (d) 2 Ω

Electric Power

When discussing work and energy, we defined power as the rate at which work is done, or the ratio of energy to time.

$$\text{Power} \equiv \frac{\text{Energy}}{\text{time}}$$

It is easy to see that the power consumed by an electrical device depends on the voltage across it. We might, for example, connect two or three flashlight bulbs in series and try to make them light exactly as bright as one bulb connected to one battery. Since voltage in a series circuit is divided between resistances, we would find it necessary to boost the voltage of our power source in proportion to the number of bulbs we wish to light. Increasing the voltage of the power source, by such means as putting more batteries in series, would bring up the brightness of the bulbs, so that each one would dissipate energy at the same rate as one bulb connected across one battery. We can see from this argument that power dissipated in an electrical circuit is proportional to the voltage across a circuit.

$$P \propto V$$

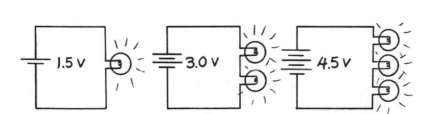

Figure 12-15

The power dissipated by light bulbs connected in a series circuit is proportional to the voltage across them.

We could have connected the bulbs in parallel, of course, and we would not have had to raise the voltage to make them all light as bright as one. The current drain on the battery would be greater, however, and the battery would wear out more quickly. Additional batteries could be connected in parallel to the original battery to keep the current drain per battery constant, but this would not be essential to our argument except to emphasize the fact that the power dissipated by an electrical circuit is also proportional to the current in it.

$$P \propto I$$

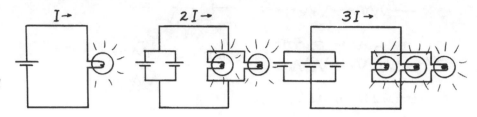

Figure 12-16

The power dissipated by light bulbs connected in parallel is proportional to the current they draw.

Since power is proportional to both voltage and current, it is also proportional to their product.

$$P \propto VI$$

It happens that by a very clever choice of units for both voltage and current our forefathers have managed to make the proportionality constant come out to be unity (1), so we can say,

$$\boxed{P = VI}$$

As mentioned earlier, voltage is formally defined as potential energy per unit charge. All our forefathers had to do to make things come out right was to define a volt as one joule per coulomb, since one ampere (the unit for current or flow rate of charge) is already defined as one coulomb per second. One volt times one ampere then works out to be one watt, which is indeed appropriate for power.

$$P = (1 \text{ V})(1 \text{ A})$$

$$= \left(1 \frac{\text{J}}{\text{C}}\right)\left(1 \frac{\text{C}}{\text{s}}\right)$$

$$= 1 \frac{\text{J}}{\text{s}}$$

$$= 1 \text{ W}$$

CHECK QUESTION

A 60-W light bulb is rated at 120 V. How much current must there be at this voltage to dissipate 60 W of power?

Answer: 0.5 A

Compound Circuits

The rules for adding resistors in series and parallel, as well as the rule for power dissipated in an electrical circuit, can be applied to a whole network of electrical components. You can find the equivalent resistance of a complex circuit by looking at the resistance of each section and adding them together. Ohm's law applied at various stages of looking at the basic parts of the circuit will tell you the current in each component as well as the voltage across it. The power rule will then yield the power dissipated by each component.

Suppose, for example, that you wish to connect two 8-Ω speakers in series to your radio so that they will produce nice loud music in your living room.[2] Suppose further that your neighbors have asked you to place small auxiliary speakers in their bedrooms so that they can listen to the high notes as well as the booming bass notes that inevitably come through the walls. You decide to connect these two auxiliary speakers, also 8-Ω resistance, in parallel, but you connect that circuit in series with a 12 Ω power resistor to reduce the output of the little speakers before connecting the whole mess in parallel with your living room speakers. You wonder how much power will be dissipated by one of the little speakers when your radio is putting out 40 W.

The first thing to do is draw a picture of the whole circuit, as in Figure 12-17a. We could then draw an equivalent circuit, Figure 12-17b, in which the two 8-Ω auxiliary speakers (in parallel) are represented as a single 4-Ω resistor.

$$\frac{1}{8\Omega} + \frac{1}{8\Omega} - \frac{1}{4\Omega}$$

Of course, this equivalent 4-Ω resistor is still connected in series with the 12-Ω power resistor. This circuit, shown in Figure 12-17b, can in turn be represented by a still simpler equivalent circuit, as in Figure 12-17c. In this figure the rule for adding resistors in series is used to represent the two 8-Ω main speakers as a single 16-Ω resistor, and the 12-Ω power resistor and the 4-Ω equivalent resistor are represented as another 16-Ω resistor. This circuit can be further simplified by looking at the two 16-Ω branches

[2]Audio speakers are rated in ohms of **impedance**. We can think of this impedance as the same thing as resistance. Impedance is actually a generalization of the concept of resistance for circuits involving alternating current.

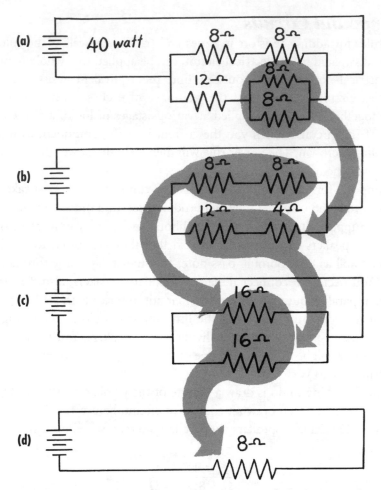

Figure 12-17

Successive equivalent circuits used to analyze an arrangement of 8-W speakers and a 12-W power resistor.

as a single 8-Ω resistor connected to the radio, since by the rule for adding resistors in parallel

$$\frac{1}{R_t} = \frac{1}{R_1} + \frac{1}{R_2}$$

$$= \frac{1}{16\Omega} + \frac{1}{16\Omega} = \frac{2}{16\Omega}$$

$$R_t = \frac{16\Omega}{2} = 8\Omega$$

We can now apply the rule for power to find the current in the speaker network. We recall that power is voltage times current:

$$P = VI$$

Although we don't know either the voltage or current at this point, we do know their ratio. Ohm's law tells us that voltage is current times resistance.

$$V = IR$$

We can use this relationship to eliminate voltage from our power relationship:

$$P = (IR)I$$
$$P = I^2R$$

Solving for current,

$$I = \sqrt{\frac{P}{R}}$$

and inserting our data, we find that the total current in the speaker circuit is 2.24 A.

$$I_t = \sqrt{\frac{40\ W}{8\Omega}}$$
$$= \sqrt{\frac{40\ V \cdot A}{8\ V/A}}$$
$$= \sqrt{5A^2}$$
$$= 2.24\ A \cong 2\ A$$

At this point, we could apply Ohm's law to find the voltage across the speaker circuit and then deduce the current through each branch, but in this special case where the branches are of equal resistance it is easy to see that the current will split up into two equal 1.12-A currents in each branch. The current in the lower branch will then again split up into two 0.56-A currents through the two 8-Ω auxiliary speakers. We can then put this result back into the general solution for power we just derived in terms of current squared and resistance to get the result we were looking for, namely the power dissipated by each little speaker:

$$P = I^2R$$
$$= (0.56\ A)^2(8\Omega)$$
$$= (0.31)(8)(A^2)\left(\frac{V}{A}\right)$$
$$= 2.5\ W$$

The power dissipated in your neighbors' bedrooms would be somewhat less than in your living room. Their speakers would dissipate 5 W of sound energy, and 15 W would go into the 12-Ω resistor—20 W altogether. From the symmetry of your speaker circuit, the other half of the 40 W from the radio will go into your main speakers.

PROBLEM SET 12

SUMMARY OF CONCEPTS

The problems in this chapter are related to electrical forces, voltage, current, resistance, and power.

The electrical force between two charges q_1 and q_2 is directly proportional to the product of the charges and inversely proportional to the square of the distance, d, between them.

$$F = k\frac{q_1 q_2}{d^2}$$

The proportionality constant k is equal to 9.0×10^9 N $\cdot$ m^2/C^2 when charge is measured in coulombs. The **coulomb**, abbreviated C, may be thought of as the charge on a very large number of protons, namely 6.25×10^{18}, which is equal but opposite to the charge on the same number of electrons.

The rate of charge flow, called **current**, and abbreviated I

$$I \equiv \frac{q}{t}$$

is measured in amperes. The **ampere** is therefore defined as a coulomb per second.

$$1 \text{ A} = \frac{1 \text{ C}}{1 \text{ s}}$$

Voltage may be thought of as the electrical pressure that induces in a wire or resistor. For many wires, the voltage, V, is directly proportional to the current, I:

$$V = IR$$

The proportionality constant R is called the **resistance** of the wire, measured in **ohms**, abbreviated Ω, which is a volt per ampere.

The resistance of two wires connected in series is the sum of the resistances of the two separate wires:

$$R_{\text{total}} = R_1 + R_2$$

The resistance of two wires connected in parallel is the reciprocal of the sum of the reciprocals of the two separate resistances.

$$\frac{1}{R_{\text{total}}} = \frac{1}{R_1} + \frac{1}{R_2}$$

Electric power, which like all power is defined as the ratio of energy to time or the rate of doing work, turns out to be in the product of voltage, V, and current, I:

$$\text{Power} = VI$$

Electric power is measured in watts (W). A watt is a joule per second.

Try to do the Examples yourself before looking at the Sample Solutions.

EXAMPLE PROBLEMS

12-1 In demonstrating the concept of electrical attraction to a friend, you inflate a rubber balloon and charge it by rubbing it against your shirt and show how the force of attraction pulls your shirt away from your body.

Given: A force of 0.12 N is produced when the balloon is held 1.0 cm from the shirt, but this force is shown to decrease when the distance is increased to 2.0 cm to simulate a heart throb. The charge on an electron is 1.60×10^{-19} C. ($k = 9.0 \times 10^9$ N · m²/C²).

Find: (a) How much electrical charge was transferred between the shirt and the balloon to produce the force observed at the smaller distance?
(b) How many electrons does this represent?
(c) What force is produced at the larger distance?

BACKGROUND

This problem illustrates the Coulomb force between electrical charges and demonstrates the vast number of electrons in a tiny amount of charge. Assume that the charges are located at two points that are separated by the given distances. We will see, in the next chapter, how to deal with charges that are distributed over two surfaces.

12-2 An automobile starter draws a heavy current and will soon discharge your battery if something is wrong with your engine and it doesn't start right away. The generator will recharge the battery, once the engine does start, by passing the same total amount of charge through the battery the other direction but at a much slower rate.

Given: Your automobile is hard to start. You operate your starter for a total of 30 s (at ten-second intervals), drawing an average of 300 A. The engine finally starts and the generator replaces this charge at the rate of 15 A.

Find: (a) How much total charge flows through the starter?
(b) How long does it take the generator to replace this charge?

BACKGROUND

This problem illustrates the definition of current. Most people encounter the unit of charge flow, the ampere, in their daily lives, but they do not usually think of it in terms of the charge transferred. One time when the actual amount of charge becomes important is when the charge must be replaced—as in recharging a battery.

12-3 An unwise workman is operating an electric drill without a three-wire grounded extension cord. Fortunately, he gets a little electric shock as a warning before he works up a sweat. The current through his body is less than the bad range, the range where the current starts to take control of his body. At that point, which we will call the "let-go" current, he finds it difficult to let go of the drill.

Given: Dirt on an insulating surface inside the electric drill permits enough leakage current to flow so that the metal casing is at 90 V relative to ground. The workman's body has a resistance of 2×10^4 Ω before he works up a sweat. The "let-go" current for this workman is 10 mA.

Find: (a) How much current flows through his body, no sweat?
(b) At what electrical resistance does the current through his body enter the bad range?

BACKGROUND

This problem is an example of Ohm's law. The voltage across the workman's body is proportional to the current passing through. Leakage currents across dirty insulators in electric appliances are a major problem. They have killed many people. Even new appliances with three-wire cords and plugs, which provide an extra ground, can be lethal if the safety ground is removed so that the plug can be used with an older outlet. People who remove that third prong probably don't understand the magnitude of the problem.

12-4 The ignition coil in an older automobile is connected in series with the points, which act like a switch to interrupt the current through the coil at the proper time. Even when closed, however, the points have considerable resistance and are usually made of the hard metal tungsten to resist pitting due to sparks.

Given: In a twelve-volt auto ignition system, the coil is designed to pass 3.0 A of current when the voltage across it is 10 V, with the assumption that the voltage drop across the points will be about 2 V.

Find: (a) What is the resistance of the points?
(b) What is the resistance of the coil?
(c) What is the total resistance of the circuit, including coil and points?

BACKGROUND

This problem is an example of two electrical resistors connected in a series circuit. The electrical resistance of ignition points increases with age as the points become pitted by the tiny electric spark that takes place each time they are opened. The points should be changed every 20,000 miles or so for this reason.

12-5 A light fixture holds two light bulbs and is wired so that they are connected in parallel to the line voltage.

Given: A sixty watt lamp and a seventy-five-watt lamp are connected in parallel to 115 V. The sixty watt lamp passes a current of 0.52 A, and the seventy-five-watt lamp passes 0.65 A.

Find: (a) How much current is passed by both lamps together?
(b) What is the resistance of the sixty-watt lamp?
(c) What is the resistance of the seventy-five-watt lamp?
(d) What is the total resistance of both lamps together?

BACKGROUND

This problem illustrates the rule for adding resistances in parallel. Instead of the voltage adding as for a series circuit, the current flowing through the two parallel paths is the sum of the current through each separate path. When you plug appliances into the wall, they are connected in parallel.

12-6 You decide to have a coffee party in a dormitory room and invite as many friends as you can without blowing out a fuse when they all plug in their immersion heaters at the same time. Everyone has the same kind of immersion heater for boiling water because the campus bookstore only handles one kind.

Given: You look on your immersion heater and discover that it is rated for 300 W at 115 V. The fuse panel shows that the wall outlets in your room can handle 15 A altogether.

Find: (a) How much current does your immersion heater draw?
(b) How many friends can you invite?

BACKGROUND

This problem is an example of electric power expressed in terms of voltage and current. From the power rating of an electric appliance, you can figure out the current it draws and estimate the number of appliances you can use at the same time on a particular circuit.

12-7 A Wheatstone bridge is a common circuit used in electrical measurements. In this bridge, two branches, each consisting of two series-connected resistors, are connected in parallel. This array is then connected in series to a current limiting resistor.

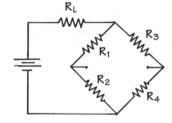

Given: $R_1 = 6\ \Omega$, $R_2 = 8\ \Omega$, $R_3 = 12\ \Omega$, $R_4 = 16\ \Omega$, $R_L = 10\ \Omega$.

Find: (a) What is the resistance of the branch containing R_1 and R_2?
(b) What is the resistance of the branch containing R_3 and R_4?
(c) What is the resistance of the whole circuit?

BACKGROUND

This problem illustrates a compound electrical circuit in which resistors are connected in series and parallel.

In the usual application, the Wheatstone bridge is balanced by adjusting the values of resistances in the two arms so that there is no voltage between the two corners, which represent the junctions between R_1 and R_2 on the one hand and between R_3 and R_4 on the other hand. It is easy to show that the ratio of R_1 to R_2 is equal to the ratio of R_3 to R_4 if there is a null reading, or zero voltage, between these two junctions. This circuit can be used to measure an unknown resistance by putting it into a bridge with known resistances or to detect the change of resistance of a temperature-, pressure-, or light-sensitive resistor.

Solutions to EXAMPLE PROBLEMS

Given:

A force of 0.12 N is produced when the balloon is held 1.0 cm from the shirt, but this force is shown to decrease when the distance is increased to 2.0 cm to simulate a heart throb. The charge on an electron is 1.60×10^{-19} C.
($k = 9.0 \times 10^9 \cdot N \cdot m^2/C^2$).

Find:

(a) How much electrical charge was transferred between the shirt and the balloon to produce the force observed at the smaller distance?

(b) How many electrons does this represent?

(c) What force is produced at the larger distance?

(a) $F = k\dfrac{q_1 q_2}{r^2}$

$= k\dfrac{q_1(-q_1)}{r^2}$

$= -k\dfrac{q_1^2}{r^2}$

$q = r\sqrt{\dfrac{-F}{k}}$

$= (1.0 \ cm)\sqrt{\dfrac{-(-0.12 \ N)}{(9.0 \times 10^9 \ Nm^2/C^2)}}$

$= \boxed{3.7 \times 10^{-8} C}$

(b) $q = 3.7 \times 10^{-8} C\left(\dfrac{1 \ electron}{1.60 \times 10^{-19} C}\right)$

$= \boxed{2.3 \times 10^{11}}$

(c) $F = -k\dfrac{q_1^2}{r^2}$

$= \left(9.0 \times 10^9 \ \dfrac{Nm^2}{C^2}\right)\left(-\dfrac{(3.7 \times 10^{-8})^2}{(2.0 \ cm)^2}\right)$

$= \boxed{-0.03 \ N}$

DISCUSSION

The force between two charges is directly proportional to the product of the charges and inversely proportional to the square of the distance r between them. Assuming that the two charges are equal and opposite, since the positive charge on the shirt was formed by transferring a certain number of electrons from the balloon to the shirt, the product of the two charges is $-q^2$. Part (a) is worked out by solving for q and inserting the data.

Part (b) is just a conversion factor to illustrate the point that it takes a huge number of electrons to produce this little eighth newton force at a distance of a centimeter. The number of electrons turns out to be nearly a quarter million times a million electrons.

Part (c) shows that even this force falls off to a few hundredths of a newton at two centimeters, thus illustrating the inverse square law for the Coulomb force.

Given:
Your automobile is hard to start. You operate your starter for a total of 30 s (at ten-second intervals), drawing an average of 300 A. The engine finally starts and the generator replaces this charge at the rate of 15 A.

Find:
(a) How much total charge flows through the starter?

(b) How long does it take the generator to replace this charge?

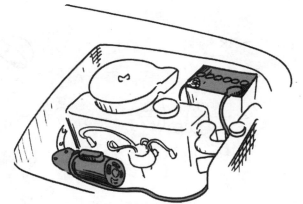

(a) $I = \dfrac{q}{t}$

$q = It$

$= (300 \text{ A})(30 \text{ s})\left(\dfrac{C/s}{A}\right)$

$= \boxed{9 \times 10^3 \text{ C}}$

(b) $I = \dfrac{q}{t}$

$t = \dfrac{q}{I}$

$= \dfrac{(9 \times 10^3 \text{ C})}{(15 \text{ A})}\left(\dfrac{A}{C/s}\right)\left(\dfrac{min}{60s}\right)$

$= \boxed{10 \text{ min}}$

DISCUSSION

The amount of charge drawn by the starter motor in 30 s is found in part (a) by solving the definition of current for charge and inserting the given current and time. The units are straightened out by using a conversion factor based on the fact that an ampere of current is defined as a coulomb of charge per second.

Part (b) is found by solving the definition of current for time. The result is that it takes 20 times as long to charge the battery as it took to discharge it, since the generator only produces one-twentieth of the amount of current that the starter motor draws.

SAMPLE SOLUTION 12-3

Given:

Dirt on an insulating surface inside the electric drill permits enough leakage current to flow so that the metal casing is at 90 V relative to ground. The workman's body has a resistance of $2 \times 10^4 \, \Omega$ before he works up a sweat. The "let-go" current for this workman is 10 mA.

Find:

(a) How much current flows through his body, no sweat?

(b) At what electrical resistance does the current through his body enter the bad range?

(a) $V = IR$

$I = \dfrac{V}{R}$

$= \dfrac{(90 \text{ V})}{(2 \times 10^4 \Omega)}\left(\dfrac{\Omega}{V/A}\right)$

$= 4.5 \times 10^{-3} A = \boxed{4.5 \text{ mA}}$

(b) $V = IR$

$R = \dfrac{V}{I}$

$= \dfrac{(90 \text{ V})}{(10 \text{ mA})}\left(\dfrac{mA}{10^{-3} A}\right)\dfrac{\Omega}{V/A}$

$= \boxed{9 \times 10^3 \Omega}$

DISCUSSION

The current is found in part (a) by solving Ohm's law and inserting the data. The resulting current, 4.5 mA, will give a good stout shock. Solving for resistance, as in part (b), gives a value several times greater than the actual resistance of a person's body after they have become moist with perspiration. That third wire, the grounding wire, in a three-wire extension cord is a life saver.

The ignition coil in an older automobile is connected in series with the points, which act like a switch to interrupt the current through the coil at the proper time. Even when closed, however, the points have considerable resistance and are usually made of the hard metal tungsten to resist pitting due to sparks.

(a) $V = IR$

$$R_p = \frac{V_p}{I}$$

$$= \frac{(2\ V)}{(3.0\ A\ )}\left(\frac{\Omega}{V/A}\right)$$

$$= \boxed{0.67\ \Omega}$$

3.0 A →

$V_{points} = 2\ V$ $R_{points} = ?$

12 V

$V_{coil} = 10\ V$ $R_{coil} = ?$

Given:

In a twelve-volt auto ignition system, the coil is designed to pass 3.0 A of current when the voltage across it is 10 V, with the assumption that the voltage drop across the points will be about 2 V.

Find:

(a) What is the resistance of the points?

(b) What is the resistance of the coil?

(c) What is the total resistance of the circuit, including coil and points?

(b) $R_c = \dfrac{V_c}{I}$

$$= \frac{(10\ V)}{(3.0\ A\ )}\left(\frac{\Omega}{V/A}\right) = \boxed{3.3\ \Omega}$$

(c) $R_{total} = R_p + R_c$

$$= (0.67\ \Omega) + (3.3\ \Omega) = \boxed{4.0\ \Omega}$$

DISCUSSION

Resistance in parts (a) and (b) is found from the voltage and current, using Ohm's law. The resistance of the series circuit can be found in part (c) by using the rule for adding resistance in series. The total resistance of two resistors connected in series is the sum of the two resistances. An alternative approach would be to use the total voltage and current in the same general solution as found from part (a) and part (b).

$$R_{total} = \frac{V_{total}}{I} = \frac{12\ V}{3\ A} = 4\ \Omega$$

The result is the same.

SAMPLE SOLUTION 12-5

Given:

A sixty watt lamp and a seventy-five-watt lamp are connected in parallel to 115 V. The sixty watt lamp passes a current of 0.52 A, and the seventy-five-watt lamp passes 0.65 A.

Find:

(a) How much current is passed by both lamps together?

(b) What is the resistance of the sixty-watt lamp?

(c) What is the resistance of the seventy-five-watt lamp?

(d) What is the total resistance of both lamps together?

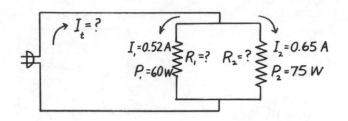

(a) $I_t = I_1 + I_2$

$= (0.52 \text{ A}) + (0.65 \text{ A})$

$= \boxed{1.17 \text{ A}}$

(b) $R_1 = \dfrac{V_1}{I_1}$

$= \dfrac{(115 \text{ V})}{(0.52 \text{A})}$

$= \boxed{221 \ \Omega}$

(c) $R_2 = \dfrac{V_2}{I_2}$

$= \dfrac{(115 \text{ V})}{(0.65 \text{ A})}$

$= \boxed{177 \ \Omega}$

(d) $\dfrac{1}{R_t} = \dfrac{1}{R_1} + \dfrac{1}{R_2}$

$= \dfrac{1}{(221 \ \Omega)} + \dfrac{1}{(177 \ \Omega)}$

$R_t = \boxed{98 \ \Omega}$

DISCUSSION

The current in part (a) is the sum of the currents in the separate branches. The resistance of the two separate paths can be calculated from the voltage and current, using Ohm's law as shown in parts (b) and (c).

The total resistance of the combined parallel circuit can be found from the rule for adding resistors in parallel. The reciprocal of the total resistance is the sum of the reciprocals of the resistances of the branches. Instead of inserting the data as the second step, as shown in part (d), a general solution could be obtained by solving for R_{total} instead of its reciprocal. The result is the sum divided by the product of the branch resistances. An alternative to using the parallel resistance rule would be to calculate the resistance from Ohm's law, using the total current as found in part (a).

$$R_{total} = \frac{V}{I_{total}} = \frac{115 \text{ V}}{1.17 \text{ A}} = 98 \ \Omega$$

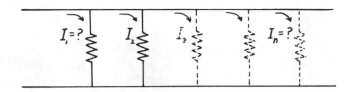

Given:
You look on your immersion heater and discover that it is rated for 300 W at 115 V. The fuse panel shows that the wall outlets in your room can handle 15 A altogether.

Find:
(a) How much current does your immersion heater draw?

(b) How many friends can you invite?

(a) $\quad P = VI$

$$I_1 = \frac{P_1}{V}$$

$$= \frac{(300\ W)}{(115\ V)}\left(\frac{V \cdot A}{W}\right)$$

$$= \boxed{2.61\ A}$$

(b) $\quad I_t = I_1 + I_2 + \ldots + I_n$

$\quad I_t = nI_1 \qquad$ since $I_1 = I_2 = I_3 = \ldots = I_n$

$$n = \frac{I_t}{I_1}$$

$$= \frac{(15\ A)}{(2.61\ A)} = 5.75$$

$$= \boxed{5}$$

DISCUSSION

The electric power rule states that power is the product of voltage and current. Solving for current, as in part (a), the power is divided by the voltage and the units are handled with a conversion factor based on the fact that a watt is a volt times an amp.

The total current in a parallel circuit is the sum of the currents in each branch. This parallel circuit rule is used in part (b) to calculate the number of branches, each carrying the same current, which will add up to current limit of the fuse. The total current is just the number of branches times the average current flowing in each branch. Solving for the number of branches and inserting the data shows that something over five heaters could be connected. The result must be rounded to the lowest integer, however, since six heaters will blow the fuse.

Given:
$R_1 = 6\,\Omega$, $R_2 = 8\,\Omega$, $R_3 = 12\,\Omega$,
$R_4 = 16\,\Omega$, $R_L = 10\,\Omega$,

Find:

(a) What is the resistance of the branch containing R_1 and R_2?

(b) What is the resistance of the branch containing R_3 and R_4?

(c) What is the resistance of the whole circuit?

(a) $R_{1,2} = R_1 + R_2$

$\quad = (6\,\Omega) + (8\,\Omega)$

$\quad = \boxed{14\,\Omega}$

(b) $R_{3,4} = R_3 + R_4$

$\quad = (12\,\Omega) + (16\,\Omega)$

$\quad = \boxed{28\,\Omega}$

(c) $R_t = R_L + R_{1,2,3,4}$

$\quad$ where $\dfrac{1}{R_{1,2,3,4}} = \dfrac{1}{R_{1,2}} + \dfrac{1}{R_{3,4}}$

$\quad R_t = R_L + \dfrac{1}{\frac{1}{R_{1,2}} + \frac{1}{R_{3,4}}}$

$\quad = 10\,\Omega + \dfrac{1}{\frac{1}{14\,\Omega} + \frac{1}{28\,\Omega}}$

$\quad = \boxed{19\,\Omega}$

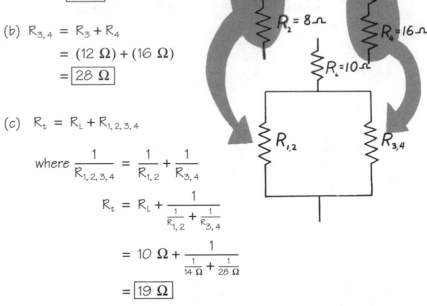

DISCUSSION

The resistances of the two branches are found by using the series resistance rule, as shown in parts (a) and (b).

These two combined resistances are used in a simplified parallel circuit having as its two branches the results of parts (a) and (b). The reciprocal of the total resistance representing all four resistances, R_1, R_2, R_3, and R_4, is the sum of the reciprocals of the two branches according to the parallel resistance rule. Then this resistance, which is the reciprocal of the sum of the reciprocals, is added to the load resistance, R_L, by the series resistance rule.

Even the most complicated circuit can be handled with the series and parallel resistance rule, since those are the only two ways of connecting two resistors together. Just start small and build up like building blocks.

ESSENTIAL PROBLEMS

Be sure you can do the Example Problems (without peeking at the Sample Solutions) before working these Essential Problems.

12-8 Two fog droplets have a close encounter on a typical cold, foggy, summer evening in downtown San Francisco. They never get close enough to touch because of those invigorating positive ions folks sometimes talk about. Actually, the positive ions are few and far between. Of the hundred or so millions of water molecules in the droplet, ever so few have a positive charge.

Given: The droplets, both having the same positive charge, are repelled from each other by a force of 2.36×10^{-16} N when they are 41 μm apart, but this force decreases when they are 200 μm apart. The charge on an electron is 1.60×10^{-19} C. ($k = 9.0 \times 10^9$ N · m²/C²)

Find: (a) How much electrical charge is on each droplet?
(b) How many extra protons, unmatched by an electron, are in each droplet?
(c) What force is produced at the larger distance?

12-9 Automobile batteries are rated in terms of the total amount of charge, the charge that will completely discharge them if it flows from one terminal to the other. The unit used in the auto parts trade to measure battery capacity is the amp-hour, which is over three thousand times our friend, the SI coulomb.

Given: A typical battery is supposed to deliver 60 A for 3600 s before becoming completely discharged. A typical home battery charger delivers 3.0 A when charging such a battery.

Find: (a) How many coulombs of charge can the fully charged battery cause to flow before it becomes completely discharged?
(b) How long would it take a home battery charger to recharge a completely discharged battery?

12-10 Jerry Hurt has ungrounded wall sockets in his Victorian flat. That makes it difficult to use his new electric drill because the three-wire plug won't fit in the socket. Rather than upgrade his house wiring he is tempted to cut off the third grounding prong, a common and all too frequently fatal error in judgment.

Given: Leakage current brings the metal casing of Jerry's drill to 170 V relative to ground. Jerry has a resistance of $1.6 \times 10^4 \, \Omega$ before he works up a good sweat. Jerry's heart will stop if the current through his body gets up to 25 mA.

Find: (a) How much current flows through his body, no sweat?

(b) At what electrical resistance will the current through his body stop Jerry's heart?

12-11 Someone tells you that you can magnetize a steel nail by winding insulated wire around it and connecting the stripped ends of the wire to a flashlight battery. You have a flashlight, some wire and a nail, so you decide to try this. To limit the current and avoid running the battery down too quickly, you connect the wire, coiled around the nail, in series with the flashlight bulb.

Given: When the current through the circuit is 55 mA, the voltage across the coil is 0.35 V and the voltage across the flashlight bulb is 2.45 V.

Find: (a) What is the resistance of the coil?

(b) What is the resistance of the flashlight bulb?

(c) What is the combined resistance of the coil and the flashlight bulb?

12-12 High beams in many automobile headlights are separate lamp bulbs that can be turned on in addition to the low beam lamps.

Given: A high beam lamp and a low beam lamp are connected in parallel to the 13.5 V available in an automobile electrical system. The low beam lamp passes a current of 0.60 A, and the high beam lamp passes 0.75 A.

Find: (a) How much current is passed by both lamps together?

(b) What is the resistance of the low beam lamp?

(c) What is the resistance of the high beam lamp?

(d) What is the total resistance of both lamps together?

12-13 Mike Yu has been asked to bring coffee for a graduation party and decides to make it ahead of time. He fills a large number of flasks, borrowed from the chemistry department, with coffee and plans to reheat them at the party with hot plates borrowed from the physics department undergraduate lab.

Given: The hot plates are rated for 240 W at 120 V. The fuse panel shows that the wall outlet where coffee is to be served can handle 20 A.

Find: (a) How much current does one hot plate draw?
(b) How many hot plates can Mike operate from this outlet at a time?

12-14 One way to improve that old monaural radio in your ancient car is to add speakers from junk radios and tape recorders. Clever series and parallel combinations provide an effective resistance close to that of the original speaker. Two speakers pumping out good music for folks in the front seat are connected in series, as are two speakers for the back seat. Front and back seat pairs are then connected in parallel. This speaker circuit works so well that you add a power resistor in series with the whole thing to keep your brother-in-law from playing it too loud.

Given: The front seat speakers are 7 Ω and 8 Ω in series. The back seat speakers are 6 Ω and 4 Ω also in series. Both pairs are connected in parallel. The circuit is then connected in series with 2 Ω power resistor.

Find: (a) What is the effective resistance of the front seat branch?
(b) What is the effective resistance of the back seat branch?
(c) What is the effective resistance of the whole circuit?

MORE INTERESTING PROBLEMS

12-15 Hellen Angel has built an electric motorcycle because of her concern about air pollution. Hellen counts her watts, but her friends talk horsepower.

Given: Hellen's motorcycle runs on a 18 V battery that can provide 80 A for 3600 s without hurting the battery. Her motor draws current at 32 A when accelerating. She knows that 1 kilowatt = 1.34 horsepower.

Find: (a) What is the total charge that the battery can provide without hurting it?
(b) For what total time can Hellen accelerate her motorcycle on one charging?
(c) What is the horsepower equivalent of her motor?
(d) How many joules of energy are available from her battery?

12-16 Electric golf carts run on deep cycle batteries designed to be discharged to a greater extent than in gas car batteries. You should not use more than ten percent of the total charge available in a typical car battery or you will shorten its life. The design of the electrodes in a gas car battery is for high current and light weight. Golf cart batteries, on the other hand, can be discharged fifty percent without hurting them.

Given: A golf cart operates on a 24 V battery that can intermittently provide 65 A for 4200 s and still be recharged like new. The golf cart motor draws current at 22 A when being driven between holes. (1 kilowatt = 1.34 horsepower).

Find: (a) What is the total charge that the battery can provide without hurting it?
(b) For what total time can the cart be driven between rechargings?
(c) What is the horsepower equivalent of the motor?
(d) How many joules of energy are available without hurting the battery?

12-17 Chlorine gas is manufactured by passing an electric current through molten salt in an electrolysis cell. It takes two electrons to produce one molecule of this diatomic gas.

Given: A current of 125 A is passed through an electrolysis cell in the production of chlorine. One mole of any gas occupies 22.4 liters at atmospheric pressure and contain Avogadro's number, 6.02×10^{23}, of molecules. It takes 3.20×10^{-19} C of charge to produce one molecule of chlorine.

Find: (a) How many molecules are in 1 liter of chlorine at atmospheric pressure?

(b) How much electrical charge must pass through the electrolysis cell to produce 1 liter of gas?

(c) How much time does it take to produce 1 liter of chlorine in this cell?

12-18 Do not stick the bare ends of an electrical cord into an ordinary dill pickle and then plug them into the house current. This is a dangerous, crazy thing to do and it probably produces poisonous pickle gas. We don't know. That's chemistry. We are not responsible for chemistry. When you don't do this, also *please* don't hold the pickle in your hand and stand in a puddle of water. That's physics. You would be killed—electrocuted to death—before you smelled the pickle gas. DON'T DO IT.

Given: A current of 20 A producing pickle gas would blow the fuse. Like all gasses, one mole of this gas would occupy 22.4 liters at atmospheric pressure and contain Avogadro's number, 6.02×10^{23}, of molecules. It reportedly takes 6.40×10^{-19} C of charge to produce one molecule of dangerous poisonous pickle gas[3].

Find: (a) How many molecules are in 1 liter of pickle gas at atmospheric pressure?

(b) How much electrical charge must reportedly pass through the pickle to produce 1 liter of pickle gas?

(c) If the fuse did not blow, how much time would it take to produce 1 liter of dangerous poisonous pickle gas?

[3]These reports are only by people who have never tried it. People who have tried it are not talking. Comedy magicians Pen and Teller, in their delightful book *How To Play With Your Food* (Penn Jillette and Teller, Villard Books, New York, 1992), warn you not to build The Incredibly Dangerous Glowing Pickle Machine if "you don't know everything about electricity." They claim that it will glow a ghostly yellow and say it is the most beautifully goofy science thing you'll ever see. We know a physics professor who always does the glowing pickle at his annual party for physics majors and graduate students. Each year he seems freshly surprised when the house lights go off because the pickle blows a fuse.

12-19 Some people believe a deterrent to burglary is to leave their front porch lights on all the time. Other people believe the benefits are outweighed by the cost of the electrical energy. Many astronomers want the lights off because they hate light pollution and even question the crime deterrence of light.

Given: Your front porch light contains a 60-W bulb. You look on your electric bill and find that you were charged $117.62 for 913 "KWH" of electrical energy one 30-day month, where a "KWH" is a kW times an hour.

Find: (a) How much did 1 "KWH" of energy cost you that month?
(b) How many kilowatt-hours would a 60-W bulb use in a month?
(c) How much would it cost you to leave the bulb on for the whole month?

12-20 You can save energy by using more efficient light bulbs. In the long run, you also save money. Replacing an ordinary light bulb with a compact fluorescent bulb is a big initial investment, but the fluorescent bulb will last about thirteen times as long and consume less than a third of the energy for the same amount of light.

Given: A 20 W compact fluorescent bulb is advertised as a replacement for a standard seventy-five watt bulb. You look at an old electric bill and find that you were charged $104.96 for 827 "KWH" of electrical energy one 30 day month, where a "KWH" is a kW times an hour.

Find: (a) How much did 1 "KWH" of energy cost you that month?
(b) How many kilowatt-hours would a compact florescent bulb use in a month?
(c) How much would it cost you to leave the bulb on for the whole month?

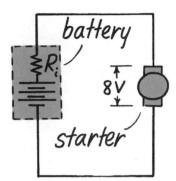

12-21 Your headlights will dim if you leave them on while you start your car. That is internal resistance. The voltage your battery will deliver, without the lights or other load, is called the emf of the battery. The terminal voltage of the battery with a load is reduced because of the internal resistance. At least that is one way of looking at it. As an automobile battery runs down, to a first approximation the emf remains constant while the internal resistance increases.[4]

Given: The voltage of a particular automobile battery drops from 12 to 8 V while the starter is drawing 200 A of current.

Find: (a) What is the effective resistance of the starter motor?
(b) What is the internal resistance of the battery?

12-22 Professor Marvin Dejong explains battery operation in terms of academic discipline. Batteries have something in them that produces voltage. We call it emf, short for electromotive force, but Marvin has penetrated this mysterious fifth force of nature and refers it as the "chemical force." He assumes it is static. We all know chemistry is not static, but it is a useful approximation. We call it the *Marvin approximation*.[5] Batteries also have an internal resistance that increases as time goes on.

Given: The voltage of a particular flashlight battery drops from 3.1 V to 2.8 V when lighting a lamp that draws a current of 2 A.

Find: (a) What is the effective resistance of the lamp?
(b) What is the internal resistance of the battery?

[4]This mental model, with constant emf and internal resistance that increases as the battery runs down, is good for explaining how the battery can have enough charge to operate your car radio but be dead when it comes to starting the engine. It is, however, only approximately true. The emf of a lead-acid battery does change somewhat as a function of the state of charge.

[5]Marvin Dejong, Physics Department, School of the Ozarks, Point Lookout, MO 65726. A prolific author and highly respected physics teacher. Call him at (417) 334-6411 and ask him to explain his delightful "little man" theory.

12-23 Ellen Wonderful likes to watch TV while fixing breakfast. She has a TV set and toaster plugged into the same wall outlet. Appliances plugged into the same outlet are connected in parallel.

Given: Ellen reads labels. She notices that the toaster draws 10.5 A. She becomes concerned that she has been operating it while the TV set is on. She therefore reads the label on the TV and finds that it is rated at 120 W. Both the TV and the toaster are rated for 120 V.

Find: (a) How much power does the toaster use?
(b) How much current does the TV set use?
(c) What is the current drain of the toaster and TV set together?

12-24 You need to buy Martha a present for her tenth birthday. If, instead of at the toy store, you spend your cash in the elecronics store, your real gift will be the quality time spent in showing her how to do safe and simple electrical experiments. First you show how easy it is to light up an LED by connecting it in series with a current limiting resistor. Then you connect it in parallel with an incandescent lamp.

Given: You pick a resistor such that the current through the LED is limited to 25 mA. The incandescent lamp produces 0.60 W. Both are powered by a 9.0 V battery.

Find: (a) How much power does the LED and resistor use?
(b) How much current does the incandescent lamp use?
(c) What is the current drain on the battery?

12-25 To demonstrate the magnetic field around a current-carrying wire, a length of fourteen gauge house wire is connected to a low voltage power supply with the voltage turned down. Then the voltage is turned up until the wire starts to get hot.

Given: It requires 1.14 V to push 30 A through a length of ordinary house wire at room temperature. But then the wire heats up, and its resistance increases by 4×10^{-3} Ω

Find: (a) What is the resistance of the wire at room temperature?
(b) What is the resistance of the wire when it gets hot?
(c) What will be the current passing through the wire when hot assuming the voltage remains constant?

12-26 Your friend Martha also needs some insulated hook-up wire, 22 or 24 gauge, from the electronics store. Wrap a couple of meters of it into a coil, and then suspend a couple of paper clips in the middle of the coil.[6] Strip the insulation from the ends of the wire and attach them to a battery. The paper clips will become magnetized and will jump apart. Very cute! Then put a refrigerator magnet under the coil, and the coil will jump. Reverse the connections and ....Well, you must try it. As the wire heats up, of course, the resistance will increase. See if you can tell.

Given: It requires 2.84 V to push 8.5 A through your coil at room temperature. When the wire heats up, its resistance increases by 4×10^{-2} Ω

Find: (a) What is the resistance of the wire at room temperature?
(b) What is the resistance of the wire after it heats up?
(c) What will be the current passing through the wire after it heats up assuming the voltage remains constant?

[6] *Safe and Simple Electrical Experiments*, Rudolf Graf, Dover, New York, 1964. Graf suggests using a broom handle as a form.

ANSWERS

Solutions to **ESSENTIAL PROBLEMS**

12-8   (a) 6.6×10^{-18} C; (b) 41 protons; (c) 9.9×10^{-18} N

12-9   (a) 2.16×10^5 C; (b) 20 hours

12-10   (a) 10.6 mA; (b) 6.8 k Ω

12-11   (a) 6.4 Ω; (b) 44.5 Ω; (c) 50.9 Ω

12-12   (a) 1.35 A; (b) 22.5 $\Omega \cong 23$ Ω;(c) 18.0 Ω; (d) 10.0 Ω

12-13   (a) 2.0 A; (b) 10

12-14   (a) 15 Ω; (b) 10 Ω; (c) 8 Ω

Solutions to **MORE INTERESTING PROBLEMS,** *odd answers*

12-15   (a) 2.88×10^5 C; (b) 150 min; (c) 0.77 hp; (d) 5.2×10^6 J

12-17   (a) 2.69×10^{22} molecules; (b) 8600 C $\cong 8.60 \times 10^3$ C; (c) 68.8 s

12-19   (a) 12.9¢ (= 0.129$); (b) 43.2 kW · hr; (c) $5.56

12.21   (a) 4×10^{-2} Ω; (b) 2×10^{-2} Ω

12-23   (a) 1.26 kW; (b) 1.0 A; (c) 11.5 A

12-25   (a) 38 m Ω; (b) 42 m Ω; (c) 27 A

CHAPTER 13 *Electro-magnetic Interactions*

Electromagnetic Fields In The Age Of Steam

WHEN THE POWER GRID GOES DOWN on a summer day in California, society suspends normal operations. Stores and banks close. The whole population of the central valley get in their cars. Without air conditioning, they must abandon their homes and head for the hills. In parts of California, it is not possible to survive without air conditioning. Life in parts of California depends on electrical power.

The technology of electrical power is largely a twentieth century accomplishment, but that technology rests on a nineteenth century understanding of the relationship between electricity and magnetism. That understanding was put into place by people who lived in the age of steam. The cutting edge of their technology was ruled by fluid pressure and fluid energy.

We have already seen the influence of the fluid model in our present thinking about voltage, current, and resistance. To go further, to understand the power grid, we need a somewhat more sophisticated idea that embraces both electricity and magnetism. It is through that idea, the idea of fields, that electricity and magnetism interact.

Near the beginning of the nineteenth century, the great English chemist Sir Humphry Davy went to visit Count Alessandro Volta, the Italian physician, who had invented the electric battery. In Sir Humphry's hands, the Italian physician's battery had become a powerful research tool, and the two noble gentlemen had much to discuss.

Sir Humphry brought along his lab assistant, Michael Faraday, mostly to deal with his suitcases. As it turned out, it was the lab assistant who became the central figure in the understanding of electricity and magnetism. It was

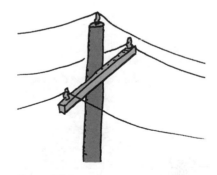

Figure 13-1

Life in parts of California depends on electrical power.

Faraday who came up with the field concept, and it was he who remained as the central figure as he and his colleagues built the operating theory of the modern generator.

Without benefit of formal education, the young Michael Faraday had learned a great deal about scientific inquiry from the great English chemist, but Sir Humphery had seen fit to teach him very little about mathematics. Although field theory had been put into a beautifully elegant mathematical form by the end of the nineteenth century, that is not how most early pioneers in electrical power thought about it. They thought like Faraday in terms of a conceptual understanding of lines of force. While you may find these ideas sophisticated and well adapted to a mathematical treatment, it may help to remember that the basic concepts are, at their core, non-mathematical.

Since our twentieth century understanding of such things as light and radio waves is built on the elegantly mathematical form of field theory, we will take a conceptual look at where the theory leads us at the end of this chapter. Along the way, we will use algebra as a tool in our understanding, but that should be to simplify—not to complicate. In trying to think like Faraday, we should keep in mind that this non-mathematical genius of the nineteenth century lived in the age of steam. Fluid theory guided his thoughts as it may guide ours.

The Origin of the Field Concept

The person to first see a relationship between electricity and magnetism was the Danish professor Hans Christian Oersted in 1820. The story goes that Professor Oersted was doing a physics demonstration showing his class that electricity and magnetism, while clearly similar, are totally independent of each other. That was what people believed until that moment.

The demonstration failed! Oersted saw that a large current through a wire near a compass needle tends to turn the needle perpendicular to the wire.

Before then, most electrical phenomena were related to electrostatics. Count Volta's battery made currents flow through wires, but they were relatively weak currents. As people like Davy improved battery technology, larger currents were available. These new batteries made possible Oersted's important observation.

Michael Faraday, by then a young journeyman in the field of chemical research, was intrigued by the Danish physics professor's report. Faraday, with his training in chemistry, thought of this phenomenon in terms of a reversible reaction. He set out to do the reverse. He started looking for a way to get electricity from magnetism.

Figure 13-2

A large current through a wire near a compass needle tends to turn the needle perpendicular to the wire.

HIGHLIGHT
Michael Faraday

Apprenticed as a bookbinder, Michael Faraday (1791–1867) started his scientific career as a lab assistant to Sir Humphry Davy (1778–1829), one of the most prolific chemists of all time.

There is a story that Sir Humphry, who was Director of the Royal Institution in London at the time, was browsing in a bookshop one day and came across a beautifully bound copy of notes, the subject of which was his own popular lectures given for the public. When he learned that the notes had been prepared by the apprentice in the shop, he hired the young man on the spot.

While it makes a good story, the truth is that young Michael Faraday so valued his notes on the Davy lectures that he would never have offered them for sale. It is the case that his notes were so impressive that Sir Humphry hired him in spite of his lack of scientific training and, in effect, discovered him. Indeed, many of Sir Humphry's friends and admirers came to believe that of all the great chemist's many discoveries, Michael Faraday may have been the most important of all.

In the tradition of the sorcerer's apprentice, Faraday became at least his master's equal, and succeeded him as director of the Royal Institution. A person of great intelligence who lived well and modestly, he gained the highest respect of all who knew him.[1] Today, we honor Faraday by naming the unit of capacitance after him.

[1] Faraday's modesty, amiability, and intelligence were such that, when Sir Humphrey would take him traveling, treating him as little more than a servant, people were impressed as much with one as the other. Said one, "We admired Davy, we *loved* Faraday." See the biography by Sylvanus Thompson, *Michael Faraday: His Life and Work*

Much was already known about electricity. Over four decades earlier, in about 1777, Charles Coulomb had established the inverse square law for electrical force, which can be written, as in the last chapter,

$$F = k\frac{q_1 q_2}{d^2}$$

As we saw in the last chapter, Coulomb's law was developed in analogy to Newton's theory of universal gravitation developed in Chapter 5.

$$F = G\frac{m_1 m_2}{d^2}$$

Faraday was familiar with these relationships on a conceptual level, but his lack of mathematical training led him to give little weight to their algebraic form. For him, they seemed to leave the phenomenon itself simply unexplained. For Faraday, the key to understanding the phenomenon lay in understanding the geometry of the space *between* the objects upon which the forces act, whether they be charges, masses, or magnetic poles.

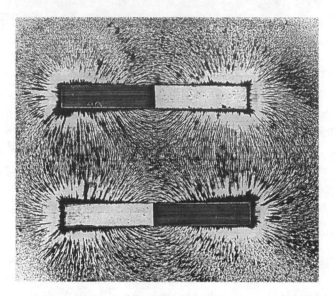

Figure 13-3

If the magnets have opposite poles they attract one another. The lines of force go from one magnet to another.

The space between two magnets, which either attract or repel each other depending on their orientation, looks and feels empty. Sprinkle iron filings between the two magnets, however, and the tiny pieces of iron will align themselves into lines, in very predictable patterns. If the magnets attract one another, the lines go from one magnet to the other. If the magnets repel one another, the lines seem to push on each other.

Faraday called these patterns "lines of force," and he generalized the concept to the space between electrical charges, as well as magnets. Using

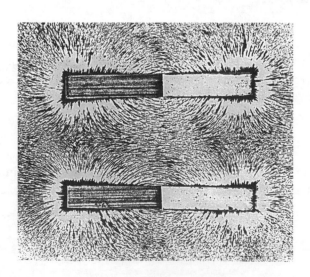

Figure 13-4

If the magnets have the same poles they repel one another, and the lines of force seem to push against one another.

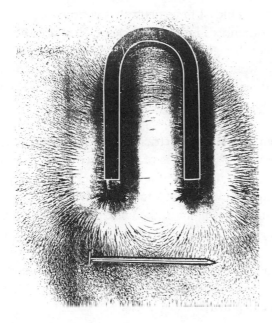

Figure 13-5

When you sprinkle iron fillings around magnets they align themselves into what Faraday called "lines of force." Here we see a horseshoe magnet and nail with iron filings. This was NOT done on a copy machine when no one was looking. We deny it.

this mental tool, Faraday was able to understand and predict both qualitatively and quantitatively the results of many experiments. Ultimately, this mental tool helped him find a way to get electricity from magnetism.

Nearly fifty years later, James Clerk Maxwell (1831–1879), possessed of a fully trained and powerful mathematical mind, turned his attention to electricity and magnetism. Before he started, however, he resolved to read no mathematical treatment on the subject until he had read Faraday's work. When he finished, as we shall see at the end of this chapter, Maxwell had turned Faraday's conceptual tool into a set of four equations that form the foundation for our modern understanding.

Force Fields

The idea of a force field is quite appealing to the imagination of science fiction authors. It usually takes the form of a hard-edged region of space that will exert a strong force on any object that enters or tries to enter. It may come as a surprise to some people to learn that force fields are actually part of our present technology. It may, on the other hand, come as a disappointment to others to learn that the hard edge of an invisible force field is not something that ordinary humans can yet experience.

We do know of hard-edged force fields, but they exist only inside the atomic nucleus. They have a very short range. The so-called strong interaction force is what holds the atomic nucleus together. All of other force fields we know of are much more gentle in their treatment of distance than those of science fiction. The most familiar of these is the gravitational field. Then there are electric and magnetic fields, which account for most of everything else in our experience.

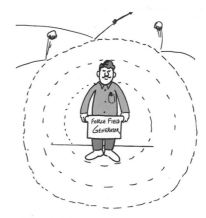

Figure 13-6

The idea of a force field is appealing to the imagination of science fiction authors.

The only other kind of force fields that we know of are also limited in their range to distances inside the nucleus. They are the so-called weak interaction forces that govern radioactive decay. In recent times, we have found that this weak force is so closely related to electric and magnetic forces that all three may be looked upon as different aspects of the same thing. People working in this area of research now speak of the "electro-weak" force.

Even though gravitation, electricity and magnetism seem like different things, there are many properties of their force fields that are remarkably similar. This is one of the beautiful things about physics. We can actually learn about things that we do not know by studying more familiar cases and comparing them. At this point in the book, we have only touched on electricity, but have spent a great deal of time with gravity. In order to study force fields, then, let us look at what we know about gravity as a force field.

As you recall, the ideas of weight and mass are closely related. Everything that has mass has weight. The ratio of these two quantities is the acceleration due to gravity.

$$g = \frac{w}{m}$$

Also as you recall, weight w is a vector since it is a type of force, and mass, m, is a scalar having only magnitude as there is no particular direction associated with it. Their ratio, $\mathbf{g}$, is a vector having the same direction as $\mathbf{w}$. The magnitudes of $\mathbf{g}$ in London and San Francisco are nearly the same, but their directions are nearly perpendicular.

If we look at the gravitational acceleration around the world, one way to picture the directional properties of g is draw a number of radial lines with arrows pointing in the direction that an object would fall. The lines that we draw construct a picture of what Faraday called "lines of force."

One of the nice things about this picture is that the gravitational lines of force tell us about the magnitude as well as the direction of $\mathbf{g}$. The more dense the lines are at any point in the picture, the stronger the gravitation, $\mathbf{g}$, is. Referring to Figure 13-7, the lines are close together at the surface of the earth. Farther from the earth, as the lines radiate outward, the lines are farther apart, corresponding to a weaker $\mathbf{g}$.

Another virtue to this approach is that it helps us to think in the abstract. Standing at one place on the surface of the earth, we might think of ourselves as being attracted to the center of the earth by a huge point mass down there. In truth, we are all attracted by the big island of Hawaii. If you happen to be in San Francisco, the attraction is stronger for some of us than for others. For all of us in that fair city, however, it is mostly horizontal, but the horizontal part is nearly balanced by the attraction of the islands that make up the United Kingdom pulling in the other direction. The UK is

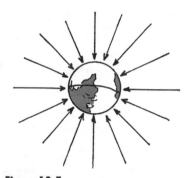

Figure 13-7

Faraday's lines of force for gravity on the planet Earth.

further but bigger. All parts of the globe pull on all of us in different directions, but the net effect is nearly the same as if all the mass of the world were located at the center.

On most days, we don't think of ourselves as being attracted to Hawaii. Nor do we think of ourselves being attracted to some imaginary point mass some six thousand kilometers beneath our feet. Instead we think of the space around us as being filled with something that pulls on things. That something is a gravitational field.

Just as the force fields of science fiction exert a force on any object that enters their region of space, the gravitational fields of planets and stars may be thought of as a property of their region of space such that any object that enters that space will experience a force.

Electric Fields

Aside from the gravitational field of the earth, which you experience constantly, the next most familiar force field is electrical. There are a few times when you actively notice electric fields, for example when you comb your hair on a dry day or take clothes out of the dryer, you are likely to see or feel static electric forces acting over a distance. Actually, you experience electrical forces every minute. All chemical and molecular forces are electrical. Without electrical fields you would literally fall apart.

We define the electric field, traditionally abbreviated with a capitol **E**, in the same way that we defined the gravitational field **g**. In the gravitational case, the force is weight and weight was proportional to mass. So we took the ratio of gravitational force to mass.

$$\mathbf{g} = \frac{\mathbf{w}}{m}$$

In the electrical case, the force is the electrical force of attraction or repulsion between two objects, and we have seen that this force is proportional to the charge on one of the objects.

$$\mathbf{E} = \frac{\mathbf{F}}{q_o}$$

In words, the electric field **E** is defined as the ratio of the electrical force **F** acting on an object to the amount of charge q_o on that object. The units of the electric field are newtons per coulomb, or N/C.

We think of electric fields, like gravitational fields, as existing in empty space. With this in mind, a good way to think of an electric field is in terms of the effect it has on other things. In the above equation, then, the electric field is defined by the amount of mechanical force that the field exerts on something with a charge. To describe the electric field at a particular place, we use a mental tool called a **test charge**, q_0. A test charge is a very small charge, small enough that it's charge does not affect the electric field by its presence.

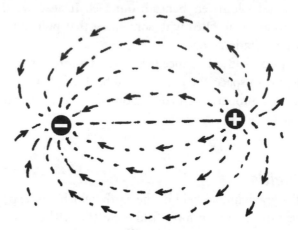

Figure 13-8

We use the convention that an electric field comes out of a positively charged object and goes onto a negatively charged object.

In real life, as discussed in the last chapter, the smallest electrical charge is the electron. Depending on the situation, we may want to let our imaginary test charge be smaller than the charge of an electron. This is sometimes necessary in a weaker field, since the charge must be so small that it's presence does not change the field (and the results).

In the above relationship, we define our test charge as being positive. In that way, we can put a positive direction of an electric field coming out of a positively charged object and going into a negatively charged object.

This relationship is often turned around to find the force that an electric field would place on a charged object. In this case, the equation is written

$$\mathbf{F} = \mathbf{E}q_o$$

If q_0 is positive, $\mathbf{F}$ and $\mathbf{E}$ are in the same direction, and if q_0 is negative, $\mathbf{F}$ is in the opposite direction from $\mathbf{E}$.

The idea of an electric field is a powerful mental tool for dealing with charge that is spread out over space. Say your charge is spread out evenly over the surface of a metal sphere. This is a very typical situation. Just as we can draw lines of force representing earth's gravitational field to simplify our thinking so that we can pretend that all of the mass of the earth is located at its center, the electric field simplifies our thinking for spread out or unevenly distributed charges. In this example, where the charge is distributed around the outside of the sphere, the electric field outside the sphere will be the same as if all the charge were at the center.

Of course, in the gravitational case, the only interesting objects are spherical. They have to be. The only interesting objects are big, such as planets, moons and stars. Any object big enough to produce a substantial

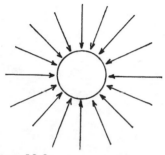

Figure 13-9

Electric field lines for a negatively charged sphere.

gravitational field is also big enough to collapse of its own weight into a spherical shape.

In the electrical case, it is easy to have charge distributions that are non-spherical. That is where the idea of an electric field becomes a powerful tool for reasoning. Say you had two metal balls with equal but opposite charges. If the balls are far enough away from each other that the electrical force between them is very weak, the charges will distribute themselves evenly over the spherical surfaces. You can use the distances between the centers of the balls as the distance d in Coulomb's law.

$$F = k\frac{q_1 q_2}{d^2}$$

If you bring the balls closer together, the charges on the two balls will begin to attract each other. The charges will move around on the spherical surfaces, and concentrate themselves on the sides of the ball nearest to the other set of charges. The distance d, the effective distance between the two charges, is now less than the distance between the centers of the balls.

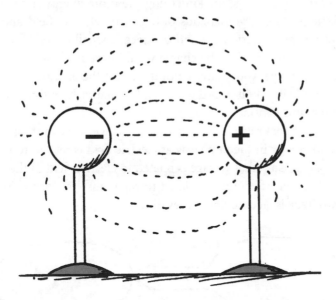

Figure 13-10

The effective distance between the charges on two conducting balls is less than the distance between the centers of the balls.

Although the exact solution of this problem requires sophisticated mathematics, a very good approximate solution requires only geometry. By visualizing Faraday's lines of force, we can see how the charges must move around on the surface of the metal. Because the charges are free to move around in the metal, we can predict the shape of the electric field just outside the metal balls. It has to be perpendicular to the surface of the ball. Any parallel component of the electric field will tend to cause charges to move. The only static situation is arrived at when the charges are in a position such that there is no component of the electric field parallel to the surface.

Potential Energy

Just as the idea of an electric field helps in understanding forces acting on charged objects, the idea of voltage, encountered in the last chapter, turns out to be an even more powerful tool for understanding the potential energy of charged objects. We are now in a position to see how voltage is related to the ideas of electric field, energy and force.

Remember from Chapter 8 that potential energy is the energy stored in an object when work is done on it to pull it away from gravity. This energy is transformed into other energy—kinetic energy—when the object is allowed to fall and the gravitational field pulls the object back to earth.

Although we all experience the same gravitational field, some of us experience more weight than others. Those of us blessed with great weight also enjoy greater potential energy when perched at a certain height above the floor. Thus the old saying, "The bigger they are, the harder they fall."

Just as an electric field may be thought of as something that gives force to an electric charge, it is also appropriate to think of it as something that can give potential energy to a charge where there is a distance for the charge to move. The bigger it is, the harder it falls electrically.

The greater the charge on something the more energy it has—but the ratio of energy to charge depends only on the electric field and the distance. For a given electric field and distance, a small charge would have a small potential energy. Half the charge would have half the energy. A hundredth of the charge would have a hundredth of the energy. A billionth the charge would have a billionth the energy. No matter how small the charge, the ratio remains constant.

The limit of that ratio is how we define voltage. We can think of voltage as related to energy in the same way as electric field is related to force. That means, as a corollary, that voltage is related to electric field in the same way work, the transfer of energy, is related to force. Just as work is force times distance, voltage is electric field times distance.

Figure 13-11

Voltage is related to the idea of force through either of the ideas of work or electric field. If you multiply by distance, to get work or energy, and then divide by charge, you come to the same place in your thoughts as if you first divide by charge, to get electric field, and then multiply by distance.

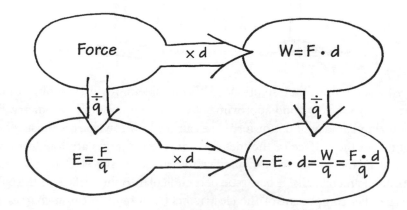

These two equivalent ways of thinking about voltage are reflected in the units we use in dealing with electrical matters. The volt can be expressed as a joule per coulomb.

$$1 \text{ V} = 1\left(\frac{\text{J}}{\text{C}}\right)$$

It could also be expressed in terms of units for electric field and distance if only we had units for electric field. Rather than to have separate units for electric field, however, we find it more convenient to express the concept in terms of either force or voltage. For example, the dielectric strength of air, E_{max}, which is the strongest electric field you can have before you get a spark, may be expressed in either joules per coulomb or volts per meter.

$$E_{max} = 3 \times 10^6 \left(\frac{\text{N}}{\text{C}}\right) = 3 \times 10^6 \left(\frac{\text{V}}{\text{m}}\right)$$

There are some further refinements we could worry about, and deal with using calculus. For the moment, however, as long as the charge is small enough that we can ignore its own electric field and the force is constant and in the same direction as the distance, we need not worry about these refinements.

One of the virtues of this idea of voltage is that, like energy, it is a scalar. That's good because, when you go to add two voltages, you only have their magnitudes to worry about. Electric fields, on the other hand, are vectors and add like forces. It is therefore reasonably difficult to calculate the electric field of several charges spread out over space, but it is far easier to calculate the voltage due to the these charges.

Magnetic Fields

As mentioned earlier, Michael Faraday got his inspiration for his idea of electric fields from the magnetic lines of force that appear when you sprinkle iron filings onto a magnet. Put a pair of magnets with opposite poles near each other, and you can see by the lines of force that they attract each other. As we have seen, the lines of force graphically pull between the poles. Turn one magnet end for end, and you can see by the lines of force that they repel. The lines of force graphically push against each other.

It is conventional to mark magnets with an uppercase N on one end and an uppercase S on the other. That is because, if free to move, a compass will tend to line up with the earth's magnetic field. The end of the magnet marked with an N should point north. It is not proper to call these ends of the magnet "North" and "South," however. The earth's north pole, if you were to go there and mark it with the same convention as for magnets, would be marked with an S since it attracts the N ends of magnets. To avoid claiming that "the earth's north pole is a south pole," we should be a little more careful in our language. The N pole of a magnet is a "north *seeking* pole." That way we can say "the earth's north pole is a south *seeking* pole." Careful use of language can at least minimize confusion.

Compass needles are sometimes made in the form of magnetized metal arrows tending to point north. We think of the magnetic field of a magnet as pointing in the direction that such a compass needle would point. This is a convention, an arbitrary choice, but it is a convention that holds general agreement. We therefore think of the magnetic field as coming out of the N pole, or north seeking pole, of a magnet and going into the S pole, or south seeking pole.

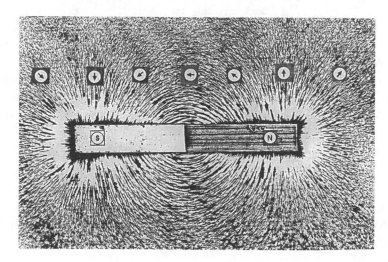

Figure 13-12

The convention is that the magnetic field lines come from the N and go to the S pole of a magnet.

Inside the magnet, the magnetic field returns to form closed loops. That is how we interpret the fact that a broken magnet turns into two magnets. Every magnet has two poles of equal strength. Nobody has ever found a magnet with only one pole, and they have searched for such things. The first person to find a so-called magnetic monopole will almost certainly win a Nobel prize. Many believe that will never happen and that the absence of magnetic monopoles is a fundamental law of physics. Maybe, maybe not—but in the absence of magnetic monopoles, all magnetic field lines form closed loops.

There are many similarities between electric fields and magnetic fields. The force lines, for example, will be the same for a magnet with two poles and for two oppositely charged forces the same distance apart. A main difference between the two is that every charged object, by itself, is surrounded by an electric field radiating outward. Since there is no such thing as a magnetic monopole, however, magnetic lines of force always loop back on one another—they never radiate outward to infinity.

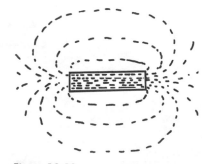

Figure 13-13

Magnets always have closed loops.

Electromagnets

Run current through a wire and, as Oersted discovered while giving a physics lecture, a compass near the wire will tend to turn at right angles to the wire.[2] Further investigation showed that the magnetic field encircled the wire.

This was the beginning of the important discovery that an electric current is always surrounded by a magnetic field. In fact, any moving electric charge produces a magnetic field.

Using the right hand as a convention, we take the direction of the magnetic field to be the direction in which the fingers of the right hand will point if you were to take hold of the wire with your thumb pointing in the direction of the current. We call this the "right hand rule." It is consistent with the right hand rule used in the mechanics of our culture.

There is a story about how many college deans it takes to change a light bulb. The story goes that some pranksters manufactured a light bulb with a left hand thread just to prove some point or other. One version of the story has the college dean getting it right on the first try. He turned it the wrong way on the first try. (An unlikely story to those of us who know and love our college administration.) The moral of the story is that right hand rule is embedded in our culture. Any *reasonable* person knows which way to turn a light bulb. It's an arbitrary choice, but one that holds common agreement.

If you wrap the wire around, say, a nail, the nail will become magnetized. Anyone who hasn't tried this should. A dozen turns of wire with current from a single cell of a flashlight battery is enough to turn a nail into a strong enough magnet to pick up paper clips. The direction of the magnetic field can be predicted with the right hand rule. If the current travels around the nail in the same direction as the fingers of your right hand, your thumb will point toward the N pole of the magnetized nail.

If you try this with, say, a pencil, you will still get a magnetic field. It will be much weaker, but the nail is not essential to producing a magnetic field. A coil of wire wrapped into the shape of a cylinder has an magnetic field all by itself. This coiled wire is called a solenoid. Looking at the magnetic field of a solenoid is what gives us the idea that the magnetic field of a permanent magnet forms closed loops with the field going one way outside the magnet and returning inside the magnet.

Andre-Marie Ampere (1775–1836), in whose honor we name the unit for current, saw in the magnetic field of a solenoid evidence of the atomic theory of matter. He suggested that a permanent magnet held its magnetism by virtue of tiny circulating currents on the atomic scale. He suggested that a permanent magnet is, in effect, an electromagnet. This idea is based

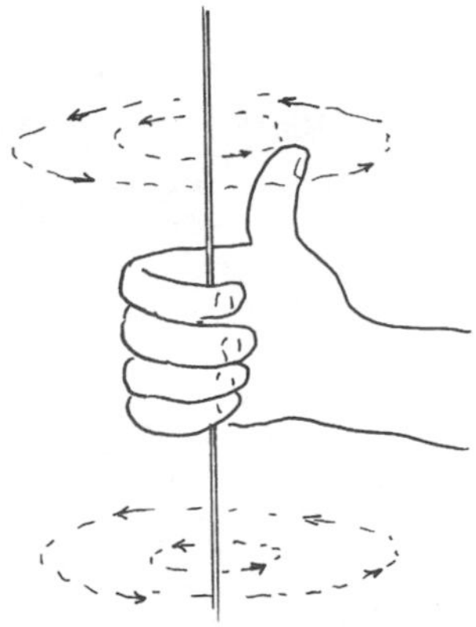

---

[2]When Frank Openheimer created San Francisco's Exploratorium, he took great pains to incorporate optical illusions. His point, and he often made it, was that many discoveries can be made by careful observation. In Oersted's day, others must have seen what he saw. The difference was that Oersted noticed.

Figure 13-14

Wrap your right hand around a charge-carrying wire in such a way that your thumb is pointing in the direction of the electric current. Your fingers will be pointing in the direction of the magnetic field.

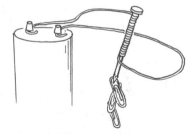

Figure 13-15

You can create an electromagnet by wrapping a current-carrying wire around a nail.

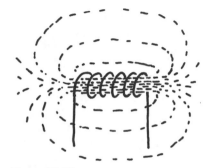

Figure 13-16

A solenoid.

on the above assumption that a moving electric charge always produces a magnetic field. Ampere was right about matter being made of atoms, and it turns out that this is a good way to understand permanent magnets, even with our deeper understanding of atoms in the present day. Now we think that atoms do not actually have charges moving around inside them, but they certainly act as if they do.

HIGHLIGHT
Geomagnetism

Compasses are made of magnetized needles which are allowed to float freely so that they can align themselves with the earth's magnetic field. Because of this, most people tend to think of the earth as having a sort of bar magnet through the middle, with two distinct poles. The earth's magnetic field, however, is less consistent and less predictable than one might think.

The earth's magnetic field is actually quite uneven and changes over time. There is evidence that the poles have switched position at least once, and the magnetic field does indeed slide around relatively quickly. In fact, in eighty years or so there will be a measurable difference in the direction of the magnetic field. Generally, people can get around just fine using a compass, but there are provisions made for those occasions when accuracy is important. Maps give not only north and south, but somewhere usually give the information of how far different the official "north" is from what a compass would read.

The earth, then, does not act as a bar magnet. The question of the cause of the earth's magnetic field puzzled physicists for a long time. The current thinking on the subject is that the moving liquid that sloshes around (slowly) under the earth's core acts as a conductive layer, which, as it passes through the magnetic field of it's own making, has charge moving around in it. The net result of this is that there is, in effect, a layer under the earth of electric charges moving around. Moving charges create a magnetic field, and so is born the magnetic field of the earth.

Magnetic Forces

Originally, magnets were interesting to people because of the way they interacted with other magnets, particularly the big magnet we call the earth. Early work on this subject was done by William Gilbert (1544–1603), the court physician to Queen Elizabeth I. In the year 1600, he published in London his great work on magnetism, correctly describing the main features of the earth's magnetic field and showing why compasses point north. London, after all, is a sea port, and Elizabethan England had a more than passing interest in matters of navigation. Gilbert's work, however, was of theoretical as well as practical value. We owe much of what we know about the forces between magnetic poles, and indeed the idea itself of magnetic poles, to this renaissance scholar.

In our electrically driven society, however, we are more interested in how electricity and magnetism interact. Oersted's discovery in 1820 that a compass needle tends to set itself a right angles to an electric current was the first hint that electricity and magnetism do interact. Electric motors, generators, and most modern appliances are driven by this discovery.

In our modern society, some of us spend too much of our lives watching the end of an electron beam paint pictures on a phosphor screen. That electron beam, the beam in a television tube, makes a familiar example for discussing how magnetic forces act on charges. The electrons, having been

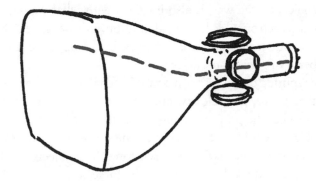

Figure 13-17

Deflection of eletron beam in a television picture tube.

shot out from the back of the tube toward the screen by electric fields, are then guided to different parts of the screen by magnetic fields created by coils of wire called, naturally enough, "deflection coils." This deflection has a number of features that we should appreciate.

For one, the charges need to be moving. A charged particle at rest in a static magnetic field feels no magnetic force at all. Only when it is moving relative to the field does it experience a force, and what's more, it needs to be moving perpendicular to the field. In the television set, the electrons are moving toward you, as you watch the phosphor screen, and the direction of the magnetic fields that deflect them are up and down and side to side.

Another feature is that the direction of the force is perpendicular to both the magnetic field and the direction the charge is moving. The field that deflects the electron beam from side to side is directed up and down. The field that deflects the electron beam up and down is a second field, and its direction is from side to side.

In this respect, magnetic forces are different from gravitational and electrical forces. In the gravitational case, the force is always in the direction of the field. In the electrical case, the force is either in the same or opposite direction, depending on whether the charge is positive or negative. But in

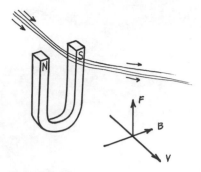

Figure 13-18

The force deflecting a beam of positive particles is perpendicular to the magnetic field and to the velocity.

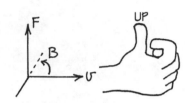

Figure 13-19

Right hand rule

the magnetic case, the force is never in the same or opposite direction. It's off to the side.

Whether it is off to the side in one direction or the other, however, depends on what is positive and what is negative. Suppose, for example, a beam of positive ions were directed between the poles of a magnet. Let us assume that the ions are going from left to right and that the N pole of the magnet is closer to you than the S pole. In that case, the direction of the force on the ions will be upward.

If you point the fingers of your right hand in the direction of velocity and hold your hand such that your fingers can curl into the direction of the magnetic field, then your thumb will point in the direction of the force. This convention is called the "right hand rule." The relative directions of things are chosen by Nature, but our names for things are chosen by people. The common convention is to choose the direction of the magnetic field to be such that v rotated into B will be in the direction of F following the same convention as a right-handed threaded lightbulb.

This choice, by the way, is also involved with what we decide to call a positive ion. Electrons, by human definition, are negative. The beam of electrons in your television set is deflected in the opposite direction from that of a beam of positive particles.

The magnitude of the force on a moving charge is proportional to three things—the charge, the velocity, and the strength of the magnetic field. You might expect the proportionality with charge. All else being equal, two charged objects would each experience the same magnetic force, and so together they experience twice as much force. Ten objects would have ten time the charge and produce ten times the force, and so forth. Mathematically, we can say the force **F** is proportional to charge q.

$$\mathbf{F} \propto q$$

As it happens, the force is also directly proportional to velocity. Twice the velocity, twice the force. Ten times the velocity, ten times the force. Mathematically, we can express this as

$$\mathbf{F} \propto \mathbf{v}$$

We can combine these two proportionalities by saying

$$\mathbf{F} \propto q\mathbf{v}$$

or by saying that the ratio of force to the product of charge and velocity is constant in a given magnetic field. That, in fact, is how magnetic field is usually defined.

The force and the velocity are not in the same direction, which makes the math a bit more complex. For now, we can drop the vector notation and just look at the magnitude of the fields in the equations. We can deal with the direction as a separate issue, using the right hand rule.

The magnitude of the magnetic field, abbrieviated B, is the proportionality constant in the above equation and is written

$$B = \frac{F}{q_O(v\sin\theta)}$$

Theta, θ, is the angle that the incoming charge makes with the magnetic field. This means that ($v \sin \theta$) is the component of the velocity perpendicular to the magnetic field. As we discussed above, the part of the charge travelling perpendicular to the magnetic field is the only part that affects it.

This (sin θ) factor was, by the way, one of the first things noticed about magnetic forces acting on moving charges. When J. J. Thompson, who discovered the electron, tried deflecting a beam of positive ions with a magnetic field, he made the electric and magnetic fields parallel. The shape of the traces puzzled him, and he only gradually came to realize that the magnetic force was only acting on the ions which happened to be moving perpendicular to the field.

The SI unit for magnetic field is the tesla, named after Nikola Tesla (1870–1943), the Croatian born American inventor. A tesla is, by definition, the magnetic field that will cause one newton of force to be exerted on a charge of one coulomb moving at one meter per second perpendicular to the direction of the field.

$$1 \text{ T} \equiv \frac{(1 \text{ N})}{(1 \text{ C})\left(1\frac{\text{m}}{\text{s}}\right)}$$

Mass Spectrometers and the Fusion Torch

It was Thompson's student, Francis William Aston (1877–1945), who deliberately set the magnetic field perpendicular to the ion's velocity. In so doing, he turned Thompson's positive ray method into one of the most powerful research techniques in the early part of the twentieth century. He invented the mass spectrometer.

We know that atoms of a certain element can have different masses. We say they are different isotopes, and we understand the difference in terms of their having different numbers of neutrons. The reason we know as much as we do about isotopes is because of Aston's mass spectrometer.

A material, such as uranium, can be vaporized in an oven and ionized by a strong electric field. A beam of positive ions passing through a magnetic field will then be bent into a circular path with a radius of curvature that depends on the mass of the ion. This approach has already been used as part of the process for obtaining weapons grade uranium. If nuclear

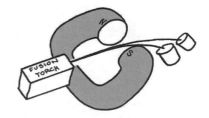

Figure 13-20

A fusion torch may someday be used to separate material on a large scale.

fusion ever becomes commercially feasible, one application may be to separate solid waste with a magnetic field.

Suppose uranium ions, for example, are traveling at 8.0×10^6 m/s in a magnetic field of 0.44 T. Since the charge on an uranium nucleus is 92 times the charge on an electron, which is 1.6×10^{-19} C, the force can be found from the equation for magnetic fields

$$B = \frac{F}{q(v\sin\theta)}$$

solved for F

$$F = qvB\sin\theta$$

becomes

$$F = (92)(1.6 \times 10^{-19}\text{C})(8 \times 10^6\text{m/s})(0.44\text{ T})\sin(90°)$$

which comes to

$$F = 5.2 \times 10^{-11}\text{ N}$$

This might not seem like much force, but if the ion travels 10 cm at this speed, the impulse, acting for a time of 1.25×10^{-8} s,

$$\text{Impulse} = F\Delta t$$

becomes

$$\text{Impulse} = (5.2 \times 10^{-11}\text{ N})(1.25 \times 10^{-8}\text{ s})$$

which comes to

$$\text{Impulse} = 6.5 \times 10^{-19}\text{ N} \cdot \text{s}$$

and is comparable to the momentum, since the mass of a uranium ion is 3.95×10^{-25} kg.

$$mv = (3.95 \times 10^{-25}\text{ kg})(8 \times 10^6\text{ m/s}) = 3.2 \times 10^{-18}\text{ kg} \cdot \text{m/s}$$

Induction of Voltage

Electrical power for the power grid (from which we all get electricity delivered to our houses) is entirely produced by passing wires through a magnetic field. First discovered by Joseph Henry (1797–1878), an American physics teacher, in 1829, the way to get electricity from magnetism was independently discovered in England by Michael Faraday. Faraday's greatness was such that he supported Henry's priority in this important discovery. We now honor both Faraday and Henry by naming two of the most commonly used units in electricity and magnetism after them.

If a wire passes through a magnetic field, positive charge in the wire is pushed in one direction along the wire while negative charge is pushed in the opposite direction. In most wires, it is the negative electrons that are free to move, although we frequently think of current as flowing in the direction that positive charges would flow.

Figure 13-21

Passing a wire through a magnetic field produces a voltage.

The amount of voltage depends on the length of the wire and the relative velocity between it and the magnetic field. An easy way to think of it is in terms of Faraday's lines of force. He defined something called **magnetic flux** which is the product of magnetic field **B** and area *A*, which is taken to be perpendicular to the magnetic field. The magnetic flux is represented by the greek capital letter Phi Φ, (pronounced "fee").

$$\Phi = BA$$

In terms of Faraday's magnetic flux, the maximum voltage available, which is called the electromotive force ξ, is just the rate of change of flux.

$$\xi = -\frac{\Delta\Phi}{\Delta t}$$

In this way of looking at it, it doesn't matter whether it is the magnetic field or the area that changes. Either will produce a voltage. For example, generators in older automobiles produced electrical power for the ignition system and battery by spinning a coil of wire, called a rotor, in a stationary magnetic field produced by a stationary electromagnet called a stator. The voltage regulator adjusted the strength of the stationary magnetic field so that the voltage produced by the rotor would match the voltage needed by the battery. As the rotor turned, the area of the magnetic field passing through the spinning coil would change, producing the voltage that was sent to the battery through stationary carbon brushes rubbing against contacts that spun with the rotor.

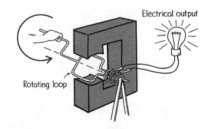

Figure 13-22

As the coil spins, the area the of the field that is perpendicular to the magnetic field changes.

Generators have been replaced in modern automobiles with alternators that do the same job more efficiently. The voltage regulator again adjusts the strength of a magnetic field, but this time the rotor is the electromagnet. As the rotor spins, the magnetic field through the stators not only changes but reverses, producing alternating current. Since the battery needs direct current, this alternating current must be sent through semiconductor rectifiers to do the same job that had been done by carbon brushes and rubbing contacts.

In the days of vacuum tubes, before semiconductor power rectifiers, only police cars had alternators. Vacuum tube rectifiers were expensive and had a short service life, but they needed them because police cars sit for long periods with their radios on and their engines at idle. Ordinary cars in

HIGHLIGHT
Maxwell's Equations (Optional)

The basic theory linking electric and magnetic fields was devised by James Clerk Maxwell (1831–1879) in 1864. Setting out to summarize what was then known about these fields, he found that he could do it with four equations. Two of the equations describe how electric and magnetic fields radiate from any point in space. The concept is easy in the case of magnetic fields—they don't radiate. As we mentioned earlier, magnetic fields always form closed loops.

The fact that magnetic field lines never radiate from any point in space can be expressed using a mathematical function called the divergence. This is a little higher level of mathematics than we want to get into right now, but the concept is fairly straight forward. Fortunately, the language used in mathematics is fairly descriptive of the concept. If all the magnetic field going into any point in space also comes out, the divergence is zero.[3]

$$\text{Divergence of magnetic field} = 0$$

For an electric field, the situation is a little more involved because electric fields radiate from charges. We think of electric fields as coming out of positive charges and going into negative charges. Using the same mathematical function, the divergence, on the electric field, expresses this fact by saying that the divergence of the electric field is proportional to charge density. That turns out to be a very general rule, and useful in all kinds of situation. For one particular situation, however, it is just as easy as the magnetic field. That situation is the description of the electric field in free space, where there is no charge. Then we can say that the divergence of the electric field is zero.

$$\text{Divergence of electric field} = 0$$

We are interested in empty space because that's what light travels through on its way to us from the sun and the stars. That is what fascinated Maxwell because he saw an imbalance in two other expressions that described how electric and magnetic fields are related. When he balanced the equations, making them symmetric, he saw how light could travel through empty space.

Another of Maxwell's equations describes in geometrical terms how an electrical field is created by a changing magnetic field. In the previous section, we saw how voltage is equal to the rate of change of Faraday's magnetic flux. Maxwell saw how to express this very relationship in terms of how a changing magnetic field causes an electric field to curl around in a closed loop. We, today,

---

[3]The divergence uses a vector differential operator called the del operator ∇, which is essentially a three-dimensional derivative. The divergence of a vector field is the dot product of the del operator with the field: $\text{div} = \nabla \cdot (\text{Field})$.

the old days had fewer demands on their electrical systems. Changing the angle was good enough to keep the battery charged. With the semiconductor revolution, all cars have alternators with changing magnetic fields to power everything from the ignition system to the rear window defroster.

call this mathematical function the "curl," and a very descriptive name it is.[4] We say that the amount that an electric field curls around a point in space is equal to the rate of change of the magnetic field through that point in space.

$$\text{Curl of electric field } = -\frac{\Delta B}{\Delta t}$$

The thing that puzzled Maxwell was that the only way people knew how to create a magnetic field was to pass a current through a wire. Maxwell knew that he could write an equation describing the amount that magnetic field curls around a point in space as proportional to the current density through that point. That expression would be equivalent to Oersted's discovery that a compass will tend to turn sideways to a current in a wire. The puzzling thing is that there is no current density in empty space, so the curl of magnetic field should be zero. His intuition told him that, if the curl of electric field depends on how fast the magnetic field changes, then the curl of magnetic field should depend on how fast the electric field changes.

$$\text{Curl of magnetic field} \propto \frac{\Delta E}{\Delta t}$$

Maxwell called this changing electric field a "displacement current" because he saw it in terms of increasing and decreasing polarization of the "aether," the theoretical substance that scientists thought filled empty space. When he worked out the value of the proportionality constant, he knew he was on to something. It turns out that the proportionality constant is none other than the square of the speed of light in a vacuum, c.

$$\text{curl } B = c^2 \frac{\Delta E}{\Delta t}$$

From these four equations, Maxwell was able to show that his aether, or what we think of as free space, can transmit electromagnetic waves at the speed of light. This theory predicted the discovery three decades later of things other than visible light, such as radio waves, X-rays, and infrared radiation that also travel at this speed.

---

[4]The curl uses the sane vector differential del operator ∇, used by the divergence, but it is the vector cross product of the the del operator with the field: curl $= \nabla \times$ (Field).

SUMMARY OF CONCEPTS

Weight, the force of gravity, is proportional to mass. The ratio of these two quantities is the acceleration due to gravity.

$$\mathbf{g} = \frac{\mathbf{w}}{m}$$

If we consider this mass to be very small, this ratio can also be thought of in terms of gravitational field. The electrical equivalant, the ratio of force to charge is electric field,

$$\mathbf{E} = \frac{\mathbf{F}}{q_o}$$

which is also defined in the limit where the charge approaches zero. Such a charge, an infinitessimally small positive charge, is known as a test charge q_O.

Voltage may be defined as electric field times distance

$$V = \mathbf{E} \cdot \mathbf{d}$$

as well as the ratio of work W, or energy, to test charge q_o.

$$V = \frac{W}{q_o}$$

In terms of units, a volt is equal to one joule per coulomb of potential.

$$1V = \frac{1\,J}{1\,C}$$

The magnetic fields act only on charges when they are in motion. The direction of the force is proportional to the charge. It is also proportional to and perpendicular to both the velocity **v** of the charge and to the direction of the magnetic field **B**. The magnitude of the force is therefore given by the equation:

$$F = Bq_o(v\sin\theta)$$

where B is the magnitude of the magnetic field, v how fast the test charge, q_0, is moving and θ is the angle the travelling charge makes with the magnetic field.

We take the direction of magnetic field to be such that direction of the force will be in the direction of the thumb of the right hand when the fingers are pointed into the direction of the velocity such that they can be curled into the direction of the magnetic field. We take the magnitude of magnetic field to be such that, when measured in teslas (T), 1 coulomb (C) of charge moving at 1m/s in a magnetic field of 1T will experience a force of 1 newton (N).

$$1\,T = \frac{1\,N}{C\left(\dfrac{m}{s}\right)} = \frac{N \cdot s}{C \cdot m}$$

Voltage is induced in a loop of wire when the amount of magnetic flux Φ changes, where magnetic flux is defined as the product of magnetic field **B** and the area A perpendicular to the magnetic field.

$$\Phi = \mathbf{B}A$$

The rate of change of the flux through the loop gives the maximum voltage, known as the electromotive force ξ,

$$\xi = -\frac{\Delta\Phi}{\Delta t}$$

EXAMPLE PROBLEMS

Try to do the Examples yourself before looking at the Sample Solutions.

13-1 The most volcanically active object in our solar system seems to be Io (pronounced EE-oh), the closest moon to Jupiter. Volcanic eruptions on Io are all the more spectacular because of Io's small gravitational field. Boulders and rocks thrown up from the surface travel to great heights before falling back.

Given: A boulder with a mass of 5,000 kg, thrown upward from the surface of Io, has a weight of 8800 N. A rock, also thrown upward from the surface of Io, has a mass of 1.14 kg.

Find: (a) What is the gravitational field at the surface of Io?
(b) What is the weight of the rock?

BACKGROUND

Our limited experience, being confined to live out our lives on the surface of the earth as most of us are, leads us to think of the gravitational field of the earth as a natural constant. The closest thing to our experience with other gravitational fields are in the form of videos of astronauts on the surface of the moon, but someday ordinary folks may visit such strange places.

Io would be a very interesting place to visit. The gravitational field is similar to our own moon, and think of all those fire fountains you could watch. Might get hot though.

13-2 The electric field of a Van de Graff generator can drive a puffed rice fountain. Put the puffed rice in a plastic foam cup and set it on top of the generator's sphere. It's a beautiful sight to behold, but get out the broom.

Given: A grain of puffed rice having a charge of 83 pC (pico coulombs) experiences a force of 26 μN. A smaller grain of puffed rice has a charge of only 38 pC.

Find: (a) How great is the electric field near the generator's sphere?
(b) How great is the force on the smaller grain?

BACKGROUND

Use only puffed rice. For some reason, puffed wheat doesn't work. Most schools have a Van de Graff generator. If you can get your hands on one, this is a demonstration that you must do.

13-3 The dielectric strength of plastic insulation keeps you safe when you handle the power cord to that electric drill in your tool box. There is danger, however, in handling power cords when the insulation gets old and cracked. The dielectric strength of air is less than that of plastic.

Given: The "hot" wire in the electric cord you are holding has a peak voltage of 170 V relative to your hand. The plastic insulation on that wire has a dielectric strength of 45 MV/m and is 0.83 mm thick. The dielectric strength of air is 3 MV/m.

Find: (a) How great is the electric field in the insulation?
(b) What is the minimum distance your finger can come to the metal wire before a spark jumps through the air?
(c) What is the minimum distance your finger can come to the metal wire before a spark jumps through the insulation?

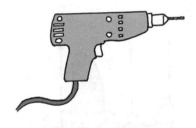

BACKGROUND

*Your 120 V wall socket gives you more voltage than you might think. We call the 120 V a **rms** value, which stands for root mean square or average value. It is the peak value that is likely to give you a shock. The ratio of rms to peak value is one-half the square root of two, or 0.707.*

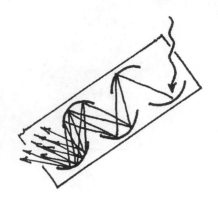

13-4 A photomultiplier tube is a vacuum tube that is sensitive to very low levels of light. A photon of light coming into the tube hits a metal plate and knocks out a single electron. That electron is accelerated by a voltage toward a second metal plate, gaining enough kinetic energy along the way to knock out two other electrons, which, in turn, are accelerated toward a third meal plate, and so on.

Given: Each stage of a photomultiplier tube accelerates electrons through a voltage of 180 V. Electrons have a charge of 1.60×10^{-19} C and a mass of 9.11×10^{-31} kg.

Find: (a) How much kinetic energy does each electron gain in being accelerated from one stage to the next?

(b) What is the minimum velocity each electron can have as it hits each plate?

BACKGROUND

Vacuum tube technology still lives! Solid state is rapidly replacing hollow state, but here is a place where tubes can still be found.

Part (b) asks for the minimum velocity because some electrons have an initial velocity as they are liberated from the metal plate.

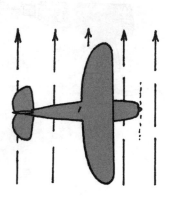

13-5 An airplane flying east along the earth's equator picks up a slight positive charge due to friction with the air through which it passes. The magnetic field of the earth at the equator points due north, producing a slight force on the airplane.

Given: The airplane having a mass of 3.3×10^5 kg, a charge of 2.2 μC and a velocity of 250 m/s due east, is flying through a magnetic field of magnitude 5.5×10^{-5} T .

Find: (a) What is the magnitude of the magnetic force acting on the airplane?

(b) What is the direction of the magnetic force acting on the airplane?

(c) What is the ratio of the magnetic force to the gravitational force acting on the airplane?

BACKGROUND

This is a moderate amount of charge for an airplane aloft. Sometimes you will see a glow from the tips of wings when flying at night. The charge would then be a few orders of magnitudes greater than the charge in this example.

13-6 The horizontal and vertical deflection coils in a television set produce magnetic fields that combine to push the electron beam back and forth, up and down.

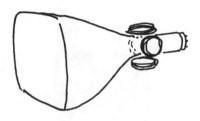

Given: An electron having a charge of 1.60×10^{-19} C and a velocity of 5.6×10^7 m/s is deflected horizontally by a magnetic field having a strength of 0.23 T and deflected vertically by a magnetic field having a strength of 0.32 T.

Find: (a) What is the magnitude of the resulting magnetic field?
   (b) What is the direction, measured from the horizontal, of the resulting magnetic field?
   (c) How great is the force acting on the electron?

BACKGROUND

Two separate deflection coils are used in television sets because the sweep rate is hundreds of times faster in the horizontal than in the vertical. The magnetic fields they produce add by the rules of vector addition. In this way, they add like forces but they are more abstract since they are mutually perpendicular to both force and velocity.

13-7 Some day electrical energy may be generated by sending a high-speed plasma beam, made up of electrons and positive ions, through a magnetic field. Given the right conditions, the electrons will go one way and the positive ions will go the other.

Given: An electron and an ion are traveling at 2.5×10^7 m/s perpendicular to a magnetic field of 6×10^{-3} T. The electron has a mass of 9.11×10^{-31} kg and a charge of 1.60×10^{-19} C. The ion has a mass 3.27×10^{-25} kg of and a charge of 1.26×10^{-17} C.

Find: (a) How great is the force acting on the electron?
   (b) What is the radius of curvature of the electron's path?
   (c) How great is the force acting on the ion?
   (d) What is the radius of curvature of the ion's path?

BACKGROUND

A plasma is like a neutral gas but made up of a mixture of electrons and positive ions. That romantic candle flame on your dinner table is a plasma, but not nearly hot enough for this application. In experimental generators, called magnetohydrodynamic generators, a magnet has indeed been used to separate a plasma beam into positive and negative parts. The temperatures, however, are like those needed for a fusion reactor. We can do it as an experiment, but not on a commercial scale.

13-8 You tear apart the speaker from a junk radio to see how it works. You find a coil of wire attached to the paper speaker cone. The coil of wire fits into a gap between the poles of a permanent magnet. Moving the coil changes the flux passing through it, which produces a voltage.

Given: The coil has 45 turns of wire, each with an area of 2.2×10^{-4} m². It is in the gap of a magnet having a strength of 0.25 T. It takes you 38 ms of time to pull the coil out of the gap of the magnet.

Find: (a) How much magnetic flux passes through each loop of the coil?
(b) What is the emf generated by each loop of the coil?
(c) What is the total voltage produced in pulling the coil from the gap of the magnet?

BACKGROUND

The coil of wire fits into a gap between the poles of a permanent magnet and is intended to exert force on the speaker cone in response to an electrical signal. As the current increases and decreases, so does the force on the paper cone. The paper cone pushes air back and forth, making sound.

As is usually the case with things that turn electrical energy into mechanical energy, a speaker can be used backward. The same mechanism can change mechanical energy into electrical energy. Intercom systems frequently let a speaker serve double duty as a microphone. It's not high fidelity, but you can sure hear what they are saying at the other end.

If you have two junk radios or tape players, you might be able to use the speaker from one as a microphone to drive the amplifier and speaker of the other. Touch various solder connections near where the volume control is attached, and you can usually hear a hum or snap that means you're into the amplifier circuit somewhere. Low voltage battery powered junk equipment won't hurt you. Don't try this with old vacuum tube amplifiers. Also stay away from junk television sets except as parts. There are thousands of volts on a television picture tube.

Solutions to EXAMPLE PROBLEMS

(a) $w = mg$

$g = \dfrac{w}{m}$

$= \dfrac{(8,800 \text{ N})}{(5,000 \text{ Kg})}$

$= \boxed{1.76 \text{ N/kg}}$

(b) $w = mg$

$= (1.14 \text{kg})(1.76 \text{ N/kg})$

$= \boxed{2.00 \text{ N}}$

DISCUSSION

The gravitational field on the surface of Io is about ten percent greater than on the surface of our moon.

SAMPLE SOLUTION 13-1

Given:
A boulder with a mass of 5,000 kg, thrown upward from the surface of Io, has a weight of 8800 N. A rock, also thrown upward from the surface of Io, has a mass of 1.14 kg.

Find:
(a) What is the gravitational field at the surface of Io?

(b) What is the weight of the rock?

SAMPLE SOLUTION 13-2

Given:

A grain of puffed rice having a charge of 83 pC (pico coulombs) experiences a force of 26 μN. A smaller grain of puffed rice has a charge of only 38 pC.

Find:

(a) How great is the electric field near the generator's sphere?

(b) How great is the force on the smaller grain?

(a) $F = Eq$

$E = \dfrac{F}{q}$

$= \dfrac{(26\ \mu N)}{(83\ pC)}$

$= \boxed{3.13 \times 10^{5}\,N/C}$

(b) $F = Eq$

$= (3.13 \times 10^{5}\,N/C)(38\ pC)$

$= \boxed{12\ \mu N}$

DISCUSSION

We have picked numbers to be consistent with an electric field about one tenth of the size needed to produce a spark. The force doesn't seem like much, but it's enough to send a grain of puffed rice into a nice parabolic path.

(a) $V = Ed$

$E = \dfrac{V}{d}$

$= \dfrac{(170V)}{(0.83 \text{ mm})}$

$= \boxed{2.05 \times 10^{5} \text{ V/m}}$

(b) $d = \dfrac{V}{E}$

$= \dfrac{(170V)}{(3 \text{ MV/m})}$

$= \boxed{0.057 \text{ mm}}$

(c) $d = \dfrac{V}{E}$

$= \dfrac{(170V)}{(45 \text{ MV/m})}$

$= \boxed{0.0038 \text{ mm}}$

SAMPLE SOLUTION 13-3

Given:
The "hot" wire in the electric cord you are holding has a peak voltage of 170 V relative to your hand. The plastic insulation on that wire has a dielectric strength of 45 MV/m and is 0.83 mm thick. The dielectric strength of air is 3 MV/m.

Find:
(a) How great is the electric field in the insulation?

(b) What is the minimum distance your finger can come to the metal wire before a spark jumps through the air?

(c) What is the minimum distance your finger can come to the metal wire before a spark jumps through the insulation?

DISCUSSION

The distance a spark will jump through air is quite small at line voltage in the United States. At higher line voltage, the distance is proportionally greater.

The real danger in using a power cord with old cracked insulation is, however, that moisture from your sweaty hands will get into the cracks and serve as a conduction path.

Replace those old cords. Your life is worth more than a few bucks.

SAMPLE SOLUTION 13-4

Given:

Each stage of a photomultiplier tube accelerates electrons through a voltage of 180 V. Electrons have a charge of 1.60×10^{-19} C and a mass of 9.11×10^{-31} kg.

Find:

(a) How much kinetic energy does each electron gain in being accelerated from one stage to the next?

(b) What is the minimum velocity each electron can have as it hits each plate?

(a) $V = \dfrac{W}{q}$

$W = (180V)(1.60 \times 10^{-19}C)\dfrac{1\ J/C}{1\ V}$

$= \boxed{2.88 \times 10^{-17}\ J}$

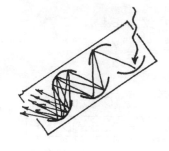

(b) $E_k = \frac{1}{2}mv^2$

$v = \sqrt{\dfrac{2E_k}{m}}$

$= \sqrt{\dfrac{2(2.88 \times 10^{-17}\ J)}{(9.11 \times 10^{-31}\ kg)}}$

$= \boxed{7.95 \times 10^6\ m/s}$

DISCUSSION

The velocity, found from the kinetic energy, is still less than a tenth the velocity of light. We will see that relativistic considerations can be ignored for speeds this small.

(a) $F = qvB$

$= (2.2 \text{ μC})(250 \text{ m/s})5.5 \times 10^{-5}\text{T}$

$= \boxed{3.03 \times 10^{-8}\text{N}}$

(b)

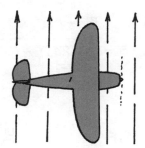

(c) $\dfrac{F_B}{F_W} = \dfrac{qvB}{mg}$

$= \dfrac{(3.03 \times 10^{-8}\text{N})}{(3.3 \times 10^{5}\text{kg})(9.8 \text{ N/kg})}$

$= \boxed{9.4 \times 10^{-15}}$

SAMPLE SOLUTION 13-5

Given:
The airplane having a mass of
3.3×10^5 kg, a charge of 2.2 μC and
a velocity of 250 m/s due east, is flying
through a magnetic field of magnitude
5.5×10^{-5} T

Find:
(a) What is the magnitude of the magnetic force acting on the airplane?

(b) What is the direction of the magnetic force acting on the airplane?

(c) What is the ratio of the magnetic force to the gravitational force acting on the airplane?

DISCUSSION

The force is up when flying east with a positive charge. It would be down if you are either flying west or flying east with a negative charge. It doesn't much matter, however, because the force is so small. Even with two orders of magnitude more charge, the magnetic force would be on the order of a million million times smaller than the weight.

SAMPLE SOLUTION 13-6

Given:

An electron having a charge of $1.60 \times 10^{-19}\,C$ and a velocity of $5.6 \times 10^{7}\,m/s$ is deflected horizontally by a magnetic field having a strength of 0.23 T and deflected vertically by a magnetic field having a strength of 0.32 T.

Find:

(a) What is the magnitude of the resulting magnetic field?

(b) What is the direction, measured from the horizontal, of the resulting magnetic field?

(c) How great is the force acting on the electron?

(a) $B^2 = B_v^2 + B_H^2$

$B = \sqrt{(0.23T)^2 + (0.32T)^2}$

$= \boxed{0.39T}$

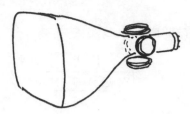

(b)

$\tan\theta = \dfrac{B_v}{B_H}$

$\theta = \tan^{-1}\dfrac{B_v}{B_H}$

$= \tan^{-1}\dfrac{0.23T}{0.32T}$

$= \boxed{35.7°}$

(c) $F = q\upsilon B$

$= (1.60 \times 10^{-19}C)(5.6 \times 10^7\,m/s)(0.39T)$

$= \boxed{3.5 \times 10^{-12}N}$

DISCUSSION

Parts (a) and (b) explore the rules of vector addition as applied to magnetic fields. In part (b), it is worth noting that it is the vertical component of magnetic field that deflects the electrons in the horizontal. That's the field that changes hundreds of times faster than the vertical deflection field—which is actually pointed in the horizontal.

The force in part (c) seems to be mighty small, being three and a half millionths of a millionth of a newton. In a real television set, however, the force is smaller still. The mass of an electron is on the order of ten to the minus 31, so the acceleration produced by this force would be huge.

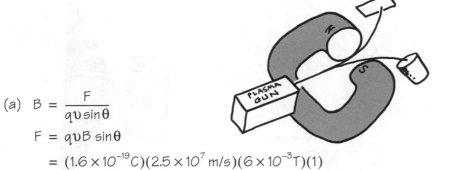

(a) $B = \dfrac{F}{q\upsilon\sin\theta}$

$F = q\upsilon B \sin\theta$

$= (1.6 \times 10^{-19} C)(2.5 \times 10^7 \, m/s)(6 \times 10^{-3} T)(1)$

$= \boxed{2.4 \times 10^{-14} N}$

(b) $F = m\dfrac{\upsilon^2}{r}$

$r = \dfrac{m\upsilon^2}{F}$

$= \dfrac{(9.11 \times 10^{-31} kg)(2.5 \times 10^7 \, m/s)^2}{2.4 \times 10^{-14} N}$

$= \boxed{2.4 \, cm}$

(c) $F = q\upsilon B \sin\theta$

$= (1.26 \times 10^{-17} C)(2.5 \times 10^7 \, m/s)(6 \times 10^{-3} T)\sin 90°$

$= \boxed{1.89 \times 10^{-12} N}$

(d) $r = \dfrac{m\upsilon^2}{F}$

$= \dfrac{(3.27 \times 10^{-25} kg)(2.5 \times 10^7 \, m/s)^2}{1.89 \times 10^{-12} N}$

$= \boxed{1.08 \, m}$

SAMPLE SOLUTION 13-7

Given:

An electron and an ion are traveling at 2.5×10^7 m/s perpendicular to a magnetic field of 6×10^{-3} T. The electron has a mass of 9.11×10^{-31} kg and a charge of 1.60×10^{-19} C. The ion has a mass 3.27×10^{-25} kg of and a charge of 1.26×10^{-17} C.

Find:

(a) How great is the force acting on the electron?

(b) What is the radius of curvature of the electron's path?

(c) How great is the force acting on the ion?

(d) What is the radius of curvature of the ion's path?

DISCUSSION

Since the velocity is perpendicular to the magnetic field, we can substitute unity for the sine term.

Using these numbers, we can see that the electrons will bend into a circular path with a radius of curvature of a few centimeters while the ion will bend the other way with a much longer radius of curvature, on the order of meters.

In this example, we are using a velocity a little less than a tenth the velocity of light so that we do not need to worry about relativistic effects. In a real fusion reactor, the speeds are relativistic. Also in a real plasma beam, the electrons go faster than the positive ions.

The ion charge and mass used are, by the way, about right for a fully ionized gold nucleus.

SAMPLE SOLUTION 13-8

Given:

The coil has 45 turns of wire, each with an area of $2.2 \times 10^{-4} \, m^2$. It is in the gap of a magnet having a strength of 0.25 T. It takes you 38 ms of time to pull the coil out of the gap of the magnet.

Find:

(a) How much magnetic flux passes through each loop of the coil?

(b) What is the emf generated by each loop of the coil?

(c) What is the total voltage produced in pulling the coil from the gap of the magnet?

(a) $\Phi \equiv BA$

$= (0.25 \, T)(2.2 \times 10^{-4} \, m^2)$

$= \boxed{5.5 \times 10^{-5} \, Tm^2}$

(b) $\xi_1 = -\dfrac{\Delta\Phi}{\Delta t}$

$= -\dfrac{5.5 \times 10^{-5} \, Tm^2}{38 \, ms}$

$= \boxed{-1.45 \, mV}$

(c) $\xi_n = nE_1$

$= (45)(-1.45 \, mV)$

$= \boxed{-65 \, mV}$

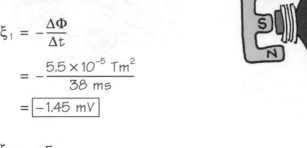

DISCUSSION

A millivolt and a half per turn is only a few tens of millivolts for the whole coil. Doesn't seem like much, but this is a reasonable signal. Amplifying a signal by a factor of a hundred is no trick at all.

ESSENTIAL PROBLEMS

Be sure you can do the Example Problems (without peeking at the Sample Solutions) before working these Essential Problems.

13-9 The planet Pluto, discovered in 1930 by Clyde Tombaugh, appears to be a big snowball about two-thirds the diameter of our Moon, but with a much smaller gravitational field. It's a great place for a jumping contest, or so they say. Visit Pluto with a friend, and you could both jump very high. Dress warmly though. It's cold.

Given: With your space suit and gear, you have a mass of 120 kg and, on Pluto, a weight of 71 N. Your friend, with more equipment, has a mass of 180 kg.

Find: (a) What is the gravitational field at the surface of Pluto?
(b) What is your friend's weight?

13-10 What holds up clouds? One might think that all those little droplets of water would just fall as rain. Slow rain maybe, but rain nonetheless. The earth, as it turns out, has an electric field and each droplet has a bit of charge. The Coulomb force between cloud droplets hold them apart and the earth's electric field holds them up.

Given: A large droplet of water in a cloud has a charge of 2.6×10^{-17} C and is supported by a force of 2.3×10^{-15} N. A smaller droplet of water has a charge of only 8.8×10^{-18} C.

Find: (a) How great is the earth's electric field holding up the cloud?
(b) How great is the force holding up the smaller droplet?

13-11 Some insulating materials are better than others for the spark plug wires in an automobile. That's high voltage that gets sent to those spark plugs. Plug wires with silicon resin insulation are top of the line, but expensive. Cut one by accident, and you should replace the whole set. Some mechanics will use a little epoxy putty to fix a mistake, but it's not as good.

Given: Voltage is delivered to the spark plugs at 20 kV. The silicon resin insulation on the plug wire has a dielectric strength of 20 MV/m and is 1.2 mm thick, except where it has been accidentally cut. The dielectric strength of epoxy putty is 15 MV/m.

Find: (a) How great is the electric field in the silicon insulation?
(b) What is the minimum thickness of silicon resin needed for insulation?
(c) What is the minimum thickness of epoxy putty needed to repair the cut?

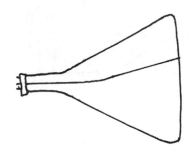

13-12 Electrons in a television picture tube are given so much kinetic energy before they slam into that phosphor screen that they light up the phosphor to provide us with a seductively bright, beautiful picture. They go so fast, in fact, that their mass begins to increase in accord with Einstein's laws of relativity. Even from classical mechanics, ignoring relativistic effects, we can see that those electrons must move along right quick.

Given: Electrons in a television picture tube are accelerated through a voltage of 5.0 kV. Electrons have a charge of 1.60×10^{-19} C and a mass of 9.11×10^{-31} kg.

Find: (a) Before slamming into the phosphor, how much kinetic energy does each electron gain?
(b) Ignoring relativistic effects, what is the velocity of the electrons?

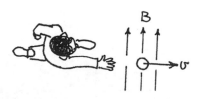

13-13 A baseball player in Ecuador throws a positively charged baseball eastward, perpendicular to the earth's magnetic field.

Given: The baseball, having a mass of 0.132 kg, a charge of 0.72 μC and a velocity of 25 m/s due east, is flying through a magnetic field of magnitude 4.5×10^{-5} T.

Find: (a) What is the magnitude of the magnetic force acting on the ball?
(b) What is the direction of the magnetic force acting on the ball?
(c) What is the ratio of the magnetic force to the gravitational force acting on the ball?

13-14 We think of the magnetic field of the earth as pointing north. Actually that's only true at the equator. In San Francisco, for example, the magnetic field points more downward than sideways. Not that you would notice by the magnetic force acting on the charge produced when you comb your hair.

Given: A hair comb having a charge of 3.2×10^{-17} C and a velocity of 5.6 m/s is deflected but little by the earth's magnetic field having horizontal component of 1.8×10^{-5} T and a vertical component of 5.4×10^{-5} T.

Find: (a) What is the magnitude of the resulting magnetic field?
(b) What is the direction, measured from the horizontal, of the resulting magnetic field?
(c) How great is the force acting on the comb?

13-15 One day, if fusion energy becomes practical, garbage dumps may become a thing of the past. Solid waste may be completely vaporized in the star-hot flame of a fusion torch and sent through a magnetic field to be separated into raw material.

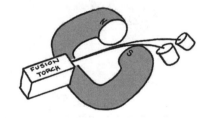

Given: An iron ion and a gold ion are traveling at 6.8×10^5 m/s perpendicular to a magnetic field of 1.4 T. The iron ion has a mass of 9.27×10^{-26} kg and a charge of 4.16×10^{-18} C. The gold ion has a mass of 3.27×10^{-25} kg and a charge of 1.26×10^{-17} C.

Find: (a) How great is the force acting on the iron ion?
(b) What is the radius of curvature of the iron ion's path?
(c) How great is the force acting on the gold ion?
(d) What is the radius of curvature of the gold ion's path?

13-16 The strength of the earth's magnetic field is easy to measure with a sensitive voltmeter or oscilloscope. Just wind a coil of wire on an aluminum bicycle wheel (such as can be found as scrap in most bicycle shops). Holding the coil with the axis in the direction of the earth's magnetic field, turn it so that the axis is perpendicular to the earth's field. The voltage produced by the coil as the flux goes to zero gives a good measure of the field strength.

Given: The coil has 150 turns of wire, each with an area of 0.33 m² It is at a place on the earth where the magnetic field strength is 5.7×10^{-5} T. It takes you 0.16 s of time to turn the axis coil from parallel to perpendicular to the earth's magnetic field.

Find: (a) How much magnetic flux passes through each loop of the coil?
(b) What is the emf generated by each loop of the coil?
(c) What is the total voltage produced in the coil?

MORE INTERESTING PROBLEMS

13-17 Hearing what causes the ocean tides, you devise a simple experiment. Ocean going folk say that the moon causes water on the surface of the earth to form tidal bulges which, as the earth turns, make two high tides and two low tides each day. The tidal bulge on the side of the earth closest to the moon is said to be formed by the stronger gravitational field of the moon on that side as compared to the other side.

Given: You hold an apple having a mass of 0.480 kg up to the moon when it is directly overhead. The moon has a mass of 7.36×10^{22} kg and is 3.78×10^8 m away. Twelve hours later, your apple is 1.28×10^7 m farther from the moon. The universal gravitational constant is 6.67×10^{-11} Nm^2/kg^2.

Find: (a) What is the gravitational force between the apple and the moon overhead?
(b) What is the gravitational field of the moon at the near side of the earth?
(c) How much smaller is the gravitational field of the moon on the far side of the earth?

13-18 Don't try this, but if you were to fall into a black hole, long before you got to where the black edge of the thing begins, you would be torn apart by tidal forces. Falling feet first, your feet would simply fall faster than your head. Your boots would become so heavy, you would probably struggle to take them off.

Given: Your right boot, still on your foot, has a mass of 2.8 kg and is 38 km from the center of a black hole having a mass of 7.36×10^{30} kg. Your left boot, having the same mass as the right, is in your hand and is 0.78 m farther from the black hole. The universal gravitational constant is, as always, G = 6.67×10^{-11} Nm^2/kg^2.

Find: (a) What is the gravitational force between the right boot and the black hole?
(b) What is the gravitational field of the black hole at your right boot?
(c) How much smaller is the gravitational field of the black hole on your left boot?

13-19 In a controversial study published in1979, Werthemer and Leeper suggested a relationship between childhood leukemia and the distance a child lives from electrical power lines. Hundreds of subsequent studies catalyzed by this report have been inconsistent and inconclusive. Most concern has been about magnetic fields rather than electric fields because electrical forces at any reasonable distance from a power line are comparable to the normal thermal fluctuations of electric field within a normal cell.

Given: Sixty meters from a five hundred kilovolt powerline, the electric field inside the body due to the power line is about 3×10^{-6} V/m, producing a force on a certain ion in the chromosome inside a cell of 9.2×10^{-25} N. Thermal fluctuations of the electric field of 2×10^{-2} V/m acting over a distance of 62 μm produce random voltage changes across the same chromosome.

Find: (a) What is the charge on that certain ion?
(b) How great is the force on the ion due to thermal fluctuations of electric field?
(c) How great are the random voltage fluctuations experienced by a chromosome?

13-20 Electric fields travel far better in water, where sharks hunt, than in air, where you and I hunt. Sharks find their lunch by feeling their prey's electric fields generated by normal muscle and nerve action. Land hunters like us feel only much stronger fields. In the surf, therefore, the shark can feel us long before we can feel it.

Given: We can feel an electric field of 15 kV/m, producing a force of 3.5×10^{-5} N per strand of hair, making our hair stand up. Sharks, on the other hand, can detect an electric field as small as 1.6×10^{-5} V/m over a distance of 25 cm from the shark's body.

Find: (a) What charge must be induced in a strand of hair?
(b) If you could sense as small an electric field as a shark, how great would be the force on the same amount of charge?
(c) How much voltage does the shark feel?

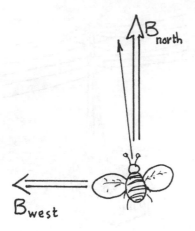

13-21 Honey bees navigate by sensing tiny changes in the earth's magnetic field. Other animals such as sea turtles and homing pigeons also use the earth's field, but honey bees seem to be the most sensitive. Or so say people who try it. They generally add an artificial magnetic field perpendicular to the earth's field, mostly changing the angle of the resultant but slightly increasing its magnitude.

Given: A honey bee can sense a change in direction of as small as 0.80° in the horizontal component of magnetic field when that component of the earth's field is 1.8×10^{-5} T.

Find: (a) How much artificial field must be added for this change in angle?

(b) What change in magnitude does this represent?

13-22 Some bacteria grow around tiny magnetic mineral crystals, allowing them to tell up from down. Down is food. Up is deadly light. People who play with bacteria like to trick them with an artificial magnetic field perpendicular to the earth's field. Bacteria, as it turns out, are pretty good sports. Not like honey bees in this respect.

Given: Bacteria can sense a change in direction of as small as 38° in the vertical component of the earth's field when that component of the earth's field is 3.8×10^{-5} T.

Find: (a) How much artificial field must be added for this change in angle?

(b) What change in magnitude does this represent?

13-23 Sharks use their electric sensing ability not only to locate their prey but to navigate. The shark's long sensing organ, called the Ampullae of Lorenzini, can feel the voltage induced as the shark crosses the earth's magnetic field. To the shark, swimming east feels different than swimming west.

Given: Sharks can detect a voltage as small as 4μV over a distance of 25 cm from the shark's body. The shark swims for a time of 2.2 s perpendicular to the earth's magnetic field of 4.8×10^{-5} T.

Find: (a) What is the smallest amount of flux change needed to produce the voltage a shark can feel?
(b) How much area must be swept out to change this amount of flux?
(c) How fast must the shark swim to feel the earth's field?

13-24 Pipe fitters who work in chemical plants are exposed to high levels of magnetic fields. Large currents in overhead bus bars produce magnetic fields that will lift a pipe wrench right off a worker's shoulder. Even so, a person would need to walk fast through one of these magnetic fields to produce a voltage that a human can feel.

Given: Human hair can sense a voltage as small as 40 V over a distance of 2.5 mm from the body. A pipe fitter walks for a time of 0.55 s perpendicular to a magnetic field of 5.8×10^{-2} T.

Find: (a) What is the smallest amount of flux change needed to produce the voltage a pipe fitter can feel?
(b) How much area must be swept out to change this amount of flux?
(c) How fast must the pipe fitter walk to feel this magnetic field?

ANSWERS

Solutions to **ESSENTIAL PROBLEMS**

13-9   (a) 0.59 N/kg; (b) 106 N

13-10   (a) 88 N/C; (b) 7.8×10^{-16} N

13-11   (a) 17 MV/m; (b) 1.0 mm; (c) 1.3 mm

13-12   (a) 8.0×10^{-16} J; (b) 4.2×10^{7} m/s

13-13   (a) 8.1×10^{-10} N; (b) up; (c) 6.26×10^{-10}

13-14   (a) 5.7×10^{-5} T; (b) 72°; (c) 1.02×10^{-20} N

13-15   (a) 4.0×10^{-12} N; (b) 1.07 cm; (c) 1.2×10^{-11} N; (d) 1.25 cm

13-16   (a) 1.88×10^{-5} Tm^2; (b) 0.118 mV; (c) 18 mV

Solutions to **MORE INTERESTING PROBLEMS,** *odd answers*

13-17   (a) 1.65×10^{-5} N; (b) 3.44×10^{-5} m/s^2; (c) 2.2×10^{-6} m/s^2

13-19   (a) 3×10^{-19} C; (b) 6×10^{-21} N; (c) 1.2×10^{-6} V

13-21   (a) 2.5×10^{-7} T; (b) 1.8×10^{-9} T

13-23   (a) 8.8×10^{-6} Tm^2; (b) 0.183 m^2; (c) 33 cm/s

CHAPTER 14 *Modern Physics*

Something Old and Something New

CLASSICAL PHYSICS IS FREQUENTLY CALLED Newtonian physics, because it is based on the laws of motion laid down by Newton. It generally refers to that body of physics that was fairly well understood by the turn of this century. At about that time, some new developments came to light that led to the theory of relativity on the one hand and atomic and nuclear physics on the other. These topics are generally referred to as *modern physics,* which is somewhat misleading, since it implies that there is a replacement for the old-fashioned kind. This is not the case. Modern physics supplements our classical understanding, adding depth and extending it into the realms of the very fast and the very tiny. At the boundaries, however, the old and the new match perfectly, leaving both intact in their own right.

We shall end this book by probing these boundaries—first in the realm of the very fast, to see how the rules of relativity are consistent with classical physics, and then in the realm of the very small, to take a look at the atom and its nucleus. Should the physics of these realms seem strange, it is because they are not a part of your everyday experience. You must stretch your imagination to understand what is right and proper in these realms of existence.

The Special Theory of Relativity

Our present understanding of things that move at velocities near the speed of light is based on a theory called the *special theory of relativity.* Some things don't happen exactly as you might expect at these speeds. The laws of classical physics, which work so well at ordinary velocities, clearly need some modification as one enters the realm of the very fast. For one thing, the laws of motion must be modified to account for nature's speed limit. We have found that nothing can be accelerated through the speed of light. Try as we may, we just can't get anything to go any faster. In fact, we cannot even get bits of matter, regardless of how tiny, up to a speed quite as great as the speed of light. The laws of classical physics do not predict this speed limit, and they need to be modified to account for it.

To understand how things should behave at high velocities, imagine that you are going nearly that fast yourself. The theory of relativity is largely based on how things look from a moving **frame of reference**. A frame of reference is an arbitrary set of axes that you use to judge the position or velocity of something. Let us examine this concept of a frame of reference to see how you already use it in your thinking.

Most people usually select a frame of reference in which most of the things around them seem stationary. Their minds make this selection so automatically that they seldom stop to question the process. For example, you might be sitting at home at your desk, and your frame of reference would be selected so that your chair, desk, bookcase, walls, ceiling, as well as most of the other objects in the room, would seem stationary. Even if you got up and walked across the room, you would probably keep using the same frame of reference. You would likely think of yourself as moving and everything else as stationary. The idea that they are actually stationary would probably be so firmly established in your mind that you would not even give it a second thought. When an earthquake strikes, people frequently panic because of the sudden and unexpected loss of their stationary frame of reference. Most people find it very disquieting to have everything upon which they judge their bearings suddenly start to move.

On the other hand, people sometimes select a frame of reference that they know full well is moving but that still makes things around them seem stationary. Commuters gathered around the bar in the club car of a train are aware of the fact that the bar, their drinks, and the people standing around them are all moving, but they still judge position and velocity relative to this moving frame of reference. Two friends standing at the bar might match coins to see who will pay for the next round of drinks. They flip the coins into the air, catch them, and slap the coins onto the backs of their hands in the time-honored gambling ritual. If you asked them to stop and think about it, they would agree that the "true path" of the coins was an arc, as the train carries both them and the coins down the track at the same time that the coins go up into the air and back down to their hands. If you don't remind them of the motion of the train, however, they would intuitively use a moving frame of reference in which the bar, their drinks, and the club car furniture all seem to be stationary.

The special theory of relativity is based on the assumption that the laws of physics all work the same in a moving frame of reference, without regard to the velocity of that frame of reference, as long as the velocity is constant.

Figure 14-1

Commuters in a club car understand relativity, at least in a limited sense.

That is just what the commuters in the club car have learned to expect, at least insofar as it applies to mechanical things around them. The coins flipped into the air go through the same motion, as far as the gamblers are concerned, without regard to the speed with which the world is flashing by outside. The bartender pours liquor the same without glancing out the window to check on the velocity of the train. Only when the train lurches or sways, would the motion of the train affect the difficulty with which people stagger across the floor.

Albert Einstein was the first to suggest that the same principle of relativity that applies to mechanical things, such as coins flipped into the air and people staggering across the floor, would apply as well to electrical and magnetic phenomena, even including the behavior of light. He saw no reason why some laws of physics, some forms of experience, would look different in a moving frame of reference if others do not. He therefore took it as a basic postulate that all forms of experience, even experiments to measure the speed of light, are relative to the frame of reference you use and would work the same in any uniformly moving frame of reference. This means that the laws of physics are no better suited to one frame of reference than to any other.

This assumption forms the basis of the *special* theory of relativity, so called because it is restricted to the special case of frames of reference moving at a constant velocity—cases in which there is no acceleration. This is the principle of relativity that applies to the train moving smoothly down the track without lurching or swaying. In this special case, the assumption is that the motion of the reference frame has *no effect* on the laws of physics as long as there is no acceleration.

After you have fully understood the special theory of relativity, you will be in a position to appreciate the **general theory of relativity**, which applies even to frames of reference that accelerate. This is the theory that applies to the train as it lurches or sways. The general theory of relativity is based on the assumption that the effects of acceleration are the same as a change in the gravitational field. For now, however, we will not get into this theory of gravitation. It will be enough to understand something about the structure of space and time implied by the special theory. All we will do in this book will be to introduce some of the basic ideas of relativity, so that you can start to think about some of the puzzles they contain.

The Structure of Space and Time

Einstein was the first to realize that the relativity experienced by the commuters gathered around the bar in the club car of the moving train is even more important than the mental picture they have formed of space and time, based on what they probably feel is common sense. Common sense tells them, for example, that they are really moving and that the world outside their window is really stationary. If they wanted to go even faster, they could take a walk toward the front of the train. Common sense tells them that their resultant velocity would be the sum of a person's walking speed and the speed of the train. This simple rule for adding velocities assumes that everybody agrees on their measurements of space and time. Common sense tells them that time is independent of space and that any two observers who see two events simultaneously will agree that they happened at the same time. These common-sense assumptions are based on common experience with things moving at ordinary velocities. Einstein realized that these common-sense assumptions would be modified if we lived our lives at velocities nearer to the speed of light.

Take the simple rule of adding velocities, for example. If the train were going at near the speed of light, say nine-tenths the speed of light, and you were to walk forward in the train at two-tenths the speed of light, the common-sense rule for adding velocities would state that your actual velocity would be eleven-tenths the speed of light. Common sense therefore seems to tell you that you can go faster than the speed of light. Since you can't, common sense must be in need of some fixing up.

Even the notion that you have an "actual" velocity assumes that one frame of reference is better than all the rest. You may like to think that the world outside your train window is stationary and that you are moving. Of course, the world isn't really stationary; it is circling the sun. Even the sun isn't stationary. We know that the sun is a member of a galaxy that is spinning on its axis. Even the galaxy is moving through space. Common sense would seem to suggest that the sum of all of these motions should add up to something we could measure. People have looked long and hard for some sort of absolute frame of reference by which we could measure how fast we are "really" going. There just doesn't seem to be any.

One of the most famous experiments in the history of physics was an experiment to measure the velocity of the earth through the universe by using the speed of light as a reference. In 1887, Albert Michelson teamed up with a young chemist named Edward Morley to build a big interferometer, which could compare the velocity of light in two perpendicular paths. Michelson had invented this interferometer and found it useful for measuring tiny distances, even to a fraction of a wavelength of light.[1] Light is split into two perpendicular paths by a half-silvered mirror and then reflected back on itself by two mirrors at the end of each path. When the light in these two paths gets back to the half-silvered mirror, it forms light and dark fringes of constructive and destructive interference, depending on whether the light waves reinforce or cancel each other. Michelson realized that the wavelength in these two paths depends on the velocity of light. He wondered if the motion of the earth around the sun wouldn't affect the apparent velocity of light in one direction more than the other. It was a

---

[1] The Michelson interferometer is still used as one of the finest measures of distance. Our standards of length, in fact, are now based on the wavelength of light as measured by this instrument.

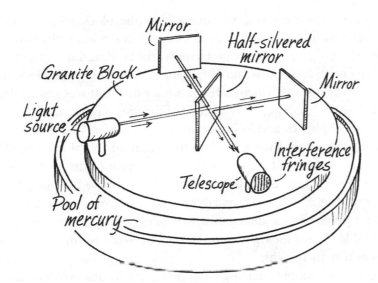

Figure 14-2

The Michelson interferometer splits up a beam of light and compares the light paths by recombining the light to produce fringes of constructive and destructive interference of the light waves.

simple matter to point one arm of his interferometer in the direction he knew the earth was going, note the position of the fringes, and then turn the apparatus around so that the other arm of the interferometer was pointed in that direction. Any shift in the fringe pattern would indicate a difference in the apparent velocity of light in these two perpendicular directions. But if there was any shift in the fringe pattern due to the velocity of the earth, he discovered, it was too small to measure with any of the interferometers he had built to that date. That was when he teamed up with Morley to build a bigger interferometer. They mounted the mirrors and everything on a large granite block, so that the distances would be as stable as possible, and then they floated the whole thing in a large pool of mercury so that there wouldn't be any chance that the granite block would bend in the slightest as they turned it about a vertical axis.

Michelson and Morley thought that they would get at least some change in the fringe pattern due to the motion of the earth. It seemed to them that light in one path would have to chase after the mirror at the end of the path, which would be receding at the velocity of the earth, until it finally is reflected and returns to the half-silvered mirror, which the light beam would now see as coming toward it—also at the velocity of the earth. Light in the other path, meanwhile, would have to take a slightly diagonal path to get back to the presumably moving half-silvered mirror. The net effect should have been small but measurable if only the motion of the earth around the sun was to be taken into account, but they thought they might see other motions as well.

In order to get the path lengths long enough to measure the tiny change they predicted, Michelson and Morley's interferometer became so sensitive that you couldn't even stand in the same room with it. The heat of your body would warm one side of the granite block and cause a noticeable shift in the fringe pattern due to thermal expansion. Michelson and Morley had to stand outside the room and manipulate the block with strings and pulleys to turn it on its axis so as to exchange the roles of the two paths. But they finally got all of these problems worked out, and then they tried it.

Much to their surprise, nothing happened. Turning the block on its axis didn't change the fringe pattern even a tiny bit. It was just as though the earth weren't moving. This was one of the most famous experiments in the history of physics, and its importance lies in the fact that it failed.

For years, people tried to explain the failure of the Michelson-Morley experiment in terms of some change in the length of the apparatus. The motion of the granite block through space was supposed to push the electron orbits out of shape or something, and that would make the granite block get smaller in the direction it was going. This change in length was supposed to be just the right amount to compensate for the change in the apparent velocity of light.

All sorts of people started getting interested in this problem at about the turn of the century, and they all tried different ways to measure the velocity of the earth through space. All attempts failed. The theories became more complicated as they tried to explain away the failure of each fresh approach. It began to look as if all the laws of nature had entered into some kind of conspiracy to prevent these people from measuring the absolute velocity of the earth.

Then in 1905, a young unknown Swiss patent examiner, Albert Einstein, published a paper that swept away all of these contrived difficulties by suggesting a whole new theoretical approach. Instead of assuming that the motion of the earth made Michelson and Morley's granite block contract, Einstein suggested that the contraction was a property of space itself. Before that, no one had thought to question the common-sense view that space and time were absolute quantities, somehow fundamental constants in all the universe. Einstein saw space and time as things to be measured. He reasoned that if every method you can think of for measuring the distance between two points gives the same result, then that is the distance that actually exists as far as you are concerned. Similarly, if you measure the time between two events, then that is the time that actually exists.

It is true that someone who was stationary with respect to the sun would get a different result if he were to measure the distance between the mirrors mounted on Michelson's granite block. This hypothetical observer would see the earth and everything on it as being in motion. To that person, it would seem that any distance measured by Michelson and Morley along the forward direction of their travel would be foreshortened by a certain factor, depending on their velocity. As we shall see, that factor turns out to be the square root of 1 minus the ratio of velocity, *v*, squared over the velocity of light, *c*, squared:

$$\sqrt{1 - \frac{v^2}{c^2}}$$

This factor is always less than 1 as long as v is something between zero and c. Thus, the person who is stationary with respect to the sun would see the length l', measured in Michelson and Morley's moving frame of reference, as being shorter than his own length, l, by an amount that would be found by multiplying by this factor:

$$l' = l\sqrt{1 - \frac{v^2}{c^2}}$$

This does not mean that the hypothetical observer's measurements are any better than Michelson's. In fact, Michelson would see the hypothetical person's measurements as foreshortened in exactly the same way. From Michelson and Morley's point of view, the hypothetical observer is the one who is moving. The motion of the earth would make a person who is stationary with respect to the sun seem to be moving in the opposite direction. Instead of agreeing with that person's view of space, the Michelson and Morley team would see any distances measured by him as being shorter by exactly the same factor. Strange as it may seem, each sees the other's distances as foreshortened—and they are both right.

Figure 14-3

A hypothetical observer who is stationary with respect to the sun would see the distance between the interferometer mirrors mounted on Michelson and Morley's granite block as being foreshortened by the motion of the apparatus.

The resolution of this seeming paradox comes from the fact that both parties also measure time differently. The person who is stationary with respect to the sun sees Michelson's clocks as running slow. If that person measures the time, t, between any two events, he will see Michelson as measuring a longer time t', between the same two events. He will see Michelson's time measurements as lengthened or dilated by the same factor by which he sees Michelson's space measurements foreshortened. Just as you multiply by this factor to get a shorter distance, you divide by it to get a longer time:

$$t' = \frac{t}{\sqrt{1 - \dfrac{v^2}{c^2}}}$$

Michelson would also see the hypothetical person's time measurements as dilated in exactly the same way. To Michelson, that person is moving, and moving clocks, as we will see, always seem to run slow. Just as each person sees the other's distances as foreshortened, they also see the other's time as expanded.

These two changes work together to bring each observer's view of things back into agreement. If only one effect occurred, one frame of reference would have to be better than the other. As we will see by considering several examples, however, the fact that both effects occur allows both observers to see the same events happening and agree on such things as cause and effect, even if they see the events from different perspectives of space and time.

The truly remarkable feature of the theory of relativity is that it points out a whole new structure of space and time. Instead of being absolute things, tied to some fundamental frame of reference, they are relative things, which can be seen equally well from different perspectives. Instead of being independent things, as time and space might seem to be if viewed only from one frame of reference, they are closely related concepts, so that a new perspective from a different frame of reference shows you more of one and less of the other.

This relativistic structure of space and time differs in several respects from our common-sense notion of the way things are. Consider, for example, our common-sense notion of simultaneity. We think of two events as being simultaneous when we see them happen at the same time. Just as it seems natural to think in terms of an absolute frame of reference in space, it is also natural to think of that frame of reference as being absolute in time. We might like to think that our idea of "at the same time" is an absolute. Just as we might like to think of ourselves as being the center of the universe in space, we might also like to believe our zero point in time is better than everyone else's. From the relativistic point of view, however, the idea of any absolute frame of reference is only a simplified mental model, too simple, in fact, to deal with reality. Just as our spatial reference frame is a mental contrivance, so is our notion of simultaneity. If two things happen at different places but at the same time as you see it, they might well seem to happen at different times as another observer would see it. If he is moving relative to you, he will have a different perspective of what is meant by "at the same time."

Let us return to the Michelson-Morley experiment as an illustration. As far as Michelson and Morley were concerned—or at least as far as they could tell from their experiment—the light beam was split up to travel two equal paths and be reflected back simultaneously by the mirrors at the ends of their paths. The interference pattern produced when the two beams returned indicated that this is exactly what did happen. To our hypothetical observer, who is stationary with respect to the sun, however, the two light beams would not bounce off the two mirrors simultaneously. The two paths would not be the same length, and the speed of light would be the same in both directions only in *his* frame of reference. He would, however, agree with Michelson and Morley on the final result. He would see these two effects as ganging up to produce the same fringe pattern.

At this point, you might well wonder who is "correct." As we consider examples in relativity, you will find that you have a strong tendency to look at the observers in moving frames of reference as being somehow mistaken in interpreting what they see. It might seem that these people should realize that *they* are the ones who are moving and that they should take that into account when they measure time. The fact is that Michelson and Morley thought that the hypothetical observer had a better view of reality than they did, since he would be more stationary. When they tried to partly compensate in their calculations for their own motion, however, they came out with a result that failed to agree with the experiment. That was because they didn't understand the unified structure of space and time. Einstein's contribution to our understanding of that structure made it unnecessary for either observer to compensate for his own motion.

If you find it difficult to put aside your common-sense view of space and time, think what it must have been like for people in Einstein's day. When his paper was published in 1905, there was a huge wave of protest. If you look under "relativity" in the card catalog of the Library of Congress, you will find that nearly half the books written on the topic are intended to show why it would never work. All these books now sit forgotten on the shelf, of historical interest only. Even in 1921, 16 years after Einstein's original paper on relativity was published, the topic was so controversial that it wasn't directly mentioned in Einstein's Nobel prize presentation.

One reason that our common-sense view of space and time has such a strong grasp on our thinking is that we have little direct experience with frames of reference that go fast enough for relativity to make a noticeable difference. You have to be going at near the speed of light relative to another observer before what you see differs much from what the observer sees. These velocities lie so far beyond what is technically possible for people to travel, even today, that it requires a stretch of the imagination even to think about going that fast.

A Thought Experiment at Near the Speed of Light

People of today have a cultural advantage over Einstein's contemporaries when it comes to thinking about frames of reference moving at high velocities. People of Einstein's day had plenty of experience with moving frames of reference on trains and the like, but the intelligentsia had a low regard for science fiction. Science fiction had not yet developed into an accepted art form. Today, however, sci-fi fans have little trouble imagining what it would be like to go at speeds near the speed of light.

To get an idea of why length is contracted in a moving frame of reference, let us imagine that a spaceship has been assigned the task of making some space surveys at velocities below the speed of light. Two special survey ships have been outfitted to make careful angular measurements on far-distant objects. Captain Epstein puts Officer Stern in charge of one of these vessels and Officer Bower in charge of the other. These officers are directed to keep their survey craft exactly 1 light-second in front of and behind the mother craft. (One light-second, the distance traveled by light in 1 s, is less than the distance from the earth to the moon.) Mr. Stern and Ms. Bower position their ships and set their instruments to measure the far-distant objects of interest. While the computer on board the mainship is comparing the readings reported from the two survey craft, Captain Epstein directs the helm officer to get underway at one-tenth the velocity of light, $0.1c$. Lieutenant Antonelli, the communications officer, signals Stern and Bower to get underway at this velocity. They both get the message at the same time and simultaneously increase their speed to $0.1c$.

Let us use some advanced science fiction technology to simplify for our imagination the problems involved in this little speed adjustment. Given the physics we understand today, it would take a little over a month to attain the velocity we are talking about at the comfortable acceleration of 9.8 m/s^2. Even at ten g of acceleration, it would take over 3 days. Using sci-fi technology, however, we can pretend that someone has been clever enough to figure out a way to do it in a few moments. It is important to ignore what goes on during the acceleration, because we are now considering only the special theory of relativity, the theory of reference frames moving at a uniform constant velocity.

Suppose we are observing this whole operation from a space station that remains in the original frame of reference. We make a graph with the position of the mainship and the two survey craft on one axis and time on the other. If the mainship started out from our starbase, the graph of its position before it got underway would be a straight line along the time axis, representing no change in position.

If we could use the vertical axis to represent time, Mr. Stern and Ms. Bower will both have positions that are shown as vertical lines before they get their first signal to speed up.

We could choose any convenient scale to represent distance and time; let us measure distance in light-seconds, so that light signals travel upward and outward at a 45° line, covering equal distances on both scales of our graph. This choice of scale tends to put time and distance on an equal footing in our thinking.

We can see that Bower and Stern get the signal to get underway simultaneously by the fact that the first two 45° lines cross Ms. Bower's and Mr. Stern's paths at the same height on the time axis. At that point, the paths bend to be inclined at a certain angle to the space axis, so that their velocity can be represented as progress along the space axis with increasing time.

We need to assume that the helm officer on the mainship delays execution of the captain's command for 1 second to allow that command to reach the survey craft so that they can all start up simultaneously. The path representing the mainship also moves in the vertical direction and bends at the same angle at the same time as the other two paths. That way, the mainship stays halfway between the two survey craft.

We can now see how distances get foreshortened in a moving frame of reference. If Captain Epstein orders both craft to speed up again, his order will reach Bower before it reaches Stern, at least in our frame of reference. We see Ms. Bower traveling toward the command ship and thus intercepting

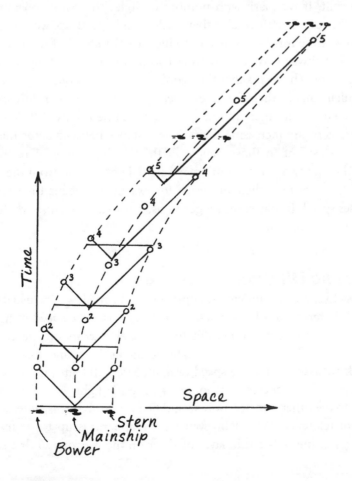

Figure 14-4

A space-time diagram shows that the distance between Ms. Bower and Mr. Stern gets smaller, because Bower always gets the signal to speed up sooner than Mr. Stern.

the signal before it has a chance to catch up to Mr. Stern. On our graph this would be represented by the 45° signal path intercepting Ms. Bower's path earlier, or lower on the time axis, than Mr. Stern's, since both craft's paths are inclined to the right. Each time the captain orders an increase in speed, Bower gets the order first and speeds up sooner than Stern. The distance between them therefore decreases the faster they go.

The same argument for the distance between ships can be made for the ships themselves and any material object on board the ships. Everything is squeezed down in the direction of travel. We can stand outside on our space station and say that this squeezing down is caused by the length of time it takes to get information from one end of an object to the other about how fast it is supposed to be going, but we could not expect people on board the moving spacecraft even to notice this squeezing effect, since it happens uniformly on everything around them. Everything, even a meter stick that they might use to measure things with, is contracted in length by the same amount.

The people on the spacecraft seem to us to have a warped perspective of time as well as space. Notice that the helmsman on the mainship must always wait for a certain length of time before increasing the speed of the mother craft if he is to keep the command ship halfway between the two survey craft. If we mark with numbered circles the points on our graph where the survey craft receive their orders and speed up, we can see that the proper time for the helmsman to increase the speed of the mainship is at a point halfway between the other two points along a straight line connecting them. The fact is that the people on the spacecraft see these points as simultaneous in their frame of reference. The helmsman just increases the speed of the mother craft at the time when he thinks both Ms. Bower and Mr. Stern are increasing theirs. The distance between ships may seem shorter to us; but to those people in the moving frame of reference, it seems that the two survey craft are exactly 1 light-second from the mother craft. The helmsman therefore waits for exactly 1 s of his time and then increases speed. It may seem longer to us, but that is because we think his watch is running slow.

The Time Dilation

It is easy to see why a moving timepiece would run slow. The usual trick is to think in terms of a **light clock**, a clock that uses the speed of light as a standard for measuring time. Put two parallel mirrors on the ends of a meter stick and send a pulse of light bouncing back and forth between them. If we assume that the speed of light, 3.00×10^8 m/s, is a constant in any frame of reference, the amount of time for light to travel from one mirror to the other 1 meter away would be $(1/3.00) \times 10^{-8}$ s, at least in that frame of reference. We might even fix the apparatus up to be used as a clock by putting only a thin layer of silver on the mirrors, so that some of the light gets through to photocells mounted behind the mirrors. One photocell would be connected to an amplifier that says "tick" every time the light pulse bounces from the mirror. The other photocell would, of course, be connected to an amplifier that says "tock."

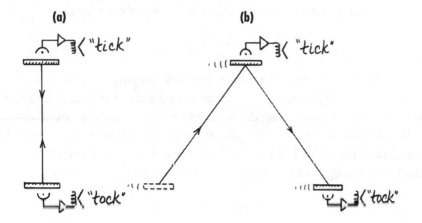

We could now define a standard time interval as the time between a "tick" and a "tock" of our light clock. Using the definition of velocity,

$$v \equiv \frac{d}{t}$$

solved for distance,

$$d = vt$$

we can relate the lengths of the light paths, as seen from two different points of view, to the time intervals as seen in those frames of reference. For a stationary clock, or at least one that you see as stationary, the time interval, t, would be defined in terms of the perpendicular distance between the mirrors and the velocity of light, c, as it travels between these mirrors. If the clock were moving, however, you would see the time interval as being longer because the distance the light would have to travel would be longer, as you see it. This longer time interval, t', would be defined in terms of the velocity of light, c, and the hypotenuse of a right triangle having as one leg the distance you originally used for a stationary clock. In fact, the moving clock runs slow by exactly the same proportion that the hypotenuse of this right triangle is greater than the leg that represents the distance between the two mirrors.

Let us label the sides of a right triangle in Figure 14-6, using a prime to indicate the time t' as measured by the moving clock and no prime to indicate the time t as measured by the stationary clock. In this diagram, the hypotenuse is the velocity of light times the primed time:

$$\text{Hypotenuse} = ct'$$

Figure 14-5

(a) A stationary light clock would measure time based on the speed of light and the distance between two mirrors. (b) A moving light clock would run slower, because the distance traveled by the light would need to be longer.

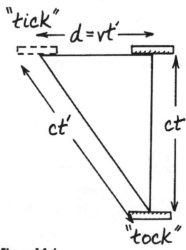

Figure 14-6

We can relate the time measured by a moving clock to that measured by a stationary clock by using the Pythagorean theorem to relate the distance between the mirrors of a light clock to the distance we see the light travel while we see the mirrors move in our frame of reference.

while the side leg is the velocity of light times the unprimed time:

$$\text{Side leg} = ct$$

As you see it, the hypotenuse is longer than the side leg because the mirrors moved a distance that is represented by the top leg of the triangle:

$$\text{Top leg} = vt'$$

The distance the mirrors move is their velocity times the primed time, t', which is the time as measured in the same frame of reference in which the light seems to travel along the hypotenuse of the right triangle.

We can now apply the Pythagorean theorem to show that the times as measured in these two frames of reference are related by the square of the hypotenuse being equal to the sum of the squares of the two sides:

$$(ct')^2 = (ct)^2 + (vt')^2$$

We can get the primed values of time on one side of the equation by subtracting $(vt')^2 = v^2(t')^2$ from both sides of the equation:

$$c^2(t')^2 - v^2(t')^2 = c^2 t^2$$

This can be solved for t' by factoring $(t')^2$ out of the two terms on the left and then isolating it on one side of the equation and taking the square root. It is traditional, however, to divide both sides of the equation by c^2 first;

$$\frac{(t')^2(c^2 - v^2)}{c^2} = t^2$$

simplify by dividing c^2 into both terms of the numerator, and then take the square root of both sides

$$t'\sqrt{1 - \frac{v^2}{c^2}} = t$$

and solve for t'

$$\boxed{t' = \frac{t}{\sqrt{1 - \dfrac{v^2}{c^2}}}}$$

This relationship is called the **relativistic time dilation**. The theory of relativity is based on the assumption that not only do moving clocks run slow but that time itself runs slow in a moving frame of reference and that length is foreshortened by the same factor.

The Relativistic Point of View

We have seen how the measurement of length might be distorted by the foreshortening of a moving meter stick, and we have seen how the measurement of time might be distorted by the slow running of a moving light clock. Neither of these two effects taken alone require the theory of relativity to understand. In fact, the relationships for length foreshortening and time dilation were known before Einstein proposed the theory of relativity, but people thought that these relationships applied only to the distortion of physical objects. They thought that the distortions of their measuring instruments seemed to conspire to make it difficult to measure their velocity relative to some fundamental frame of reference. Einstein was the first to suggest that these were more than experimental difficulties, that they were actually theoretical impossibilities. He suggested that the real reason that you couldn't measure your velocity relative to some fundamental frame of reference was that no fundamental frame of reference exists. Einstein realized that if the absence of a fundamental frame of reference were a basic fact of nature, the length foreshortening and time dilation are more than independent effects resulting from the distortion of the physical objects used to make the measurements. They become two related effects, which apply to the structure of space and time itself. If there is no such thing as an absolute frame of reference, then the structure of space and time must be such that all measurements of any physical phenomena, including the velocity of light, must come out the same in any frame of reference. Einstein's genius was to realize that the length contraction and time dilation were two aspects of a single phenomenon and that they fit together in just the right way to suggest this new view of the structure of space and time.

It is possible to get pretty well into the study of relativity and still cling unconsciously to the idea of a fundamental frame of reference and an absolute time permeating the universe. You might, for example, agree that Mr. Stern's survey craft is closer to Ms. Bower's and that they have difficulty measuring this fact because their clocks are running slow, but it might still seem that Stern should know that this is the case. Mr. Stern is a very smart guy; he should know he is the one who is moving. It is true that Stern would remember having accelerated from our frame of reference into the one he is in now. Also, he would know that his frame of reference is moving relative to us. But Mr. Stern might think of us as moving relative to *him*. As long as he remains at a constant velocity, his point of view is just as valid as ours. From his frame of reference, *our* lengths are foreshortened and *our* clocks run slow.

It seems a bit strange at first that observers moving relative to each other would both observe the same changes in space and time in the other's frame of reference. If we could look out into the universe and find another race of people like ourselves, except that they are moving at near the speed of light, we would see them living their lives in slow motion. If they were moving fast enough, their life spans might well be several times our own. They, on the other hand, would see us living in slow motion. We would live, as they see it, several times as long as they. Each race would see the other as having a longer life span. How could they both be right? If we considered only the time dilation, or only the length foreshortening, this would be a paradox. Both time and distance must be taken into account if we are to understand how people moving relative to us make measurements that differ in both respects but agree in the final result.

To see how a different view of space in a moving frame of reference must be accompanied by a different view of time, let us return to our example of the mainship and the two survey craft. Suppose the two survey craft are to fly past two space buoys in the course of their mission, and suppose they are to maintain a constant velocity of $0.8c$ (eight-tenths the speed of light). The buoys are 1 light-hour apart, as we see them. From our point of view, the mainship would take a little longer than 1 hour to travel this distance at less than the speed of light. In fact, we could easily calculate the time from the definition of velocity. Solving the definition of velocity for time,

$$v \equiv \frac{d}{t}$$

$$t = \frac{d}{v}$$

and setting the velocity equal to $0.8c$ while the distance is c times 1.00 hour,

$$t = \frac{(c)(1.00 \text{ hour})}{0.8c}$$

we see the mainship taking 1.25 hours to travel from one buoy to the next:

$$t = 1.25 \text{ hours}$$

Although we will measure 1.25 hours for the starship to travel 1.00 light-hour at eight-tenths the speed of light, we do not expect its crew to measure that much time. Their clocks are running slow. The time dilation in their frame of reference makes them measure less time than we measure for the same events. Their time as we see it is t', while their time as they see it is t. Solving the time dilation relationship for the time as they see it,

$$t' = \frac{t}{\sqrt{1 - \dfrac{v^2}{c^2}}}$$

$$t = t' \sqrt{1 - \frac{v^2}{c^2}}$$

At eight-tenths the speed of light, their clocks are running so slow that they only measure 1.25 hours, as we see it, as three-quarters of an hour:

$$t = (1.25 \text{ hours})\sqrt{1 - \frac{(0.8c)^2}{c^2}}$$

$$= (1.25 \text{ hours})\sqrt{1 - 0.64}$$

$$= (1.25 \text{ hours})\sqrt{0.36}$$

$$= (1.25 \text{ hours})(0.6)$$

$$= 0.75 \text{ hours}$$

As it turns out, the crew on the mainship agrees that the time they measure between passing one buoy and the next is 0.75 hour, but they do not see this time measurement as resulting from their clock running slow. To them, the shorter time is a natural result of a shorter distance. In their frame of reference, they are stationary and the buoys approach and pass them at 0.8c. The distance between the buoys, however, is foreshortened from their point of view. In their frame of reference, the distance between the space buoys is only 0.6 light-hour.

$$l' = l\sqrt{1 - \frac{v^2}{c^2}}$$

$$= (1.0c \cdot \text{hour})\sqrt{1 - \frac{(0.8c)^2}{c^2}}$$

$$= 0.6c \cdot \text{hour}$$

Given this shorter distance, it seems quite natural that the time necessary to travel it at 0.8c would be 0.75 hour. The definition of velocity, solved for time,

$$v \equiv \frac{d}{t}$$

$$t = \frac{d}{v}$$

gives 0.75 hour as a natural result.

$$t = \frac{0.6c \cdot \text{hour}}{0.8c} = 0.75 \text{ hour}$$

It is no accident that these numbers come out the same. The special theory of relativity is built on the premise that people in different frames of reference will agree on the same experimental results. Since the laws of physics work the same in any frame of reference, the same result should occur no matter how you look at it. Since one of the laws of physics is that the speed of light is the same universal constant in any frame of reference, people in different frames of reference will not see things happening in the same way, but they will still see the same things happening. The nice thing about the special theory of relativity is that people can not only agree on the final result but they can understand each other's points of view.

Suppose that Mr. Stern has been directed to leave a record of his distance standards on the space buoys as he passes. That way, Ms. Bower can compare her distance standards against those left by Stern when she passes the same buoys a few moments later. Mr. Stern therefore arranges two marking lasers (or other marking devices) to burn images of both ends of a meter stick at the same time on the hull of the space buoy. To make certain that the two lasers go off simultaneously, he actuates them both from a button located at the 50-cm mark halfway between the ends of the meter stick. Since the signal from the button travels the same distance to the two phasors at the ends of the stick, they both make their marks at the same time.

As long as Ms. Bower looks at these marks while she is traveling at the same speed Mr. Stern was going when he made them, there is no doubt that she will see them as 1 meter apart. Bower and Stern, after all, are not moving relative to each other. But let us look at this from the point of view of a workman who has stopped by one of the space buoys on his regularly scheduled service rounds. Suppose the workman has made whatever adjustments are necessary and has paused for a lunch break. As he sits back to enjoy his cup of coffee and watch the maneuvers of the mainship, Mr. Stern comes along in the first survey craft.

Figure 14-7

Mr. Stern's marking phasors don't go off simultaneously in the workman's frame of reference.

Much to the workman's distress, Mr. Stern fires his marking lasers at the exact point on the hull of the space buoy where the workman has left his meter stick. No harm is done, however, as the workman's meter stick is foreshortened in Mr. Stern's frame of reference. From Mr. Stern's point of view, the space buoy and everything on it are zipping past at eight-tenths the speed of light and are shorter by a factor of $\sqrt{1 - v^2/c^2}$, which turns out to be six-tenths at this velocity. Since Mr. Stern sees the workman's meter stick as only 60 cm long, he is easily able to miss it by putting his marks beyond both ends of it with plenty of room to spare at either end.

This might all seem very logical from Mr. Stern's point of view, but you might ask how the workman sees this result. When he goes over to pick up his meter stick, he will agree that Mr. Stern has been kind enough to miss it. He finds burn marks on the hull of the space buoy, but these marks are clearly more than 1 m apart. The strange thing is that the workman sees Mr. Stern's marking lasers as being closer than 1 m apart. From his point of view, Mr. Stern is zipping along at 0.8c; everything on Mr. Stern's survey craft must be foreshortened by a factor of six-tenths. He is therefore left with the problem as to how marking lasers that are closer than 1 m apart somehow leave marks that are further than 1 m apart.

This problem is resolved by the other side of the space-time distortion. The only way that marking lasers can leave burn marks that are significantly further apart than the separation between phasors is if they go off at different times. The result observed by the workman could only have occurred if the phasor toward the rear of Mr. Stern's craft went off first and then, after it had traveled for a considerable distance, the laser toward the front of the craft went off. This, in fact, is exactly the way things look in the workman's frame of reference. He sees Mr. Stern reach out and touch the button halfway between the two marking lasers, but he does not see the signal from the button reach the two lasers at the same time. The laser traveling toward the button (the one toward the rear) gets the signal first and goes off before the one that is traveling away.

The fact that the signal from the button travels in both directions with the same velocity in the workman's frame of reference makes the two events, which Mr. Stern sees as simultaneous, happen at different times. The workman can, however, see how Mr. Stern might think that the two lasers went off simultaneously. Light scattered from the first laser that goes off must travel a longer distance to catch up with Mr. Stern than light from the second. Mr. Stern sees the flashes from the two lasers at the same time and, as far as he is concerned, they went off simultaneously.

People in both frames of reference agree on the final result. Both see the burn marks miss the ends of the workman's meter stick. Mr. Stern thinks it is because the workman's meter stick is foreshortened; the workman thinks it is because Mr. Stern's marks were made at different times. Which is right depends on your preference. The Einsteinian point of view is that they are both right.

Relativistic Mechanics

Einstein's theory of relativity requires some modification of Newton's second law of motion. As objects approach the speed of light, their acceleration is no longer strictly proportional to the force applied, as stated in the classical expression of Newton's second law:

$$F = ma$$

At least, F and a are not proportional in the sense that m is constant. Newton's second law can be fixed up, however, so that it continues to work even at velocities near the speed of light if we include the same relativistic factor that seems to affect space and time. That is, we simply define a **relativistic gamma factor**, γ, which increases as the square root of one minus v squared over c squared decreases:

$$\gamma = \frac{1}{\sqrt{1 - \frac{v^2}{c^2}}}$$

As long as the velocity, v, is very small compared to the velocity of light, the square root factor is not much different from one, and the relativistic mass, which is the product of γm of the gamma factor and mass, is not much different from the **rest mass**, m_1, which is the good old constant inertial mass that we have become familiar with in classical physics. Thus, Newton's second law is little affected by this redefinition of mass until the velocity approaches that of light. When the velocity does approach the velocity of light, both the mass and the acceleration become a function of velocity. Newton's second law still holds, but only in the form that refers to the change of momentum with respect to time.

$$F = \frac{\Delta(\gamma m v)}{\Delta t}$$

At the one extreme where the velocity is small, mass is constant and we have the classical $F = ma$. At the other extreme where v approaches c, the square root factor in the denominator of the relativistic mass expression approaches zero and the mass becomes infinite. We find that this is actually consistent with our experience.

You might wonder what experience we could possibly have had that would make this relativistic mass more than a theoretical premise. It is true that our present technology is far short of being able to get the velocity of objects as large as people and spaceships up to any value where the relativistic nature of mass would have any noticeable effect on Newton's second law. It is quite common, however, for us to get little particles of mass up to relativistic speeds.

One device in which relativistic particle speeds are quite common is the ordinary television set. The electrons in the picture tube may easily be accelerated up to about half the speed of light before they crash into the phosphor screen. The light you see as you watch television is therefore energy that is released by a sudden change in the relativistic mass of these electrons.

The relativistic nature of mass has a substantial effect on the technology of particle accelerators. It is no trick at all to get electrons or other particles up to 99% of the speed of light. They do that in the first 10 ft of the 2-mile-long accelerator at Stanford University. For the rest of the 2 miles the velocity increases by only tiny amounts. First they get the particles up to 99.9% of the speed of light, then 99.99%, and then 99.999%. The accelerating force is exerted on the particles for the whole 2 miles, dumping more and more energy into the particles, but the only effect on their velocity is to add nines after the decimal point. The closer the speed of the particles comes to the speed of light, the harder it is to make them go any faster.

What happens to the energy they keep dumping into the particles? If the particles don't go much faster, where does all the energy go? The energy, it turns out, goes into increasing the mass of the particles instead of their velocity, and this increased mass has a very real effect on their inertia.

Particles that come out of an accelerator not only have increased mass in terms of how hard they smash into the target but they also have increased mass in terms of how hard they are to deflect. It is common, for example, to deflect the particles from a high-energy accelerator from one experiment to the other so that several experiments can be set up at the same time. People who operate the accelerator find that they need a stronger magnetic force to deflect the beam on a good day when everything is working just right and the particles in the beam have the maximum energy the accelerator can deliver. Thus, the relativistic mass is just as substantial as the rest mass in terms of the inertial effects it has on Newton's second law.

The idea that energy can show up in terms of increased mass led Einstein to what some people feel is one of the most important concepts of the twentieth century—the celebrated equivalence between mass and energy. The idea is that energy is proportional to mass, with the proportionality constant being the square of the speed of light.

$$E = mc^2$$

This relationship suggests more than that equal increases of energy will be accompanied by equal increases in mass. It also suggests that there is energy associated with rest mass and that the energy is huge. The speed of light is a very big number; the speed of light squared is that much bigger again.

Some feel that the discovery of this relationship between mass and energy led directly to the age of atomic energy and the big bombs. Others feel that it only laid a theoretical groundwork by suggesting that vast amounts of energy are somehow available. The fact is that we have been making use of this relationship for a long time without knowing it. When you burn a log in the fireplace, you are converting a tiny fraction of the mass of the log into energy. The only reason we didn't know that a long time ago is that the fraction is far too tiny to measure by conventional means. The forms of atomic energy that we can now make use of by such nuclear reactions as fusion and fission still only convert a tiny amount of

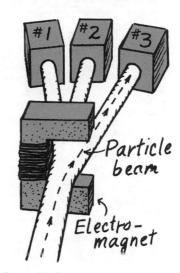

Figure 14-8

The increased relativistic mass of particles coming from an accelerator makes them harder to deflect with a magnetic field.

the available rest mass into energy, but the fraction is now large enough to measure. The only way we know of getting total mass-energy conversion is to allow anti-matter to combine with ordinary matter. However, no one has a sizable chunk of anti-matter, and no one knows where to find any either. For now, that is probably just as well. We had best first learn how to handle the tiny amounts of energy available from atomic physics.

Atomic Physics

Having looked into the realm of the very fast, it is now time to look into the realm of the very tiny. Suppose you were to commit the most heinous of all crimes—falling asleep in class. Since the lecture was exciting, exhilarating, and dynamic, we must assume you were up all night (studying). Let us imagine that you shrank in size as you fell from your seat and became smaller and smaller as you approached the floor. Instead of falling onto the floor, you would find that you fell into it. As you approached microscopic dimensions, the relatively smooth floor would open up into jagged canyons, which would become even rougher the more you fell. The walls and bottom of the canyon you would fall into would begin to take on the structure of cells and organic fibers as you approached the level of microbiology. You would begin to see long chainlike hydrocarbon molecules as you shrank to the level of biochemistry. If you could get close enough to one of these molecules to see the atoms individually, you would see something very strange indeed. You would find that the picture most people form of what an atom should look like is a gross oversimplification of reality.

A bit of historical perspective is useful in sorting out the various mental models people use in dealing with atomic structure. The planetary model most people think of when they picture an atom, where the electrons orbit the nucleus like planets going around the sun, is actually an early model first proposed by Niels Bohr just after the birth of quantum theory. We still tend to think in terms of this simple picture, even though it has been replaced several times over by models that give progressively better results but become progressively more abstract.

Quantum theory got its start just after the turn of the century, when a man named Max Planck got the idea that energy comes in little chunks, which he called quanta, just like matter comes in little chunks called atoms. He showed how this assumption would explain the frequency distribution of radiation from hot radiating bodies. Nobody took Planck seriously until Einstein, a brash unknown at the time, showed how the same assumption would explain the energy of electrons excited by light falling on a vacuum tube photocell. This was the work, by the way, for which Einstein was later given the Nobel prize. It was an important piece of work, because it established Planck's quantum hypothesis as a significant physical theory describing the fundamental nature of matter and energy. Light, which had been thought of as a wave for centuries, was shown to be broken up into

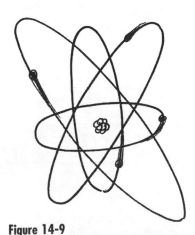

Figure 14-9

A cartoonist's concept of an atom shows electrons going around the nucleus like little planets going around a tiny sun.

little quantized chunks called photons. The wave nature of light remained, but light was now understood to have particle properties as well.

Planck's quantum hypothesis having been lent strength by Einstein's use of it to explain the photoelectric effect, Niels Bohr then used it, together with new evidence that most of the mass of an atom and all of the positive charge is located at the center, to produce the planetary model of the atom. The electrons are shown traveling around the nucleus like planets going around the sun. This model really only lasted a couple of years, but we still use it because it was the last model that was easy to visualize. Bohr's planetary model is so beautiful and forms such a clear picture in our minds that we sometimes have to remind ourselves that the atom isn't really like that miniature solar system. The biggest problem that Bohr had with his planetary model was that he couldn't explain why the orbits were stable other than to say that the energy of the electron must be quantized somehow.

The next, more sophisticated, model of the atom was invented by a Frenchman named L. V. de Broglie. It occurred to de Broglie that if light waves had particle properties, maybe particles like electrons might also have wave properties. This stroke of genius immediately explained Bohr's quantized orbits in terms of standing waves. Only those orbits could exist that met the condition that the wave would fit in the sense that it would close on itself. The de Broglie wave hypothesis was an instant success. His picture of an atom with the electron spread around the orbit as a standing wave was so popular, as a matter of fact, that it lasted even less time than Bohr's planetary model. It attracted so much attention that very little time elapsed before other people figured out significant improvements.

Within a few years two different people had independently figured out the rules of physics that should apply to electrons treated as quantized waves. One man, Erwin Schrödinger, worked out the basic equation of wave mechanics that plays the same role in quantum mechanics as Newton's second law, $F = ma$, plays in classical mechanics. The Schrödinger equation describes the behavior of a de Broglie wave in terms of a partial differential equation, which cannot really be understood without the language of calculus. Even the picture of the atom described by the Schrödinger wave equation is a bit difficult to comprehend. As is frequently the case, the more sophisticated you make a model to better describe reality, the more difficult that model is to comprehend. This particular model of the atom has been very successful and has lasted for several decades, but it is so mathematically abstract that we can only paint a hazy picture of it here.

Instead of seeing an electron as located at a particular point in space, or even as being spread out around an orbit, the Schrödinger picture sees the electron's position as being described by a wave function, usually represented by the Greek letter psi, Ψ, which is spread out in three-dimensional space as a sort of cloud. This wave function cloud has greater or less density here and there, depending on the energy and angular momentum

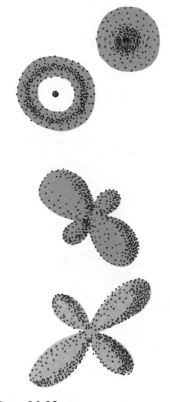

Figure 14-10

Wave function solutions to Schrödinger's equation can be represented by clouds having different density shapes.

of the electron, but it does not have sharply defined edges. The physical meaning of this wave function, Ψ, is a bit abstract, as it happens to be a complex function having both real and imaginary parts, but the absolute square of the function, when evaluated at a particular point in space, turns out to be the probability of finding the electron at that point.

Different wave functions, Ψ, which form solutions to Schrödinger's wave equation for different energies and angular momenta of the electrons, can be represented by density clouds having different shapes. Some are round and centered on the nucleus with a finite probability of finding the electron in the center of the nucleus itself. Some are doughnut-shaped, with a hole in the middle. Some are clouds with two or more bulges.

A few people may still want to cling to the idea that there is such a thing as an electron in the form of a hard little particle, like a miniature BB, somewhere in that cloud of probability. This view is shot down by the fact that some wave functions have two density bulges separated by a null plane. There is equal probability of finding the electron in either part of the cloud above or below the plane, but there is exactly zero probability of finding the electron in the plane itself. A proponent of the pure particle theory of the electron would therefore have to figure out a way for the same electron to spend an equal amount of time traveling about above and below the plane without ever crossing over. It would seem that the electron just cannot be a simple particle.

Models such as particles and waves are pictures we form in our minds based on our everyday experiences with objects large enough to see and feel. These models may be dear to our hearts, but they simply fail to describe reality in atomic dimensions. Even the clouds that we have been using to describe the density of an electron's wave function are just models to help us picture what is going on. We should keep in mind the fact that electrons don't really look like that at all. What do they look like? The answer is that they don't look like anything you have ever seen. Electrons are so small that it doesn't even make sense to talk about looking at one in the conventional sense. To look at it would be to bounce several photons off it, but the very act of bouncing even just one photon off an electron will cause it to be long gone from the position in which it was "seen" by- the photon. To ask what an electron looks like is something like asking what an earthquake would look like. You can see what it does and you can locate it in a general way, but you can't actually see it.

At the same time that Schrödinger was working out his wave mechanics, Werner Heisenberg was working out an equivalent set of quantum mechanical rules, starting from, of all things unanswerable questions. Heisenberg's so-called matrix mechanics are a set of rules that produce exactly the same results as Schrödinger's wave mechanics, but they are based on Heisenberg's **uncertainty principle**. Heisenberg pointed out that it is impossible to measure simultaneously the position and momentum of a particle to greater than a certain precision. Based on Planck's quantum hypothesis and de Broglie's wave-particle duality for an electron, Heisenberg figured out that the ultimate precision in measurement would in theory be when the uncertainty in the momentum, $\Delta(mv)$, times the uncertainty in the position, Δx, would be about equal to Planck's constant h.

$$\Delta(mv) \cdot \Delta x \cong h$$

Heisenberg also figured out that the same uncertainty between momentum and position would hold between energy and time. If you know the energy of a particle to within an uncertainty, ΔE, then the smallest uncertainty of time, Δt, for which you know this energy is when the product of the uncertainties is approximately equal to Planck's constant.

$$\Delta E \cdot \Delta t \cong h$$

Heisenberg took these uncertainties as more than a limit of technology; he took them as a limit to knowledge itself. Starting from this theory of knowledge, he developed a calculus of observable quantities representing dynamic variables by matrices. The mathematics is just as much fun as that used by Schrödinger, and the result turns out to be the same. People who deal with particles all the time find it convenient to be conversant with both approaches.

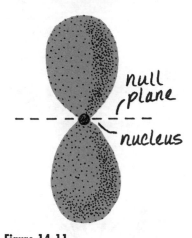

Figure 14-11

Some wave functions have two density concentrations separated by a null plane.

Nuclear Physics

One of the jolliest results of the quantum mechanics developed by Schrödinger and Heisenberg in dealing with atomic structure turned out to be that it worked nearly as well for nuclear structure. Nuclear structure, however, seems to be a great deal more complex than that of the outer atom, and we still haven't got it all figured out yet. The more we look at the nucleus, the more complex it seems to be. Still, there is a great deal that we do know and are able to put to practical use.

We know, for example, that there are two kinds of stable particles in the nucleus, and we know that they both have almost the same amount of mass. One kind has a positive charge and the other is neutral. The ones with a positive charge are called **protons**; the ones that are neutral are called **neutrons**. They both have a mass of about two-thousand times the mass of an electron, although the neutron has a tiny bit more mass than the proton. They both tend to stick to one another with about the same amount of force, sort of like candy-coated popcorn of two different colors in a popcorn ball. Because of their similarities, both kinds of particles are called **nucleons**.

The protons, with their positive charge, serve to hold the negatively charged electrons in their orbits. Each proton has the power to hold one electron in an electrically neutral atom, since protons have exactly the same size charge as an electron, but of opposite sign. The number of protons in the nucleus therefore determines the chemical properties of the atom by determining the possible structures of the electron orbits that can occur.

The number of neutrons in the nucleus has no direct effect on the chemistry of the atom. Neutrons serve, however, to help hold the nucleus together. Without neutrons, protons would have too much electrical charge to stick together. There is a strong force of attraction between nucleons but not strong enough to hold two naked protons together. You need a neutron or two around to lend attractive force and help stick two protons together. The more protons you have, the more neutrons you need to hold them together. For light elements, you need about equal numbers of neutrons and protons for a nucleus to stay together. For heavy elements, however, you have to throw in extra neutrons, because the strong nuclear force of attraction is very short-ranged. All the protons in a nucleus are electrically repelled by each other, even on opposite sides of the nucleus, but the force of attraction acts over such short distances that it is effective only between neighboring nucleons, whether they be neutrons or protons. The electrical repulsion builds up in heavy elements to the point where you need to mix in a disproportionately large number of neutrons to add their attractive force for stability. Lead, for example, has nearly one and a half times as many neutrons as protons. After a certain point, however, you just cannot add in enough neutrons to stabilize the nucleus.

One of the limiting factors in the size of a nucleus is the fact that neutrons are not stable by themselves. A naked neutron will spontaneously decay into a proton plus an electron. Out of a bunch of naked neutrons, about half of them will decay in 12 minutes. A neutron seems to need protons around to keep this from happening. After the size of a nucleus reaches a certain point, the neutrons so outnumber the protons that there are not enough protons in the mix to prevent the neutrons from decaying. It is almost as if one of the neutrons in a big heavy nucleus doesn't see enough protons around and decides to "kick out" an electron and become a proton itself.

This ability of a neutron to change itself into a proton by "kicking out" an electron might make you think that maybe a neutron is really a proton with an electron inside. This is an attractive thought, but it is wrong. One of the many reasons why this cannot be true, or at least this simple, is that ordinary electrons are far too big to fit into a neutron. Only by adding a great deal of energy to an electron can you get its dimensions down to the point where it will even fit into the nucleus of an atom, let alone into a single nucleon. According to Heisenberg's uncertainty principle, the size of a particle, or the uncertainty in its position, is inversely proportional to its momentum. A neutron has a comparably large momentum because of its rest mass.

$$\Delta(mv) \cdot {}_{\Delta x} \cong h$$

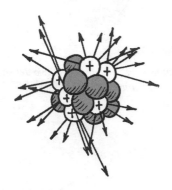

Figure 14-12

The nuclear force of attraction holds a nucleus together, but it is so short-ranged that it only acts between neighboring nucleons, whereas all the protons push on each other.

The smaller rest mass of an electron, nearly two thousand times smaller than a proton, means that the uncertainty in its position is nearly 2,000 times greater. It is therefore best to think of the neutron

$$\Delta(mv) \cdot \Delta x \cong h$$

as being a separate entity, distinct from a proton and an electron until it somehow makes the change. Someday, we hope to find some simple theory to explain what these different kinds of little boys and girls are made of, but we just don't understand it yet. What we do know is that it's not all as simple as sugar and spice.

Alpha, Beta, and Gamma Rays

Early investigators found out that radioactive substances seemed to emit three distinct kinds of rays. They named them with the first three letters of the Greek alphabet in the order with which they could be stopped by progressively heavier shielding. The easiest to stop are alpha rays. They can be stopped by a reasonably heavy piece of paper. Beta rays go right through a piece of paper but are stopped by a fairly thin sheet of metal. Gamma rays are the hardest of all to stop and require lead or other heavy shielding to contain them.

Alpha rays are easy to stop because they are big, highly charged, particles. They are what amounts to helium nuclei that have "boiled off" the nucleus of a heavy unstable atom. A helium nucleus is a pair of protons held together with the help of a pair of neutrons. If the heavy atom has too many protons to be stabilized even by an extra number of neutrons, it is possible for excess energy to get into a pair of protons near the surface of the nucleus, and they will leave with a pair of neutrons like a molecule evaporating from a drop of hot liquid. It is always a pair of neutrons and protons, because this configuration is more stable and requires less energy to escape than a single proton leaving the nucleus by itself. The reason for this has to do with the wave nature of neutrons and protons, together with the fact that they act like little magnets. Their magnetic moments, or spins, cause them to pair up and stick together like magnets holding each other by the north and south poles. This pairing is not strong enough to hold two lonesome protons together, but a pair of neutrons added in makes a very stable configuration indeed. So when they come out, they all come out together.

An alpha particle is easy to stop, because it has two positive charges that produce such a strong electric field that it shakes apart just about every molecule it comes near, leaving a trail of positive and negative ions. This takes so much energy that the particle is soon brought to a stop. Even traveling through nothing but air, an alpha particle comes to a stop after only a few centimeters. Even a piece of paper will stop an alpha particle. After it stops, it grabs up a couple of electrons to become nothing more than a humble helium atom.

Beta particles are a little harder to stop, because they carry only a single charge. They are actually nothing more than high-speed electrons that have somehow been "kicked out" of the nucleus whenever a neutron changes into a proton. It makes up for its small mass by having a large velocity—which is another reason that it doesn't lose its energy in as short a distance as does an alpha particle. It doesn't stay around long enough to ionize many of the atoms that it comes near. It requires more of a direct hit before it gives up some of its energy.

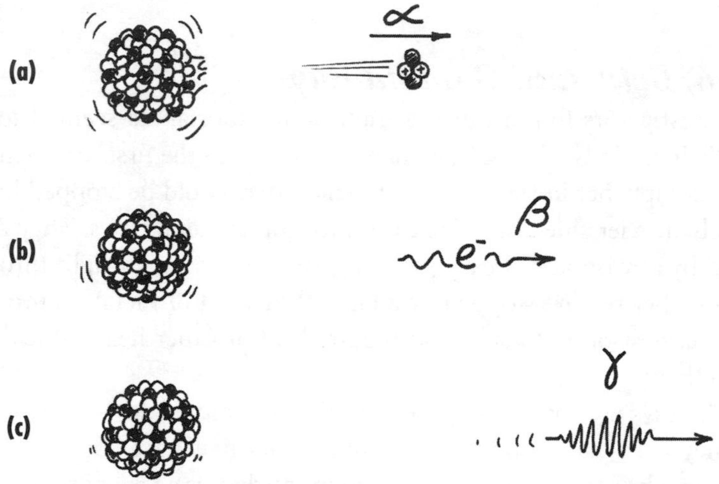

Figure 14-13

(a) An alpha particle is a helium nucleus, two protons and two neutrons, given off by a heavy nucleus. (b) A beta particle is a high-speed electron given off when a neutron turns into a proton. (c) A gamma ray is an energetic photon of electromagnetic radiation.

Gamma rays are the hardest of the three to stop, because they don't have a charge of any kind. Like visible light, they are simply photons of electromagnetic radiation, but they are of much higher energy and frequency. Visible light is emitted when electrons jump from one orbit to another having a lower energy. Gamma rays are emitted when nucleons seem to do a similar sort of thing inside the nucleus. The energy states of the nucleus are widely separated, so that the photons of "light" emitted each carry a large amount of energy.

The relationship between the energy of a photon and its frequency forms the basis of Planck's original hypothesis, which started the whole quantum theory. It was Planck's idea that the energy, E, in a photon is just directly proportional to the frequency f of the radiation. The proportionality constant h is called **Planck's constant** in honor of the man who got the whole quantum theory started.

$$E = hf$$

Planck's constant has a value of 6.63×10^{-34} J · s.

The frequency of gamma radiation is in the range of 10^{20} Hz to 10^{25} Hz, which is five to ten orders of magnitude greater than the frequency range of visible light (10^{14} Hz to 10^{15} Hz). This high frequency gives gamma rays so much energy that they travel right through light materials just like X rays do. (X rays are in the frequency range of 10^{17} Hz to 10^{20} Hz.) All the energy in a photon must be given up in a single interaction or else none of it can be. That is the very idea of a quantum, an indivisible unit of energy. The gamma ray photon therefore keeps right on going until it makes a direct enough hit with a massive enough chunk of matter that it can give up all its energy in a single interaction. That is the reason that gamma rays are so hard to stop. It is also the reason why the most efficient shielding materials against gamma radiation are the heavy metals, whose large nuclei make good targets.

CHECK QUESTIONS

1. How much energy is in a visible light photon whose frequency is 5×10^{14} Hz?

Answer: 3×10^{-19} J

2. What is the frequency of a gamma ray photon whose energy is 7×10^{-12} J?

Answer: 1×10^{22} Hz

Alpha, beta, and gamma rays were only the first kinds of radiation identified by earlier investigators. After these three were identified and named for the order of difficulty in stopping them, other kinds of radiation were discovered that are even harder to stop. One of the most fantastic of these is made of tiny particles called **neutrinos**, which have zero rest mass and no charge. They have so little interaction with matter that a neutrino can go right into the center of the sun with a good chance that it will come out the other side without having been deflected. They are so difficult to detect that people looked for them for 26 years before they found hard evidence that they really exist. Fortunately, we are so transparent to them that they don't bother us much.

Neutron Activation

Another form of radiation that is hard to stop, and which can be quite destructive, is made up of neutrons released by many nuclear reactions. Neutrons are difficult to shield against because they are electrically neutral. They don't even feel the strong electric fields that occupy most of the volume of an atom, but must bump right up against a nucleus if they are to have any interaction at all. Neutron radiation therefore passes right through heavy shielding material, making only occasional collisions with the tiny nuclei of the atoms in the material, and even these collisions are mostly glancing blows in which the neutrons keep on going. The potential destructive nature of neutron radiation comes from the fact that neutrons are radioactive themselves and that they make the materials they pass through radioactive as well. We have mentioned the fact that naked neutrons tend to turn themselves into protons by "kicking out" an electron or beta particle. This beta decay of the neutrons forms only part of the problem, however. The other part of the radiation hazard comes from the tendency of neutrons to change the very nature of the material through which they pass. Since they are electrically neutral, they don't even need to be going particularly fast to bump right up against the nucleus of an atom and upset its balance in one of several ways. If the incoming neutron either sticks to or knocks out a proton, the whole character of the nucleus is changed.

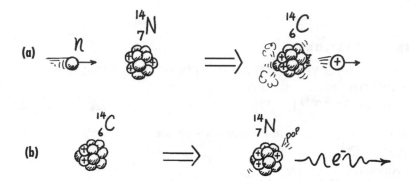

Figure 14-14

(a) Neutron activation of nitrogen 14 takes place when a neutron comes along and bumps out a proton, changing the nucleus to carbon 14. (b) Carbon 14 is radioactive and beta decays back into nitrogen with a half-life of 5,770 years.

Only certain combinations of neutrons and protons go together to form a stable nucleus. We have already mentioned the fact that the number of protons in a nucleus is the thing that determines the chemical character of the atom by determining the positive charge of the nucleus. The number of protons is therefore called the **atomic number** of the atom and dictates which element it is. Any atom having six protons in the nucleus will have the chemical properties of carbon, for example, while any atom having seven protons will be nitrogen. An atom can have different numbers of neutrons in the nucleus, however, and still remain the same element. The different number of neutrons will give the atoms of a certain element different amounts of mass but will have little effect on the chemical properties. The number of neutrons will determine whether the nucleus will be stable or not.

Atoms of the same element having different numbers of neutrons in the nucleus are called different **isotopes** of the element. Many elements have several stable isotopes as well as a few unstable ones. As we mentioned before, the isotopes of the lighter elements that tend to be most stable are the ones in which there are about the same number of neutrons as protons. The most common isotope of carbon, for example, has six neutrons and six protons. That is called carbon 12, because there are 12 nucleons in the nucleus all together. The total number of neutrons plus protons is called the **mass number** of the atom.

The usual way of identifying a particular isotope is to express its mass number as a superscript before the chemical symbol. For example, carbon 12 would be expressed with a 12 as a superscript:

$$\text{mass number} \longrightarrow {}^{12}_{6}\text{C} \longleftarrow \text{atomic number}$$

Sometimes it is convenient also to express the atomic number as a subscript before the chemical symbol, although it is redundant. Anything having six protons in the nucleus is carbon by definition.

The effect of neutron radiation can be to alter either the mass number or the atomic number of atoms in the path of the neutrons. If a neutron sticks in the nucleus of a target atom, the mass number increases by 1. If a neutron comes in and knocks out a proton, the mass number remains unchanged but the atomic number decreases by 1. Either way, a new isotope is formed, which may or may not be stable.

A natural example of the production of radioactive isotopes by free neutrons is the production of carbon 14, ${}^{14}_{6}\text{C}$, which has been going on in the upper reaches of our atmosphere since the birth of our planet. Cosmic radiation produces neutrons naturally in the upper atmosphere, and these neutrons bombard nitrogen nuclei to produce a certain low-level concentration of radioisotopes. The most abundant isotope of nitrogen is nitrogen 14, ${}^{14}_{7}\text{N}$, which has seven neutrons and seven protons. In the natural form of neutron activation, a neutron comes into the nucleus and "knocks out" a proton, leaving the nucleus richer by one neutron but poorer by one proton. This leaves the mass of the nucleus essentially unchanged, but it decreases its positive charge by one unit. The original nitrogen 14, with an atomic number of 7, is thereby changed into carbon 14, with an atomic number of 6.

$$ {}^{14}_{7}\text{N} + {}^{1}_{0}\text{n} \Rightarrow {}^{14}_{6}\text{C} + {}^{1}_{1}\text{p} $$

The neutron can be expressed as ${}^{1}_{0}\text{n}$, since it has zero charge and hence an atomic number of zero and a mass number of 1. The proton can be expressed as ${}^{1}_{1}\text{p}$, since it has one unit of charge and one unit of mass. Notice that the total charge as well as the total mass remain unaltered by the reaction. The superscripts and subscripts both add up to the same number on both sides of the reaction. Carbon 14 produced in this way has two more neutrons than carbon 12, the most abundant isotope of carbon, but it behaves chemically the same. The two extra neutrons make carbon 14 slightly unstable; it beta decays slowly back into nitrogen 14. One of the neutrons turns into a proton by emitting an electron, or beta particle.

$$ {}^{14}_{6}\text{C} \Rightarrow {}^{14}_{7}\text{N} + {}^{0}_{-1}\text{e} $$

There is also a neutrino emitted in this reaction, but we will not bother to show it in our expression because it has no effect on our simple calculations. Neutrinos have zero charge and zero rest mass. We do need to show the electron however. An electron is shown as $_{-1}^{0}e$ with an atomic number of −1, since it has one unit of minus charge, and a mass number of zero, since its loss does not change the total number of neutrons and protons in the nucleus. Here again, the sum of the superscripts is the same on both sides of the reaction, as is the sum of the subscripts. The total charge and the number of nucleons are both unchanged by the reaction.

In a natural sample of carbon, 98.89% is found to be $_{6}^{12}C$, the isotope with an equal number of neutrons and protons. Almost all the rest is $_{6}^{13}C$, an isotope with one more neutron, which is also stable. Only a trace amount of naturally occurring carbon turns out to be the radioactive isotope $_{6}^{14}C$, and all of it that is found was produced by natural neutron activation of nitrogen in the upper atmosphere by the cosmic radiation process. We know that none of the carbon 14 found today was present when the earth was formed, because we know the rate of the beta decay process is too rapid for $_{6}^{14}C$ to last that long. Any carbon 14 that was present when the earth was formed has long since turned into nitrogen 14.

The small amount of radioactive isotopes produced by cosmic radiation does not present much of a radiation hazard, but the huge amounts of neutron radiation from artificial nuclear reactions is one of the major problems of reactor design. The shielding material itself becomes radioactive and changes its chemical and structural properties as a result of the neutron bombardment. The working components of the reactor must be made of materials that can stand neutron bombardment for significant periods of time. Management of the radioactive waste that comes from the replacement of shielding and other parts of the reactor is still another problem that comes from working with neutrons.

Neutron activation can, on the other hand, be a useful tool. It is most frequently used for measuring even tiny amounts of certain elements in a sample. The radioactive decay of the isotopes in the sample after it has been bombarded with neutrons becomes a signature of the elements that were originally in the sample. For example, sodium has 11 protons in the nucleus and just a slightly larger number of neutrons. The only naturally occurring isotope of sodium is sodium 23, $_{11}^{23}Na$, with 12 neutrons in the nucleus. If a sample containing sodium is irradiated with neutrons, some of the sodium atoms will capture a neutron and become the unstable isotope sodium 24. A gamma ray, γ, is also produced in this reaction.

$$_{0}^{1}n + _{11}^{23}Na \Rightarrow _{11}^{24}Na + \gamma$$

Sodium 24 decays with a half-life of 15 hours to yield gamma rays and beta particles of certain well-known energies.

$$_{11}^{24}\text{Na} \Rightarrow {}_{12}^{24}\text{Mg} + {}_{-1}^{0}\text{e} + \gamma$$

Thus, the presence of simultaneous beta and gamma rays at a rate that decreases by half every 15 hours is a sure sign that sodium was present in the sample before it was subjected to neutrons. At least 70 elements can be activated in this way to produce radioactive isotopes that decay in a recognizable way.

Half-Life and Carbon Dating

The existence of radioactive carbon 14 in the atmosphere means that you eat the stuff all the time. You can't get away from it. The carbon produced by neutron bombardment of nitrogen in the atmosphere turns into carbon dioxide and gets into growing plants. A certain amount of the oatmeal you eat in the morning is carbon 14. If you eat eggs instead, the chicken who laid the egg ate the oatmeal, and you still get the same ratio of carbon 14 to the other isotopes of carbon in everything you eat. The only way to avoid putting this radioactive substance into your body is to quit eating.

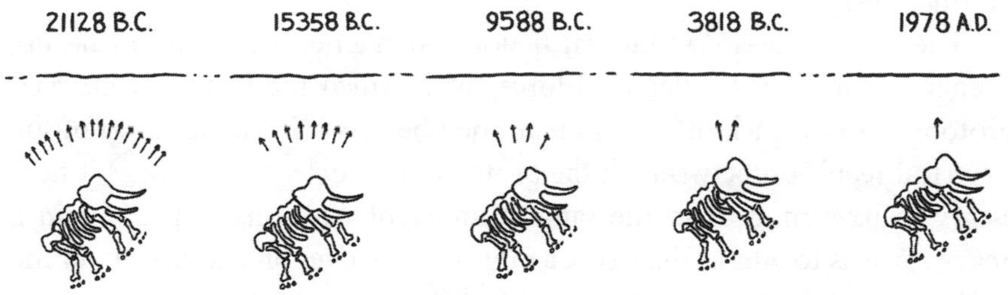

Figure 14-15

The rate of decay of carbon 14 decreases by half every 5,770 years.

Looking at the bright side, you will eventually quit eating. The carbon 14 in your body will then eventually all decay, and you won't have anything more to worry about. It will take about 5,770 years for half the carbon 14 to go away. After another 5,770 years, half the remaining will decay. You will then be down to one-fourth of what you started with. After another 5,770 years, you'll be down to one-eighth. The amount present always goes down by a factor of two each time this period of time passes.

It is impossible to guess when any one radioactive nucleus will decay. The chances of its decay are independent of what has gone on in the past. Every instant of time is therefore a new ball game, so to speak. On the other hand, you can guess how long it will take for half of a whole bunch of radioactive nuclei to decay. You may not know which nucleus will be the next to go, but you do know that the rate of decay is proportional to the number that are left. When half are gone, the rate of decay will be half as great. The period of time for this to happen is called the **half-life** of the reaction. It is a known constant under most conditions for most nuclear decay reactions.

From the rate of decay of carbon 14 it is possible to tell when an animal had its last meal to within about 30 years. This fact has been useful to archaeologists and others interested in dating old organic remains. All that needs to be done is to figure out by how many factors of 2 the concentration of carbon 14 is smaller than the concentration in a fresh sample. The number of these factors of 2 is the number of half-lives of age of the sample.

Nuclear Fission

There are two nuclear reactions which, given the state of our present technology, have the potential of giving us large amounts of power. One is nuclear fusion and the other is nuclear fission. It may seem remarkable that both reactions produce energy, since they are, in some ways, opposite kinds of reactions. Nuclear fusion is when the nuclei of two light elements are fused together to form a heavier nucleus; nuclear fission is when a heavy nucleus breaks apart to form nuclei of lighter elements. We will first consider these two reactions separately and then see how they can both produce energy.

The energy released in nuclear fission, when a heavy nucleus breaks up, is energy stored in the electrical forces with which the positively charged protons push on each other. As mentioned before, it is the build-up of the electrical repulsion between all the protons in a nucleus that makes it necessary to have more than the same number of neutrons as protons in a heavy nucleus to add in their short-ranged attractive force to help hold the nucleus together. After the number of protons gets up to a certain point, you just can't put enough neutrons into the nucleus to make it stable. Unstable heavy nuclei can reach a more stable configuration by alpha decay or beta decay or by breaking apart in nuclear fission.

The break-up of a heavy nucleus results in the release of neutrons as well as the lighter nuclei called fission fragments. The lighter nuclei don't need as many neutrons to balance the repulsive force between their protons, and so there are neutrons left over. These extra neutrons are important in the use of fission to produce nuclear power.

Nuclear fission occurs spontaneously in heavy elements, but it takes place so slowly that the power release is at too low a level to be of much use. The neutrons released by the fission of one nucleus can be used, however, to artificially induce the fission of another unstable nucleus. As long as at least one neutron released by one fission induces the fission of another nucleus, a high rate of energy release will be propagated in what is called a chain reaction. If more than one neutron on the average induces a new fission, the rate of energy release will increase. An atomic bomb makes a fine example of such a runaway chain reaction. The trick in producing power in a nuclear reactor is to absorb just the right amount of neutrons, so that the chain reaction continues at a constant level, while keeping the neutron flux up to the point that the chain reaction can perpetuate.

Getting the neutron flux up to the point where a chain reaction is possible turns out to be quite a trick in itself. The neutrons need to be moving rather slowly or else you need a lot more of them. Most of the neutrons coming directly from a nuclear fission are moving so fast that they pass through nuclei they hit with only a small probability of an interaction. These are not hard little bullets that are sure to have an effect on everything they hit, but wave-particles, which have a smaller chance of interaction the shorter the time that they stay in any one place. This problem is solved in nuclear reactors by providing a moderator, a light element in which neutrons can bounce around and slow down without being absorbed. Slow neutrons that have lost enough kinetic energy so that they are in thermal equilibrium with their surroundings are called **thermal neutrons**.

Another problem encountered in getting a large enough neutron flux to sustain a chain reaction is that only the light isotope of uranium, $^{235}_{92}U$, fissions properly under neutron bombardment. The heavy isotope of uranium, $^{238}_{92}U$, absorbs thermal neutrons but doesn't fission. Thus, the heavy isotope soaks up the neutrons, but the only thing that happens to it is that it beta decays into plutonium $^{239}_{94}Pu$ after a while. This is a problem, because 99.3% of natural uranium is the heavy isotope. The light isotope, which will fission, only makes up 0.7% in a natural mix. Refining uranium to get a larger concentration of the fissionable light isotope is, fortunately, a difficult task. The isotopes cannot be separated by chemical means, because they are both chemically uranium. Gas diffusion plants, which separate by the slight difference in mass, are very difficult to build and operate.

This difficulty turns out to be fortunate because it is relatively easy to build an atomic bomb once you have enough of the enriched mixture of the light isotope, ^{235}U. All you have to do is to get enough of it together in a small enough volume and hold it there while the neutron flux builds up. It is probably a good thing that this is as hard to do as it is.

There is another way to get around the neutron absorption of uranium ^{235}U, and that is to use plutonium instead of uranium. While plutonium does not occur in nature, it does result from the operation of a uranium reactor. The heavy isotope ^{238}U absorbs thermal neutrons and then beta decays into plutonium, and plutonium is fissionable in a chain reaction. What is more, plutonium can be chemically separated from uranium. It is produced as a by-product of refining spent reactor fuel.

Reactors can even be designed to optimize the production of plutonium from ^{238}U. Such reactors are called breeder reactors. It happens that the most efficient operation of a breeder reactor requires the operation of the reactor with little or no moderator to slow down the neutrons. Such a reactor is called a fast breeder reactor, because it operates on vast quantities of fast neutrons rather than on smaller amounts of slow neutrons. Unfortunately, a fast neutron reactor is more unstable and difficult to control than a slow neutron reactor.

Nuclear Fusion

Another nuclear reaction that can release a great deal of energy is the fusing together of light elements to make heavier ones. This is the reaction that is powering our sun. Hydrogen goes through several nuclear transformations to become helium. Several steps are necessary, because it is almost impossible to get four protons together and keep them there long enough for the fusion reaction to take place. In fact, it is a formidable task just to get two protons together. Two protons must be coming together at a tremendous velocity if they are to overcome their electrical repulsion to get close enough to interact. The temperature at which these kinds of velocities occur is on the order of many millions of degrees Kelvin. These temperatures occur in the stars, but they are very difficult for earthlings to produce.

Red stars as well as white stars that operate up to about the temperature of our sun all produce energy in just this way. Two protons come together long enough for one to turn itself into a neutron and emit a positron, which is an elementary particle much like an electron except that it has a positive charge. The resulting nucleus is deuterium $^2_1$H, a heavy isotope of hydrogen. A proton can then combine with this deuterium nucleus to form a helium 3 nucleus, $^3_2$He . Helium 3 is the light isotope of helium. The usual helium nucleus has two protons and two neutrons, but the helium 3 nucleus has two protons and only one neutron. It really wants another neutron. If two helium 3 nuclei come together, one of them will take both neutrons and the two protons in the other will then fly apart with a tremendous release of energy.

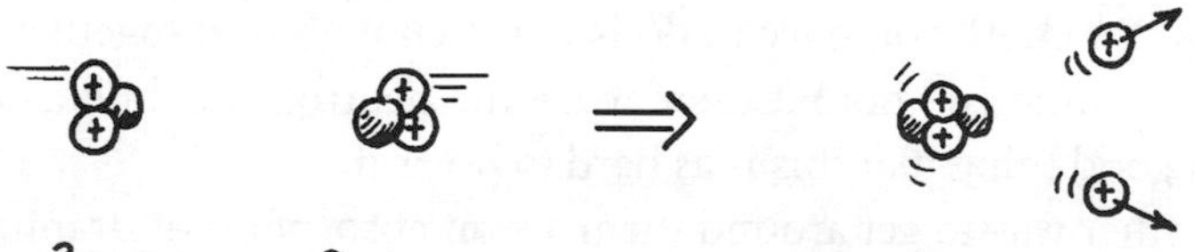

Figure 14-16

Two helium 3 nuclei come together and react to produce a helium 4 nucleus and two protons.

$$^3_2\text{He} + ^3_2\text{He} \longrightarrow ^4_2\text{He} + 2\,^1_1\text{H} + energy$$

Each step of the process releases energy, but the last step releases slightly more energy than all the other steps leading to two helium 3 nuclei taken together. This set of steps is collectively called the **proton-proton cycle**. It starts with ordinary hydrogen nuclei, or protons, and ends with a helium nucleus and two left-over protons.

Hot stars, hotter than our own sun, probably get their energy from a different set of steps called the **carbon cycle**. It involves protons interacting with a carbon nucleus as it changes back and forth between different isotopes of carbon and nitrogen to finally produce a helium nucleus and the original carbon. Carbon acts as a nuclear catalyst in this reaction.

Controlled nuclear fusion may be the energy source of the future if we can figure out a way to hold things together long enough at a high enough temperature to get out enough energy to make it pay. All of our efforts so far give less energy than we have had to put in to contain the reaction. It now looks like we will be able to get energy out, but not without a large number of problems still to be solved.

For one thing, the proton-proton cycle and the carbon cycle are still far beyond our capability. The only fusion reactions we are now working on are ones that generate vast numbers of neutrons as a by-product. As is the case with nuclear fission reactors, neutron activation of the working components and shielding material is a major problem of reactor design. Controlled nuclear fusion has many such problems to be solved before it can answer our energy needs of the future. For now, the only fusion power plant we can rely on is our sun.

You may wonder what will happen when the sun runs out of hydrogen to fuse into helium. This won't happen for a while; the sun has enough hydrogen to last for some tens of billions of years. Even after it runs out of hydrogen, however, it will keep on going. The fusion of hydrogen into helium is but one of many fusion reactions that can power a star.

After the sun runs out of hydrogen, the gravitational collapse of the helium that has been produced will raise its pressure and temperature at the center of the sun to the point where helium will start fusing into heavier elements. The heavier elements will then fuse into still heavier elements, and so on until the element iron is reached. Every fusion process liberates energy until we come to iron. Every fusion process after iron absorbs energy instead of giving it off. Therefore, after the sun develops an iron core, the next fusion process will be quite different. The energy produced by fusion reactions up to this point always tended to fight off the gravitational collapse, but fusion reactions past that point will actually favor gravitational collapse. Our sun will implode suddenly. The core will become immensely more dense; our sun will become a neutron star in a sudden event, called a supernova, and the mantle of various heavy elements will be blown away. We believe that all the known elements that make up our planet were formed in the interior of an aging star and blown away in that cataclysmic event when the star imploded in a supernova. The elements up to iron were formed before the supernova; the elements past iron were formed in that cataclysmic event itself and just happened to get blown away with the rest of the mantle. You can therefore look upon yourself as literally being made of star dust. Those stars in your eyes are really stars in your eyes.

Mass Defect

We are now in a position to answer two puzzles about nuclear reactions and nuclear structure. One is how you can get energy from both nuclear fission and nuclear fusion. It does seem strange that energy is released in an atomic bomb by allowing the nuclei of heavy elements to break apart into lighter elements and in a hydrogen bomb by allowing nuclei of light elements to combine into heavier elements. The other puzzle is one that bothered early investigators as well as present students who first study a chart of the elements. The atomic masses of the elements seem to bear some relationship to the atomic number, but the relationship doesn't seem to be exact. The relationship improves once we understand isotopes and the need for increasing numbers of neutrons to stabilize heavy nuclei, but the mass still doesn't seem to be exactly proportional to the number of nucleons in the nucleus. This defect, as it was called, in the mass relationship is caused by the same phenomenon that allows you to get energy both from fission and fusion.

Looking at Table 13-1, we see that the atomic mass is almost the same number as the mass number of the isotope, but not exactly. For atoms lighter than carbon, the number giving the atomic mass (in atomic mass units) is a tiny bit larger than the number of nucleons in the nucleus. At about that point, it dips below and becomes a tiny bit smaller. The mass stays below the mass number until about the element 86 where it crosses over and becomes larger again.

These tiny differences in the fifth, sixth, and seventh places turn out to be very important. They mean that neutrons and protons do not always have the same mass. The whole key to nuclear power resides in the fact that this tiny discrepancy in mass represents huge amounts of energy. The proportionality constant, in fact, is none other than the speed of light squared:

$$E = mc^2$$

We could make a graph of the numbers in the list to see what the relationship looks like, but what we are really interested in is the ratio of the

Figure 14-17

A graph of the mass per nucleon shows that the neutrons and protons in the iron nucleus seem to have less mass than those in other elements.

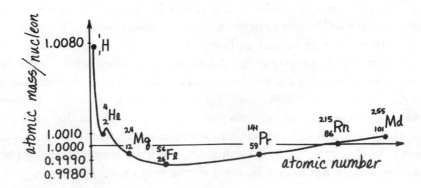

atomic mass to the number of nucleons in each nucleus. This ratio gives us the average mass of neutrons and protons in that nucleus. A graph of this ratio shows a low point at element 26, iron, $_{26}^{56}$Fe which has an atomic mass of 55.9349. Dividing this mass by 56, the number of neutrons and protons in this isotope of iron, we get a ratio of 0.998837 atomic mass units per nucleon. All other isotopes have more mass per nucleon.

Elements lighter than iron therefore lose mass when they combine by nuclear fusion to produce nuclei closer in atomic number to iron. Elements heavier than iron lose mass when they break apart or fission to approach iron. The decrease in atomic mass per nucleon in either process is accounted for by energy liberated in the reaction.

Atomic Number	Atomic Isotope	Mass	Atomic Number	Atomic Isotope	Mass
1	^{1}H	1.007825	28	^{60}Ni	59.9332
2	^{4}He	4 00260	32	^{74}Ge	73.9219
3	^{7}Li	7 01600	37	^{85}Rb	84.9117
4	^{9}Be	9.01218	41	^{93}Nb	92.9060
5	^{11}B	11.00931	47	^{107}Ag	106.9051
6	^{12}C	12.00000	51	^{121}Sb	120.9038
7	^{14}N	14.00308	53	^{127}I	126.9006
8	^{16}O	15.99491	59	^{141}Pr	140.9074
9	^{19}F	18.99840	66	^{161}Dy	160.9266
10	^{20}Ne	19.99244	72	^{180}Hf	179.9468
11	^{23}Na	22.9898	76	^{190}Os	189.9586
12	^{24}Mg	23.98504	82	^{208}Pb	207.9766
13	^{27}Al	26.98153	86	^{215}Rn	214.9987
14	^{28}Si	27.97693	86	^{216}Rn	216.0002
15	^{31}P	30.99376	88	^{220}Ra	220.0110
16	^{32}S	31.9721	92	^{235}U	235.0439
20	^{40}Ca	39.9626	92	^{238}U	238.0508
25	^{55}Mn	54.9381	94	^{239}Pu	239.0522
26	^{56}Fe	55.9349	98	^{250}Cf	250.0766
27	^{59}Co	58.9332	100	^{250}Fm	250.0795
			101	^{255}Md	255.0906

Table 14-1

A list of some isotopes shows that their masses are nearly, but not exactly proportional to the number of nucleons.

Elementary Particle Research

We have been speaking of electrons, protons, and neutrons as though they are the only things that go into making up an atom. That is a useful model for understanding basic atomic and nuclear structure, but, as usual, there is more to it than that. Neutrons and protons seem to be made up of more elementary things called **quarks**.

One of the most active fields of modern research is in trying to understand the particles that come out of a target when you send in a beam of high-energy particles. The number and diversity of particles that came out, as we have improved our techniques for sending particles in, has led us in recent years to a better understanding of the rules of nature. The so called **standard model** of elementary particles, puts the number of quarks at six with certain properties and six more antiquarks with symetrically opposite properties.

One result of particle study has been the discovery of antiparticles, which have symmetrical properties to ordinary particles. A positron, for example, has the same mass and spin as an electron but the opposite charge. Positrons don't last very long in the presence of ordinary matter. Their positive charge causes them to be attracted to an electron, whereupon the electron and positron combine to annihilate each other. Their rest masses are converted into energy in a huge burst of electromagnetic radiation. The same thing is true for the other anti-particles. They tend to annihilate ordinary particles.

We have found anti-protons and anti-neutrons, which are symmetrical to protons and neutrons but which are annihilated and turned into energy when they find their opposite numbers. We have seen very energetic photons of electromagnetic energy produce anti-particles and particles in pairs. It may be that the whole universe was created in this way. Maybe half the stars we see are made of anti-matter. Instead of protons and neutrons in the nucleus, there may be anti-protons and anti-neutrons. Instead of electrons in orbit about the nucleus, there may be positrons. Anti-matter would look just like ordinary matter. You wouldn't be able to tell that it was anti-matter without touching it. If you shook hands with an antimatter person, however, it would be the last friendly gesture you would ever make before being literally annihilated.

High-energy particle research makes use of wonderful particles that can do tricks for us that ordinary particles can't. A muon, for example, acts very much like an electron in terms of electric charge and everything else except that it is about 200 times as massive. It only lives for about 2.2×10^{-6} s, but that is plenty of time for it to do a rather remarkable thing. Replacing an electron in an atom, a muon will quickly decay into its lowest energy orbit. Because of its large mass, the inner orbits of a muon are actually smaller than the nucleus itself. We can therefore learn more about the charge distribution of the nucleus by studying the energy levels of a muon as it decays to orbits inside the nucleus.

It might seem a bit strange that a muon spends most of its short life in an orbit that is actually inside the nucleus without even bumping into the nucleons in the same way that nucleons bump into each other. The muon doesn't even seem to feel the strong interaction force that acts between neutrons and protons. Muons, like electrons, only interact electrically and by the weak interaction that governs most nuclear decay processes. Particles that interact by one kind of force can be transparent in some ways to particles that interact by another.

Just as the rules of behavior for things that move very fast differ from your ordinary experience with objects moving at ordinary velocities, the rules of behavior of these elementary particles differ markedly from what you might expect, based on your experience with ordinary-size objects. Still, it is remarkable that our understanding of the rules that govern ordinary objects that we can see has led us, little by little, to understand better by analogy the different rules that govern things that we have not seen and believe that we cannot ever hope to see.

PROBLEM SET 14

SUMMARY OF CONCEPTS

The special theory of relativity, applying to frames of reference moving at a constant velocity with no acceleration, is based on the fact that moving meter sticks must get short as their velocity approaches the velocity of light and that moving clocks must run slow as they approach the same speed. This **length contraction**

$$l' = l\sqrt{1 - \frac{v^2}{c^2}}$$

(handwritten: moving, $< 1 \Rightarrow l' < l$)

and **time dilation**

$$t' = \frac{t}{\sqrt{1 - \frac{v^2}{c^2}}}$$

(handwritten: $< 1 \Rightarrow t' > t$)

are experimental facts resulting from the finite nature of the velocity of light,

$$c = 3 \times 10^8 \, \text{m/s}$$

The special theory of relativity says that the laws of physics are the same in any frame of reference and that the velocity of light is also constant in any frame of reference. This implies that the length contraction and time dilation apply to the actual structure of space and time. Several results predicted by this theory have stood up to experimental test. Among these are the fact that Newton's second law, $F = (mv)/t$ remains valid even at speeds near the speed of light as long as the mass, m, is understood to be the **relativistic mass**

$$m = \frac{m_0}{\sqrt{1 - \frac{v^2}{c^2}}}$$

(handwritten: $< 1 \Rightarrow m > m_0$)

which increases as v approaches c. This relativistic mass results in the Einsteinian mass-energy relationship.

$$E = mc^2$$

Atomic and nuclear physics apply to dimensions so small that the wave nature of a particle puts limits on our knowledge of its momentum, (mv), and position, x, as well as its energy, E, and time, t. Heisenberg's uncertainty principle states that the product of our uncertainties in our knowledge of these variables is limited by the wave nature of the particle to about the size of Planck's constant:

$$\Delta(mv) \cdot \Delta x \cong h$$
$$\Delta E \cdot \Delta t \cong h$$

Planck's constant, h, is the proportionality constant between energy, E, and frequency, f, of electromagnetic radiation that got the whole theory of quantum mechanics started.

$$E = hf$$

The particle nature of waves, suggested by Einstein when he proposed that energy of electromagnetic radiation comes in little quantum bundles called photons whose energy is proportional to the frequency of the radiation, was what eventually suggested the idea that particles also have a wave nature. The wave-particle nature of both waves and particles is not apparent until you get to small dimensions, because Planck's constant is a very small number ($h = 6.63 \ 10^{-34}$ J $\cdot$ s).

A nucleus can decay by emitting an **alpha particle,** which is a helium nucleus consisting of two neutrons and two protons; a **beta particle**, which is a high-speed electron; or a **gamma ray**, which is an energetic photon. Alpha decay decreases the mass number by 4 and the atomic number by 2. Beta decay changes a neutron into a proton, leaving the mass number unchanged but increasing the atomic number by 1. Gamma decay leaves both mass number and atomic number unchanged.

$$_a^b X \Rightarrow {}_{a-2}^{b-4} Y + {}_2^4 \alpha$$
$$_a^b X \Rightarrow {}_{a+1}^{b} Y + {}_{-1}^{0} \beta$$

EXAMPLE PROBLEMS

Try to do the Examples yourself before looking at the Sample Solutions.

14-1 You are watching a large man-made earth satellite zip across the night sky in a low circular orbit and start wondering how fast it would have to be going to relativistically contract a measurable fraction of its original length.

Given: The satellite is 30.0 m long and is traveling at 8×10^3 m/s, but you imagine that it somehow speeds up to the point where it will be only 29.9 m long.

Find: (a) How fast would the satellite have to travel to contract to the length that you imagine?
(b) How much time would a clock on the satellite lose in an hour of earth time because of traveling at this imaginary high velocity?
(c) By how much does the length of the satellite actually contract?

BACKGROUND

This is an exercise in using the length contraction and time dilation. It gives you an opportunity to calculate the length contraction for the fastest large object you are likely to see near the surface of the earth.

The calculation in part (c) involves taking the square root of a number that is so close to unity (1) that a slide rule is of little use. If you don't have a calculator that takes square roots of this many digits, try a hand calculation squaring a number which is very close to 1. Then try another number that gives a result even closer to the number whose square root you want. This is called the method of successive approximations.

14-2 A science fiction spacecraft is somehow able to combine a small amount of anti-matter with an equal amount of ordinary matter and convert all the energy into its own kinetic energy.

Given: The rest mass of the spacecraft is 5×10^6 kg. It converts a total mass of 2.0 kg into its own kinetic energy.

Find: (a) How much energy, in joules, is produced?

(b) If that energy all goes into the kinetic energy of the spacecraft, how fast does the spacecraft go?

(c) By how much does the relativistic mass of the spacecraft increase as a result of its increase in kinetic energy?

BACKGROUND

This problem illustrates the concept of relativistic mass and the Einsteinian mass-energy relation. You will also need to remember that kinetic energy is one-half the mass times the square of the velocity.

We know of no way that a spacecraft could propel itself forward by giving itself all the kinetic energy expended. In rocket propulsion, most of the energy goes into the exhaust gas. Here we assume that science fiction technology has overcome this problem.

14-3 Looking through a microscope at a 10-micron animal swimming about in a drop of water, you wonder if these dimensions are small enough to bring the uncertainty principle into play. Planck's constant is equal to 6.63×10^{-34} J · s.

Given: The microscopic animal has a mass of about 10^{-12} kg and a velocity of about 10^{-4} m/s. You estimate the uncertainty in these measurements to be 10^{-13} kg and 10^{-5} m/s.

Find: (a) What is the uncertainty in the animal's momentum from your estimates?

(b) What uncertainty in position is predicted by the uncertainty principle?

(c) What uncertainty in time is associated with the uncertainty in the animal's kinetic energy?

BACKGROUND

This exercise provides an opportunity to see if Heisenberg's uncertainty principle has much of an effect on observations made with an optical microscope. Remember that momentum is mass times velocity, and kinetic energy is one-half mass times the square of velocity.

14-4 A low-power laser produces a very bright spot of light, the power content of which is measured in milliwatts. You decide to estimate the number of photons per second in the laser beam. You know that Planck's constant $h = 6.6 \times 10^{-34}$ J · s.

Given: The laser beam is rated at a quarter of a milliwatt, 2.5×10^{-4} W, for light having a frequency of 4×10^{14} Hz.

Find: (a) How much energy does one photon of this light have?

(b) How many photons of this energy come from the laser in 1 s to produce the rated power?

BACKGROUND

This is an exercise in using Planck's constant and Einstein's quantum theory of radiation. Remember that power is defined as energy per unit time. This exercise gives you an opportunity to see how many photons go into making a bright spot of light.

14-5 In the thorium series of natural radioactivity, thorium 232 undergoes a series of alpha and beta decays to become eventually a stable isotope of lead.

Given: Thorium, $^{232}_{90}$Th, has an atomic number of 90 and a mass number of 232. It alpha decays with a half-life of about 10^{10} years into an isotope of radium. The radium then beta decays, with a half-life of 6.7 years into actinium, which again beta decays into another isotope of thorium. This second isotope of thorium is much less stable than thorium 232 and alpha decays with a half-life of 1.9 years.

Find: (a) What is the mass number of radium, $^{?}_{88}$Ra , into which $^{232}_{90}$Th decays?

(b) What are the mass numbers of actinium (Ac) and thorium resulting from beta decay?

(c) What is the next isotope in the chain resulting from alpha decay of thorium?

BACKGROUND

This problem in nuclear transformation involves remembering the difference between an alpha particle and a beta particle. The isotope resulting in part (c) then alpha decays again into radon, which alpha decays once more into polonium. Two more alpha decays and two more beta decays can then be seen to result in lead 208, $^{208}_{82}$Pb.

Solutions to EXAMPLE PROBLEMS

(a) $\quad l' = l\sqrt{1 - \dfrac{v^2}{c^2}}$

$\left(\dfrac{l'}{l}\right)^2 = 1 - \dfrac{v^2}{c^2}$

$\dfrac{v^2}{c^2} = 1 - \left(\dfrac{l'}{l}\right)^2$

$v = c\sqrt{1 - \left(\dfrac{l'}{l}\right)^2}$

$\quad = c\sqrt{1 - \left(\dfrac{29.9 \text{ m}}{30.0 \text{ m}}\right)^2}$

$\quad = c\sqrt{1 - (0.9967)^2}$

$\quad = c(0.08158)$

$\quad = \boxed{2.45 \times 10^7 \text{ m/s}}$

(b) $\quad t' = \dfrac{t}{\sqrt{1 - v^2/c^2}}$

$\quad = \dfrac{1.0 \text{ hr}}{\sqrt{1 - \dfrac{(2.45 \times 10^7 \text{ m/s})^2}{(3.0 \times 10^8 \text{ m/s})^2}}}$

$\quad = 1.0033 \text{ hr}$

$\Delta t = t' - t = 0.0033 \text{ hr} = \boxed{12 \text{ s}}$

(c) $\quad l' = l\sqrt{1 - v^2/c^2}$

$\quad = (30 \text{ m})\sqrt{1 - \dfrac{(8 \times 10^3 \text{ m/s})^2}{(3 \times 10^8 \text{ m/s})^2}}$

$\quad = (30 \text{ m})\sqrt{1}$

$\quad = (30 \text{ m})(1)$

$\Delta l = l - l'$

$\quad = 30 \text{ m} - (29.99999999 \text{ m})$

$\quad = \boxed{1.1 \times 10^{-8} \text{ m}}$

SAMPLE SOLUTION 14-1

Given:
The satellite is 30.0 m long and is traveling at 8×10^3 m/s, but you imagine that it somehow speeds up to the point where it will be only 29.9 m long

Find:
(a) How fast would the satellite have to travel to contract to the length that you imagine?

(b) How much time would a clock on the satellite lose in an hour of earth time because of traveling at this imaginary high velocity?

(c) By how much does the length of the satellite actually contract?

DISCUSSION

The velocity found in part (a) is far greater than the velocity the satellite would actually have for a circular orbit. In fact, it is about 2,000 times the escape velocity of a space vehicle.

The fractional change in time, namely 0.9967, turns out to be the same as the fractional change in length, namely 29.9 m divided by 30.0 m, because a dilation of one is associated with a contraction of the other by the same rule of relativity.

The change in length found in part (c) for the velocity of an actual orbiting satellite would be about 2% as long as the wavelength of yellow light.

SAMPLE SOLUTION 14-2

Given:

The rest mass of the spacecraft is 5×10^6 kg. It converts a total mass of 2.0 kg into its own kinetic energy.

Find:

(a) How much energy, in joules, is produced?

(b) If that energy all goes into the kinetic energy of the spacecraft, how fast does the spacecraft go?

(c) By how much does the relativistic mass of the spacecraft increase as a result of its increase in kinetic energy?

(a) $E = mc^2$

$= (2.0 \text{ kg})(3.0 \times 10^8 \text{ m/s})^2$

$= 2.0 \times 3.0^2 \times 10^{2 \times 8} \text{ kg m}^2/\text{s}^2 \left(\dfrac{1 \text{ J}}{1 \text{ N} \cdot \text{m}}\right) \dfrac{1 \text{ N}}{1 \text{ kg m/s}^2}$

$= \boxed{1.8 \times 10^{17} \text{ J}}$

(b) $E_k = \frac{1}{2}mv^2$

$v = \sqrt{\dfrac{2 E_k}{m}}$

$= \sqrt{\dfrac{2(1.8 \times 10^{17} \text{ J})}{5 \times 10^6 \text{ kg}} \dfrac{\text{kg m}^2/\text{s}^2}{\text{J}}} = 2.6832816 \times 10^5 \text{ m/s}$

$\cong \boxed{2.7 \times 10^5 \text{ m/s}}$

(c) $m = \dfrac{m_0}{\sqrt{1 - v^2/c^2}}$

$= \dfrac{(5 \times 10^6 \text{ kg})}{\sqrt{1 - \dfrac{(2.7 \times 10^5 \text{ m/s})^2}{(3.0 \times 10^8 \text{ m/s})^2}}} = (5.000002 \times 10^6 \text{ kg})$

$\Delta m = m - m_0 = (5.000002 \times 10^6 \text{ kg}) - (5.000000 \times 10^6 \text{ kg}) = \boxed{2 \text{ kg}}$

DISCUSSION

You can see that $E = mc^2$ at least has the correct units, since a joule of energy is a newton of force times a meter, while a newton is defined as that amount of force that accelerates 1 kg at 1 m/s².

Even the huge amount of energy that would be liberated by combining a kilogram of anti-matter with the same amount of ordinary matter would, according to part (b), only speed a 5,000-metric-ton spacecraft up to about a thousandth the velocity of light.

The increase in relativistic mass of the spacecraft found in part (c) is exactly equal to the mass converted into energy found in part (a). The mass-energy relationship works both ways.

(a) Momentum including uncertainty

$$= (m + \Delta m)(\upsilon + \Delta \upsilon)$$

$$= 10^{-12-4} \text{ kg m/s} \pm (10^{-12-5} \pm 10^{-13-4}) \text{ kg m/s} \pm 10^{-13-5} \text{ kg m/s}$$

$$= 10^{-13-5} \text{ kg m/s} \pm (2 \times 10^{-17} \text{ kg m/s})$$

where $\Delta(m\upsilon) = \boxed{2 \times 10^{-17} \text{ kg m/s}}$

(b) $\Delta(m\upsilon)\Delta x \cong h$

$$\Delta x \cong \frac{h}{\Delta(m\upsilon)}$$

$$= \frac{6.63 \times 10^{-34} \text{ J}}{2 \times 10^{-17} \text{ kg m/s}} \frac{\text{kg m2/s2}}{\text{J}}$$

$$= \boxed{3 \times 10^{-17} \text{m}}$$

(c) Energy including uncertainty

$$= \tfrac{1}{2}(m \pm \Delta m)(\upsilon \pm \Delta \upsilon)^2$$

$$= \tfrac{1}{2}(10^{-12} \text{ kg} \pm 10^{-13} \text{ kg})(10^{-4} \text{ m/s} \pm 10^{-5} \text{ m/s})^2$$

$$= \tfrac{1}{2}10^{-12} \text{ kg}(10^{-4} \text{ m/s}^2) \pm \tfrac{1}{2}(10^{-12} \text{ kg})2(10^{-4} \text{ m/s})(10^{-5} \text{ m/s})$$

$$\pm \tfrac{1}{2}(10^{-13} \text{ kg})10^{-4} \text{ m/s}^2 \pm (\text{insignificantly small terms})$$

$$= \tfrac{1}{2} \cdot 10^{-20} \text{ kg m/s}^2 \pm 10^{-21} \text{ kg m/s}^2 \pm \tfrac{1}{2} \cdot 10^{-21} \text{ kg m/s}^2$$

$$= E \pm \Delta E$$

where ΔE = "Plus or minus" terms above

$$= 1.5 \times 10^{-21} \text{ kg m/s}^2 = 1.5 \times 10^{-21} \text{ J}$$

$\Delta E \Delta t \cong h$

$$\Delta t \cong \frac{h}{\Delta E} \cong \frac{(6.63 \times 10^{-34} \text{ J} \cdot \text{s})}{(1.5 \times 10^{-21} \text{ J})} \cong \boxed{4 \times 10^{-13} \text{ s}}$$

Given:
The microscopic animal has a mass of about 10^{-12} kg and a velocity of about 10^{-4} m/s. You estimate the uncertainty in these measurements to be 10^{-13} kg and 10^{-5} m/s.

Find:
(a) What is the uncertainty in the animal's momentum from your estimates?

(b) What uncertainty in position is predicted by the uncertainty principle?

(c) What uncertainty in time is associated with the uncertainty in the animal's kinetic energy?

DISCUSSION

The method of finding the error, done as a special case in part (a), is done in general in calculus as the "chain rule." The result is as follows:

$$\Delta(mv) = m(\Delta v) + (\Delta m)v$$

$$\Delta(\tfrac{1}{2}mv^2) = \tfrac{1}{2}(\Delta m)v^2 + \tfrac{1}{2}m(2v\Delta v)$$

Those students who haven't had calculus can do part (c) by finding the uncertainty in the energy as a special case, as shown in part (a).

    The resulting uncertainties in position and time found in parts (b) and (c) are far too small to be observed. The uncertainty principle therefore comes into play only at dimensions much smaller than are observable with an optical microscope. Quantum effects are therefore not noticeable on these dimensions.

SAMPLE SOLUTION 14-4

Given:

The laser beam is rated at a quarter of a milliwatt, 2.5×10^{-4} W, for light having a frequency of 4×10^{14} Hz.

Find:

(a) How much energy does one photon of this light have?

(b) How many photons of this energy come from the laser in 1 s to produce the rated power?

(a) $E = hf$

$$= (6.6 \times 10^{-34} \text{ J} \cdot \text{s})(4 \times 10^{14} \text{ Hz})\left(\frac{\text{cycle/s}}{\text{Hz}}\right)$$

$$= \boxed{2.6 \times 10^{-19} \text{ J}}$$

(b) $P \equiv \dfrac{E}{t}$

$$= \frac{n\, E_{per\ photon}}{t}$$

$$n = \frac{Pt}{E_{per\ photon}}$$

$$= \frac{(2.5 \times 10^{-4} \text{ W})(1 \text{ s})}{(2.6 \times 10^{-19} \text{ J/photon})} \frac{\text{J/s}}{\text{W}}$$

$$= \boxed{9 \times 10^{14} \text{ Photons}}$$

DISCUSSION

The energy in part (a) is found from Planck's hypothesis that the energy, E, per photon in electromagnetic radiation is proportional to the frequency, f, with Planck's constant being the proportionality constant. The unit for frequency, the hertz, can be converted to cycles per second, with the cycles being merely understood in the final answer. The result is understood to be the energy per photon.

The result of part (a) is the energy used in part (b) with the units joule per photon being explicitly stated when inserted into the general solution. This general solution is obtained from the definition of power where the total energy is assumed to be the number of photons, n, times the energy per photon, as found in part (a).

(a) $^{232}_{90}\text{Th} \Rightarrow {}^{x}_{88}\text{Ra} + {}^{4}_{2}\alpha$

   $232 = x + 4$

   $x = 232 - 4 = 228$

   hence $^{?}_{88}\text{R}$ is $\boxed{^{228}_{88}\text{Ra}}$

(b) $^{228}_{88}\text{Ra} \Rightarrow {}^{x}_{89}\text{Ac} + {}^{0}_{-1}\beta \Rightarrow {}^{x}_{90}\text{Th} + {}^{0}_{-1}\beta$

   $288 = x + 0$

   $x = 228$

   hence $\boxed{^{228}_{89}\text{Ac}}$ and $\boxed{^{228}_{90}\text{Th}}$ are the resulting isotopes

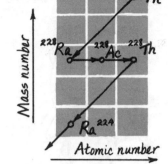

(c) $^{228}_{90}\text{Th} \Rightarrow {}^{y}_{z}x + {}^{4}_{2}\alpha$

   $\left.\begin{array}{l} 228 = y + 4 \\ 90 = z + 2 \end{array}\right\}$ hence: $\begin{array}{l} y = 228 - 4 = 224 \\ z = 90 - 2 = 88 \end{array}$

   Thus $^{y}_{z}x$ is $\boxed{^{224}_{88}\text{Ra}}$

DISCUSSION

Both the mass numbers and atomic numbers add up on both sides of the reaction in part (a). The atomic number of radium is 88, two less than that of thorium, because the alpha particle carries away two protons. This mass number, 228, does not change in beta decay, since a beta particle has no mass number. Thorium again alpha decays into radium in part (c), since we already know from part (a) that the element having atomic number 88 is called radium. The rest of the thorium series is

$$^{224}_{88}\text{RA} \Rightarrow {}^{220}_{86}\text{Rn} + {}^{4}_{2}\alpha \Rightarrow {}^{216}_{84}\text{Po} + {}^{4}_{2}\alpha$$

$$\Rightarrow {}^{212}_{82}\text{Pb} + {}^{4}_{2}\alpha \Rightarrow {}^{212}_{83}\text{Bi} + {}^{0}_{-1}\beta \begin{array}{c} \nearrow \\ \\ \searrow \end{array} \begin{array}{c} {}^{208}_{81}\text{Tl} + {}^{4}_{2}\alpha \\ \\ {}^{216}_{84}\text{Po} + {}_{-1}\beta \end{array} \begin{array}{c} \nearrow \\ \\ \nearrow \end{array} {}^{208}_{82}\text{Pb} + \begin{array}{c} \alpha \\ \beta \end{array}$$

SAMPLE SOLUTION 14-5

Given:
Thorium, $^{232}_{90}\text{Th}$ has an atomic number of 90 and a mass number of 232. It alpha decays with a half-life of about 10^{10} years into an isotope of radium. The radium then beta decays, with a half-life of 6.7 years into actinium, which again beta decays into another isotope of thorium. This second isotope of thorium is much less stable than thorium 232 and alpha decays with a half-life of 1.9 years.

Find:
(a) What is the mass number of radium, $^{?}_{88}\text{Ra}$, into which $^{232}_{90}\text{Th}$ decays?

(b) What are the mass numbers of actinium (Ac) and thorium resulting from beta decay?

(c) What is the next isotope in the chain resulting from alpha decay of thorium?

Be sure you can do the Example Problems (without peeking at the Sample Solutions) before working these Essential Problems.

14-6 Superman is a big six foot two husky fellow when at rest, but he is considerably shorter when flying at near the speed of light.

Given: Superman is 1.88 m tall when at rest, but loses some of that when traveling at 0.5c, half the speed of light relative to you. When he goes still faster, his height is reduced to 1.21 m (4 ft tall) .

Find: (a) What is Superman's height when traveling at half the speed of light?
(b) How much does his watch lose in an hour of your time at this speed?
(c) How fast is he going when his height is reduced still more?

14-7 Ms. Atom is chasing bad men in her atomic-powered car. The fuel gauge in her car tells her how much mass is converted into energy.

Given: Ms. Atom's car has a mass of 1.2×10^3 kg and converts 1.0 g of matter into kinetic energy.

Find: (a) How much energy, in joules, is produced?
(b) How fast is Ms. Atom's car going if, starting from rest, all that energy is converted into the kinetic energy of the atomic car?
(c) By how much does the relativistic mass of the car increase as a result of this increase in velocity?

14-8 The molecules in a flask of oxygen at 0° C under atmospheric pressure have an average velocity of about 460 m/s. The atoms in the molecule have an equilibrium separation of 1.12×10^{-10} m and a combined mass of 5×10^{-26} kg. ($h = 6.63 \times 10^{-34}$ J · s).

Given: A molecule has a momentum of 2.4×10^{-23} kg · m/s, give or take about 10%, and an average kinetic energy of 5×10^{-21} J, with about the same uncertainty.

Find: (a) What is the uncertainty in the molecule's momentum?
(b) What uncertainty in position is predicted by the uncertainty principle?
(c) What uncertainty in time is associated with the molecule's uncertainty in kinetic energy?

14-9 A mobile ham radio transmitter is tuned into the 10–m band and is sending out photons of energy in every direction. ($h = 6.63 \times 10^{-34}$ J · s)

Given: The radio transmitter is tuned to a frequency of 3.3×10^8 Hz and is operating at a power of 100 W.

Find: (a) How much energy does 1 photon of this frequency radio wave have?
(b) How many photons of this energy are radiated in 1 s?

14-10 In the uranium series of natural radioactivity, uranium 238 undergoes a series of alpha and beta decays to eventually become a stable isotope of lead, $^{206}_{82}$Pb .

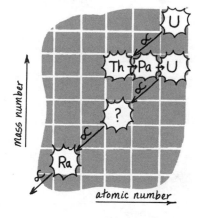

Given: Uranium $^{238}_{92}$U has an atomic number of 92 and a mass number of 238. It alpha decays with a half-life of about 10^9 years into an isotope of thorium. The thorium (Th) then beta decays with a half-life of about 24 days into an isotope of protactinium (Pa). The protactinium then beta decays into another isotope of uranium, which again alpha decays.

Find: (a) What is the mass number of the thorium isotope, $^{?}_{90}$Th, into which the uranium 238 decays?
(b) What are the isotopes of the Pa and U atoms resulting from beta decay?
(c) What is the next isotope in the chain resulting from alpha decay?

14-11 Besides the thorium series and uranium series, there is only one other series of natural radioactivity. It is called the actinium series and starts with the less abundant isotope of Uranium, $^{235}_{92}U$. Like the other natural radioactive series, it ends in a stable isotope of lead.

Given: Uranium 235 alpha decays with a half-life of 103 years into an isotope of thorium. The thorium beta decays into an isotope of protactinium, which then alpha decays into actinium.

Find: (a) What isotope of thorium results from alpha decay of $^{235}_{92}U$?
(b) What isotope of actinium results from alpha decay of Pa?

14-12 You are on a spaceship going at near the speed of light traveling from one star to another several light-years distant. To you it seems that the distance between these stars is foreshortened, because you see them moving relative to you at near the speed of light.

Given: You are traveling at eight-tenths the speed of light, $0.8c$. The distance between the stars is 2.5 light-years in a frame of reference in which they are stationary, which you decide to call the star frame.

Find: (a) How much star frame time does it take you to make the journey in the star frame?
(b) Your clock runs slow because you are traveling so fast. How much time does your clock measure for the journey?
(c) How far does it seem to be between stars as measured in your frame of reference?
(d) How long does it seem to you for the distance between the stars to zip past going at the velocity you see the stars moving at?

14-13 A 5-W flashlight bulb emits a large number of photons of visible light and even a larger amount of heat energy photons. ($h = 6.63 \times 10^{-34}$ J · s)

Given: Thirty per cent of the energy going into a 5-W flashlight bulb is radiated as visible light with an average wavelength of 5×10^{-7} m. The flashlight is left on for 1 min.

Find: (a) How much energy does 1 photon of the average wavelength have?

(b) If all the visible light were of this wavelength, how many photons would be emitted in 1 min?

(c) What is the mass equivalent of this many photons of this energy?

14-14 The size of individual nuclear power plants is expected to increase from the 200-megawatt experimental plants to something between 3,000 and 3,500 megawatts.

Given: A 3×10^9 · W nuclear power plant operates for 180 days on one set of fuel rods. Each uranium nucleus liberates about 1.5×10^{-13} J of energy when it fissions.

Find: (a) How much energy is produced on one set of fuel rods?

(b) What is the mass equivalence of this much energy?

(c) How many uranium nuclei must fission to produce this much energy?

14-15 The energy liberated in a chemical explosion produces a tiny change in mass, in accordance with the Einsteinian mass-energy relationship.

Given: One kilogram of dynamite liberates 5.4×10^6 J of energy when it explodes.

Find: (a) What is the mass equivalence of this much energy?

(b) What percentage of the total mass is converted to energy?

ANSWERS

Solutions to **ESSENTIAL PROBLEMS**

14-6   (a) 1.63 m; (b) 0.13 hours = 8 min; (c) $0.765c = 2.3 \times 10^8$ m/s

14-7   (a) 9×10^{13} J; (b) approx 3.9×10^5 m/s; (c) 1×10^{-3} kg = 1 g

14-8   (a) 2.4×10^{-24} kg · m/s; (b) $\cong 3 \times 10^{-10}$ m; (c) $\cong 1.3 \times 10^{-12}$ s

14-9   (a) 2.19×10^{-25} J; (b) 4.6×10^{26} photons

14-10   (a) 234; (b) $^{234}_{91}$Pa and $^{234}_{92}$U ; (c) $^{230}_{90}$Th

Solutions to **MORE INTERESTING PROBLEMS,** *odd answers*

14-11   (a) $^{231}_{90}$Th; (b) $^{227}_{89}$Ac

14-12   (a) 3.1 years; (b) 1.9 years; (c) 1.5 light-years; (d) 1.9 years

14-13   (a) 4.0×10^{-19} J/photon; (b) 2.3×10^{20} photons; (c) 1.0×10^{-15} kg

14-14   (a) 4.7×10^{16} J; (b) 0.5 kg; (c) 3×10^{29} nuclei

14-15   (a) 6.0×10^{-11} kg; (b) 6×10^{-9} %

Algebra Review

Algebra is the next level of abstraction beyond arithmetic. In arithmetic you learn certain rules of logic regarding a set of things called real numbers. If you take a 2 and add it to a 7, you always get a 9:

$$2 + 7 = 9$$

If you take a 2, however, and multiply it by a 7, you always get a 14:

$$2 \times 7 = 14$$

The usefulness of these rules is that the real numbers to which they apply can stand for anything. Once you know that two apples added to seven apples always gives you nine apples, you also know that two dollars added to seven dollars will give you nine dollars.

In algebra, we let certain symbols, such as x, y, z, a, or t, stand for real numbers that are either unknown or of no importance at the moment. The great secret of algebra is that you can let a symbol stand for anything and still the rules and operations of arithmetic hold the same for them as for the real numbers. If you don't know what something is, you just fake it. You put down an x or y and pretend that you know what you are talking about. This is a valuable trick, and if you follow the correct rules of logic you will come out all right in the end.

For example, if someone gave you two dollars and you later count your money to find that you have nine dollars all together, it would be easy to figure out how much money you had to start with. The number 2, added to whatever it was, let's call it x, is 9:

$$2 + x = 9$$

If you subtract 2 from both sides of the equation, you see that x must have been equal to 7 if the equality was true in the first place:

$$x = 9 - 2 = 7$$

Getting the unknown isolated on one side of the equation so that you know what it is in terms of other symbols and numbers is called **solving** the equation. Certain operations are permissible in solving an equation, because they do not change the equality. One of these Great Truths of algebra is that you can move an additive term from one side of the equation to the other by changing its sign. This First Great Truth is an almost direct application of the more general rule that you can do almost anything you like to an equation as long as you do the same thing to both sides. In the example, we subtract 2 from both sides of the equation:

$$2 + x = 9$$
$$-2 = -2$$
$$x = 9 - 2 = 7$$

Most people would say that that is all there is to it; it seems obvious to most people that subtracting 2, or adding −2, to $(2 + x)$ gives x as an answer. People who know about forms of algebra that are more sophisticated than the simple kind that we will use in this course have to think about associative and distributive laws to persuade themselves that this is the case, but we will not worry about such things. We should note, however, that the addition and subtraction rule works as well for algebraic symbols as it does for the real numbers they represent.

EXERCISES

Solve the following equations for x. (Cover answers and then compare.)

1. $5 + x = 10$
2. $x + 7 = y$
3. $x - 3 = 7$
4. $-4 + a = x + y$
5. $5 + x = a - 7$
6. $a + x = 2a - y$
7. $5a + y = 12a + x + y$
8. $5 + 3x = 7 + 2x$
9. $12x + y = 7x$
10. $x + 3y = 4x + 3a$

Answer: 1. $x = 5$; 2. $x = y - 7$; 3. $x = 10$; 4. $x = -4 + a - y$; 5. $x = a - 12$; 6. $x = a - y$; 7. $x = -7a$; 8. $x = 2$; 9. $x = \frac{-y}{5}$; 10. $x = y - a$

Multiplication and Division Rule

The preceding Exercises 9 and 10 involve dividing both sides of the equation by the same number to get x all by itself on one side of the equation. In Exercise 9, for example, subtracting $7x$ and y from both sides of the equation gives $5x$ equal to minus y. This expression can be solved for x by dividing both sides of the equation by 5;

$$\frac{5x}{5} = \frac{-y}{5}$$

Since 5 divides into itself leaving unity (1), the answer is $-y$ divided by 5.

$$x = \frac{-y}{5}$$

The Second Great Truth of algebra is that you can take something out of the numerator on one side of the equation and put it into the denominator on the other side. The reverse works equally well. You can also move something from the denominator on one side to the numerator on the other. If we were solving for y in the above expression, for example, we could go backwards to the preceding expression to get $-y$ equal to $5x$. There is one exception. Division by zero is undefined in the sense that the reciprocal of an infinitesimally small number is an infinitely large number.

Exercise 10 illustrates a point of caution that must be kept in mind when using the Second Great Truth. Subtracting $4x$ and $3y$ from both sides of the equation, and then changing the signs of both sides, gives

$$3x = 3y - 3a$$

Dividing both sides by 3 gives

$$\frac{3x}{3} = \frac{3y - 3a}{3}$$
$$x = y - a$$

One Infernal Pit which must be avoided is to divide the 3 into only one of the two terms on the right side of the equation instead of into both terms. Whenever we perform an operation on an equation, we must perform that operation on every member of the equation. Many people who would know better if only they would stop to think about it have stepped into this First Infernal Pit of algebra, and it has pulled them right down.

EXERCISES

Solve the following equations for x. (Cover answers and then compare.)

1. $12x = 24y$

2. $ax = 2y$

3. $\frac{x}{3} = 3$

4. $\frac{x}{2} = y + 3$

5. $2y + 3x = 6a$

6. $4x + 3y = 3x + 5a$

7. $7(x + y) = a$

8. $\frac{1}{4}x + a = x + 6y$

9. $\frac{1}{x} = y + \frac{1}{a}$

10. $\frac{1}{a} = \frac{1}{x} + \frac{1}{y}$

Answer: 1. $x = 2y$; 2. $x = \frac{2y}{a}$; 3. $x = 9$; 4. $x = 2y + 6$; 5. $x = 2a - \frac{2y}{3}$;

6. $x = 5a - 3y$; 7. $x = \frac{a - 7y}{7} = \frac{a}{7} - y$; 8. $x = \frac{4a}{3} - 8y$; 9. $x = \frac{a}{ay + 1}$; 10. $x = \frac{ay}{y - a}$

Fractions

In several of the preceding exercises, the unknown was part of an algebraic fraction. The $\frac{1}{4}x$ in Exercise 8, for example, is subtracted from x to give $\frac{3}{4}x$ after rearranging the expression to get the unknown on one side:

$$x - \frac{1}{4}x = a - 6y$$

As in adding any fractions, a common denominator must be found. In this case, the terms containing x have 4 as a common denominator.

$$\frac{4x}{4} - \frac{x}{4} = a - 6y$$

$$\frac{3x}{4} = a - 6y$$

The unknown, x, is then isolated by multiplying both sides of the equation by the reciprocal of $\frac{3}{4}$, namely $\frac{4}{3}$:

$$x = \frac{4}{3}(a - 6y) = \frac{4a}{3} - 8y$$

Exercise 9 also requires finding a common denominator. A common denominator must be found in the first step of this exercise when it comes to adding y and $1/a$. In this case, a itself is the common denominator:

$$\frac{1}{x} = \frac{ay}{a} + \frac{1}{a}$$

The two terms on the right can then be added, since they are fractions with the same denominator:

$$\frac{1}{x} = \frac{ay + 1}{a}$$

Some people would be tempted to let the a in the denominator cancel the a in the numerator but they must put down such temptation and recognize it as the same First Internal Pit of algebra discussed in the last section

$$\frac{\not{a}\, y + 1}{\not{a}} \quad \text{WRONG!}$$

If you divide a into one term of the numerator, you must divide it into both, and that would put you back at the original statement of the problem in Exercise 9.

People who mistakenly believe that you can cancel the a in the denominator have probably forgotten the difference between an *additive term* and a *multiplicative factor* in an algebraic expression. Additive components of an algebraic expression connected with a plus or minus sign are called **terms**, while multiplicative components are called **factors**. Thus, in an expression such as $4x^2 + 3ay + 2axy$, there are three terms, which are $4x^2$, $3ay$, and $2axy$ all added together. In the last term, 2 and the symbols a, x, and y are all factors. It is okay to divide something into one factor of an expression, but not one term of an expression. Thus,

$$\frac{2axy}{a} = 2xy \quad \text{but} \quad \frac{4x^2 + 2axy}{a} \neq 4x^2 + 2xy$$

The proper final step in solving Exercise 9 is to take the reciprocal of both sides of the equation, just as in Exercise 10. This gets the unknown out of the denominator of the fraction. The first step in Exercise 10, however, is to get the term containing the unknown, x, alone on one side of the equation even though x is in the denominator.

$$\frac{1}{x} = \frac{1}{a} - \frac{1}{y}$$

Then, after adding the two terms on the right with the same common denominator, we are ready to take the reciprocal.

$$\frac{1}{x} = \frac{y}{ay} - \frac{a}{ay} = \frac{y-a}{ay}$$

Taking the reciprocal of both sides is perfectly legal, as it may be looked upon as a multiple application of the Second Great Truth of algebra, whereby things are moved from the denominator on one side to the numerator on the other,

$$x = \frac{ay}{y-a}$$

Taking the reciprocal of both sides may, on the other hand, be looked upon as an application of a Third Great Truth of algebra, which is that both sides of an equation can be raised to the same exponential power. In this case, the exponential power is −1.

Exponents

By an exponential power, we mean the number of times something is multiplied by itself. Thus x squared is x to the second power; x cubed is x to the third power, and so on.

$$x \cdot x = x^2 \quad \text{and} \quad x \cdot x \cdot x = x^3$$

In this case, x is called the **base** and 2 or 3 is called the **exponent**. The rule for multiplying two things together, if they have the same base, is to add their exponents. Thus, x squared times x cubed is equal to x to the fifth power.

$$x^2 \cdot x^3 = (x \cdot x)(x \cdot x \cdot x) = x^{2+3} = x^5$$

Application of this rule to division leads to the concept of a negative exponent. Dividing by x cubed is the same as subtracting its exponent.

$$\frac{x^2}{x^3} = \frac{x \cdot x}{x \cdot x \cdot x} = x^{2-3} = x^{-1}$$

By x^{-1} we mean the reciprocal of x, or $1/x$.

Another application of this rule to division leads to the fact that anything raised to the zero power equals 1. That is what happens when you divide something by itself. For example, x^2 divided by x^2 equals unity (1).

$$\frac{x^2}{x^2} = x^{2-2} = x^0 = 1$$

One very useful application of exponents is in expressing very large or very small numbers in terms of powers of 10. A million, for example, is 10 to the sixth power:

$$1,000,000 = 10^6$$

The speed of light, which is two hundred ninety-nine million seven hundred ninety-three thousand meters per second, can be expressed in power-of-ten notation as

$$c = 299,793,000 \text{ m/s} = 2.99793 \times 10^8 \text{ m/s}$$

This power-of-ten notation is therefore useful for placing the decimal point without using zeros, which may or may not be significant in terms of information content. In the speed of light, for example, the last three zeros are not significant. To nine significant figures, the speed of light is 299,793,456 m/s.

Another use of exponents is in expressing square and cube roots when solving algebraic equations for an unknown that is raised to an exponential power. The exponent of the root, in this case, turns out to be a fraction rather than a whole number.

$$\sqrt{x} = x^{1/2}$$

The logic of this notation comes from the rule for multiplying. Squaring both sides of the equation gives x to the first power just as it should:

$$(\sqrt{x})^2 = (\sqrt{x})(\sqrt{x}) = x^{1/2}x^{1/2} = x^{1/2+1/2} = x^1 = x$$

As an example of the fractional exponent notation, we can solve the Pythagorean theorem for one of the sides. The Pythagorean theorem, you may recall, states that the square of the hypotenuse, h, of a right triangle is equal to the sum of the squares of the sides x and y.

$$h^2 = x^2 + y^2$$

Solving this expression for x^2,

$$x^2 = h^2 - y^2$$

and then taking the square root of both sides of the equation gives the solution in terms of the side x.

$$x = (h^2 - y^2)^{1/2}$$

This brings us to the edge of the Second Infernal Pit of algebra, into which the unwary sometimes fall. Some people might be tempted to let the square root of the difference $(h^2 - y^2)$ be equal to the difference of the square roots. While it is true that the exponential power of a product is equal to the product of the exponential powers, it is *not* true that the same rule holds for sums and differences. People fall into this Second Infernal Pit for the same reason that they fall into the first. They fail to distinguish additive terms and multiplicative factors.

$$(A \cdot B)^2 = A^2 \cdot B^2 \quad \text{but} \quad (A + B)^2 \neq A^2 + B^2$$

The rule that does hold for the square of a sum of two terms will be discussed in the next section.

EXERCISES

Solve for x. (Cover answers and then compare.)

1. $\frac{x}{m} + 3m = 4m$

2. $\frac{x}{r} - 5 = r + 2$

3. $x^2 + y^2 = 5y^2$

4. $\frac{x^2 - a^2}{y} = \frac{2y}{a}$

5. $5x^4 = 3x^4 + 4a^3$

Evaluate:

6. $x^2 x^5 =$

7. $\frac{x^2}{x^5} =$

8. $x^4(xa)^{-2} =$

9. $\frac{(6c^2d)^2(a^4d)}{(2acd)} =$

10. $x^2 + x^5 =$

Answers: 1. $x = m^2$; 2. $x = r^2 + 7r$; 3. $x = 2y$; 4. $x = \left(\frac{2y^2}{a} + a^2\right)^{1/2}$;

5. $x = (2a^3)^{1/4}$; 6. x^7; 7. x^{-3}; 8. x^2a^{-2}; 9. $18a^3c^3d^2$; 10. $x^2 + x^5$

Operations with Polynomials

Exercise 10 in the preceding section might be thought of as a trick question in that many people will try to add the exponents in the same way as in Exercise 6. If you stand back and look at it, you will see that this is just another edge of that First Infernal Pit of algebra that people keep sliding into. We again need to draw a distinction between multiplying two things together, as in Exercise 6, and adding them, as in Exercise 10. The two factors x^2 and x^5 are part of the same term in Exercise 6, since they are multiplied together. They are separate terms in Exercise 10, since they are added.

An algebraic expression having only one term is frequently called a **monomial** to distinguish it from an expression having two or more terms, which is called a **polynomial**. A polynomial with two terms is called a **binomial**, and one with three terms is called a **trinomial**, and so on, but all polynomials have more than one term.

Polynomials differ from monomials in the way in which they are multiplied or divided by things. Consider multiplying the polynomial $(x^2 + y)$ by the number 2, for example:

$$2(x^2 + y) = 2x^2 + 2y$$

Two times anything is just two of those things added together. Two times the quantity x^2 plus y is just $2x^2$ plus $2y$. The same would be true of any number times the polynomial.

$$n(x^2 + y) = nx^2 + ny$$

Thus, a monomial times a polynomial is the monomial times each term of the polynomial.

The equality works both ways. The polynomial in Exercise 10, for example, could be produced by multiplying x^2 times $(1 + x^3)$.

$$x^2(1 + x^3) = x^2 + x^5$$

In fact, $x^2(1 + x^3)$ could be used as an answer to Exercise 10, but it is not particularly better or simpler than the original statement $x^2 + x^5$. There are some applications, however, in which it might be desirable to "factor out" common members of a polynomial and express it as the product of that factor times a simpler polynomial. Consider the following expression, for example:

$$6xy^2 + 2xy^3 = 2xy^2(3 + y)$$

Whether the factored form of the expression, $2xy^2(3 + y)$, or the unfactored form, $6xy^2 + 2xy^3$, is better depends on the problem. When you see a polynomial having common factors, you only need to be aware of the fact that you can factor them out if need be.

We can extend the rule for multiplying a monomial by a polynomial to the case of multiplying polynomials together by looking again at the above development. There we considered multiplying 2 times a polynomial and then multiplying n times that same polynomial. We might consider multiplying the polynomial $(2 + n)$ times that polynomial. Since $(2 + n)$ means two of anything added to n of anything, we can just add the two above expressions to get

$$(2 + n)(x^2 + y) = 2x^2 + 2y + nx^2 + ny$$

Thus, the rule for multiplying polynomials together is to multiply each term of one polynomial by each term of the other and add the results.

Here again, the equality runs both ways; there are many cases in which an algebraic expression is best expressed in a factored form rather than the "multiplied out" form. It all depends on the particular problem. Consider, for example, the following expression:

$$\frac{2x^2 + 2y + nx^2 + ny}{x^3 + xy}$$

This expression can be simplified by factoring out common terms in both numerator and denominator:

$$\frac{2(x^2 + y) + n(x^2 + y)}{x(x^2 + y)}$$

Dividing the $(x^2 + y)$ in the denominator into both terms of the numerator gives a much simpler expression:

$$\frac{2 + n}{x}$$

There are two polynomials that you should be able to recognize as expressible in factored form. They are the polynomials resulting from the square of a sum and the product of the sum and difference of two terms.

$$(x + y)^2 = x^2 + 2xy + y^2$$
$$(x + y)(x - y) = x^2 - y^2$$

When you see the polynomials on the right, you should know that they can be expressed in terms of the products on the left.

EXERCISES

Multiply out the following expressions. (Cover answers and then compare.)

1. $a(x + y) =$

2. $x^2(1 + x^3) =$

3. $\frac{5y^2}{x^2}(3y + x^3) =$

4. $(x + 1)(a + b) =$

5. $(x + 1)(x^2 + x) =$

Factor the following polynomials.

6. $ax + ay =$

7. $4y^2 + y^4 =$

8. $mx + my + 4x + 4y =$

9. $x^2 + 6x + 9 =$

10. $y^2 - 4 =$

Answers: 1. $x = ax + ay$; 2. $x = x^2 + x^5$; 3. $x = 15y^3x^{-2} + 5y^2x$;

4. $x = xa + xb + a + b$; 5. $x = x^3 + 2x^2 + x$; 6. $a(x + y)$; 7. $y^2(4 + y^2)$;

8. $(m + 4)(x + y)$; 9. $(x + 3)^2$; 10. $(y + 2)(y - 2)$

Solving Equations Having Polynomial Fractions

Our review of algebra has led us to consider three Great Truths of algebra, or rules of logic that allow us to move things around in an equation to isolate one unknown in terms of the others. The First Great Truth allows us to move a whole additive or subtractive term from one side to another by changing the sign. The Second Great Truth allows us to move a factor from one side to the other by changing it between numerator and denominator. The Third Great Truth allows us to raise both sides of an equation to the same exponential power. We have also considered two Infernal Pits, or mistakes of reasoning which some people occasionally fall into. These errors in reasoning are based on a failure to distinguish between terms and factors in an algebraic expression. With three rules of reason to guide us and only two pitfalls to avoid, we can do all the algebra needed in this course. We can even do algebra more complex than we need at present.

One example of these more difficult problems is an algebraic equation having both monomials and polynomials as parts of a fraction. It is useful to work problems of this nature at this point because they summarize the algebraic operations that you will need to master in the first part of the course. Consider, for example, the following equation to be solved for x:

$$\frac{x^2 + y^2}{2xy} = \frac{x}{y^2}$$

We see that the unknown we are solving for is on both sides of the equation and that it is in both the numerator and denominator on the left. The best first step is frequently to get the unknown out of the denominator. Multiplying both sides by the denominator on the left, or using the Second Great Truth of algebra,

$$x^2 + y^2 = \frac{x(2xy)}{y^2}$$

which is the same as

$$x^2 + y^2 = \frac{2x^2}{y}$$

Using the First Great Truth of algebra to get terms containing the unknown on one side of the equation,

$$x^2 - \frac{2x^2}{y} = -y^2$$

Factoring Out x^2,

$$x^2\left(1 - \frac{2}{y}\right) = -y^2$$

Or, using negative exponent notation, we can write the same thing as follows:

$$x^2(1 - 2y^{-1}) = -y^2$$

The Second Great Truth of algebra can then be used to solve for x^2:

$$x^2 = \frac{-y^2}{1 - 2y^{-1}}$$

We could just take the square root of both sides at this point and be done with it, but it is nice to try to simplify the fraction on the right. Multiplying both numerator and denominator by $-y$ gets rid of the negative exponent in the denominator.

$$x^2 = \frac{-y^2(-y)}{-y + 2}$$

$$= \frac{y^3}{2 - y}$$

Then, taking the square root of both sides, we get the solution. By the Third Great Truth of algebra,

$$x = \frac{y^{3/2}}{(2-y)^{1/2}}$$

Thus, this one problem makes use of all three algebra rules we are reviewing and confronts us with both pitfalls that people should learn to avoid. You might find it instructive to go back over this example to see if you can find where the pitfalls lie.

EXERCISES

Solve for x. (Cover answers and then compare.)

1. $\dfrac{x^2 + y^2}{2z} = \dfrac{z}{y}$

2. $\dfrac{x + 2y + a}{3x} = a$

3. $a = \dfrac{x+y}{x-y}$

4. $\dfrac{x^2 + 4x + 4}{x + 2} = 4$

5. $\dfrac{(x+1)(x^2-1)}{(x-1)(x+1)} = 1$

Answers: 1. $x = \left(\dfrac{2z^2}{y} - y^2\right)^{1/2}$; 2. $x = \dfrac{2y+a}{3a-1}$; 3. $x = y\dfrac{a+1}{a-1}$;

4. $x = 2$; 5. $x = 0$

Trigonometry Review

Trigonometry is the study of angles. There are several ways of measuring an angle. One is to draw a circle around the apex and then divide the circle up into some number of parts and call them **degrees**.

The number of parts can be any arbitrary number. We obtained the convention of using 360° from the ancient Mesopotamians. It seems likely that they got that number by noting that there are about 360 days in the year. No one knows why they did not use $364 \frac{1}{4}$. It probably has to do with the fact that they had a numbering system based on 60.

Another approach would be to measure an angle by taking the ratio of the arc length to the radius of the circle.

$$\sphericalangle \theta = \frac{C}{r}$$

This ratio is called the radian measure of the angle. Since the arc length of a complete circle is 2π times the radius, one complete circle, or 360°, is equal to 2π radians.

$$360° = \frac{2\pi r}{r} = 2\pi \text{ radians}$$

We do not have to draw a circle in order to measure an angle. Instead, we can drop a perpendicular from one side of the angle to the other, forming a right triangle.

Right triangles have two legs and a hypotenuse. The hypotenuse, labeled h, is the side opposite the right angle. One of the two remaining

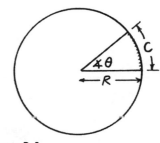

Figure B-1

An angle can be measured with a circle divided into parts.

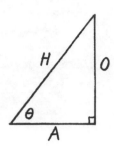

Figure B-2

An angle can also be measured using a right triangle. One leg will be opposite the angle and the other adjacent.

sides is **adjacent** to the angle θ and the other is **opposite**. Let us label them *a* and *o*, respectively. The angle can now be measured in terms of the ratios of the sides of the triangle. The ratio of the opposite over the hypotenuse depends only on the size of the angle, not on the size of the triangle we draw. A larger triangle would have a larger leg opposite the angle, but it would also have a proportionally larger hypotenuse. The ratio of the opposite to the hypotenuse would be the same in the large triangle as it is in the small one.

The ratio of the opposite to the hypotenuse is a *signature* of the angle θ, called the **sine** of the angle.

$$\sin \theta \equiv \frac{o}{h}$$

The ratio of the adjacent leg to the hypotenuse is also independent of the size of the triangle. It is the *co-signature*, called the **cosine**, of the angle.

$$\cos \theta \equiv \frac{a}{h}$$

The only other ratio left is that of the opposite over the adjacent leg. This ratio is called the **tangent** of the angle.

$$\tan \theta \equiv \frac{o}{a}$$

(Its name comes from the fact that this ratio is used in finding the slope of a line that is tangent to a curve at a particular point.)

Figure B-3

The ratio of the leg opposite the angle to the hypotenuse is independent of the size of the triangle. It depends only on the angle.

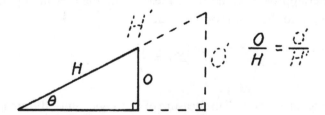

A good way to remember the sine and the cosine is to visualize that part of the triangle with which they are associated. We will often be given the hypotenuse and wish to know the lengths of the legs. Multiplying the definitions of sine and cosine by the hypotenuse, we find that the opposite side equals hypotenuse times the sine of the angle; and the adjacent side equals the hypotenuse times the cosine of the angle.

Eventually you will think sine when you see the side of a right triangle opposite an angle, and you will think cosine when you see the side adjacent. The tangent is then the ratio of the sine to the cosine.

The sum of all three angles in any triangle is 180°. Since one of the angles is a right triangle, equal to 90°, the other two add up to the remaining 90°. The other angle in the triangle, unmentioned until now, is called the **complement** of θ.

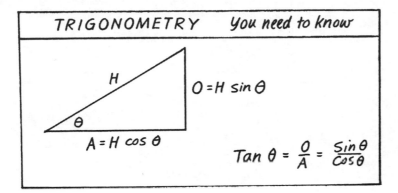

Figure B-4

The basic trigonometric definitions give the legs of a right triangle in terms of the hypotenuse. The side opposite the angle is the hypotenuse times the sine, and the side adjacent is the hypotenuse times the cosine. The tangent is the ratio of the sine to the cosine, which is the ratio of the legs they are associated with.

The side of the triangle opposite 0 is adjacent to that other angle. This fact is the basis of the trigonometric identity that says, **The sine of an angle is the cosine of its complement**.

$$\sin \theta \ = \ \cos(90° - \theta)$$

When you look at the side of a right triangle, look at it in relation to an angle. The side is the hypotenuse times the sine of one of the angles. It is also the hypotenuse times the cosine of the other angle.

Index